W9-CAZ-793

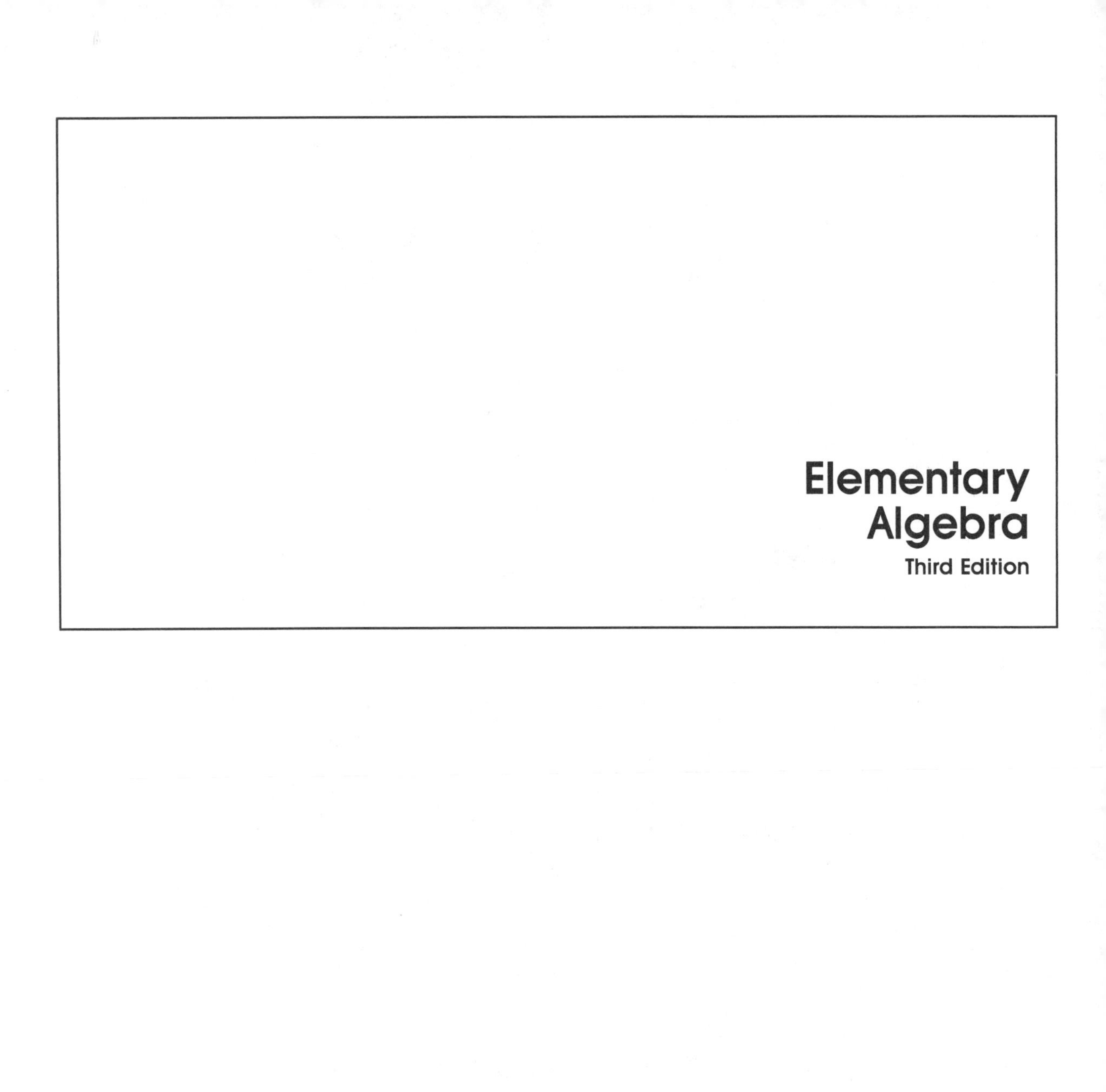

Elementary Algebra

Third Edition

ELEMENTARY ALGEBRA

THIRD EDITION

Charles P. McKeague

Cuesta College

Harcourt Brace Jovanovich, Publishers

and its subsidiary, Academic Press

San Diego New York Chicago Austin

London Sydney Tokyo Toronto

ISBN: 0-15-520958-2
Library of Congress Catalog Card Number: 85-48017
Printed in the United States of America

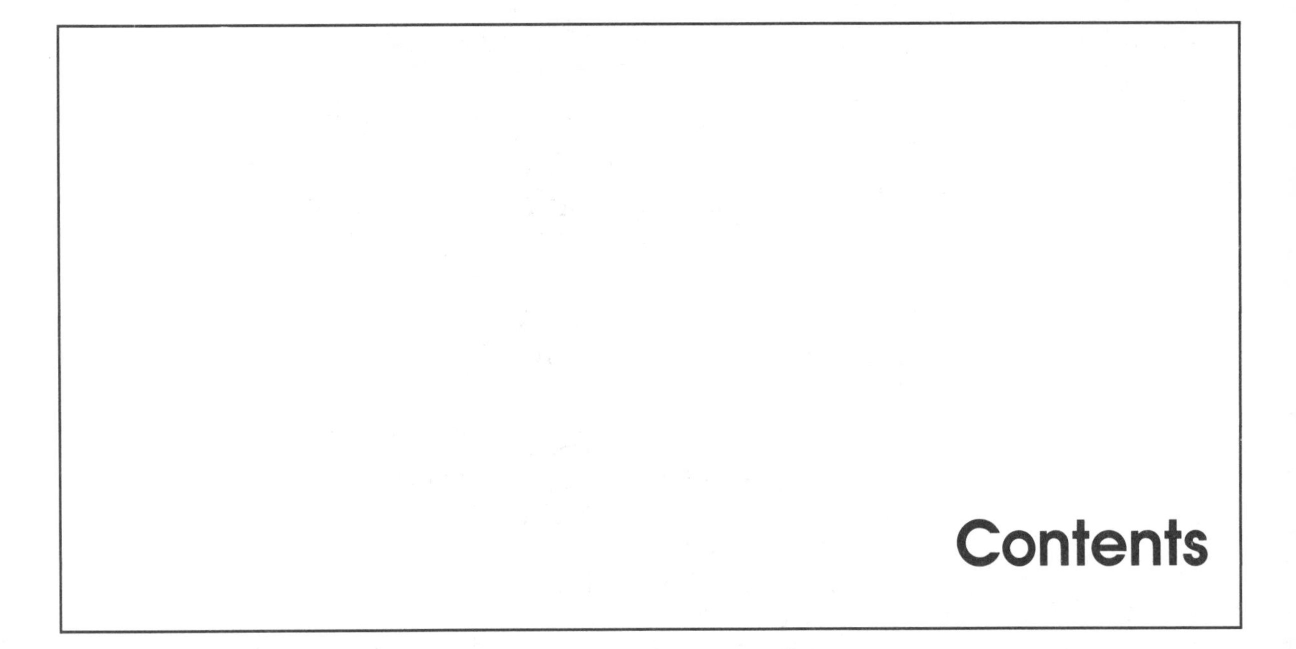

Contents

9 Additional Topics 369

Preface to the Instructor

This third edition of *Elementary Algebra* retains the same basic format and style as the second edition. It is designed to be used in a lecture situation for students with no prior background in algebra.

Organization of the Text The book begins with a preface to the student explaining what study habits are necessary to ensure success in an introductory algebra course.

The rest of the book is divided into chapters. The first eight chapters cover material usually taught in an introductory algebra class. Chapter 9 contains additional topics that I cover if time permits.

Each chapter is divided into sections, each of which can be discussed in a 45–50 minute class session. Following each section is a problem set. As was the case in the second edition, I have incorporated the following five ideas into each problem set:

1. *Drill:* There are enough problems in each set to ensure student proficiency in the material.
2. *Progressive Difficulty:* The problems increase in difficulty as the problem set progresses.
3. *Odd-Even Similarities:* Each pair of consecutive problems is similar. Since the answers to the odd-numbered problems are listed in the back of the book, the similarity of each odd-even pair of

problems gives your students a chance to check their work on an odd-numbered problem and then try a similar even-numbered problem.

4. *Application Problems:* My students are always curious about how the algebra they are learning can be applied, but at the same time many of them are afraid of word problems. I have found that they are more likely to put some time and effort into trying application problems if they don't have to work an overwhelming number of them at one time and if they work on them every day. For these reasons, I have placed a few application problems toward the end of almost every problem set in the book.
5. *Review Problems:* As was the case in the second edition, each problem set, beginning with Chapter 2, contains a few review problems. When possible, these problems review material that will be needed in the next section; otherwise, they review material from the previous chapter. (For example, the problems at the end of the problem sets in Chapter 5 review material from Chapter 4.)

Each chapter of the book ends with a chapter summary and a chapter test. The chapter summary lists the new properties and definitions found in the chapter, along with some of the common mistakes students make. As in the second edition, the margins in the chapter summaries contain examples that clarify the topics being reviewed. Each chapter test contains a representative sample of the problems covered in the chapter. All answers for these chapter tests are included in the back of the book.

Changes in the Third Edition

In General There are a number of changes in this edition that may not be apparent at first glance. Many of the sections have been rewritten so that the progression of topics proceeds more smoothly. In some cases, a new example or two has been added or the wording of the explanations rewritten. Although you may not notice these changes, your students will find the book easier to read. The more noticeable changes are listed below.

Sequence of Topics The definition of positive integer exponents has been included with the material in Section 1.1. Subsets of the real numbers are now part of Section 1.2, rather than being in a separate section. The section on formulas has been completely rewritten and placed in Section 2.5. The material on the slope of a line and the equation of a line has been taken from Chapter 9 and placed in Sections 3.3 and 3.4. Ratio and proportion are now covered at the end of Chapter 6 rather than in Chapter 9. Scientific notation has been incorporated into the first three sections of Chapter 4, rather than being concentrated in a single section. The section on solving compound inequalities has been moved from Chapter 2 to the beginning of Chapter 9. Similarly, the section on graphing linear inequalities in two variables has been moved from Chapter 3 to Section 9.2. At the request of some reviewers, I have moved the section that covers long division with polynomials to the end of Chapter 4.

New Topics Three new sections are introduced in this edition: Section 5.6 covers factoring in general. (It is actually a summary of all the methods of factoring previously covered.) Section 9.3 explains the distance formula and gives an introduction to circles. Section 9.5 covers factoring the sum and difference of two cubes. Factoring by grouping is new and has been included with the material on factoring out the greatest common factor in Section 5.2.

Review Problems The review problems introduced in the previous edition have been expanded in this edition. They still cover material that will be used in the next section, when appropriate. However, they also review material from the previous chapter. That is, the review problems at the end of the problem sets in Chapter 3 cover material from Chapter 2. Those in Chapter 4 cover material from Chapter 3, and so on through Chapter 8. If you give tests on two or more chapters at a time, you will find that this new organization of the review problems helps your students remember what they did in previous chapters. As was the case in the second edition, each section in Chapter 9 ends with a set of review problems from one of the first eight chapters of the book. Thus, while covering the last chapter, your class will also be reviewing for the final exam.

Problem Sets A number of the problem sets have been rewritten and most of them have been lengthened. You will find more challenging problems than you did in the second edition, and many more application problems.

Word Problems There are more than three times as many word problems in this edition as in the previous edition and they are spread throughout the book rather than being concentrated in a few sections. Also, many of the word problems are now explained with the use of tables.

Instructor's Resource Manual The manual that accompanies this edition now includes a set of supplementary problems for each chapter of the book. These supplementary problems can be reproduced and given to students who need more practice with problem solving. They can also be used as extensive chapter reviews before testing. As in the second edition, the manual also includes multiple tests for each chapter of the book, three final exams, and answers to the even-numbered text problems.

Acknowledgments

There are many people to thank for their assistance with this revision. Amy Barnett, my editor at Academic Press, did a fine job of coordinating the reviews of the manuscript and keeping all my other projects in line while I completed this one. Sheila Korman and the production staff at Academic Press did an incredible job of keeping the book on schedule. The answers to all the problems in the text were checked by Frank Gunnip. As always, Frank was a great help to me. My wife Diane, and my children, Patrick and Amy, are a constant source of encouragement.

Finally, I want to thank the people who reviewed this third edition for their suggestions and contributions. I was very lucky to have such a fine group of instructors contribute their thoughts.

Jeffrey C. Barnett
Fort Hayes State University

Jean Berdon
Canada College

Donley A. Chandler
El Paso Community College

Larry Curnutt
Bellevue Community College

Louise Dyson
Clark College

John Fujii
Merritt College

Martha Jordan
Okaloosa-Walton Junior College

Patricia Newell
Edison Community College

Allen Newhart
Parkersburg Community College

Orlan Ohlhausen
Mountain View College

Dana Piens
Rochester Community College

M. Pillai
Simmons College

Balbir Singh
College of San Mateo

Edward Specht
Indiana University

Charlene Sprankel
Richland Community College

Linda Whitener
University of Louisville

Preface to the Student

I often find my students asking themselves the question "Why can't I understand this stuff the first time?" The answer is, "You're not expected to." Learning a topic in algebra isn't always accomplished the first time around. There are many instances when you will find yourself reading over new material a number of times before you can begin to work problems. That's just the way things are in algebra. If you don't understand a topic the first time you see it, it doesn't mean there is something wrong with you. Understanding algebra takes time. The process of understanding requires reading the book, working problems, and getting your questions answered.

Here are some questions that are often asked by students starting an elementary algebra class.

How much math do I need to know before taking algebra?

You should be able to do the four basic operations (addition, subtraction, multiplication, and division) with whole numbers, fractions, and decimals. Most important is your ability to work with whole numbers. If you are a bit weak at working with fractions because you haven't worked with them in a while, don't be too concerned; we will review fractions as we progress through the book. I have had students who eventually did very well in algebra, even though they were initially unsure of themselves when working with fractions.

What is the best way to study?

The best way to study is to study consistently. You must work problems every day. A number of my students spend an hour or so in the morning working problems and reading over new material and then spend another hour in the evening working problems. The students I have that are successful in algebra are the ones who find a schedule that works for them and then stick to it. They work problems every day.

If I understand everything that goes on in class, can I take it easy on my homework?

Not necessarily. There is a big difference between understanding a problem someone else is working and working the same problem yourself. There is no substitute for working problems yourself. The concepts and properties are understandable to you only if you yourself work problems involving them.

If you have decided to be successful in algebra, then the following list will be important to you.

How to be Successful in Algebra

1. **Attend all class sessions on time.** There is no way to know exactly what goes on in class unless you are there. Missing class and then expecting to find out what went on from someone else is not the same as being there yourself.
2. **Read the book.** It is best to read the section that will be covered in class before you go to class. Reading in advance, even if you do not understand everything you read, is still better than going to class with no idea of what will be discussed.
3. **Work problems every day and check your answers.** The key to success in mathematics is working problems. The more problems you work, the better you will become at working them. The answers to the odd-numbered problems are listed in the back of the book. So, when you have finished an assignment, be sure to compare your answers with the ones in the back of the book. If you have made a mistake, find out what it was.
4. **Do it on your own.** Don't be misled into thinking that someone else's work is your own. Having someone else show you how to work a problem is not the same as working the same problem yourself. It is okay to get help when you are stuck. As a matter of fact, it is a good idea. Just be sure that you do the work yourself.
5. **Review every day.** After you have finished the problems your instructor has assigned, take another 15 minutes and review a section you did previously. The more you review, the longer you will retain the material you have learned.
6. **Don't expect to understand every new topic the first time you see it.** Sometimes you will understand everything you are doing,

and sometimes you won't. That's just the way things are in mathematics. Expecting to understand each new topic the first time you see it can lead to disappointment and frustration. Remember, the process of understanding algebra takes time. It requires that you read the book, work problems, and get your questions answered.

7. **Spend as much time as it takes for you to master the material.** There is no set formula for the exact amount of time you need to spend on algebra to master it. You will find out as you go along what is or isn't enough time for you. If you end up spending two or more hours on each section in order to master the material there, then that's how much time it takes; trying to get by with less will not work.
8. **Relax.** It's probably not as difficult as you think.

Chapter 1 contains some of the most important material in the book. It also contains some of the easiest material to understand. Be sure that you master it. Your success in the following chapters depends upon how well you understand Chapter 1. Here is a list, in order of importance, of the ideas you must know after having completed Chapter 1.

1. You *must* know how to add, subtract, multiply, and divide positive and negative numbers. There is no substitute for consistently getting the correct answers.
2. You *must* understand and recognize when the commutative, associative, and distributive properties are being used. These properties are used continuously throughout the book to create other properties and definitions. They are the fundamental properties upon which our algebraic system is built.
3. You should know the major classifications of numbers. That is, you should know the difference between whole numbers, integers, rational numbers, and real numbers.

If the material in Chapter 1 seems familiar to you, you may have a tendency to skip over it lightly. Don't do it. Look at the above list as you proceed through the chapter and make sure you understand each topic. By doing so, you will be off to a good start and increase your chance for success with the rest of the material in the book.

Section 1.1 Notation and Symbols

Since much of what we do in algebra involves comparison of quantities, we will begin by listing some symbols used to compare mathematical quantities. The comparison symbols fall into two major groups: equality symbols and inequality symbols.

We will let the letters a and b stand for (represent) any two mathematical quantities.

Comparison Symbols

Equality:	$a = b$	a is equal to b (a and b represent the same number)
	$a \neq b$	a is not equal to b
Inequality:	$a < b$	a is less than b
	$a \nless b$	a is not less than b
	$a > b$	a is greater than b
	$a \ngtr b$	a is not greater than b
	$a \geq b$	a is greater than or equal to b
	$a \leq b$	a is less than or equal to b

The symbols for inequality, $<$ and $>$, always point to the smaller of the two quantities being compared. For example, $3 < x$ means 3 is smaller than x. In this case, we can say "3 is less than x" or "x is greater than 3"; both statements are correct. Similarly, the expression $5 > y$ can be read as "5 is greater than y" or as "y is less than 5" because the inequality symbol is pointing to y, meaning y is the smaller of the two quantities.

Next we consider the symbols used to represent the four basic operations—addition, subtraction, multiplication, and division.

Operation Symbols

Addition:	$a + b$	The *sum* of a and b
Subtraction:	$a - b$	The *difference* of a and b
Multiplication:	$a \cdot b$, $(a)(b)$, $a(b)$, $(a)b$, ab	The *product* of a and b
Division:	$a \div b$, a/b, $\frac{a}{b}$, $b\overline{)a}$	The *quotient* of a and b

When we encounter the word *sum,* the implied operation is addition. To find the sum of two numbers, we simply add them. *Difference* implies subtraction, *product* implies multiplication, and *quotient* implies division. Notice also that there is more than one way to write the product or quotient of two numbers.

Grouping Symbols

Parentheses ( ) and brackets [ ] are the symbols used for grouping numbers together. (Occasionally braces { } are also used for grouping, although they are usually reserved for set notation, as we shall see.)

The following examples illustrate the relationship between the symbols for comparing, operating, and grouping and the English language.

▼ Examples

Mathematical expression	*English equivalent*
1. $4 + 1 = 5$	The sum of 4 and 1 is 5
2. $8 - 1 < 10$	The difference of 8 and 1 is less than 10
3. $2(3 + 4) = 14$	Twice the sum of 3 and 4 is 14
4. $3x \geq 15$	The product of 3 and x is greater than or equal to 15
5. $\frac{y}{2} = y - 2$	The quotient of y and 2 is equal to the difference of y and 2 ▲

The last type of notation we need to discuss is the notation that allows us to write repeated multiplications in a more compact form—*exponents*. In the expression 2^3, the 2 is called the *base* and the 3 is called the *exponent*. The exponent 3 tells us the number of times the base appears in the product. That is,

$$2^3 = 2 \cdot 2 \cdot 2 = 8$$

The expression 2^3 is said to be in exponential form, while $2 \cdot 2 \cdot 2$ is said to be in expanded form. Here are some additional examples of expressions involving exponents.

▼ Examples Expand and multiply.

6. $5^2 = 5 \cdot 5 = 25$ Base 5, exponent 2

7. $2^5 = 2 \cdot 2 \cdot 2 \cdot 2 \cdot 2 = 32$ Base 2, exponent 5

8. $10^3 = 10 \cdot 10 \cdot 10 = 1{,}000$ Base 10, exponent 3 ▲

The symbols for comparing, operating, and grouping are to mathematics what punctuation symbols are to English. These symbols are the punctuation symbols for mathematics.

Consider the following sentence:

Paul said John is tall.

It can have two different meanings, depending on how it is punctuated.

1. "Paul," said John, "is tall."
2. Paul said, "John is tall."

Without the punctuation we do not know which meaning is intended. It is ambiguous without punctuation.

Let's take a look at a similar situation in mathematics. Consider the following mathematical statement:

$$5 + 2 \cdot 7$$

If we add the 5 and 2 first and then multiply by 7, we get an answer of 49. On the other hand, if we multiply the 2 and the 7 first and then add 5, we are left with 19. We have a problem that seems to have two different answers, depending on whether we add first or multiply first. We would like to avoid this type of situation. That is, every problem like $5 + 2 \cdot 7$ should have only one answer. Therefore, we have developed the following rule for the order of operations.

Rule (Order of Operations)

When evaluating a mathematical expression, we will perform the operations in the following order, beginning with the expression in the innermost parentheses or brackets first and working our way out.

1. Simplify all numbers with exponents, working from left to right if more than one of these expressions is present.
2. Then do all multiplications and divisions left to right.
3. Perform all additions and subtractions left to right.

▼ **Examples** Simplify each expression using the rule for order of operations.

9. $5 + 8 \cdot 2 = 5 + 16$ Multiply $8 \cdot 2$ first

$= 21$

	Step	Reason
10.	$2 \cdot 7 + 3(6 + 4) = 2 \cdot 7 + 3 \cdot 10$	Do what is in the parentheses first
	$= 14 + 30$	Multiply left to right
	$= 44$	Add
11.	$2[5 + 2(6 + 3 \cdot 4)] = 2[5 + 2(6 + 12)]$	Simplify within the innermost grouping symbols first
	$= 2[5 + 2(18)]$	
	$= 2[5 + 36]$	Next, simplify inside the brackets
	$= 2[41]$	
	$= 82$	
12.	$10 + 12 \div 4 + 2 \cdot 3 = 10 + 3 + 6$	Multiply and divide left to right
	$= 19$	Add left to right

Notice in Example 12 that we divided 12 by 4 and multiplied 2 times 3 before we did any addition. The rule for order of operations indicates that we always multiply and divide before we add when simplifying expressions like the one in Example 12.

	Step	Reason
13.	$4 \cdot 2^3 - 2 \cdot 3^2 = 4 \cdot 8 - 2 \cdot 9$	First, simplify each number with an exponent
	$= 32 - 18$	Then, multiply left to right
	$= 14$	Finally, subtract
14.	$2^4 + 3^3 \div 9 - 4^2 = 16 + 27 \div 9 - 16$	Simplify numbers with exponents
	$= 16 + 3 - 16$	Then, divide
	$= 19 - 16$	Finally, add and subtract left to right
	$= 3$	▲

What to Do Now

Simplifying expressions using the rule for order of operations, as illustrated in the preceding examples, is a very important skill in mathematics. You need to master it, along with the other concepts presented in this section. As you work through the problems in the problem set that follows, make it your goal to use them as tools to help you understand what you have read in this section. Above all, don't just hurry through the problems in order to finish an assignment given by your instructor. Take enough time so that when you are finished, you are sure you understand the mathematics behind the problems.

For most people taking algebra for the first time, reading through this section and looking over the examples is not enough. Real understanding of this material requires working problems, making mistakes, and correcting those mistakes. You may be worried about how you are going to do in algebra. Right now your success in algebra depends only on mastering the material in this section, so don't worry, just start working problems.

Problem Set 1.1

For each mathematical expression below, write an equivalent expression in English. Include the words sum, difference, product, and quotient when possible.

1. $7 + 8 = 15$
2. $6 + 3 = 9$
3. $7 < 10$
4. $12 < 15$
5. $8 - 3 \neq 6$
6. $10 - 5 \neq 15$
7. $2x < 20$
8. $3x > 20$
9. $x + 1 = 5$
10. $y + 2 = 10$

For each expression below, write an equivalent expression in symbols.

11. The sum of x and 5 is 14.
12. The difference of x and 4 is 8.
13. The product of 5 and y is less than 30.
14. The product of 8 and y is greater than 16.
15. The sum of x and 3 is equal to the difference of 5 and y.
16. The difference of x and 4 is equal to the sum of $2x$ and 1.
17. The product of 3 and y is less than or equal to the sum of y and 6.
18. The product of 5 and y is greater than or equal to the difference of y and 16.
19. The quotient of x and 3 is equal to the sum of x and 2.
20. The quotient of x and 2 is equal to the difference of x and 4.

Mark the following statements true or false.

21. $16 < 17$
22. $18 < 15$
23. $10 = 19$
24. $11 = 21$
25. $3 + 2 < 5$
26. $5 + 1 > 6$
27. $11 \neq 10$
28. $9 \neq 8$

Expand and multiply.

29. 3^2
30. 4^2
31. 7^2
32. 9^2
33. 2^3
34. 3^3
35. 4^3
36. 5^3
37. 2^4
38. 3^4
39. 10^2
40. 10^4
41. 11^2
42. 111^2

Use the rule for order of operations to simplify each expression as much as possible.

43. $2 \cdot 3 + 5$

44. $8 \cdot 7 + 1$

45. $2(3 + 5)$

46. $8(7 + 1)$

47. $5 + 2 \cdot 6$

48. $8 + 9 \cdot 4$

49. $(5 + 2) \cdot 6$

50. $(8 + 9) \cdot 4$

51. $5 \cdot 4 + 5 \cdot 2$

52. $6 \cdot 8 + 6 \cdot 3$

53. $5(4 + 2)$

54. $6(8 + 3)$

55. $8 + 2(5 + 3)$

56. $7 + 3(8 - 2)$

57. $(8 + 2)(5 - 3)$

58. $(7 + 3)(8 - 2)$

59. $20 + 2(8 - 5) + 1$

60. $10 + 3(7 + 1) + 2$

61. $5 + 2(3 \cdot 4 - 1) + 8$

62. $11 - 2(5 \cdot 3 - 10) + 2$

63. $8 + 10 \div 2$

64. $16 - 8 \div 4$

65. $3 + 12 \div 3 + 1 \cdot 6$

66. $18 + 6 \div 2 + 3 \cdot 4$

67. $(5 + 3)(5 - 3)$

68. $(7 + 2)(7 - 2)$

69. $5^2 - 3^2$

70. $7^2 - 2^2$

71. $(4 + 5)^2$

72. $(6 + 3)^2$

73. $4^2 + 5^2$

74. $6^2 + 3^2$

75. $3 \cdot 10^2 + 4 \cdot 10 + 5$

76. $6 \cdot 10^2 + 5 \cdot 10 + 4$

77. $2 \cdot 10^3 + 3 \cdot 10^2 + 4 \cdot 10 + 5$

78. $5 \cdot 10^3 + 6 \cdot 10^2 + 7 \cdot 10 + 8$

79. $3[5 + 2(4 + 2 \cdot 3)]$

80. $4[6 + 2(8 - 2 \cdot 3)]$

81. $(10 + 8) \div 3^2$

82. $(9 + 7) \div 2^3$

83. $3^4 + 4^2 \div 2^3 - 5^2$

84. $2^5 + 6^2 \div 2^2 - 3^2$

85. On Monday Bob buys 10 shares of a certain stock. On Tuesday he buys 4 more shares of the same stock. If the stock splits Wednesday, then he has twice the number of shares he had on Tuesday. Write an expression using parentheses and the numbers 2, 4, and 10, to describe this situation.

86. Patrick has a collection of 25 baseball cards. He then buys 3 packs of gum, and each pack contains 5 baseball cards. Write an expression using the numbers 25, 3, and 5 to describe this situation.

87. A gambler begins an evening in Las Vegas with \$50. After an hour she has tripled her money. The next hour she loses \$14. Write an expression using the numbers 3, 50, and 14 to describe this situation.

88. A flight from Los Angeles to New York has 128 passengers. The plane stops in Denver, where 50 of the passengers get off and 21 new passengers get on. Write an expression containing the numbers 128, 50, and 21 to describe this situation.

Section 1.2 Real Numbers

In this section we will get an idea of what real numbers are. In order to do this we will draw what is called the *real number line*. We first draw a straight line and label a convenient point on the line with 0. Then we mark off equally spaced distances in both directions from 0. Label the points to the right of 0 with the numbers 1, 2, 3, . . . (the dots mean "and so on"). The points to the left of 0 we label, in order, -1, -2, -3, ... Here is what it looks like:

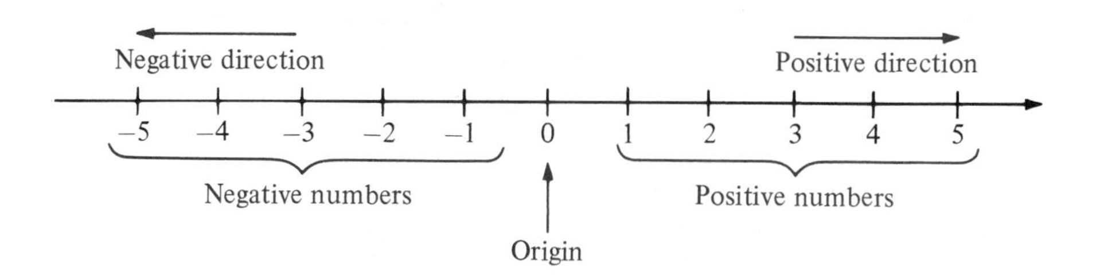

The numbers increase in value going from left to right. If we "move" to the right, we are moving in the positive direction. If we move to the left, we are moving in the negative direction. When we compare two numbers on the number line, the number on the left is always smaller than the number on the right. For instance, -3 is smaller than -1 since it is to the left of -1 on the number line.

Note If there is no sign ($+$ or $-$) in front of a number, the number is assumed to be positive ($+$).

▼ **Example 1** Locate and label the points on the real number line associated with the numbers -3.5, $-1\frac{1}{4}$, $\frac{1}{2}$, $\frac{3}{4}$, 2.5

Solution We draw a real number line from -4 to 4 and label the points in question.

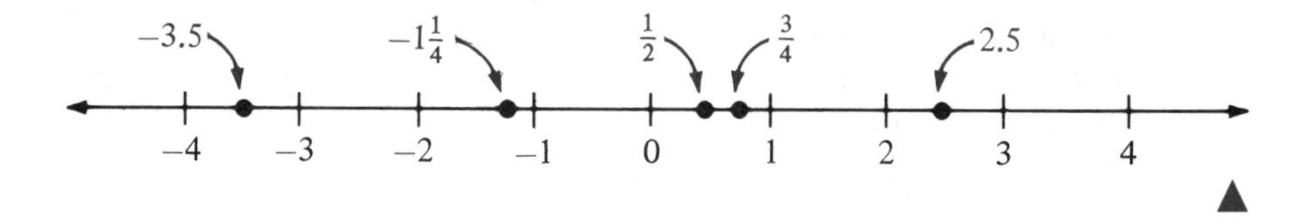

▲

DEFINITION The number associated with a point on the real number line is called the *coordinate* of that point.

In the preceding example, the numbers $\frac{1}{2}$, $\frac{3}{4}$, 2.5, -3.5, and $-1\frac{1}{4}$ are the coordinates of the points they represent.

DEFINITION The numbers that can be represented with points on the real number line are called *real numbers.*

Real numbers include whole numbers, fractions, decimals, and other numbers that are not as familiar to us as these. The numbers within the set of real numbers can be classified as *counting numbers, whole numbers, integers, rational numbers,* and *irrational numbers.* Each of these is said to be a *subset* of the real numbers.

The counting numbers are the numbers with which we count. They are the numbers 1, 2, 3, and so on. The notation we use to specify a group of numbers like this is *set notation.* We use the symbols { and } to enclose the members of the set.

$$\text{Counting numbers} = \{1, 2, 3, \ldots\}$$

▼ **Example 2** Which of the numbers in the following set are not counting numbers?

$$\{-3, 0, \frac{1}{2}, 1, 1.5, 3\}$$

Solution The numbers -3, 0, $\frac{1}{2}$, and 1.5 are not counting numbers. ▲

The whole numbers include the counting numbers and the number 0.

$$\text{Whole numbers} = \{0, 1, 2, \ldots\}$$

The set of integers include the whole numbers and the opposites of all the counting numbers.

$$\text{Integers} = \{\ldots -3, -2, -1, 0, 1, 2, 3 \ldots\}$$

▼ **Example 3** Which of the numbers in the following set are not integers?

$$\{-5, -1.75, 0, \frac{2}{3}, 1, \pi, 3\}$$

Solution The only numbers in the set that are not integers are -1.75, $\frac{2}{3}$, and π. ▲

Rational numbers include any number that can be written as the ratio of two integers. That is, rational numbers are numbers that can be put in the form

$$\frac{\text{integer}}{\text{integer}}$$

▼ **Example 4** Show why each of the numbers in the following set is a rational number.

$$\{-3, -\frac{2}{3}, 0, .333\ldots, .75\}$$

Solution The number -3 is a rational number because it can be written as the ratio of -3 to 1. That is,

$$-3 = \frac{-3}{1}$$

Similarly, the number $-\frac{2}{3}$ can be thought of as the ratio of -2 to 3, while the number 0 can be thought of as the ratio of 0 to 1.

Any repeating decimal, such as .333 . . . (the dots indicate that the 3's repeat forever) can be written as the ratio of two integers. In this case, .333 . . . is the same as the fraction $\frac{1}{3}$.

Finally, any decimal that terminates after a certain number of digits can be written as the ratio of two integers. The number .75 is equal to the fraction $\frac{3}{4}$ and is therefore a rational number. ▲

Irrational numbers are all the other numbers on the number line that cannot be written as the ratio of two integers. Because they cannot be written as the ratio of two integers, we use different kinds of notation for them. Some irrational numbers you may have seen before are π, $\sqrt{2}$, and $\sqrt{3}$.

Fractions on the Number Line

As we proceed through Chapter 1, from time to time we will review some of the major concepts associated with fractions (rational numbers). The number line can be used to visualize fractions. Recall that for the fraction $\frac{a}{b}$, a is called the numerator and b is called the denominator. The denominator indicates the number of equal parts in the interval from 0 to 1 on the number line. The numerator indicates how many of those parts we have. For the fraction $\frac{3}{4}$, the numerator is 3 and the denominator is 4. If we want to visualize $\frac{3}{4}$ on the number line, we divide the interval from 0 to 1 into 4 equal parts, and then count 3 of them.

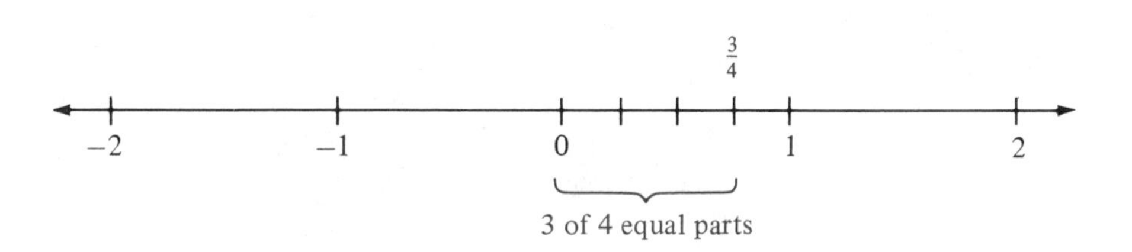

Absolute Values and Opposites

Representing numbers on the number line lets us give each number two important properties: a direction from zero and a distance from zero. The direction from zero is represented by the sign in front of the number. (A number without a sign is understood to be positive.) The distance from zero is called the absolute value of the number, as the following definition indicates.

DEFINITION The *absolute value* of a real number is its distance from zero on the number line. If x represents a real number, then the absolute value of x is written $|x|$.

▼ **Examples** Write each expression without absolute value symbols.

5. $|5| = 5$ The number $+5$ is 5 units from zero

6. $|-5| = 5$ The number -5 is 5 units from zero

7. $|-\frac{1}{2}| = \frac{1}{2}$ The number $-\frac{1}{2}$ is $\frac{1}{2}$ units from zero

8. $|-3.5| = 3.5$ The number -3.5 is 3.5 units from zero ▲

The absolute value of a number is *never* negative. It is the distance the number is from zero without regard to which direction it is from zero. When working with the absolute value of sums and differences, we must simplify the expression inside the absolute value symbols first, and then find the absolute value of the simplified expression.

▼ **Examples** Simplify each expression.

9. $|8 - 3| = |5| = 5$

10. $|3 \cdot 2^3 + 2 \cdot 3^2| = |3 \cdot 8 + 2 \cdot 9| = |24 + 18| = |42| = 42$

11. $|9 - 2| - |8 - 6| = |7| - |2| = 7 - 2 = 5$ ▲

Another important concept associated with numbers on the number line is that of opposites. Here is the definition.

DEFINITION Numbers the same distance from zero but in opposite directions from zero are called *opposites*.

▼ **Examples** Give the opposite of each number.

	Number	*Opposite*	
12.	5	-5	5 and -5 are opposites
13.	-3	3	-3 and 3 are opposites
14.	$\frac{1}{4}$	$-\frac{1}{4}$	$\frac{1}{4}$ and $-\frac{1}{4}$ are opposites
15.	-2.3	2.3	-2.3 and 2.3 are opposites ▲

Each negative number is the opposite of some positive number and each positive number is the opposite of some negative number. The opposite of a negative number is a positive number. In symbols, if a represents a positive number, then

$$-(-a) = a$$

Opposites always have the same absolute value. When you add any two opposites, the result is always zero:

$$a + (-a) = 0$$

Reciprocals and Multiplication with Fractions

The last concept we want to cover in this section is the concept of reciprocals. Understanding reciprocals requires some knowledge of multiplication with fractions. To multiply two fractions we simply multiply numerators and multiply denominators.

▼ **Example 16** Multiply: $\frac{3}{4} \cdot \frac{5}{7}$.

Solution The product of the numerators is 15 and the product of the denominators is 28:

$$\frac{3}{4} \cdot \frac{5}{7} = \frac{3 \cdot 5}{4 \cdot 7} = \frac{15}{28}$$ ▲

▼ **Example 17** Multiply: $7\left(\frac{1}{3}\right)$.

Solution The number 7 can be thought of as the fraction $\frac{7}{1}$:

$$7\left(\frac{1}{3}\right) = \frac{7}{1}\left(\frac{1}{3}\right) = \frac{7 \cdot 1}{1 \cdot 3} = \frac{7}{3}$$ ▲

Note In past math classes you may have written fractions like $\frac{7}{3}$ (improper fractions) as mixed numbers, such as $2\frac{1}{3}$. In algebra, it is usually better to leave them as improper fractions. However, if you do use mixed number notation, remember, $2\frac{1}{3}$ means $2 + \frac{1}{3}$. So, $-2\frac{1}{3}$ would mean $-2 - \frac{1}{3}$.

▼ **Example 18** Expand and multiply: $(\frac{2}{3})^3$.

Solution Using the definition of exponents from the previous section, we have

$$\left(\frac{2}{3}\right)^3 = \frac{2}{3} \cdot \frac{2}{3} \cdot \frac{2}{3} = \frac{8}{27}$$ ▲

We are now ready for the definition of reciprocals.

DEFINITION Two numbers whose product is 1 are called *reciprocals*.

▼ **Examples** Give the reciprocal of each number.

	Number	*Reciprocal*	
19.	5	$\frac{1}{5}$	Because $5(\frac{1}{5}) = \frac{5}{1}(\frac{1}{5}) = \frac{5}{5} = 1$
20.	2	$\frac{1}{2}$	Because $2(\frac{1}{2}) = \frac{2}{1}(\frac{1}{2}) = \frac{2}{2} = 1$
21.	$\frac{1}{3}$	3	Because $\frac{1}{3}(3) = \frac{1}{3}(\frac{3}{1}) = \frac{3}{3} = 1$
22.	$\frac{3}{4}$	$\frac{4}{3}$	Because $\frac{3}{4}(\frac{4}{3}) = \frac{12}{12} = 1$ ▲

Although we will not develop multiplication with negative numbers until later in the chapter, you should know that the reciprocal of a negative number is also a negative number. For example, the reciprocal of -4 is $-\frac{1}{4}$. Likewise, the reciprocal of $-\frac{2}{3}$ is $-\frac{3}{2}$. Their products are 1, but we will have to wait until we develop multiplication with negative numbers to see why.

Every real number, except zero, has a reciprocal. Any time we multiply by 0, the result is 0. (It can never be 1.) This is a special property 0 has. In symbols it looks like this:

For any number a,

$$a \cdot 0 = 0 \cdot a = 0$$

Multiplying by 0 always results in 0.

Problem Set 1.2

Draw a number line that extends from -5 to $+5$. Label the points with the following coordinates

1. 5 **2.** -2 **3.** -4 **4.** -3
5. 1.5 **6.** 2.25 **7.** -3.5 **8.** -1.5
9. $\frac{9}{4}$ **10.** $\frac{8}{3}$

Given the numbers in the set $\{-3, -2.5, 0, 1, \frac{3}{2}, \pi, \sqrt{5}\}$ list all the

11. whole numbers **12.** integers
13. rational numbers **14.** irrational numbers

Given the numbers in the set $\{-10, -8, \frac{2}{3}, \pi, 4.5, 9\}$, list all the

15. whole numbers **16.** integers
17. rational numbers **18.** irrational numbers

For each of the following numbers, give the opposite, the reciprocal, and the absolute value. (Assume all variables are nonzero.)

19. 10 **20.** 8 **21.** $\frac{3}{4}$ **22.** $\frac{5}{7}$
23. $\frac{11}{2}$ **24.** $\frac{15}{3}$ **25.** -3 **26.** -5
27. $-\frac{2}{5}$ **28.** $-\frac{3}{8}$ **29.** x **30.** a

Place one of the symbols $<$ or $>$ between each of the following to make the resulting statement true.

31. $-5 \quad -3$
32. $-8 \quad -1$
33. $-3 \quad -7$
34. $-6 \quad 5$
35. $|-4| \quad -|-4|$
36. $3 \quad -|-3|$
37. $7 \quad -|-7|$
38. $-7 \quad |-7|$
39. $-\frac{3}{4} \quad -\frac{1}{4}$
40. $-\frac{2}{3} \quad -\frac{1}{3}$
41. $-\frac{3}{2} \quad -\frac{3}{4}$
42. $-\frac{8}{3} \quad -\frac{17}{3}$

Simplify each expression.

43. $|8 - 2|$
44. $|6 - 1|$
45. $|5 \cdot 2^3 - 2 \cdot 3^2|$
46. $|2 \cdot 10^2 + 3 \cdot 10|$
47. $|7 - 2| - |4 - 2|$
48. $|10 - 3| - |4 - 1|$
49. $10 - |7 - 2(5 - 3)|$
50. $12 - |9 - 3(7 - 5)|$
51. $15 - |8 - 2(3 \cdot 4 - 9)| - 10$
52. $25 - |9 - 3(4 \cdot 5 - 18)| - 20$

Identify the following statements as true or false.

53. $(\frac{3}{5})(\frac{5}{3}) = 1$
54. $(\frac{4}{3})(\frac{3}{4}) = 1$
55. $-10(0) = 1$
56. $0 \cdot 2 = 1$
57. $-(-3) = -3$
58. $-(-\frac{1}{2}) = \frac{1}{2}$
59. $-(-15) = 15$
60. $-(-3) = 3$
61. $-|-5| = 5$
62. $-|-8| = -8$
63. $-|-9| = -9$
64. $-|-3| = 3$
65. $-3 + 3 = 6$
66. $-5 + 5 = 0$

Multiply the following.

67. $\frac{2}{3} \cdot \frac{4}{5}$
68. $\frac{1}{4} \cdot \frac{3}{5}$
69. $\frac{1}{2}(3)$
70. $\frac{1}{3}(2)$
71. $\frac{1}{4}(5)$
72. $\frac{1}{5}(4)$
73. $\frac{4}{3} \cdot \frac{3}{4}$
74. $\frac{5}{7} \cdot \frac{7}{5}$
75. $6(\frac{1}{6})$
76. $8(\frac{1}{8})$
77. $3 \cdot \frac{1}{3}$
78. $4 \cdot \frac{1}{4}$

Expand and multiply.

79. $(\frac{3}{4})^2$
80. $(\frac{5}{6})^2$
81. $(\frac{2}{3})^3$
82. $(\frac{1}{2})^3$
83. $(\frac{1}{10})^4$
84. $(\frac{1}{10})^5$

85. A football team gains 6 yards on one play and then loses 8 yards on the next play. To what number on the number line does a loss of 8 yards correspond? The total yards gained or lost on the two plays corresponds to what negative number?

86. A woman has a balance of $20 in her checking account. If she writes a check for $30, what negative number can be used to represent the new balance in her checking account?

Section 1.3 Addition of Real Numbers

We begin this section by defining addition of real numbers in terms of the real number line. Once we have the idea of addition from the number line, we can summarize our results by writing a rule for addition of real numbers.

Since real numbers have both a distance from zero (absolute value) and a direction from zero (direction), we can think of addition of two numbers in terms of distance and direction from zero.

Let's look at a problem we know the answer to. Suppose we want to add the numbers 3 and 4. The problem is written $3 + 4$. To put it on the number line, we read the problem as follows:

1. The 3 tells us to "start at the origin and move 3 units in the positive direction."
2. The + sign is read "and then move."
3. The 4 means "4 units in the positive direction."

To summarize, $3 + 4$ means to start at the origin, move 3 units in the *positive* direction and then 4 units in the *positive* direction.

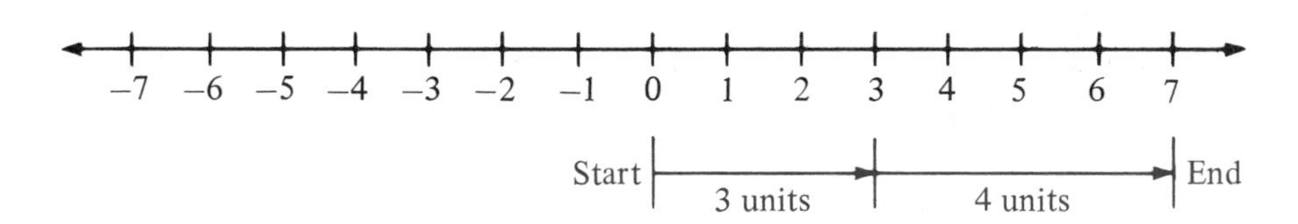

We end up at 7, which is the answer to our problem: $3 + 4 = 7$.

Let's try other combinations of positive and negative 3 and 4 on the number line.

▼ **Example 1** Add $3 + (-4)$.

Solution Starting at the origin, move 3 units in the *positive* direction and then 4 units in the *negative* direction.

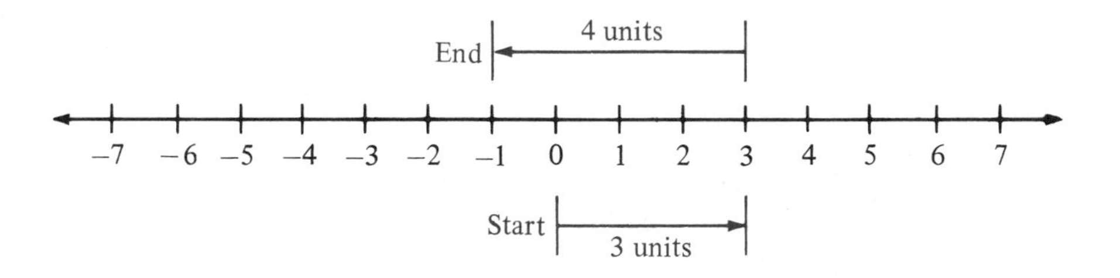

We end up at -1; therefore $3 + (-4) = -1$. ▲

▼ **Example 2** Add $-3 + 4$.

Solution Starting at the origin, move 3 units in the *negative* direction and then 4 units in the *positive* direction.

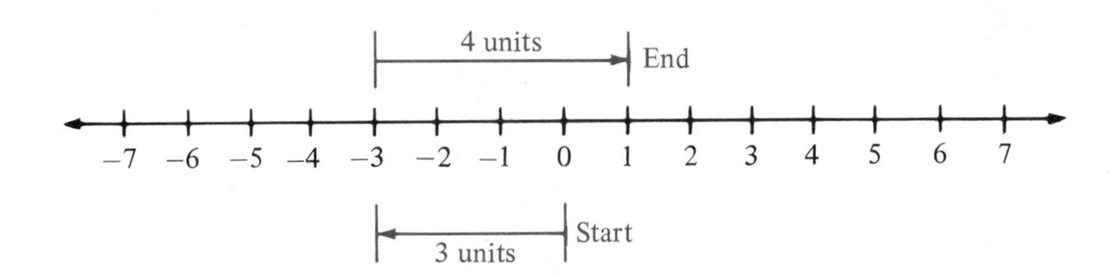

We end up at $+1$; therefore $-3 + 4 = 1$. ▲

▼ **Example 3** Add $-3 + (-4)$.

Solution Starting at the origin, move 3 units in the *negative* direction and then 4 units in the *negative* direction.

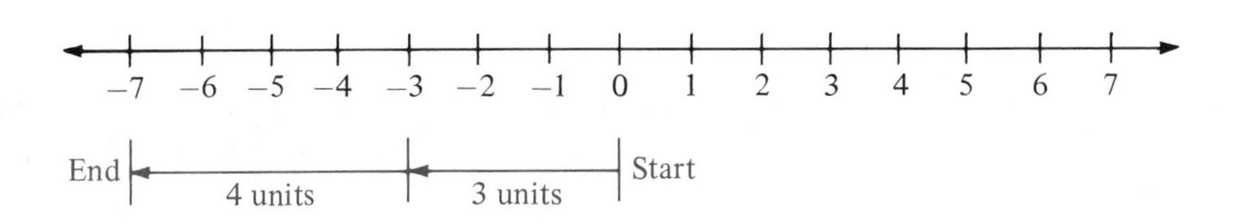

We end up at -7; therefore $-3 + (-4) = -7$. Here is a summary of what we have just completed:

$$3 + 4 = 7$$
$$3 + (-4) = -1$$
$$-3 + 4 = 1$$
$$-3 + (-4) = -7$$ ▲

Let's do four more additional problems on the number line and then summarize our results into a rule we can use to add any two real numbers.

▼ **Example 4** Add $5 + 7 = 12$.

Solution

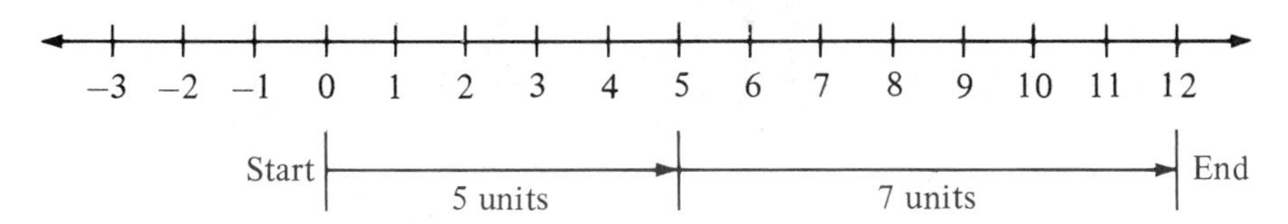

▼ **Example 5** Add $5 + (-7) = -2$.

Solution

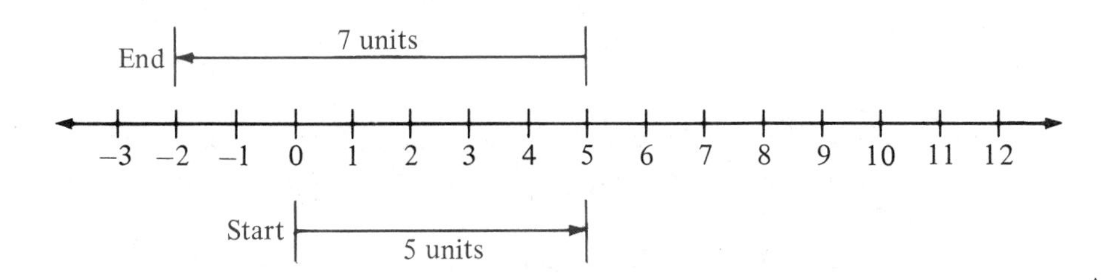

▲

▼ **Example 6** Add $-5 + 7 = 2$.

Solution

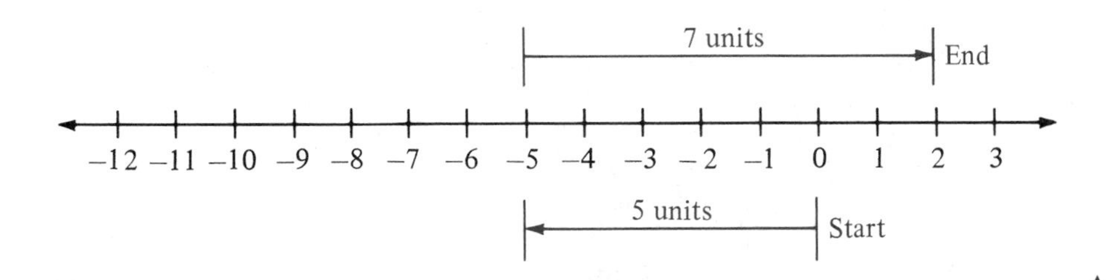

▲

▼ **Example 7** Add $-5 + (-7) = -12$.

Solution

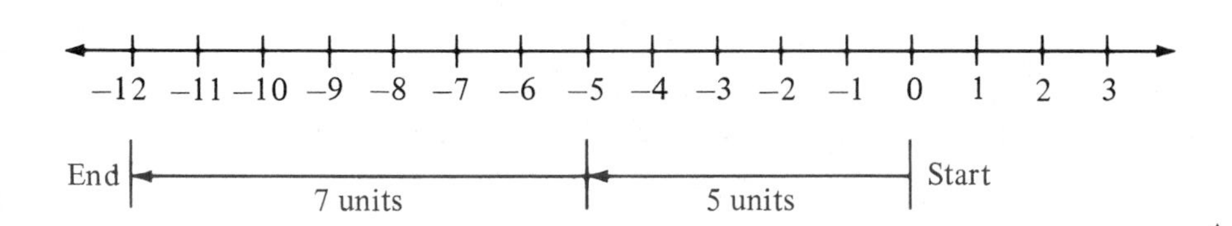

▲

If we look closely at the results of the preceding addition problems, we can see that they support (or justify) the following rule:

Rule To add two real numbers with

1. the *same* sign: Simply add their absolute values and use the common sign. (Both numbers positive, the answer is positive. Both numbers negative, the answer is negative.)
2. *different* signs: Subtract the smaller absolute value from the larger. The answer will have the sign of the number with the larger absolute value.

This rule covers all possible combinations of addition with real numbers. You must memorize it. After you have worked a number of problems, it will seem almost automatic.

Here are some more examples. Check to see that the answers to them are consistent with the rule for addition of real numbers.

▼ **Example 8** Add all combinations of positive and negative 10 and 13.

Solution Rather than work these problems on the number line, we use the rule for adding positive and negative numbers to obtain our answers:

$$10 + 13 = 23$$
$$10 + (-13) = -3$$
$$-10 + 13 = 3$$
$$-10 + (-13) = -23$$

▲

▼ **Example 9** Add all combinations of positive and negative 12 and 17.

Solution Applying the rule for adding positive and negative numbers, we have:

$$12 + 17 = 29$$
$$12 + (-17) = -5$$
$$-12 + 17 = 5$$
$$-12 + (-17) = -29$$

▲

▼ **Example 10** Add $-3 + 2 + (-4)$.

Solution Applying the rule for order of operations, we add left to right:

$$\begin{aligned} -3 + 2 + (-4) &= -1 + (-4) \\ &= -5 \end{aligned}$$

▲

▼ **Example 11** Add $-8 + [2 + (-5)] + (-1)$.

Solution Adding inside the brackets first, and then left to right, we have:

$$\begin{aligned} -8 + [2 + (-5)] + (-1) &= -8 + (-3) + (-1) \\ &= -11 + (-1) \\ &= -12 \end{aligned}$$

▲

▼ **Example 12** Simplify $(-8 + 4) + [3 + (-5)] + (-3)$.

Solution We begin by simplifying inside the parentheses and brackets. Then, we add, left to right.

$$\begin{aligned} (-8 + 4) + [3 + (-5)] + (-3) &= -4 + (-2) + (-3) \\ &= -6 + (-3) \\ &= -9 \end{aligned}$$

▲

▼ **Example 13** Simplify $-10 + 2(-8 + 11) + (-4)$.

Solution First, we simplify inside the parentheses. Then, we multiply. Finally, we add, left to right.

$$\begin{aligned} -10 + 2(-8 + 11) + (-4) &= -10 + 2(3) + (-4) \\ &= -10 + 6 + (-4) \\ &= -4 + (-4) \\ &= -8 \end{aligned}$$

▲

The next example contains absolute value. Absolute value works in the same way that parentheses do when following the rule for order of operations. That is, we simplify inside the absolute value symbols first and then find the absolute value of the simplified expression. Here is the example:

▼ **Example 14** Simplify $|-8 + 3| + |-5 + (-2)|$.

Solution It would be a mistake to find the absolute value of each number inside the absolute value symbols first. We must begin by simplifying inside each set of absolute values until we have just a single number left inside each. We can then find the absolute value of that number.

$$\begin{aligned} |-8 + 3| + |-5 + (-2)| &= |-5| + |-7| \\ &= 5 + 7 \\ &= 12 \end{aligned}$$

▲

Problem Set 1.3

1. Add all positive and negative combinations of 3 and 5. (Look back to Examples 8 and 9.)
2. Add all positive and negative combinations of 6 and 4.
3. Add all positive and negative combinations of 15 and 20.
4. Add all positive and negative combinations of 18 and 12.

Work the following problems. You may want to begin by doing a few on the number line.

5. $6 + (-3)$
6. $7 + (-8)$
7. $13 + (-20)$
8. $15 + (-25)$
9. $18 + (-32)$
10. $6 + (-9)$
11. $-6 + 3$
12. $-8 + 7$
13. $-30 + 5$
14. $-18 + 6$
15. $-6 + (-6)$
16. $-5 + (-5)$
17. $-9 + (-10)$
18. $-8 + (-6)$
19. $-10 + (-15)$
20. $-18 + (-30)$

Work the following problems using the rule for addition of real numbers. You may want to refer back to the rule for order of operations.

21. $5 + (-6) + (-7)$
22. $6 + (-8) + (-10)$
23. $-7 + 8 + (-5)$
24. $-6 + 9 + (-3)$
25. $5 + [6 + (-2)] + (-3)$
26. $10 + [8 + (-5)] + (-20)$
27. $[6 + (-2)] + [3 + (-1)]$
28. $[18 + (-5)] + [9 + (-10)]$
29. $20 + (-6) + [3 + (-9)]$
30. $18 + (-2) + [9 + (-13)]$
31. $-3 + (-2) + [5 + (-4)]$
32. $-6 + (-5) + [-4 + (-1)]$
33. $(-9 + 2) + [5 + (-8)] + (-4)$
34. $(-7 + 3) + [9 + (-6)] + (-5)$
35. $[-6 + (-4)] + [7 + (-5)] + (-9)$
36. $[-8 + (-1)] + [8 + (-6)] + (-6)$
37. $(-6 + 9) + (-5) + (-4 + 3) + 7$
38. $(-10 + 4) + (-3) + (-3 + 8) + 6$

The problems that follow involve some multiplication. Be sure that you work inside the parentheses first, then multiply, and finally, add left to right.

39. $-5 + 2(-3 + 7)$
40. $-3 + 4(-2 + 7)$
41. $9 + 3(-8 + 10)$
42. $4 + 5(-2 + 6)$
43. $-10 + 2(-6 + 8) + (-2)$
44. $-20 + 3(-7 + 10) + (-4)$
45. $2(-4 + 7) + 3(-6 + 8)$
46. $5(-2 + 5) + 7(-1 + 6)$

For the following problems, simplify inside each set of absolute value symbols first, then take the absolute value of the expression that results. Finally, simplify as much as possible.

47. $|-5 + (-3)|$

48. $|-1 + (-9)|$

49. $|-3 + 2| + |-4 + 1|$

50. $|-7 + 3| + |-3 + 8|$

51. $|5 + (-4)| + |-3 + (-2)|$

52. $|9 + (-3)| + |-5 + (-4)|$

The following problems will give you experience working with parentheses and negative signs. After you have worked these problems, see if you can state a rule using the letters a and b to summarize your results. Simplify each expression.

53. $-3 + (-4)$

54. $-2 + (-5)$

55. $-(3 + 4)$

56. $-(2 + 5)$

57. $-8 + (-2)$

58. $-6 + (-1)$

59. $-(8 + 2)$

60. $-(6 + 1)$

61. $-10 + (-15)$

62. $-20 + (-10)$

63. $-(10 + 15)$

64. $-(20 + 10)$

Recall that the word "sum" indicates addition. Write the numerical expression that is equivalent to each of the following phrases and then simplify.

65. The sum of 5 and 9

66. The sum of 6 and -3

67. Four added to the sum of -7 and -5

68. Six added to the sum of -9 and 1

69. The sum of -2 and -3 increased by 10

70. The sum of -4 and -12 increased by 2

Answer the following questions.

71. What number do you add to -8 to get -5?

72. What number do you add to 10 to get 4?

73. The sum of what number and -6 is -9?

74. The sum of what number and -12 is 8?

75. The temperature at noon is 12 degrees below zero. By 1:00 it has risen 4 degrees. Write an expression using the numbers -12 and 4 to describe this situation.

76. On Monday a certain stock gains 2 points. On Tuesday it loses 3 points. Write an expression using positive and negative numbers with addition to describe this situation and then simplify.

77. On three consecutive hands of draw poker a gambler wins $10, loses $6, and then loses another $8. Write an expression using positive and negative numbers and addition to describe this situation and then simplify.

78. You know from your past experience with numbers that subtracting 5 from 8 results in 3. ($8 - 5 = 3$.) What addition problem that starts with the number 8 gives the same result?

Section 1.4 Subtraction of Real Numbers

In the last section we spent some time developing the rule for addition of real numbers. Since we want to make as few rules as possible, we can define subtraction in terms of addition. By doing so, we can then use the rule for addition to solve our subtraction problems.

> **Rule** To subtract one real number from another, simply add its opposite.
>
> Algebraically, the rule is written like this: If a and b represent two real numbers, then it is always true that
>
> $$\underbrace{a - b}_{\text{To subtract } b} = \underbrace{a + (-b)}_{\text{add the opposite of } b}$$

This is how subtraction is defined in algebra. This definition of subtraction will not conflict with what you already know about subtraction, but it will allow you to do subtraction using negative numbers.

▼ **Example 1** Subtract all possible combinations of positive and negative 7 and 2.

Solution

$$\left.\begin{aligned} 7 - 2 &= 7 + (-2) = 5 \\ -7 - 2 &= -7 + (-2) = -9 \end{aligned}\right\} \quad \text{Subtracting 2 is the same as adding } -2$$

$$\left.\begin{aligned} 7 - (-2) &= 7 + 2 = 9 \\ -7 - (-2) &= -7 + 2 = -5 \end{aligned}\right\} \quad \text{Subtracting } -2 \text{ is the same as adding } +2$$

▲

Notice that each subtraction problem is first changed to an addition problem. The rule for addition is then used to arrive at the answer.

We have defined subtraction in terms of addition and still obtain answers consistent with the answers we are used to getting with subtraction. Moreover, we can now do subtraction problems involving both positive and negative numbers.

As you proceed through the following examples and the problem set, you will begin to notice shortcuts you can use in working the problems. You will not always have to change subtraction to addition of the opposite to be able to get answers quickly. Use all the shortcuts you wish as long as you consistently get the correct answers.

▼ **Example 2** Subtract all combinations of positive and negative 8 and 13.

Solution

$$8 - 13 = 8 + (-13) = -5$$
$$-8 - 13 = -8 + (-13) = -21$$

Subtracting $+13$ is the same as adding -13

$$8 - (-13) = 8 + 13 = 21$$
$$-8 - (-13) = -8 + 13 = 5$$

Subtracting -13 is the same as adding $+13$ ▲

▼ **Examples** Simplify each expression as much as possible.

3. $7 + (-3) - 5 = 7 + (-3) + (-5)$ Begin by changing all subtractions to additions

$= 4 + (-5)$

$= -1$ Then add left to right

4. $8 - (-2) - 6 = 8 + 2 + (-6)$ Begin by changing all subtractions to additions

$= 10 + (-6)$

$= 4$ Then add left to right

5. $-2 - (-3 + 1) - 5 = -2 - (-2) - 5$ Do what is in the parentheses first

$= -2 + 2 + (-5)$

$= -5$ ▲

The next two examples involve multiplication and exponents as well as subtraction. Remember, according to the rule for order of operations, we evaluate the numbers containing exponents and multiply before we subtract.

▼ **Example 6** Simplify $2 \cdot 5 - 3 \cdot 8 - 4 \cdot 9$.

Solution First, we multiply left to right and then, we subtract.

$$\begin{aligned} 2 \cdot 5 - 3 \cdot 8 - 4 \cdot 9 &= 10 - 24 - 36 \\ &= -14 - 36 \\ &= -50 \end{aligned}$$

▲

▼ **Example 7** Simplify $3 \cdot 2^3 - 2 \cdot 4^2$.

Solution We begin by evaluating each number that contains an exponent. Then, we multiply before we subtract:

$$\begin{aligned} 3 \cdot 2^3 - 2 \cdot 4^2 &= 3 \cdot 8 - 2 \cdot 16 \\ &= 24 - 32 \\ &= -8 \end{aligned}$$

▲

Our next example contains some absolute values. Since we treat absolute value symbols in the same way we treat grouping symbols, we simplify all expressions between absolute value symbols first. Next, we obtain the absolute value of the simplified expression. Finally, we simplify what remains, if possible.

▼ **Example 8** Simplify $|3 - 4| - |-2 - 8|$.

Solution Since we can only evaluate an absolute value when we have a single number between the absolute value symbols, we simplify inside each pair of absolute value symbols first:

$$\begin{aligned} |3 - 4| - |-2 - 8| &= |-1| - |-10| \\ &= 1 - 10 \\ &= -9 \end{aligned}$$

▲

We end this section by considering some examples of subtraction problems that are written in words.

▼ **Example 9** Subtract 7 from -3.

Solution First, we write the problem in terms of subtraction. We then change to addition of the opposite:

$$\begin{aligned} -3 - 7 &= -3 + (-7) \\ &= -10 \end{aligned}$$

▲

▼ **Example 10** Subtract -5 from 2.

Solution Subtracting -5 is the same as adding $+5$:

$$\begin{aligned} 2 - (-5) &= 2 + 5 \\ &= 7 \end{aligned}$$

▲

▼ **Example 11** Find the difference of 9 and 2.

Solution Written in symbols, the problem looks like this:

$$9 - 2 = 7$$

The difference of 9 and 2 is 7.

▲

▼ **Example 12** Find the difference of 3 and -5.

Solution Subtracting -5 from 3 we have:

$$\begin{aligned} 3 - (-5) &= 3 + 5 \\ &= 8 \end{aligned}$$

▲

Subtraction and Taking Away

For some people taking algebra for the first time, subtraction of positive and negative numbers can be a problem. These people may believe that $-5 - 9$ should be -4 or 4, not -14. If this is happening to you, you are probably thinking of subtraction in terms of taking one number away from another. Thinking of subtraction in this way works well with positive numbers if you always subtract the smaller number from the larger. In algebra, however, we encounter many situations other than this. The definition of subtraction, that $a - b = a + (-b)$, clearly indicates the correct way to use subtraction. That is, when working subtraction problems, you should think "addition of the opposite," not "take one number away from another." To be successful in algebra, you need to apply properties and definitions exactly as they are presented here.

Problem Set 1.4

The following problems are intended to give you practice with subtraction of positive and negative numbers. Remember, in algebra, subtraction is not taking one number away from another. Instead, subtracting a number is equivalent to adding its opposite.

Subtract.

1. $5 - 8$
2. $6 - 7$
3. $3 - 9$
4. $2 - 7$
5. $5 - 5$
6. $8 - 8$
7. $-8 - 2$
8. $-6 - 3$
9. $-4 - 12$
10. $-3 - 15$
11. $-6 - 6$
12. $-3 - 3$
13. $-8 - (-1)$
14. $-6 - (-2)$
15. $15 - (-20)$
16. $20 - (-5)$
17. $-4 - (-4)$
18. $-5 - (-5)$

Simplify each expression by applying the rule for order of operations.

19. $3 - 2 - 5$
20. $4 - 8 - 6$
21. $9 - 2 - 3$
22. $8 - 7 - 12$
23. $-6 - 8 - 10$
24. $-5 - 7 - 9$
25. $-22 + 4 - 10$
26. $-13 + 6 - 5$
27. $10 - (-20) - 5$
28. $15 - (-3) - 20$
29. $8 - (2 - 3) - 5$
30. $10 - (4 - 6) - 8$
31. $7 - (3 - 9) - 6$
32. $4 - (3 - 7) - 8$
33. $5 - (-8 - 6) - 2$
34. $4 - (-3 - 2) - 1$
35. $-(5 - 7) - (2 - 8)$
36. $-(4 - 8) - (2 - 5)$
37. $-(3 - 10) - (6 - 3)$
38. $-(3 - 7) - (1 - 2)$
39. $16 - [(4 - 5) - 1]$
40. $15 - [(4 - 2) - 3]$
41. $5 - [(2 - 3) - 4]$
42. $6 - [(4 - 1) - 9]$
43. $21 - [-(3 - 4) - 2] - 5$
44. $30 - [-(10 - 5) - 15] - 25$

The following problems involve multiplication and exponents. Use the rule for order of operations to simplify each expression as much as possible.

45. $2 \cdot 8 - 3 \cdot 5$
46. $3 \cdot 4 - 6 \cdot 7$
47. $3 \cdot 5 - 2 \cdot 7$
48. $6 \cdot 10 - 5 \cdot 20$
49. $5 \cdot 9 - 2 \cdot 3 - 6 \cdot 2$
50. $4 \cdot 3 - 7 \cdot 1 - 9 \cdot 4$
51. $3 \cdot 8 - 2 \cdot 4 - 6 \cdot 7$
52. $5 \cdot 9 - 3 \cdot 8 - 4 \cdot 5$
53. $2 \cdot 3^2 - 5 \cdot 2^2$
54. $3 \cdot 7^2 - 2 \cdot 8^2$
55. $4 \cdot 3^3 - 5 \cdot 2^3$
56. $3 \cdot 6^2 - 2 \cdot 3^2 - 8 \cdot 6^2$

For each of the following problems, simplify inside each pair of absolute value symbols first and then simplify the resulting expressions as much as possible.

57. $|4 - 8| - |2 - 3|$
58. $|5 - 4| - |6 - 9|$
59. $|8 - 4| - |-2 - 6|$
60. $|3 - 8| - |-7 - 4|$
61. $|-3 - 4| - |-2 - 5|$
62. $|-1 - 7| - |-3 - 6|$

Rewrite each of the following phrases as an equivalent expression in symbols and then simplify.

63. Subtract 4 from -7.
64. Subtract 5 from -19.
65. Subtract -8 from 12.
66. Subtract -2 from 10.
67. Subtract -7 from -5.
68. Subtract -9 from -3.
69. Subtract 17 from the sum of 4 and -5.
70. Subtract -6 from the sum of 6 and -3.

Recall that the word "difference" indicates subtraction. The difference of a and b is $a - b$, in that order. Write a numerical expression that is equivalent to each of the following phrases and then simplify.

71. The difference of 8 and 5.
72. The difference of 5 and 8.
73. The difference of -8 and 5.
74. The difference of -5 and 8.
75. The difference of 8 and -5.
76. The difference of 5 and -8.

Answer the following questions.

77. What number do you subtract from 8 to get -2?
78. What number do you subtract from 1 to get -5?
79. What number do you subtract from 8 to get 10?
80. What number do you subtract from 1 to get 5?

81. A man with $1500 in a savings account makes a withdrawal of $730. Write an expression using subtraction that describes this situation.
82. The temperature inside a space shuttle is 73 degrees before reentry. During reentry the temperature inside the craft increases 10 degrees. Upon landing it drops 8 degrees. Write an expression using the numbers 73, 10, and 8 to describe this situation. What is the temperature inside the shuttle upon landing?

83. A man who has lost \$35 playing roulette in Las Vegas wins \$15 playing blackjack. He then loses \$20 playing the wheel of fortune. Write an expression using the numbers -35, 15, and 20 to describe this situation and then simplify it.

84. An airplane flying at 10,000 feet lowers its altitude by 1500 feet to avoid other air traffic. Then, it increases its altitude by 3000 feet to clear a mountain range. Write an expression that describes this situation and then simplify it.

Section 1.5 Properties of Real Numbers

In this section we will list all the facts (properties) you know from past experience are true about numbers in general. We will give each property a name so we can refer to it later in the book. Mathematics is very much like a game. The game involves numbers. The rules of the game are the properties and rules we are developing in this chapter. The goal of the game is to extend the basic rules to as many new situations as possible.

You know from past experience with numbers that it makes no difference in which order you add two numbers. That is, $3 + 5$ is the same as $5 + 3$. This fact about numbers is called the *commutative property of addition.* We say addition is a commutative operation. Changing the order of the numbers does not change the answer.

There is one other basic operation that is commutative. Since 3(5) is the same as 5(3), we say multiplication is a commutative operation. Changing the order of the two numbers you are multiplying does not change the answer.

For all the properties listed in this section, a, b, and c represent real numbers.

Commutative Property of Addition

In symbols: $a + b = b + a$

In words: Changing the *order* of the numbers in a sum will not change the result.

Commutative Property of Multiplication

In symbols: $a \cdot b = b \cdot a$

In words: Changing the *order* of the numbers in a product will not change the result.

▼ **Examples**

1. The statement $5 + 8 = 8 + 5$ is an example of the commutative property of addition.
2. The statement $2 \cdot y = y \cdot 2$ is an example of the commutative property of multiplication.
3. The expression $5 + x + 3$ can be simplified using the commutative property of addition:

$5 + x + 3 = x + 5 + 3$	Commutative property of addition
$= x + 8$	Addition ▲

The other two basic operations, subtraction and division, are not commutative. The order in which we subtract or divide two numbers makes a difference in the answer.

Another property of numbers that you have used many times has to do with grouping. You know that when we add three numbers it makes no difference which two we add first. When adding $3 + 5 + 7$, we can add the 3 and 5 first and then the 7 or we can add the 5 and 7 first and then the 3. Mathematically, it looks like this: $(3 + 5) + 7 = 3 + (5 + 7)$. This property is true of multiplication as well. Operations that behave in this manner are called *associative* operations. The answer will not change when we change the association (or grouping) of the numbers.

Associative Property of Addition

In symbols: $a + (b + c) = (a + b) + c$
In words: Changing the *grouping* of the numbers in a sum will not change the result.

Associative Property of Multiplication

In symbols: $a(bc) = (ab)c$
In words: Changing the *grouping* of the numbers in a product will not change the result.

The following examples illustrate how the associative properties can be used to simplify expressions that involve both numbers and variables.

▼ **Examples** Simplify.

4.

$4 + (5 + x) = (4 + 5) + x$	Associative property of addition
$= 9 + x$	Addition

5. $5(2x) = (5 \cdot 2)x$ — Associative property of multiplication

$= 10x$ — Multiplication

6. $\frac{1}{5}(5x) = \left(\frac{1}{5} \cdot 5\right)x$ — Associative property of multiplication

$= 1x$ — Multiplication

$= x$

▲

The associative and commutative properties apply to problems that are either all multiplication or all addition. There is a third basic property that involves both addition and multiplication. It is called the *distributive property* and looks like this:

> **Distributive Property**
>
> *In symbols:* $a(b + c) = ab + ac$
>
> *In words:* Multiplication *distributes* over addition.

You will see as we progress through the book that the distributive property is used very frequently in algebra. To see that the distributive property is true, compare the following expressions:

5(3 + 4)	5(3) + 5(4)
5(7)	15 + 20
35	35

In both cases the result is 35. Since the results are the same, the two original expressions must be equal. Stated mathematically, $5(3 + 4) = 5(3) + 5(4)$. We can either add the 3 and the 4 first and then multiply that sum by 5, or we can multiply the 3 and the 4 separately by 5 and then add the results. In either case, we get the same answer.

Since subtraction is defined in terms of addition, it is also true that the distributive property applies to subtraction as well as addition. That is, for any three real numbers a, b, and c,

$$a(b - c) = ab - ac$$

Here are some examples that illustrate how we use the distributive property:

▼ **Examples** Apply the distributive property to each expression and then simplify the result.

7. $2(x + 3) = 2(x) + 2(3)$ Distributive property
$= 2x + 6$ Multiplication

8. $5(2x - 8) = 5(2x) - 5(8)$ Distributive property
$= 10x - 40$ Multiplication

Notice in this example that multiplication distributes over subtraction as well as addition.

9. $4(x + y) = 4x + 4y$ Distributive property

10. $5(2x + 4y) = 5(2x) + 5(4y)$ Distributive property
$= 10x + 20y$ Multiplication

11. $\frac{1}{2}(3x + 6) = \frac{1}{2}(3x) + \frac{1}{2}(6)$ Distributive property
$= \frac{3}{2}x + 3$ Multiplication

12. $4(2a + 3) + 8 = 4(2a) + 4(3) + 8$ Distributive property
$= 8a + 12 + 8$ Multiplication
$= 8a + 20$ Addition ▲

Special Numbers

In addition to the three properties mentioned so far, we want to include in our list two special numbers that have unique properties. They are the numbers zero and one.

Additive Identity Property

There exists a unique number 0 such that

In symbols: $a + 0 = a$ and $0 + a = a$
In words: Zero preserves identities under addition. (The identity of the number is unchanged after addition with 0.)

Multiplicative Identity Property

There exists a unique number 1 such that

In symbols: $a(1) = a$ and $1(a) = a$
In words: The number 1 preserves identities under multiplication. (The identity of the number is unchanged after multiplication by 1.)

Additive Inverse Property

For each real number a, there exists a unique number $-a$ such that

In symbols: $a + (-a) = 0$
In words: Opposites add to 0.

Multiplicative Inverse Property

For every real number a, except 0, there exists a unique real number $\frac{1}{a}$ such that

In symbols: $a\left(\frac{1}{a}\right) = 1$

In words: Reciprocals multiply to 1.

Of all the basic properties listed, the commutative, associative, and distributive properties are the ones we will use most often. They are important because they will be used as justifications or reasons for many of the things we will do in the future.

The following example illustrates how we use the preceding properties. Each line contains an algebraic expression that has been changed in some way. The property that justifies the change is written to the right.

▼ **Examples** State the property that justifies the given statement.

13. $x + 5 = 5 + x$ Commutative property of addition

14. $(2 + x) + y = 2 + (x + y)$ Associative property of addition

15. $6(x + 3) = 6x + 18$ Distributive property

16. $2 + (-2) = 0$ Additive inverse property

17. $3(\frac{1}{3}) = 1$ Multiplicative inverse property

18. $(2 + 0) + 3 = 2 + 3$ 0 is the identity element for addition

19. $(2 + 3) + 4 = 3 + (2 + 4)$ Commutative and associative properties of addition

20. $(x + 2) + y = (x + y) + 2$ Commutative and associative properties of addition ▲

As a final note on the properties of real numbers we should mention that although some of the properties are stated for only two or three real numbers, they hold for as many numbers as needed. For example, the distributive property holds for expressions like $3(x + y + z + 5 + 2)$. That is,

$$3(x + y + z + 5 + 2) = 3x + 3y + 3z + 15 + 6$$

It is not important how many numbers are contained in the sum, only that it is a sum. Multiplication, you see, distributes over addition, whether there are two numbers in the sum or two hundred.

Problem Set 1.5

State the property or properties that justify the following.

1. $3 + 2 = 2 + 3$
2. $5 + 0 = 5$
3. $4(\frac{1}{4}) = 1$
4. $10(0.1) = 1$
5. $4 + x = x + 4$
6. $3(x - 10) = 3x - 30$
7. $2(y + 8) = 2y + 16$
8. $3 + (4 + 5) = (3 + 4) + 5$
9. $(3 + 1) + 2 = 1 + (3 + 2)$
10. $(5 + 2) + 9 = (2 + 5) + 9$
11. $(8 + 9) + 10 = (8 + 10) + 9$
12. $(7 + 6) + 5 = (5 + 6) + 7$
13. $3(x + 2) = 3(2 + x)$
14. $2(7y) = (7 \cdot 2)y$
15. $x(3y) = 3(xy)$
16. $a(5b) = 5(ab)$
17. $4(xy) = 4(yx)$
18. $3[2 + (-2)] = 3(0)$
19. $8[7 + (-7)] = 8(0)$
20. $7(1) = 7$

Each of the following problems has a mistake in it. Correct the right-hand side.

21. $3(x + 2) = 3x + 2$
22. $5(4 + x) = 4 + 5x$
23. $9(a + b) = 9a + b$
24. $2(y + 1) = 2y + 1$
25. $3(0) = 3$
26. $5(\frac{1}{5}) = 5$
27. $3 + (-3) = 1$
28. $8(0) = 8$
29. $10(1) = 0$
30. $3 \cdot \frac{1}{3} = 0$

Use the associative property to rewrite each of the following expressions and then simplify the result. (See Examples 4, 5, and 6.)

31. $4 + (2 + x)$
32. $5 + (6 + x)$
33. $(x + 2) + 7$
34. $(x + 8) + 2$
35. $3(5x)$
36. $5(3x)$
37. $9(6y)$
38. $6(9y)$
39. $\frac{1}{2}(3a)$
40. $\frac{1}{3}(2a)$
41. $\frac{1}{3}(3x)$
42. $\frac{1}{4}(4x)$
43. $\frac{1}{2}(2y)$
44. $\frac{1}{7}(7y)$
45. $\frac{3}{4}(\frac{4}{3}x)$
46. $\frac{3}{2}(\frac{2}{3}x)$
47. $\frac{6}{5}(\frac{5}{6}a)$
48. $\frac{2}{5}(\frac{5}{2}a)$

Apply the distributive property to each of the following expressions. Simplify when possible.

49. $8(x + 2)$
50. $5(x + 3)$
51. $8(x - 2)$
52. $5(x - 3)$
53. $4(y + 1)$
54. $4(y - 1)$
55. $3(6x + 5)$
56. $3(5x + 6)$
57. $2(3a + 7)$
58. $5(3a + 2)$
59. $9(6y - 8)$
60. $2(7y - 4)$
61. $\frac{1}{2}(3x - 6)$
62. $\frac{1}{3}(2x - 6)$
63. $\frac{1}{3}(3x + 6)$
64. $\frac{1}{2}(2x + 4)$
65. $3(x + y)$
66. $2(x - y)$
67. $8(a - b)$
68. $7(a + b)$
69. $6(2x + 3y)$
70. $8(3x + 2y)$
71. $4(3a - 2b)$
72. $5(4a - 8b)$
73. $\frac{1}{2}(6x + 4y)$
74. $\frac{1}{3}(6x + 9y)$
75. $4(a + 4) + 9$
76. $6(a + 2) + 8$
77. $2(3x + 5) + 2$
78. $7(2x + 1) + 3$
79. $7(2x + 4) + 10$
80. $3(5x + 6) + 20$

81. While getting dressed for work, a man puts on his socks and puts on his shoes. Are the two statements "putting on your socks" and "putting on your shoes" commutative?

82. Are the statements "put on your left shoe" and "put on your right shoe" commutative?

83. A skydiver flying over the jump area is about to do two things: jump out of the plane and pull the rip cord. Are the two events "jump out of the plane" and "pull the rip cord" commutative? That is, will changing the order of the events always produce the same result?

84. Give an example of two events in your daily life that are commutative.

85. Give an example that shows that division is not a commutative operation. That is, find two numbers for which changing the order of division gives two different answers.

86. Simplify the expression $10 - (5 - 2)$ and the expression $(10 - 5) - 2$ to show that subtraction is not an associative operation.

Section 1.6 Multiplication of Real Numbers

From our experience with counting numbers, we know that multiplication is simply repeated addition. That is, $3(5) = 5 + 5 + 5$. We will use this fact, along with our knowledge of negative numbers, to develop the rule for multiplication of any two real numbers. The following examples illustrate multiplication with three of the possible combinations of positive and negative numbers.

▼ **Examples** Multiply.

1. Two positives: $3(5) = 5 + 5 + 5$
$= 15$ Positive answer

2. One positive: $3(-5) = -5 + (-5) + (-5)$
$= -15$ Negative answer

3. One negative: $-3(5) = 5(-3)$ Commutative property
$= -3 + (-3) + (-3) + (-3) + (-3)$
$= -15$ Negative answer

4. Two negatives: $-3(-5) = ?$

With two negatives, $-3(-5)$, it is not possible to work the problem in terms of repeated addition. (It doesn't "make sense" to write -5 down a -3 number of times.) The answer is probably $+15$ (that's just a guess), but we need some justification for saying so. We will solve a different problem and in so doing get the answer to the problem $(-3)(-5)$.

Here is a problem we know the answer to. We will work it two different ways.

$$-3[5 + (-5)] = -3(0) = 0$$

The answer is zero. We can also work the problem using the distributive property.

$$\begin{aligned} -3[5 + (-5)] &= -3(5) + (-3)(-5) \quad \text{Distributive property} \\ &= -15 \quad + \quad ? \end{aligned}$$

Since the answer to the problem is 0, our ? must be $+15$. (What else could we add to -15 to get 0? Only $+15$.) ▲

Note You may have to read the explanation for Example 4 several times before you understand it completely. The purpose of the explanation in Example 4 is simply to justify the fact that the product of two negative numbers is a positive number. If you have no trouble believing that, then it is not so important that you understand everything in the explanation.

Here is a summary of the results we have obtained from the first four examples:

Original numbers have		The answer is
the same sign	$3(5) = 15$	positive
different signs	$3(-5) = -15$	negative
different signs	$-3(5) = -15$	negative
the same sign	$-3(-5) = 15$	positive

By examining Examples 1 through 4 and the preceding table, we can use the information there to write the following rule. This rule tells us how to multiply any two real numbers.

> **Rule** To multiply any two real numbers, simply multiply their absolute values. The sign of the answer is
>
> 1. *positive* if both numbers have the same sign (both + or both −).
> 2. *negative* if the numbers have opposite signs (one +, the other −).

In the following examples, illustrate how we use the preceding rule to multiply real numbers.

▼ **Examples** Multiply.

5. $-8(-3) = 24$
6. $-10(-5) = 50$
7. $-4(-7) = 28$

If the two numbers in the product have the same sign, the answer is positive.

8. $5(-7) = -35$
9. $-4(8) = -32$
10. $-6(10) = -60$

If the two numbers in the product have different signs, the answer is negative. ▲

In the following examples, we combine the rule for order of operations with the rule for multiplication to simplify expressions. Remember, the rule for order of operations specifies that we are to work inside the parentheses first and then simplify numbers containing exponents. After this, we multiply and divide, left to right. The last step is to add and subtract, left to right.

▼ **Examples** Simplify as much as possible.

11. $-5(-3)(-4) = 15(-4)$
$= -60$

12. $4(-3) + 6(-5) - 10 = -12 + (-30) - 10$ Multiply
$= -42 - 10$ Add
$= -52$ Subtract

13. $(-2)^3 = (-2)(-2)(-2)$ Definition of exponents
$= -8$ Multiply, left to right

14. $-3(-2)^3 - 5(-4)^2 = -3(-8) - 5(16)$ Exponents first
$= 24 - 80$ Multiply
$= -56$ Subtract

15. $6 - 4(7 - 2) = 6 - 4(5)$ Inside parentheses first

$= 6 - 20$ Multiply

$= -14$ Subtract

16. $-3(5 - 9) + 5(-4 - 2) = -3(-4) + 5(-6)$ Inside parentheses first

$= 12 + (-30)$ Multiply

$= -18$ Add

17. $2|-9 + 2| - 5|-6 + 8| = 2|-7| - 5|2|$ Simplify inside absolute values

$= 2(7) - 5(2)$ Find absolute values

$= 14 - 10$ Multiply

$= 4$ Subtract ▲

Multiplying Fractions

Previously we mentioned that to multiply two fractions we multiply numerators and multiply denominators. We can apply the rule for multiplication of positive and negative numbers to fractions in the same way we apply it to other numbers. We multiply absolute values: the product is positive if both fractions have the same sign and negative if they have different signs. Here are some examples.

▼ **Examples** Multiply.

18. $-\frac{3}{4}\left(\frac{5}{7}\right) = -\frac{3 \cdot 5}{4 \cdot 7}$ Different signs give a negative answer

$= -\frac{15}{28}$

19. $-6\left(\frac{1}{2}\right) = -\frac{6}{1}\left(\frac{1}{2}\right)$ Different signs give a negative answer

$= -\frac{6}{2}$

$= -3$

20. $-\frac{2}{3}\left(-\frac{3}{2}\right) = \frac{2 \cdot 3}{3 \cdot 2}$ Same signs give a positive answer

$= \frac{6}{6}$

$= 1$

21. $-2\left(-\frac{1}{2}\right) = \frac{2}{1}\left(\frac{1}{2}\right)$ Same signs give a positive answer

$= \frac{2}{2}$

$= 1$

22. $\left(-\frac{5}{6}\right)^2 = \left(-\frac{5}{6}\right)\left(-\frac{5}{6}\right)$ Definition of exponents

$= \frac{25}{36}$ Multiply (same signs, positive answer) ▲

Notice in Examples 20 and 21 that the product is 1. This means that the two original numbers we multiplied in each problem were reciprocals.

We can use the rule for multiplication of real numbers, along with the associative property, to multiply expressions that contain numbers and variables.

▼ **Examples** Apply the associative property and then multiply.

23. $-3(2x) = (-3 \cdot 2)x$ Associative property

$= -6x$ Multiplication

24. $6(-5y) = [6(-5)]y$ Associative property

$= -30y$ Multiplication

25. $-2\left(-\frac{1}{2}x\right) = \left[(-2)\left(-\frac{1}{2}\right)\right]x$ Associative property

$= 1x$ Multiplication

$= x$ Multiplication ▲

The following examples show how we can use both the distributive property and multiplication with real numbers.

▼ **Examples** Apply the distributive property to each expression.

26. $-2(a + 3) = -2a + (-2)(3)$ Distributive property

$= -2a + (-6)$ Multiplication

$= -2a - 6$

27. $-3(2x + 1) = -3(2x) + (-3)(1)$ Distributive property

$= -6x + (-3)$ Multiplication

$= -6x - 3$

28.
$$
\begin{aligned}
-\frac{1}{3}(2x - 6) &= -\frac{1}{3}(2x) - \left(-\frac{1}{3}\right)(6) && \text{Distributive property}\\
&= -\frac{2}{3}x - (-2) && \text{Multiplication}\\
&= -\frac{2}{3}x + 2
\end{aligned}
$$

29.
$$
\begin{aligned}
-4(3x - 5) - 8 &= -4(3x) - (-4)(5) - 8 && \text{Distributive property}\\
&= -12x - (-20) - 8 && \text{Multiply}\\
&= -12x + 20 - 8 && \text{Definition of subtraction}\\
&= -12x + 12 && \text{Subtraction}
\end{aligned}
$$
▲

Problem Set 1.6

Use the rule for multiplying two real numbers to find each of the following products.

1. $7(-6)$
2. $8(-4)$
3. $-7(3)$
4. $-5(4)$
5. $-8(2)$
6. $-16(3)$
7. $-3(-1)$
8. $-7(-1)$
9. $-11(-11)$
10. $-12(-12)$

Use the rule for order of operations to simplify each expression as much as possible.

11. $-3(2)(-1)$
12. $-2(3)(-4)$
13. $-3(-4)(-5)$
14. $-5(-6)(-7)$
15. $-2(-4)(-3)(-1)$
16. $-1(-3)(-2)(-1)$
17. $(-7)^2$
18. $(-8)^2$
19. $(-3)^3$
20. $(-2)^4$
21. $-2(2 - 5)$
22. $-3(3 - 7)$
23. $-5(8 - 10)$
24. $-4(6 - 12)$
25. $(4 - 7)(6 - 9)$
26. $(3 - 10)(2 - 6)$
27. $(-3 - 2)(-5 - 4)$
28. $(-3 - 6)(-2 - 8)$
29. $-3(-6) + 4(-1)$
30. $-4(-5) + 8(-2)$
31. $2(3) - 3(-4) + 4(-5)$
32. $5(4) - 2(-1) + 5(6)$
33. $4(-3)^2 + 5(-6)^2$
34. $2(-5)^2 + 4(-3)^2$
35. $7(-2)^3 - 2(-3)^3$
36. $10(-2)^3 - 5(-2)^4$
37. $6 - 4(8 - 2)$
38. $7 - 2(6 - 3)$
39. $9 - 4(3 - 8)$
40. $8 - 5(2 - 7)$
41. $4(2 + 3) - 5(4 + 7)$
42. $3(5 - 7) - 4(2 - 3)$
43. $2(9 - 8) - 7(5 - 6)$
44. $5(3 - 6) - 4(2 - 5)$

45. $3|-3 + 10| - 4|-3 + 7|$
46. $5|-4 + 5| - 6|-5 + 8|$
47. $-4|3 - 8| - 6|2 - 5|$
48. $-8|2 - 7| - 9|3 - 5|$
49. $7 - 2[-6 - 4(-3)]$
50. $6 - 3[-5 - 3(-1)]$
51. $7 - 3[2(-4 - 4) - 3(-1 - 1)]$
52. $5 - 3[7(-2 - 2) - 3(-3 + 1)]$
53. $8 - 6[-2(-3 - 1) + 4(-2 - 3)]$
54. $4 - 2[-3(-1 + 8) + 5(-5 + 7)]$

Multiply the following fractions. (See Examples 18–21.)

55. $-\frac{2}{3} \cdot \frac{5}{7}$
56. $-\frac{6}{5} \cdot \frac{2}{7}$
57. $-8(\frac{1}{2})$
58. $-12(\frac{1}{3})$
59. $-\frac{3}{4}(-\frac{4}{3})$
60. $-\frac{5}{8}(-\frac{8}{5})$
61. $-3(-\frac{1}{3})$
62. $-5(-\frac{1}{5})$
63. $(-\frac{3}{4})^2$
64. $(-\frac{2}{5})^2$
65. $(-\frac{2}{3})^3$
66. $(-\frac{1}{2})^3$

Find the following products. (See Examples 23, 24, and 25.)

67. $-2(4x)$
68. $-8(7x)$
69. $-7(-6x)$
70. $-8(-9x)$
71. $-\frac{1}{3}(-3x)$
72. $-\frac{1}{5}(-5x)$
73. $-4(-\frac{1}{4}x)$
74. $-2(-\frac{1}{2}x)$

Apply the distributive property to each expression and then simplify the result. (See Examples 26–29.)

75. $-4(a + 2)$
76. $-7(a + 6)$
77. $-\frac{1}{2}(3x - 6)$
78. $-\frac{1}{4}(2x - 4)$
79. $-3(2x - 5) - 7$
80. $-4(3x - 1) - 8$
81. $-5(3x + 4) - 10$
82. $-3(4x + 5) - 20$
83. Five added to the product of 3 and -10 is what number?
84. If the product of -8 and -2 is decreased by 4, what number results?
85. Write an expression for twice the product of -4 and x and then simplify it.
86. Write an expression for twice the product of -2 and $3x$ and then simplify it.
87. What number results if 8 is subtracted from the product of -9 and 2?
88. What number results if -8 is subtracted from the product of -9 and 2?

Section 1.7 Division of Real Numbers

The last of the four basic operations is division. We will use the same approach to define division as we used for subtraction. That is, we will define division in terms of rules we already know.

Recall that we developed the rule for subtraction of real numbers by defining subtraction in terms of addition. We changed our subtraction problems to addition problems and then added to get our answers. Since we already have a rule for multiplication of real numbers, and division is the inverse operation of multiplication, we will simply define division in terms of multiplication.

We know that division by the number 2 is the same as multiplication by $\frac{1}{2}$. That is, 6 divided by 2 is 3, which is the same as 6 times $\frac{1}{2}$. Similarly, dividing a number by 5 gives the same result as multiplying it by $\frac{1}{5}$. We can extend this idea to all real numbers with the following rule.

Rule If a and b represent any two real numbers (b cannot be 0), then it is always true that

$$a \div b = \frac{a}{b} = a\left(\frac{1}{b}\right)$$

Division by a number is the same as multiplication by its reciprocal. Since every division problem can be written as a multiplication problem and since we already know the rule for multiplication of two real numbers, we do not have to write a new rule for division of real numbers. We will simply replace our division problem with multiplication and use the rule we already have.

▼ **Examples** Write each division problem as an equivalent multiplication problem and then multiply.

1. $\frac{6}{2} = 6\left(\frac{1}{2}\right) = 3$ The product of two positives is positive

2. $\frac{6}{-2} = 6\left(-\frac{1}{2}\right) = -3$

3. $\frac{-6}{2} = -6\left(\frac{1}{2}\right) = -3$

The product of a positive and a negative is a negative (Examples 2 and 3)

4. $\frac{-6}{-2} = -6\left(-\frac{1}{2}\right) = 3$ The product of two negatives is positive ▲

The second step in these examples is used only to show that we *can* write division in terms of multiplication. (In actual practice we wouldn't write $\frac{6}{2}$ as $6(\frac{1}{2})$.) The answers, therefore, follow from the rule for multiplication. That is, like signs produce a positive answer, and unlike signs produce a negative answer.

Here are some more examples. This time we will not show division as multiplication by the reciprocal. We will simply divide. If the original numbers have the same signs, the answer will be positive. If the original numbers have different signs, the answer will be negative.

▼ **Examples** Divide.

5. $\frac{12}{6} = 2$ Like signs give a positive answer

6. $\frac{12}{-6} = -2$ Unlike signs give a negative answer

7. $\frac{-12}{6} = -2$ Unlike signs give a negative answer

8. $\frac{-12}{-6} = 2$ Like signs give a positive answer

9. $\frac{15}{-3} = -5$ Unlike signs give a negative answer

10. $\frac{-40}{-5} = 8$ Like signs give a positive answer

11. $\frac{-14}{2} = -7$ Unlike signs give a negative answer ▲

Division with Fractions

We can apply the definition of division to fractions. Since dividing by a fraction is equivalent to multiplying by its reciprocal, we can divide a number by the fraction, $\frac{3}{4}$, by multiplying it by the reciprocal of $\frac{3}{4}$, which is $\frac{4}{3}$. For example,

$$\frac{2}{5} \div \frac{3}{4} = \frac{2}{5} \cdot \frac{4}{3} = \frac{8}{15}$$

You may have learned this rule in previous math classes. In some math classes, multiplication by the reciprocal is referred to as "inverting the divisor and multiplying." No matter how you say it, division by any number (except 0) is always equivalent to multiplication by its reciprocal. Here are additional examples that involve division by fractions.

▼ **Examples** Divide and write each answer in lowest terms.

12. $\frac{2}{3} \div \frac{5}{7} = \frac{2}{3} \cdot \frac{7}{5}$ Rewrite as multiplication by the reciprocal

$= \frac{14}{15}$ Multiply

13. $-\frac{3}{4} \div \frac{7}{9} = -\frac{3}{4} \cdot \frac{9}{7}$ Rewrite as multiplication by the reciprocal

$= -\frac{27}{28}$ Multiply

14. $8 \div \left(-\frac{4}{5}\right) = \frac{8}{1}\left(-\frac{5}{4}\right)$ Rewrite as multiplication by the reciprocal

$= -\frac{40}{4}$ Multiply

$= -10$ Divide 40 by 4 ▲

The last step in each of the following examples involves reducing a fraction to lowest terms. To reduce a fraction to lowest terms, we divide the numerator and denominator by the largest number that divides each of them exactly. For example, to reduce $\frac{15}{20}$ to lowest terms, we divide 15 and 20 by 5 to get $\frac{3}{4}$.

▼ **Examples** Simplify as much as possible.

15. $\frac{-4(5)}{6} = \frac{-20}{6}$ Simplify numerator

$= -\frac{10}{3}$ Reduce to lowest terms by dividing numerator and denominator by 2

16. $\frac{30}{-4-5} = \frac{30}{-9}$ Simplify denominator

$= -\frac{10}{3}$ Reduce to lowest terms by dividing numerator and denominator by 3 ▲

In the examples that follow, the numerators and denominators contain expressions that are somewhat more complicated than those we have seen thus far. To apply the rule for order of operations to these examples, we treat fraction bars the same way we treat grouping symbols. That is, fraction bars separate numerators and denominators so that each will be simplified separately.

▼ **Examples** Simplify.

17. $\frac{2(-3)+4}{12} = \frac{-6+4}{12}$ In the numerator, we multiply before we add

$= \frac{-2}{12}$ Addition

$= -\frac{1}{6}$ Reduce to lowest terms by dividing numerator and denominator by 2

18. $\dfrac{5(-4) + 6(-1)}{2(3) - 4(1)} = \dfrac{-20 + (-6)}{6 - 4}$ Multiplication before addition

$= \dfrac{-26}{2}$ Simplify numerator and denominator

$= -13$ Divide -26 by 2

19. $\dfrac{5^2 - 3^2}{-5 + 3} = \dfrac{25 - 9}{-2}$ Simplify numerator and denominator separately

$= \dfrac{16}{-2}$

$= -8$

20. $\dfrac{(3 + 2)^2}{-3^2 - 2^2} = \dfrac{5^2}{-9 - 4}$ Simplify numerator and denominator separately

$= \dfrac{25}{-13}$

$= -\dfrac{25}{13}$

21. $\dfrac{|2 - 8| - |3 - 6|}{-9 + 12} = \dfrac{|-6| - |-3|}{3}$ Simplify inside absolute value first

$= \dfrac{6 - 3}{3}$ Evaluate each absolute value

$= \dfrac{3}{3}$

$= 1$ ▲

Division with the Number 0

For every division problem there is an associated multiplication problem involving the same numbers. For example, the following two problems say the same thing about the numbers 2, 3, and 6:

Division	Multiplication
$\dfrac{6}{3} = 2$	$6 = 2(3)$

We can use this relationship between division and multiplication to clarify division involving the number 0.

First of all, dividing 0 by a number other than 0 is allowed and always results in 0. To see this, consider dividing 0 by 5. We know the answer is 0 because of the relationship between multiplication and division. This is how we write it:

$$\frac{0}{5} = 0 \quad \text{because} \quad 0 = 0(5)$$

On the other hand, dividing a nonzero number by 0 is not allowed in the real numbers. Suppose we were attempting to divide 5 by 0. We don't know if there is an answer to this problem, but if there is, let's say the answer is a number that we can represent with the letter n. If 5 divided by 0 is a number n, then

$$\frac{5}{0} = n \quad \text{and} \quad 5 = n(0)$$

This is impossible, however, because no matter what number n is, when we multiply it by 0 the answer must be 0. It can never be 5. In algebra, we say expressions like $\frac{5}{0}$ are undefined, because there is no answer to them. That is, division by 0 is not allowed in the real numbers.

The only other possibility for division involving the number 0 is 0 divided by 0. We will treat problems like $\frac{0}{0}$ as if they were undefined also. But, if you go on to take more math classes, you will find that $\frac{0}{0}$ is called an *indeterminate form,* and is, in some cases, defined.

Problem Set 1.7

Find the following quotients (divide).

1. $\frac{8}{-4}$ **2.** $\frac{10}{-5}$ **3.** $\frac{15}{-3}$ **4.** $\frac{20}{-10}$

5. $\frac{-11}{33}$ **6.** $\frac{-12}{48}$ **7.** $\frac{-48}{16}$ **8.** $\frac{-32}{4}$

9. $\frac{-27}{3}$ **10.** $\frac{-18}{9}$ **11.** $\frac{-7}{21}$ **12.** $\frac{-25}{100}$

13. $\frac{-39}{-13}$ **14.** $\frac{-18}{-6}$ **15.** $\frac{-6}{-42}$ **16.** $\frac{-4}{-28}$

17. $\frac{-14}{-35}$ **18.** $\frac{-21}{-28}$ **19.** $\frac{0}{-32}$ **20.** $\frac{0}{17}$

Divide and reduce all answers to lowest terms.

21. $\frac{4}{5} \div \frac{3}{4}$ **22.** $\frac{6}{8} \div \frac{3}{4}$

23. $-\frac{5}{6} \div (-\frac{5}{8})$ **24.** $-\frac{7}{9} \div (-\frac{1}{6})$

25. $\frac{10}{13} \div (-\frac{5}{4})$ **26.** $\frac{5}{12} \div (-\frac{10}{3})$

27. $-\frac{5}{6} \div \frac{5}{6}$ **28.** $-\frac{8}{9} \div \frac{8}{9}$

29. $-\frac{3}{4} \div (-\frac{3}{4})$ **30.** $-\frac{6}{7} \div (-\frac{6}{7})$

The following problems involve more than one operation. Simplify as much as possible.

31. $\frac{3(-2)}{-10}$ **32.** $\frac{4(-3)}{24}$

33. $\frac{-5(-5)}{-15}$

34. $\frac{-7(-3)}{-35}$

35. $\frac{-8(-7)}{-28}$

36. $\frac{-3(-9)}{-6}$

37. $\frac{27}{4-13}$

38. $\frac{27}{13-4}$

39. $\frac{20-6}{5-5}$

40. $\frac{10-12}{3-3}$

41. $\frac{-3+9}{2\cdot 5-10}$

42. $\frac{-4+8}{2\cdot 4-8}$

43. $\frac{15(-5)-25}{2(-10)}$

44. $\frac{10(-3)-20}{5(-2)}$

45. $\frac{27-2(-4)}{-3(5)}$

46. $\frac{20-5(-3)}{10(-3)}$

47. $\frac{12-6(-2)}{12(-2)}$

48. $\frac{3(-4)+5(-6)}{10-6}$

49. $\frac{5^2-2^2}{-5+2}$

50. $\frac{7^2-4^2}{-7+4}$

51. $\frac{8^2-2^2}{8^2+2^2}$

52. $\frac{4^2-6^2}{4^2+6^2}$

53. $\frac{(5+3)^2}{-5^2-3^2}$

54. $\frac{(7+2)^2}{-7^2-2^2}$

55. $\frac{(8-4)^2}{8^2-4^2}$

56. $\frac{(6-2)^2}{6^2-2^2}$

57. $\frac{-4\cdot 3^2-5\cdot 2^2}{-8(7)}$

58. $\frac{-2\cdot 5^2+3\cdot 2^3}{-3(13)}$

59. $\frac{3\cdot 10^2+4\cdot 10+5}{345}$

60. $\frac{5\cdot 10^2+6\cdot 10+7}{567}$

61. $\frac{7-[(2-3)-4]}{-1-2-3}$

62. $\frac{2-[(3-5)-8]}{-3-4-5}$

63. $\frac{6(-4)-2(5-8)}{-6-3-5}$

64. $\frac{3(-4)-5(9-11)}{-9-2-3}$

65. $\frac{3(-5-3)+4(7-9)}{5(-2)+3(-4)}$

66. $\frac{-2(6-10)-3(8-5)}{6(-3)-6(-2)}$

67. $\frac{|3-9|}{3-9}$

68. $\frac{|4-7|}{4-7}$

69. $\frac{|2-9|-|5-7|}{10-15}$

70. $\frac{|3-10|-|1-5|}{9-12}$

71. $\dfrac{|5-8|+|2-6|}{|5-8|-|2-6|}$

72. $\dfrac{|4-9|+|2-8|}{|4-9|-|2-8|}$

73. $\dfrac{5|-3+7|-6|-4+2|}{-1-5}$

74. $\dfrac{5|-2+6|-4|3-5|}{-2-4}$

Answer the following questions.

75. What is the quotient of -12 and -4?
76. The quotient of -4 and -12 is what number?
77. What number do we divide by -5 to get 2?
78. What number do we divide by -3 to get 4?
79. Twenty-seven divided by what number is -9?
80. Fifteen divided by what number is -3?
81. If the quotient of -20 and 4 is decreased by 3, what number results?
82. If -4 is added to the quotient of 24 and -8, what number results?

Chapter 1 Summary and Review

Examples We will use the margins in the chapter summaries to give examples that correspond to the topic being reviewed whenever it is appropriate.

The number(s) in brackets next to each heading indicates the section(s) in which that topic is discussed.

SYMBOLS [1.1]

$a = b$	a is equal to b	
$a \neq b$	a is not equal to b	
$a < b$	a is less than b	
$a \not< b$	a is not less than b	The inequality symbols
$a > b$	a is greater than b	always point to the
$a \not> b$	a is not greater than b	smaller quantity.

1. $2^5 = 2 \cdot 2 \cdot 2 \cdot 2 \cdot 2 = 32$

$5^2 = 5 \cdot 5 = 25$

$10^3 = 10 \cdot 10 \cdot 10 = 1000$

$1^4 = 1 \cdot 1 \cdot 1 \cdot 1 = 1$

EXPONENTS [1.1]

Exponents are notation used to indicate repeated multiplication. In the expression 3^4, 3 is the *base* and 4 is the *exponent.*

$$3^4 = 3 \cdot 3 \cdot 3 \cdot 3 = 81$$

The expression 3^4 is said to be in *exponential form,* while the expression $3 \cdot 3 \cdot 3 \cdot 3$ is in *expanded form.*

2. $10 + (2 \cdot 3^2 - 4 \cdot 2)$
$= 10 + (2 \cdot 9 - 4 \cdot 2)$
$= 10 + (18 - 8)$
$= 10 + 10$
$= 20$

ORDER OF OPERATIONS [1.1]

When evaluating a mathematical expression, we will perform the operations in the following order, beginning with the expression in the innermost parentheses or brackets and working our way out.

1. Simplify all numbers with exponents, working from left to right if more than one of these numbers is present.
2. Then do all multiplications and divisions left to right.
3. Finally, perform all additions and subtractions left to right.

SUBSETS OF THE REAL NUMBERS [1.2]

Counting numbers:	$\{1, 2, 3, \ldots\}$
Whole numbers:	$\{0, 1, 2, 3, \ldots\}$
Integers:	$\{\ldots -3, -2, -1, 0, 1, 2, 3, \ldots\}$
Rational numbers:	{all numbers that can be expressed as the ratio of two integers}
Irrational numbers:	{all numbers on the number line that cannot be expressed as the ratio of two integers}
Real numbers:	{all numbers that are either rational or irrational}

3.
a. 7 and 100 are counting numbers, but 0 and -2 are not.
b. 0 and 241 are whole numbers, but -4 and $\frac{1}{2}$ are not.
c. -15, 0, and 20 are integers.
d. -4, $-\frac{1}{2}$, 0.75, and 0.666 . . . are rational numbers.
e. $-\pi$, $\sqrt{3}$, and π are irrational numbers.
f. All the numbers listed above are real numbers.

OPPOSITES [1.2, 1.5]

Any two real numbers the same distance from zero on the number line but in opposite directions from zero are called opposites. Opposites always add to zero.

4. The numbers 3 and -3 are opposites; their sum is 0:

$$3 + (-3) = 0$$

ABSOLUTE VALUE [1.2]

The absolute value of a real number is its distance from zero on the real number line. Absolute value is never negative.

5.

$$|5| = 5$$
$$|-5| = 5$$

RECIPROCALS [1.2, 1.5]

Any two real numbers whose product is one are called reciprocals. Every real number has a reciprocal except zero.

6. The numbers 2 and $\frac{1}{2}$ are reciprocals; their product is 1:

$$2(\tfrac{1}{2}) = 1$$

ADDITION OF REAL NUMBERS [1.3]

To add two real numbers with

1. the same sign: Simply add their absolute values and use the common sign.
2. different signs: Subtract the smaller absolute value from the larger absolute value. The answer has the same sign as the number with larger absolute value.

7. Add all combinations of positive and negative 10 and 13:

$$10 + 13 = 23$$
$$10 + (-13) = -3$$
$$-10 + 13 = 3$$
$$-10 + (-13) = -23$$

8. Subtracting 2 is the same as adding -2:

$$7 - 2 = 7 + (-2) = 5$$

Subtracting -2 is the same as adding $+2$:

$$7 - (-2) = 7 + 2 = 9$$

SUBTRACTION OF REAL NUMBERS [1.4]

To subtract one number from another, simply add the opposite of the number you are subtracting. That is, if a and b represent real numbers, then

$$a - b = a + (-b)$$

9.

$$3(5) = 15$$
$$3(-5) = -15$$
$$-3(5) = -15$$
$$-3(-5) = 15$$

MULTIPLICATION OF REAL NUMBERS [1.6]

To multiply two real numbers, simply multiply their absolute values. Like signs give a positive answer. Unlike signs give a negative answer.

10.

$$\frac{-6}{2} = -6(\tfrac{1}{2}) = -3$$

$$\frac{-6}{-2} = -6(-\tfrac{1}{2}) = 3$$

DIVISION OF REAL NUMBERS [1.7]

Division by a number is the same as multiplication by its reciprocal. Like signs give a positive answer. Unlike signs give a negative answer.

PROPERTIES OF REAL NUMBERS [1.5]

	For addition	*For multiplication*
Commutative:	$a + b = b + a$	$a \cdot b = b \cdot a$
Associative:	$a + (b + c) = (a + b) + c$	$a \cdot (b \cdot c) = (a \cdot b) \cdot c$
Identity:	$a + 0 = a$	$a \cdot 1 = a$
Inverse:	$a + (-a) = 0$	$a(\frac{1}{a}) = 1$
Distributive:	$a(b + c) = ab + ac$	

COMMON MISTAKES

1. Interpreting absolute value as changing the sign of the number inside the absolute value symbols. $|-5| = +5$, $|+5| = -5$. (The first expression is correct, the second one is not.) To avoid this mistake, remember: Absolute value is a distance and distance is always measured in positive units.

2. Confusing $-(-5)$ with $-|-5|$. The first is the opposite of -5. The second expression is the opposite of the absolute value of -5. The two expressions are not the same. The first is $+5$. The second is -5.

3. Using the phrase "two negatives make a positive." This only works with multiplication. With addition, two negative numbers produce a negative answer. It is best not to use the phrase "two negatives make a positive" at all.

Chapter 1 Test

Write an equivalent statement in English. Include the words sum, difference, product, and quotient when appropriate. [1.1]

1. $3 + 4 = 7$

2. $8 - 2 < 10$

Simplify according to the rule for order of operations. [1.1]

3. $5^2 + 3(9 - 7) + 3^2$

4. $10 - 6 \div 3 + 2^3$

For each number, name the opposite, reciprocal, and absolute value. [1.2]

5. -4

6. $\frac{3}{4}$

Add. [1.3]

7. $3 + (-7)$

8. $|-9 + (-6)| + |-3 + 5|$

Subtract. [1.4]

9. $-4 - 8$

10. $9 - (7 - 2) - 4$

Match each expression on the left with the letter of the property that justifies it. [1.5]

11. $(x + y) + z = x + (y + z)$

12. $3(x + 5) = 3x + 15$

13. $5(3x) = (5 \cdot 3)x$

14. $(x + 5) + 7 = 7 + (x + 5)$

a. Commutative property of addition
b. Commutative property of multiplication
c. Associative property of addition
d. Associative property of multiplication
e. Distributive property

Multiply. [1.6]

15. $-3(7)$

16. $-4(8)(-2)$

17. $8(-\frac{1}{4})$

18. $(-\frac{2}{3})^3$

Simplify using the rule for order of operations. [1.6]

19. $-3(-4) - 8$

20. $5(-6)^2 - 3(-2)^3$

Simplify as much as possible. [1.7]

21. $7 - 3(2 - 8)$

22. $4 - 2[-3(-1 + 5) + 4(-3)]$

23. $\dfrac{4(-5) - 2(7)}{-10 - 7}$

24. $\dfrac{2(-3 - 1) + 4(-5 + 2)}{-3(2) - 4}$

Apply the associative property and then simplify. [1.5, 1.6]

25. $3 + (5 + 2x)$

26. $-4 + (2 + x)$

27. $-\dfrac{1}{3}(-3x)$

28. $-2(-5x)$

Multiply by applying the distributive property. [1.5, 1.6]

29. $2(3x + 5)$

30. $-3(x + 4)$

31. $-5(2x - 1)$

32. $-\frac{1}{2}(4x - 2)$

From the set of numbers $\{1, 1.5, \sqrt{2}, \frac{3}{4}, -8\}$ list [1.2]

33. All the integers.

34. All the rational numbers.

35. All the irrational numbers.

36. All the real numbers.

Write an expression in symbols that is equivalent to each English phrase and then simplify it.

37. The sum of 8 and -3. [1.1, 1.3]

38. The difference of -24 and 2. [1.1, 1.4]

39. The product of -5 and -4. [1.1, 1.6]

40. The quotient of -24 and -2. [1.1, 1.7]

2 Linear Equations and Inequalities

Much of what we do in algebra is concerned with solving equations. In this chapter, we will develop most of the properties necessary to solve many different types of equations. It may seem surprising, but most equations can be solved by applying just two main properties—the *addition property of equality* and the *multiplication property of equality*. To be successful in this chapter you must know the following concepts from Chapter 1:

1. Addition, subtraction, multiplication, and division of positive and negative real numbers.
2. The commutative, associative, and distributive properties.
3. That reciprocals multiply to 1 and opposites add to zero.

Section 2.1 Simplifying Expressions

As you will see in the next few sections, the first step in solving an equation is to simplify both sides as much as possible. In the first part of this section, we will practice simplifying expressions by combining what are called similar (or like) terms.

For our immediate purposes, a term is a number or a number and one or more variables multiplied together. For example, the number 5 is a term, as are the expressions $3x$, $-7y$, and $15xy$.

DEFINITION Two or more terms with the same variable part are called *similar* (or *like*) terms.

The terms $3x$ and $4x$ are similar since their variable parts are identical. Likewise, the terms $18y$, $-10y$, and $6y$ are similar terms.

To simplify an algebraic expression, we simply reduce the number of terms in the expression. We accomplish this by applying the distributive property along with our knowledge of addition and subtraction of positive and negative real numbers. The following examples illustrate the procedure.

▼ **Examples** Simplify by combining similar terms.

1. $3x + 4x = (3 + 4)x$ — Distributive property
$= 7x$ — Addition of 3 and 4

2. $7a - 10a = (7 - 10)a$ — Distributive property
$= -3a$ — Addition of 7 and -10

3. $18y - 10y + 6y = (18 - 10 + 6)y$ — Distributive property
$= 14y$ — Addition of 18, -10, and 6 ▲

When the expression we intend to simplify is more complicated, we use the commutative and associative properties first.

▼ **Examples** Simplify each expression.

4. $3x + 5 + 2x - 3 = 3x + 2x + 5 - 3$ — Commutative property
$= (3x + 2x) + (5 - 3)$ — Associative property
$= (3 + 2)x + (5 - 3)$ — Distributive property
$= 5x + 2$ — Addition

5. $4a - 7 - 2a + 3 = (4a - 2a) + (-7 + 3)$ — Commutative and associative properties
$= (4 - 2)a + (-7 + 3)$ — Distributive property
$= 2a - 4$ — Addition

6. $5x + 8 - x - 6 = (5x - x) + (8 - 6)$ — Commutative and associative properties
$= (5 - 1)x + (8 - 6)$ — Distributive property
$= 4x + 2$ — Addition ▲

Notice that in each case the result has fewer terms than the original expression. Since there are fewer terms, the resulting expression is said to be simpler than the original expression.

Simplifying Expressions Containing Parentheses

If an expression contains parentheses, it is often necessary to apply the distributive property to remove the parentheses before combining similar terms.

▼ **Example 7** Simplify the expression: $5(2x - 8) - 3$.

Solution We begin by distributing the 5 across $2x - 8$. We then combine similar terms:

$$\begin{aligned} 5(2x - 8) - 3 &= 10x - 40 - 3 && \text{Distributive property} \\ &= 10x - 43 \end{aligned}$$

▲

▼ **Example 8** Simplify: $7 - 3(2y + 1)$.

Solution By the rule for order of operations, we must multiply before we add or subtract. For that reason, it would be incorrect to subtract 3 from 7 first. Instead, we multiply -3 and $2y + 1$ to remove the parentheses and then combine similar terms:

$$\begin{aligned} 7 - 3(2y + 1) &= 7 - 6y - 3 && \text{Distributive property} \\ &= -6y + 4 \end{aligned}$$

▲

▼ **Example 9** Simplify: $5(x - 2) - (3x + 4)$.

Solution We begin by applying the distributive property to remove the parentheses. The expression $-(3x + 4)$ can be thought of as $-1(3x + 4)$. Thinking of it in this way allows us to apply the distributive property:

$$\begin{aligned} -1(3x + 4) &= -1(3x) + (-1)(4) \\ &= -3x - 4 \end{aligned}$$

The complete solution looks like this:

$$\begin{aligned} 5(x - 2) - (3x + 4) &= 5x - 10 - 3x - 4 && \text{Distributive property} \\ &= 2x - 14 && \text{Combine similar terms} \end{aligned}$$

▲

The last topic we will cover in this section will be useful later for checking solutions to equations. We will see how to find the value of an expression when the variable in the expression has been replaced by a number.

The Value of an Expression

An expression like $3x + 2$ will have a certain value depending on what number we assign to x. For instance, when x is 4, $3x + 2$ becomes $3(4) + 2$, or 14. When x is -8, $3x + 2$ becomes $3(-8) + 2$, or -22. The value of an expression is found by replacing the variable with a given number.

▼ **Examples** Find the value of the following expressions by replacing the variable with the given number.

Expression	*Value of the Variable*	*Value of the Expression*
10. $3x - 1$	$x = 2$	$3(2) - 1 = 6 - 1$ $= 5$
11. $7a + 4$	$a = -3$	$7(-3) + 4 = -21 + 4$ $= -17$
12. $2x - 3 + 4x$	$x = -1$	$2(-1) - 3 + 4(-1)$ $= -2 - 3 + (-4)$ $= -9$
13. $2x - 5 - 8x$	$x = 5$	$2(5) - 5 - 8(5)$ $= 10 - 5 - 40$ $= -35$
14. $y^2 - 6y + 9$	$y = 4$	$4^2 - 6(4) + 9$ $= 16 - 24 + 9$ $= 1$ ▲

Simplifying an expression should not change its value. That is, if an expression has a certain value when x is 5, then it will always have that value no matter how much it has been simplified as long as x is 5. If we were to simplify the expression in Example 13 first, it would look like

$$2x - 5 - 8x = -6x - 5$$

When x is 5, the simplified expression $-6x - 5$ is

$$-6(5) - 5 = -30 - 5 = -35$$

It has the same value as the original expression when x is 5.

We can also find the value of an expression that contains two variables if we know the values for both variables.

▼ **Example 15** Find the value of the expression $2x - 3y + 4$ when x is -5 and y is 6.

Solution Substituting -5 for x and 6 for y, the expression becomes

$$\begin{aligned} 2(-5) - 3(6) + 4 &= -10 - 18 + 4 \\ &= -28 + 4 \\ &= -24 \end{aligned}$$

▲

▼ **Example 16** Find the value of the expression $x^2 - 2xy + y^2$ when x is 3 and y is -4.

Solution Replacing each x in the expression with the number 3 and each y in the expression with the number -4 gives us

$$\begin{aligned} 3^2 - 2(3)(-4) + (-4)^2 &= 9 - 2(3)(-4) + 16 \\ &= 9 - (-24) + 16 \\ &= 33 + 16 \\ &= 49 \end{aligned}$$

▲

Problem Set 2.1

Simplify the following expressions.

1. $3x - 6x$
2. $7x - 5x$
3. $-2a + a$
4. $3a - a$
5. $7x + 3x + 2x$
6. $8x - 2x - x$
7. $3a - 2a + 5a$
8. $7a - a + 2a$
9. $4x - 3 + 2x$
10. $5x + 6 - 3x$
11. $3a + 4a + 5$
12. $6a + 7a + 8$
13. $2x - 3 + 3x - 2$
14. $6x + 5 - 2x + 3$
15. $3a - 1 + a + 3$
16. $-a + 2 + 8a - 7$
17. $-4x + 8 - 5x - 10$
18. $-9x - 1 + x - 4$
19. $7a + 3 + 2a + 3a$
20. $8a - 2 + a + 5a$
21. $5(2x - 1) + 4$
22. $2(4x - 3) + 2$
23. $7(3y + 2) - 8$
24. $6(4y + 2) - 7$
25. $-3(2x - 1) + 5$
26. $-4(3x - 2) - 6$
27. $5 - 2(a + 1)$
28. $7 - 8(2a + 3)$
29. $6 - 4(x - 5)$
30. $12 - 3(4x - 2)$
31. $-9 - 4(2 - y) + 1$
32. $-10 - 3(2 - y) + 3$
33. $-6 + 2(2 - 3x) + 1$
34. $-7 - 4(3 - x) + 1$
35. $3(x - 2) + 2(x - 3)$
36. $2(2x + 1) - 3(x + 4)$
37. $4(2y - 8) - (y + 7)$
38. $5(y - 3) - (y - 4)$
39. $-9(2x + 1) - 3(x + 5)$
40. $-3(3x - 2) - 4(2x + 3)$

Evaluate the following expressions when x is 2. (Find the value of the expressions if x is 2.)

41. $3x - 1$
42. $4x + 3$
43. $7x - 8$
44. $8x - 7$
45. $-2x - 5$
46. $-3x + 6$
47. $x^2 - 8x + 16$
48. $x^2 - 10x + 25$

49. $(x - 4)^2$
50. $(x - 5)^2$
51. $x^2 - 9$
52. $x^2 - 49$
53. $(x + 3)(x - 3)$
54. $(x + 7)(x - 7)$

Find the value of the following expressions when x is -5. Then, simplify the expression and check to see that it has the same value for $x = -5$.

55. $2x - 5x + 1$
56. $4x - 7x + 5$
57. $5x + 6 + 2x + 1$
58. $2x - 3 - 5x - 1$
59. $7x - 4 - x - 3$
60. $3x + 4 + 7x - 6$
61. $5(2x + 1) + 4$
62. $2(3x - 10) + 5$

Find the value of each expression when x is -3 and y is 5.

63. $4x - 3y + 8$
64. $7x - 2y + 3$
65. $-2x - y - 9$
66. $-5x - y - 10$
67. $x^2 - 2xy + y^2$
68. $x^2 + 2xy + y^2$
69. $(x - y)^2$
70. $(x + y)^2$
71. $x^2 + 6xy + 9y^2$
72. $x^2 + 10xy + 25y^2$
73. $(x + 3y)^2$
74. $(x + 5y)^2$

When we use words to describe a mathematical expression that includes parentheses, we sometimes use the phrase "the quantity" to show where the parentheses start and a comma to show where they end. For instance, the expression $2(3x + 4) - 5$ can be read "2 times the quantity $3x$ plus 4, minus 5." Write each of the following expressions in words and include the phrase "the quantity."

75. $3(5x + 1) - 6$
76. $4(2x - 6) + 7$
77. $4(2x + 3y) - 1$
78. $8(3x - 5y) + 4$
79. $4(2x + 3y - 1)$
80. $8(3x - 5y + 4)$

Review Problems Starting now, each problem set will end with a series of review problems. In mathematics, it is very important to review. The more you review, the better you will understand the topics we cover and the longer you will remember them. Also, there are times when material that seemed confusing earlier will be less confusing the second time around.

The following problems review material covered in Sections 1.6 and 1.7.

Simplify each expression.

81. $-2(4 - 9)$
82. $-6(3 - 10)$
83. $-4(-5) - 3$
84. $-2(6) - 8$
85. $3 - 5(6 - 2)$
86. $7 - 4(1 - 5)$
87. $\dfrac{2(-6) - 3(-2)}{4 - 7}$
88. $\dfrac{-3(5) - 7(-5)}{2 - 6}$

Section 2.2 Addition Property of Equality

In this section we will solve some simple equations. To solve an equation, we must find all replacements for the variable that make the equation a true statement.

DEFINITION The *solution set* for an equation is the set of all numbers that when used in place of the variable make the equation a true statement.

For example, the equation $x + 2 = 5$ has solution set $\{3\}$ because when x is 3 the equation becomes the true statement $3 + 2 = 5$, or $5 = 5$.

The important thing about an equation is its solution set. We therefore give the following definition in order to classify together all equations with the same solution set.

DEFINITION Two or more equations with the same solution set are said to be *equivalent equations*.

Equivalent equations may look different but must have the same solution set.

▼ **Examples**

1. $x + 2 = 5$ and $x = 3$ are equivalent equations, since both have solution set $\{3\}$.

2. $a - 4 = 3$, $a - 2 = 5$, and $a = 7$ are equivalent equations, since they all have solution set $\{7\}$.

3. $y + 3 = 4$, $y - 8 = -7$, and $y = 1$ are equivalent equations, since they all have solution set $\{1\}$. ▲

If two numbers are equal and we increase (or decrease) both of them by the same amount, the resulting quantities are also equal. We can apply this concept to equations. Adding the same amount to both sides of an equation always produces an equivalent equation—one with the same solution set. This fact about equations is called the *addition property of equality* and can be stated more formally as follows.

Addition Property of Equality

For any three algebraic expressions, A, B, and C,

$$\text{if} \quad A = B$$
$$\text{then } A + C = B + C$$

In words: Adding the same quantity to both sides of an equation will not change the solution set.

This property is just as simple as it seems. We can add any amount to both sides of an equation and always be sure we have not changed the solution set.

Consider the equation $x + 6 = 5$. We want to solve this equation for the value of x that makes it a true statement. We want to end up with x on one side of the equal sign and a number on the other side. Since we want x by itself, we will add -6 to both sides:

$$\begin{aligned} x + 6 + \mathbf{(-6)} &= 5 + \mathbf{(-6)} && \text{Addition property of equality} \\ x + 0 &= -1 && \text{Addition} \\ x &= -1 \end{aligned}$$

All three equations say the same thing about x. They all say that x is -1. All three equations are equivalent. The last one is just easier to read.

Here are some further examples of how the addition property of equality can be used to solve equations.

▼ **Example 4** Solve the equation $x - 5 = 12$ for x.

Solution Since we want x alone on the left side, we choose to add $+5$ to both sides:

$$\begin{aligned} x - 5 + \mathbf{5} &= 12 + \mathbf{5} && \text{Addition property of equality} \\ x + 0 &= 17 \\ x &= 17 \end{aligned}$$

▲

▼ **Example 5** Solve for a: $a + 7 = -3$.

Solution Since we want a to be by itself, we will add the opposite of 7 to both sides:

$$\begin{aligned} a + 7 + \mathbf{(-7)} &= -3 + \mathbf{(-7)} && \text{Addition property of equality} \\ a + 0 &= -10 \\ a &= -10 \end{aligned}$$

▲

▼ **Example 6** Solve for y: $y - 8 = -3$.

Solution Again, we want to isolate y, so we add the opposite of -8 to both sides:

$$\begin{aligned} y - 8 + \mathbf{8} &= -3 + \mathbf{8} && \text{Addition property of equality} \\ y + 0 &= 5 \\ y &= 5 \end{aligned}$$

▲

Sometimes it is necessary to simplify each side of an equation before using the addition property of equality. The reason we simplify both sides first is that we want as few terms as possible on each side of the equation before we use the addition property of equality. The following examples illustrate this procedure.

▼ **Example 7** Solve for x: $-x + 2 + 2x = 7 + 5$.

Solution

$x + 2 = 12$	Simplify both sides first
$x + 2 + (\mathbf{-2}) = 12 + (\mathbf{-2})$	Addition property of equality
$x + 0 = 10$	
$x = 10$	▲

▼ **Example 8** Solve: $3x - 4 - 2x = 11 - 4$.

Solution

$x - 4 = 7$	Simplify both sides first
$x - 4 + \mathbf{4} = 7 + \mathbf{4}$	Addition property of equality
$x + 0 = 11$	
$x = 11$	▲

▼ **Example 9** Solve: $4(2a - 3) - 7a = 2 - 5$.

Solution We must begin by applying the distributive property to separate terms on the left side of the equation. Following that, we combine similar terms and then apply the addition property of equality.

$4(2a - 3) - 7a = 2 - 5$	Original equation
$8a - 12 - 7a = 2 - 5$	Distributive property
$a - 12 = -3$	Simplify each side
$a - 12 + \mathbf{12} = -3 + \mathbf{12}$	Add **12** to each side
$a = 9$	Addition ▲

So far in this section we have used the addition property of equality to add only numbers to both sides of an equation. It is often necessary to add a term involving a variable to both sides of an equation, as the following examples indicate.

▼ **Example 10** Solve for x: $5x = 4x + 2$.

Solution To isolate x on the left side of the equation, we add $-4x$ to both sides:

$5x + (\mathbf{-4x}) = 4x + (\mathbf{-4x}) + 2$	Add $\mathbf{-4x}$ to both sides
$x = 2$	Simplify ▲

▼ **Example 11** Solve: $3x - 5 = 2x + 7$.

Solution We can solve this equation in two steps. First, we add $-2x$ to both sides of the equation. When this has been done, x will appear on the left side only. Second, we add 5 to both sides:

$$
\begin{aligned}
3x + (\mathbf{-2x}) - 5 &= 2x + (\mathbf{-2x}) + 7 && \text{Add } \mathbf{-2x} \text{ to both sides} \\
x - 5 &= 7 && \text{Simplify each side} \\
x - 5 + \mathbf{5} &= 7 + \mathbf{5} && \text{Add } \mathbf{5} \text{ to both sides} \\
x &= 12 && \text{Simplify each side}
\end{aligned}
$$

▲

To check your work, substitute your solution for x into the original equation.

▼ **Example 12** Check the solution for the equation in Example 11.

Solution We replace x with 12 in the equation $3x - 5 = 2x + 7$:

$$
\begin{aligned}
3(12) - 5 &\stackrel{?}{=} 2(12) + 7 \\
36 - 5 &= 24 + 7 \\
31 &= 31 && \text{A true statement}
\end{aligned}
$$

Since $x = 12$ makes the original equation a true statement, it must be the solution to the equation. ▲

Once you have become proficient at solving equations, it will not be necessary to check every solution to every equation you solve. What is important is that you know how to check your solutions when you think it is necessary—as on a test.

Many of the equations in the problem set that follows will seem very simple. You will be able to recognize solutions to some of them without doing any work. It won't always be this way. The idea here is to develop a method of solving equations. To do so, you must show all your work until you can consistently solve equations correctly using the method shown in this section.

A Note on Subtraction

Although the addition property of equality is stated for addition only, we can subtract the same number from both sides of an equation as well. Because subtraction is defined as addition of the opposite, subtracting the same quantity from both sides of an equation will not change the solution. If we were to solve the equation in Example 5 using subtraction instead of addition the steps would look like this:

$$
\begin{aligned}
a + 7 &= -3 && \text{Original equation} \\
a + 7 - \mathbf{7} &= -3 - \mathbf{7} && \text{Subtract } \mathbf{7} \text{ from each side} \\
a &= -10 && \text{Subtraction}
\end{aligned}
$$

In my experience teaching algebra, I find that students make fewer mistakes if they think in terms of addition rather than subtraction. So, you are probably better off if you continue to use the addition property just the way we have used it in the examples in this section. But, if you were curious as to whether you could subtract the same number from both sides of an equation, the answer is yes.

Problem Set 2.2

Find the solution set for the following equations. Be sure to show when you have used the addition property of equality.

1. $x - 3 = 8$
2. $x - 2 = 7$
3. $x + 2 = 6$
4. $x + 5 = 4$
5. $a + 7 = -2$
6. $a + 8 = -1$
7. $x + 5 = -3$
8. $x + 10 = -4$
9. $y + 11 = -6$
10. $y - 3 = -1$
11. $x - 5 = -2$
12. $x - 3 = -5$
13. $m - 6 = -10$
14. $m - 10 = -6$
15. $5 + x = 4$
16. $4 + x = 7$
17. $5 = a + 4$
18. $12 = a - 3$
19. $-3 = x - 8$
20. $-5 = x - 4$

Simplify both sides of the following equations as much as possible and then solve.

21. $4x + 2 - 3x = 4 + 1$
22. $5x + 2 - 4x = 7 - 3$
23. $8a - 5 - 7a = 3 + 5$
24. $9a - 18 - 8a = 20 - 7$
25. $-3 - 4x + 5x = 18$
26. $10 - 3x + 4x = 20$
27. $-11x + 2 + 10x + 2x = 9$
28. $-10x + 5 - 4x + 15x = 0$
29. $-3 + 4 = 8x - 5 - 7x$
30. $-2 + 5 = 7x - 4 - 6x$
31. $2y - 10 + 3y - 4y = 18 - 6$
32. $3y - 20 + 6y - 8y = 21$
33. $15 - 21 = 8x + 3x - 10x$
34. $23 - 17 = -7x - x + 9x$
35. $24 - 3 + 8a - 5a - 2a = 21$
36. $30 - 4 + 7a - 2 - 6a = 30$

The following equations contain parentheses. Apply the distributive property to remove the parentheses and then simplify each side before using the addition property of equality.•

37. $2(x + 3) - x = 4$
38. $5(x + 1) - 4x = 2$
39. $-3(x - 4) + 4x = 3 - 7$
40. $-2(x - 5) + 3x = 4 - 9$
41. $5(2a + 1) - 9a = 8 - 6$
42. $4(2a - 1) - 7a = 9 - 5$
43. $-(x + 3) + 2x - 1 = 6$
44. $-(x - 7) + 2x - 8 = 4$
45. $4y - 3(y - 6) + 2 = 8$
46. $7y - 6(y - 1) + 3 = 9$
47. $2(3x + 1) - 5(x + 2) = 1 - 10$
48. $4(2x + 1) - 7(x - 1) = 2 - 6$
49. $-3(2m - 9) + 7(m - 4) = 12 - 9$
50. $-5(m - 3) + 2(3m + 1) = 15 - 8$

Solve the following equations by the method used in Examples 10 and 11 in this section.

51. $4x = 3x + 2$

52. $6x = 5x - 4$

53. $8a = 7a - 5$

54. $9a = 8a - 3$

55. $2x = 3x + 1$

56. $4x = 3x + 5$

57. $3y + 4 = 2y + 1$

58. $5y + 6 = 4y + 2$

59. $2m - 3 = m + 5$

60. $8m - 1 = 7m - 3$

61. $4x - 7 = 5x + 1$

62. $3x - 7 = 4x - 6$

63. $-2x = -3x + 5$

64. $-5x = -6x + 3$

65. $4 - 2a = -7 - 3a$

66. $2 - 3a = -5 - 4a$

Review Problems The problems below review material we covered in Section 1.5. Reviewing this material will help you in the next section.

Apply the associative property to each expression and then simplify the result.

67. $3(6x)$

68. $5(4x)$

69. $\frac{1}{5}(5x)$

70. $\frac{1}{3}(3x)$

71. $8(\frac{1}{8}y)$

72. $6(\frac{1}{6}y)$

73. $-2(-\frac{1}{2}x)$

74. $-4(-\frac{1}{4}x)$

75. $-\frac{4}{3}(-\frac{3}{4}a)$

76. $-\frac{5}{2}(-\frac{2}{5}a)$

Section 2.3 Multiplication Property of Equality

In the previous section we found that adding the same number to both sides of an equation never changed the solution set. The same idea holds for multiplication by numbers other than zero. We can multiply both sides of an equation by the same nonzero number and always be sure we have not changed the solution set. (The reason we cannot multiply both sides by zero will become apparent later.) This fact about equations is called the *multiplication property of equality* and can be stated formally as follows.

> **Multiplication Property of Equality**
>
> For any three algebraic expressions A, B, and C, where $C \neq 0$,
>
> $$\text{if} \quad A = B$$
> $$\text{then } AC = BC$$
>
> *In words:* Multiplying both sides of an equation by the same nonzero number will not change the solution set.

Suppose we want to solve the equation $5x = 30$. We have $5x$ on the left side but would like to have just x. We choose to multiply both sides by $\frac{1}{5}$ since $(\frac{1}{5})(5) = 1$. Here is the solution:

$$5x = 30$$

$$\mathbf{\frac{1}{5}}(5x) = \mathbf{\frac{1}{5}}(30) \qquad \text{Multiplication property of equality}$$

$$\left(\frac{1}{5} \cdot 5\right)x = \frac{1}{5}(30) \qquad \text{Associative property of multiplication}$$

$$1x = 6$$

$$x = 6$$

We chose to multiply by $\frac{1}{5}$ because it is the reciprocal of 5. We can see that multiplication by any number except zero will not change the solution set. If, however, we were to multiply both sides by zero, the result would always be $0 = 0$, since multiplication by zero always results in zero. Although the statement $0 = 0$ is true, we have lost our variable and cannot solve the equation. This is the only restriction on the multiplication property of equality. We are free to multiply both sides of an equation by any number except zero.

Here are some more examples that use the multiplication property of equality.

▼ **Example 1** Solve for a: $-4a = 24$.

Solution Since we want a alone on the left side, we choose to multiply both sides by $-\frac{1}{4}$:

$$\mathbf{-\frac{1}{4}}(-4a) = \mathbf{-\frac{1}{4}}(24) \qquad \text{Multiplication property of equality}$$

$$\left[-\frac{1}{4}(-4)\right]a = \left(-\frac{1}{4}\right)(24) \qquad \text{Associative property}$$

$$a = -6$$

▲

▼ **Example 2** Solve for t: $-\dfrac{t}{3} = 5$

Solution Since division by 3 is the same as multiplication by $\frac{1}{3}$, we can write $-\frac{t}{3}$ as $-\frac{1}{3}t$. To solve the equation, we multiply each side by the reciprocal of $-\frac{1}{3}$, which is -3.

$$-\frac{t}{3} = 5 \qquad \text{Original equation}$$

$$-\frac{1}{3}t = 5 \qquad \text{Dividing by 3 is equivalent to multiplying by } \tfrac{1}{3}$$

$$\mathbf{-3}\left(-\frac{1}{3}t\right) = \mathbf{-3}(5) \qquad \text{Multiply each side by } \mathbf{-3}$$

$$\left[-3\left(-\frac{1}{3}\right)\right]t = -3(5) \qquad \text{Associative property}$$

$$t = -15 \qquad \text{Multiplication}$$

▲

▼ **Example 3** Solve: $\frac{2}{3}y = 4$.

Solution We can multiply both sides by $\frac{3}{2}$ and have $1y$ on the left side:

$$\mathbf{\frac{3}{2}}\left(\frac{2}{3}y\right) = \mathbf{\frac{3}{2}}(4) \qquad \text{Multiplication property of equality}$$

$$\left(\frac{3}{2}\cdot\frac{2}{3}\right)y = \frac{3}{2}(4) \qquad \text{Associative property}$$

$$y = 6 \qquad \text{Simplify: } \tfrac{3}{2}(4) = \tfrac{3}{2}(\tfrac{4}{1}) = \tfrac{12}{2} = 6$$

▲

Notice in Examples 1 through 3 that if the variable is being multiplied by a number like -4 or $\frac{2}{3}$, we always multiply by the number's reciprocal, $-\frac{1}{4}$ or $\frac{3}{2}$, to end up with just x on one side of the equation.

▼ **Example 4** Solve: $5 + 8 = 10x + 20x - 4x$.

Solution Our first step will be to simplify each side of the equation:

$$13 = 26x \qquad \text{Simplify both sides first}$$

$$\mathbf{\frac{1}{26}}(13) = \mathbf{\frac{1}{26}}(26x) \qquad \text{Multiplication property of equality}$$

$$\frac{13}{26} = x \qquad \text{Multiplication}$$

$$\frac{1}{2} = x \qquad \text{Reduce to lowest terms}$$

▲

In the next three examples we will use both the addition property of equality and the multiplication property of equality. As a general rule, we use the addition property of equality before the multiplication property of equality.

▼ **Example 5** Solve for x: $6x + 5 = -13$.

Solution We begin by adding -5 to both sides of the equation:

$$6x + 5 + (\mathbf{-5}) = -13 + (\mathbf{-5}) \qquad \text{Add } \mathbf{-5} \text{ to both sides}$$

$$6x = -18 \qquad \text{Simplify}$$

$$\frac{1}{6}(6x) = \frac{1}{6}(-18) \qquad \text{Multiply both sides by } \tfrac{1}{6}$$

$$x = -3$$

▲

▼ **Example 6** Solve for x: $5x = 2x + 12$.

Solution We begin by adding $-2x$ to both sides of the equation:

$$5x + (\mathbf{-2x}) = 2x + (\mathbf{-2x}) + 12 \qquad \text{Add } \mathbf{-2x} \text{ to both sides}$$

$$3x = 12 \qquad \text{Simplify}$$

$$\frac{1}{3}(3x) = \frac{1}{3}(12) \qquad \text{Multiply both sides by } \tfrac{1}{3}$$

$$x = 4 \qquad \text{Simplify}$$

▲

Notice that in Example 6 we used the addition property of equality first in order to combine all the terms containing x on the left side of the equation. Once this had been done, we used the multiplication property to isolate x on the left side.

▼ **Example 7** Solve for x: $3x - 4 = -2x + 6$.

Solution We begin by adding $2x$ to both sides:

$$3x + \mathbf{2x} - 4 = -2x + \mathbf{2x} + 6 \qquad \text{Add } \mathbf{2x} \text{ to both sides}$$

$$5x - 4 = 6 \qquad \text{Simplify}$$

Now we add 4 to both sides:

$$5x - 4 + \mathbf{4} = 6 + \mathbf{4} \qquad \text{Add } \mathbf{4} \text{ to both sides}$$

$$5x = 10 \qquad \text{Simplify}$$

$$\frac{1}{5}(5x) = \frac{1}{5}(10) \qquad \text{Multiply by } \tfrac{1}{5}$$

$$x = 2 \qquad \text{Simplify}$$

▲

As you can see in examples 5 through 7, both properties can be used to solve equations. The addition property is used first to combine all the terms containing the variable together on one side of the equation and terms without the variable on the other side. Once that has been accomplished, the multiplication property of equality is used to obtain just one of whatever variable is being solved for.

For the last examples in this section we will consider some simple word problems that are easy to translate into equations. They are all problems that involve percent.

▼ **Example 8** What number is 25% of 60?

Solution To solve a problem like this we let x = the number in question (that is, the number we are looking for). Then, we translate the sentence directly into an equation by using an equal sign for the word "is" and multiplication for the word "of." Here is how it is done:

$$\underbrace{\text{What number}}\ \text{is}\ 25\%\ \text{of}\ 60?$$
$$\begin{aligned} x &= .25 \cdot 60 \\ x &= 15 \end{aligned}$$

Notice that we must write 25% as a decimal in order to do the arithmetic in the problem.

The number 15 is 25% of 60. ▲

▼ **Example 9** What percent of 24 is 6?

Solution Translating this sentence into an equation as we did in Example 8, we have:

$$\underbrace{\text{What percent}}\ \text{of}\ 24\ \text{is}\ 6?$$
$$\begin{aligned} x \cdot 24 &= 6 \\ \text{or} \quad 24x &= 6 \end{aligned}$$

Next we multiply each side by $\frac{1}{24}$. (This is the same as dividing each side by 24):

$$\begin{aligned} \frac{\mathbf{1}}{\mathbf{24}}(24x) &= \frac{\mathbf{1}}{\mathbf{24}}(6) \\ x &= \frac{6}{24} \\ &= \frac{1}{4} \\ &= .25 \text{ or } 25\% \end{aligned}$$

The number 6 is 25% of 24. ▲

▼ **Example 10** 45 is 75% of what number?

Solution Again, we translate the sentence directly:

$$45\ \text{is}\ 75\%\ \text{of}\ \underbrace{\text{what number}}?$$
$$45 = .75 \cdot x$$

Next, we multiply each side by $\frac{1}{.75}$ (which is the same as dividing each side by .75):

$$\frac{\mathbf{1}}{\mathbf{.75}}(45) = \frac{\mathbf{1}}{\mathbf{.75}}(.75x)$$

$$\frac{45}{.75} = x$$

$$60 = x$$

The number 45 is 75% of 60. ▲

A Note on Division

Since division is defined as multiplication by the reciprocal, multiplying both sides of an equation by the same number is equivalent to dividing both sides of the equation by the reciprocal of that number. That is, multiplying each side of an equation by $\frac{1}{5}$ and dividing each side of the equation by 5 are equivalent operations. If we were to solve the equation $3x = 18$ using division instead of multiplication, the steps would look like this:

$3x = 18$	Original equation
$\frac{3x}{\mathbf{3}} = \frac{18}{\mathbf{3}}$	Divide each side by **3**
$x = 6$	Division

Using division instead of multiplication on a problem like this may save you some writing. On the other hand, with multiplication, it is easier to explain "why" we end up with just one x on the left side of the equation. (The "why" has to do with the associative property of multiplication.) My suggestion is that you continue to use multiplication to solve equations like the one above until you understand the process completely. Then, if you find it more convenient, you can use division instead of multiplication.

Problem Set 2.3

Solve the following equations. Be sure to show your work.

1. $5x = 10$
2. $6x = 12$
3. $7a = 28$
4. $4a = 36$
5. $-8x = 4$
6. $-6x = 2$
7. $8m = -16$
8. $5m = -25$
9. $-3x = -9$
10. $-9x = -36$
11. $-7y = -28$
12. $-15y = -30$
13. $2x = 0$
14. $7x = 0$
15. $-5x = 0$
16. $-3x = 0$

17. $\frac{x}{3} = 2$

18. $\frac{x}{4} = 3$

19. $-\frac{m}{5} = 10$

20. $-\frac{m}{7} = 1$

21. $-\frac{x}{2} = -3$

22. $-\frac{x}{3} = -2$

23. $\frac{2}{3}a = 8$

24. $\frac{3}{4}a = 6$

25. $-\frac{3}{5}x = 12$

26. $-\frac{2}{5}x = 4$

27. $-\frac{5}{8}y = -20$

28. $-\frac{7}{2}y = -14$

Simplify both sides as much as possible and then solve.

29. $-4x - 2x + 3x = 24$

30. $7x - 5x + 8x = 20$

31. $4x + 8x - 2x = 15 - 10$

32. $5x + 4x + 3x = 4 + 8$

33. $-3 - 5 = 3x + 5x - 10x$

34. $10 - 16 = 12x - 6x - 3x$

35. $18 - 3 = 8a - 5a + 2a$

36. $20 - 4 = 7a + a - 4a$

Solve each of the following equations by multiplying both sides by -1.

37. $-x = 4$

38. $-x = -3$

39. $-x = -4$

40. $-x = 3$

41. $15 = -a$

42. $-15 = -a$

43. $-y = \frac{1}{2}$

44. $-y = -\frac{3}{4}$

Solve each of the following equations using the method shown in Examples 5, 6, and 7 in this section.

45. $3x - 2 = 7$

46. $2x - 3 = 9$

47. $2a + 1 = 3$

48. $5a - 3 = 7$

49. $1 + 4x = 2$

50. $7 + 3x = -8$

51. $6x = 2x - 12$

52. $8x = 3x - 10$

53. $2y = -4y + 18$

54. $3y = -2y - 15$

55. $-7x = -3x - 8$

56. $-5x = -2x - 12$

57. $8x + 4 = 2x - 5$

58. $5x + 6 = 3x - 6$

59. $6m - 3 = m + 2$

60. $6m - 5 = m + 5$

61. $9y + 2 = 6y - 4$

62. $6y + 14 = 2y - 2$

63. There is no solution to the equation $2x - 5 = 2x + 3$. That is, there is no real number we can use in place of x to turn the equation into a true statement. What happens when you try to solve the equation?

64. The equation $5x - 2 = 5(x - 1)$ has no solution. What happens when you try to solve the equation?

65. Every real number is a solution to the equation $2x + 6 = 2(x + 3)$. Solve the equation.

66. Every real number is a solution to the equation $5(2 - x) = -5x + 10$. Solve the equation.

Translate each of the following into an equation and then solve that equation.

67. What number is 25% of 40?
68. What number is 75% of 40?
69. What number is 12% of 2000?
70. What number is 9% of 3000?
71. What percent of 28 is 7?
72. What percent of 28 is 21?
73. What percent of 40 is 14?
74. What percent of 20 is 14?
75. 32 is 50% of what number?
76. 16 is 50% of what number?
77. 240 is 12% of what number?
78. 360 is 12% of what number?

Review Problems The problems below review material we covered in Section 2.1. Reviewing this material will help you in the next section.

Simplify each expression.

79. $5(2x - 8) - 3$
80. $4(3x - 1) + 7$
81. $-2(3x + 5) + 3(x - 1)$
82. $6(x + 3) - 2(2x + 4)$
83. $7 - 3(2y + 1)$
84. $8 - 5(3y - 4)$
85. $4x - (9x - 3) + 4$
86. $x - (5x + 2) - 3$

Section 2.4 Solving Linear Equations

We will now use the material we have developed in the first three sections of this chapter to build a method for solving any linear equation.

DEFINITION A *linear equation* in one variable is any equation that can be put in the form $ax + b = 0$, where a and b are real numbers and a is not zero.

Each of the equations we will solve in this section is a linear equation in one variable. The steps we use to solve a linear equation in one variable are listed here.

To Solve a Linear Equation in One Variable

Step 1: Use the distributive property to separate terms, if necessary, and simplify both sides of the equation as much as possible.

Step 2: Use the addition property of equality to get all variable terms on one side and all constant terms on the other. A variable term is a term that contains the variable (for example, $5x$). A constant term is a term that does not contain the variable (the number 3, for example).

Step 3: Use the multiplication property of equality to get x by itself on one side.

Step 4: Check your solution in the original equation if you think it is necessary.

▼ **Example 1** Solve: $2(x + 3) = 10$.

Solution To begin, we apply the distributive property to the left side of the equation to separate terms:

$$2x + 6 = 10 \qquad \text{Distributive property}$$
$$2x + 6 + (\mathbf{-6}) = 10 + (\mathbf{-6}) \qquad \text{Addition property of equality}$$
$$2x = 4$$
$$\frac{\mathbf{1}}{\mathbf{2}}(2x) = \frac{\mathbf{1}}{\mathbf{2}}(4) \qquad \text{Multiply each side by } \tfrac{1}{2}$$
$$x = 2$$

▲

The general method of solving linear equations is actually very simple. It is based on the properties we developed in Chapter 1 and two very simple new properties. We can add any number to both sides of the equation and multiply both sides by any nonzero number. The equation may change in form, but the solution set will not. If we look back to Example 1, each equation looks a little different from each preceding equation. What is interesting and useful is that each equation says the same thing about x. They all say x is 2. The last equation, of course, is the easiest to read, and that is why our goal is to end up with x by itself.

▼ **Example 2** Solve for x: $3(x - 5) + 4 = 13$.

Solution Our first step will be to apply the distributive property to the left side of the equation:

$$3x - 15 + 4 = 13 \qquad \text{Distributive property}$$
$$3x - 11 = 13 \qquad \text{Simplify the left side}$$
$$3x - 11 + \mathbf{11} = 13 + \mathbf{11} \qquad \text{Add } \mathbf{11} \text{ to both sides}$$
$$3x = 24$$
$$\frac{\mathbf{1}}{\mathbf{3}}(3x) = \frac{\mathbf{1}}{\mathbf{3}}(24) \qquad \text{Multiply both sides by } \tfrac{1}{3}$$
$$x = 8$$

▲

▼ **Example 3** Solve for a: $5(2a - 5) = 3a - 4$.

Solution We begin by distributing the 5 across the difference $2a - 5$:

$$10a - 25 = 3a - 4 \qquad \text{Distributive property}$$
$$10a - 25 + \mathbf{25} = 3a - 4 + \mathbf{25} \qquad \text{Add } \mathbf{25} \text{ to both sides}$$
$$10a = 3a + 21$$
$$10a + (\mathbf{-3a}) = 3a + (\mathbf{-3a}) + 21 \qquad \text{Add } \mathbf{-3a} \text{ to both sides}$$
$$7a = 21$$

$$\frac{1}{7}(7a) = \frac{1}{7}(21) \quad \text{Multiply both sides by } \tfrac{1}{7}$$
$$a = 3$$

▲

▼ **Example 4** Solve: $5(x - 3) + 2 = 5(2x - 8) - 3$.

Solution In this case, we apply the distributive property on each side of the equation:

$5x - 15 + 2 = 10x - 40 - 3$	Distributive property
$5x - 13 = 10x - 43$	Simplify both sides
$5x + (\mathbf{-5x}) - 13 = 10x + (\mathbf{-5x}) - 43$	Add $\mathbf{-5x}$ to both sides
$-13 = 5x - 43$	
$-13 + \mathbf{43} = 5x - 43 + \mathbf{43}$	Add **43** to both sides
$30 = 5x$	
$\frac{1}{5}(30) = \frac{1}{5}(5x)$	Multiply both sides by $\frac{1}{5}$
$6 = x$	

▲

It makes no difference on which side of the equal sign x ends up. Most people prefer to have x on the left side because we read from left to right and it seems to sound better to say x is 6 rather than 6 is x. Both expressions, however, have exactly the same meaning.

▼ **Example 5** Solve: $7 - 3(2y + 1) = 16$.

Solution We begin by multiplying -3 times the sum of $2y$ and 1:

$7 - 6y - 3 = 16$	Distributive property
$-6y + 4 = 16$	Simplify the left side
$-6y + 4 + (\mathbf{-4}) = 16 + (\mathbf{-4})$	Add $\mathbf{-4}$ to both sides
$-6y = 12$	
$-\frac{1}{6}(-6y) = -\frac{1}{6}(12)$	Multiply both sides by $-\frac{1}{6}$
$y = -2$	

▲

▼ **Example 6** Check the solution from Example 5 in the original equation.

Solution When $y = -2$

the equation $7 - 3(2y + 1) = 16$

becomes $7 - 3[2(-2) + 1] \stackrel{?}{=} 16$

$$7 - 3(-4 + 1) = 16$$
$$7 - 3(-3) = 16$$
$$7 + 9 = 16$$
$$16 = 16 \quad \text{A true statement}$$

▲

There are two things to notice about the example that follows: first, the distributive property is used to remove parentheses that are preceded by a negative sign and second, the addition property and the multiplication property are not shown in as much detail as in the previous examples.

▼ **Example 7** Solve: $3(2x - 5) - (2x - 4) = 6 - (4x + 5)$.

Solution When we apply the distributive property to remove the grouping symbols and separate terms, we have to be careful with the signs. Remember, we can think of $-(2x - 4)$ as $-1(2x - 4)$ so that

$$-(2x - 4) = -1(2x - 4) = -2x + 4$$

It is not uncommon for students to make a mistake with this type of simplification and write the result as $-2x - 4$, which is incorrect. Here is the complete solution to our equation:

$3(2x - 5) - (2x - 4) = 6 - (4x + 5)$	Original equation
$6x - 15 - 2x + 4 = 6 - 4x - 5$	Distributive property
$4x - 11 = -4x + 1$	Simplify each side
$8x - 11 = 1$	Add $4x$ to each side
$8x = 12$	Add 11 to each side
$x = \frac{12}{8}$	Multiply each side by $\frac{1}{8}$
$x = \frac{3}{2}$	Reduce to lowest terms

▲

Problem Set 2.4

Solve each of the following equations using the four steps shown in this section.

1. $2(x + 3) = 12$
2. $3(x - 2) = 6$
3. $6(x - 1) = -18$
4. $4(x + 5) = 16$
5. $2(4a + 1) = -6$
6. $3(2a - 4) = 12$
7. $14 = 2(5x - 3)$
8. $-25 = 5(3x + 4)$
9. $-2(3y + 5) = 14$
10. $-3(2y - 4) = -6$
11. $-5(2a + 4) = 0$
12. $-3(3a - 6) = 0$
13. $6 = 3(4x + 2)$
14. $12 = 4(2x + 3)$
15. $3(t - 4) + 5 = -4$
16. $5(t - 1) + 6 = -9$
17. $4(2y + 1) - 7 = 1$
18. $6(3y + 2) - 8 = -2$
19. $4(x - 3) = 2(x + 1)$
20. $2(x - 4) = 3(x - 6)$
21. $-7(2x - 7) = 3(11 - 4x)$
22. $-3(2x - 5) = 7(3 - x)$
23. $5(3x - 1) = 4(2x + 1) - 30$
24. $4(2x + 3) = -3(x - 1) + 20$
25. $-2(3y + 1) = 3(1 - 6y) - 9$
26. $-5(4y - 3) = 2(1 - 8y) + 11$

27. $8 - 4(x + 5) = -12$
28. $7 - 3(x + 2) = 1$
29. $6 - 5(2a - 3) = 1$
30. $-8 - 2(3 - a) = 0$
31. $2x - 5 = 5 - 2(2x - 13)$
32. $4x - 1 = 7 - 3(6 - 2x)$
33. $2(t - 3) + 3(t - 2) = 28$
34. $-3(t - 5) - 2(2t + 1) = -8$
35. $5(x - 2) - (3x + 4) = 3(6x - 8) + 10$
36. $3(x - 1) - (4x - 5) = 2(5x - 1) - 7$
37. $2(5x - 3) - (2x - 4) = 5 - (6x + 1)$
38. $3(4x - 2) - (5x - 8) = 8 - (2x + 3)$
39. $-(3x + 1) - (4x - 7) = 4 - (3x + 2)$
40. $-(6x + 2) - (8x - 3) = 8 - (5x + 1)$

None of the following equations has a single number for its solution. The solution set for each one is either all real numbers or the empty set (in which case we say there are no solutions). In each case, if you try to collect all the variable terms on one side, you will eliminate the variable completely. Once the variable has been eliminated, you will be left with either a true statement or a false statement. A true statement indicates the solution set is all real numbers, in which case every real number satisfies the equation. A false statement indicates there are no solutions to the equation; any number you substitute for x will give a false statement.

Solve each equation, if possible.

41. $2(x + 3) = 2x + 6$
42. $3x - 9 = 3(x - 3)$
43. $x + 5 = x + 7$
44. $2x + 4 = 2x - 1$
45. $8 - 4x = 2(4 - 2x)$
46. $12 - 10x = 2(6 - 5x)$
47. $4x + 6 = 2(2x - 3)$
48. $2x - 14 = 2(x + 3)$

Review Problems The problems that follow review material we covered in Sections 1.2, 1.5, and 1.6. Reviewing these problems will help you understand the next section.

Multiply.

49. $\frac{1}{2}(3)$
50. $\frac{1}{3}(2)$
51. $\frac{2}{3}(6)$
52. $\frac{3}{2}(4)$
53. $\frac{5}{9} \cdot \frac{9}{5}$
54. $\frac{3}{7} \cdot \frac{7}{3}$

Apply the distributive property and then simplify each expression as much as possible.

55. $2(3x - 5)$
56. $4(2x - 6)$
57. $\frac{1}{2}(3x + 6)$
58. $\frac{1}{4}(2x + 8)$
59. $\frac{1}{3}(-3x + 6)$
60. $\frac{1}{2}(-2x + 6)$

Section 2.5 Formulas

A formula in mathematics is an equation that contains more than one variable. The equation $P = 2l + 2w$, which tells us how to find the perimeter of a rectangle, is an example of a formula.

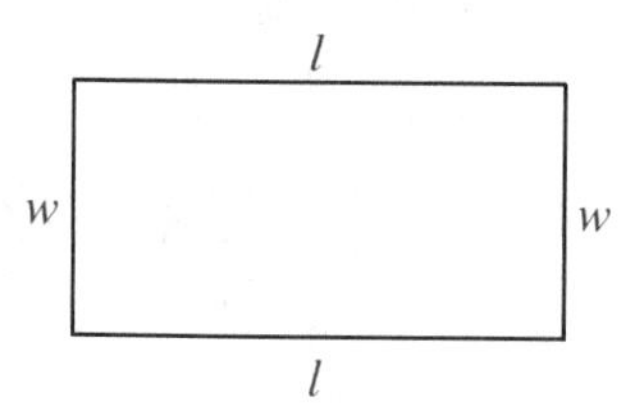

To begin our work with formulas, we will consider some examples in which we are given numerical replacements for all but one of the variables.

▼ **Example 1** The perimeter P of a rectangular livestock pen is 40 feet. If the width w is 6 feet, find the length.

Solution First we substitute 40 for P and 6 for w in the formula $P = 2l + 2w$. Then we solve for l:

When	$P = 40$ and $w = 6$	
the formula	$P = 2l + 2w$	
becomes	$40 = 2l + 2(6)$	
or	$40 = 2l + 12$	Multiply 2 and 6
	$28 = 2l$	Add -12 to each side
	$14 = l$	Multiply each side by $\frac{1}{2}$

To summarize our results, if a rectangular pen has a perimeter of 40 feet and a width of 6 feet, then the length must be 14 feet. ▲

▼ **Example 2** Find y when $x = 4$ in the formula $3x + 2y = 6$.

Solution We substitute 4 for x in the formula and then solve for y:

When	$x = 4$	
the formula	$3x + 2y = 6$	
becomes	$3(4) + 2y = 6$	
or	$12 + 2y = 6$	Multiply 3 and 4
	$2y = -6$	Add -12 to each side
	$y = -3$	Multiply each side by $\frac{1}{2}$ ▲

Our next example involves a formula that contains the number π. As a general rule, if the problem is stated with decimals, use 3.14 for π. If the problem is stated with fractions, use $\frac{22}{7}$ for π.

▼ **Example 3** A new sports car has a 6 cylinder engine with a total displacement of 3000 cubic centimeters. This means that each cylinder has a displacement of 500 cubic centimeters. If the radius of each cylinder is $\frac{9}{2}$ centimeters, find the height of each cylinder using the formula $V = \pi r^2 h$. (Use $\frac{22}{7}$ for π.)

Solution Substituting 500 for V, $\frac{9}{2}$ for r, and $\frac{22}{7}$ for π in the formula $V = \pi r^2 h$ we have:

$$500 = \frac{22}{7}\left(\frac{9}{2}\right)^2 h$$

$$500 = \frac{22}{7}\left(\frac{81}{4}\right) h$$

$$500 = \frac{1782}{28} h$$

$$500 = \frac{891}{14} h$$

$$\frac{14}{891} 500 = h$$

$$\frac{7000}{891} = h$$

If we divide 7000 by 891 we obtain 7.9 centimeters as the approximate height of the cylinder. You may want to try this example again using 4.5 for r and 3.14 for π, doing the arithmetic on a calculator. The result should be the same. ▲

In the next examples we will solve a formula for one of its variables without being given numerical replacements for the other variables.

Consider the formula for the area of a triangle:

$$A = \frac{1}{2}bh$$

h

b

where A = area, b = length of the base, and h = the height of the triangle.

Suppose we want to solve this formula for h. What we must do is isolate the variable h on one side of the equal sign. We begin by multiplying both sides by 2:

$$2 \cdot A = 2 \cdot \frac{1}{2}bh$$

$$2A = bh$$

Then we divide both sides by b:

$$\frac{2A}{b} = \frac{bh}{b}$$

$$h = \frac{2A}{b}$$

The original formula $A = \frac{1}{2}bh$ and the final formula $h = \dfrac{2A}{b}$ both give the same relationship among A, b, and h. The first one has been solved for A and the second one has been solved for h.

> **Rule** To solve a formula for one of its variables, we must isolate that variable on either side of the equal sign. All other variables and constants will appear on the other side.

▼ **Example 4** Solve $3x + 2y = 6$ for y.

Solution To solve for y we must isolate y on the left side of the equation. To begin, we use the addition property of equality to add $-3x$ to each side:

$$\begin{aligned} 3x + 2y &= 6 && \text{Original formula} \\ 3x + (\mathbf{-3x}) + 2y &= (\mathbf{-3x}) + 6 && \text{Add } \mathbf{-3x} \text{ to each side} \\ 2y &= -3x + 6 && \text{Simplify the left side} \\ \mathbf{\frac{1}{2}}(2y) &= \mathbf{\frac{1}{2}}(-3x + 6) && \text{Multiply each side by } \tfrac{1}{2} \\ y &= -\frac{3}{2}x + 3 && \text{Multiplication} \end{aligned}$$

Notice that the steps in this example are very similar to the steps in Example 2. The difference is that, in this example, we solved for y without knowing what x was. Our answer in this example gives y in terms of x. ▲

▼ **Example 5** Solve $C = \frac{5}{9}(F - 32)$ for F.

Solution This formula gives the relationship between the Fahrenheit temperature scale and the Celsius temperature scale. We begin by multiplying both sides by $\frac{9}{5}$:

$$\mathbf{\frac{9}{5}}C = \mathbf{\frac{9}{5}} \cdot \frac{5}{9}(F - 32)$$

$$\frac{9}{5}C = F - 32$$

We finish the problem by adding 32 to both sides:

$$\frac{9}{5}C + \mathbf{32} = F - 32 + \mathbf{32}$$

$$\frac{9}{5}C + 32 = F$$

or

$$F = \frac{9}{5}C + 32$$

▲

▼ **Example 6** Solve $A = \frac{1}{2}(b + B)h$ for B.

Solution The formula in this example is the formula used to find the area of a trapezoid with bases b and B, and height h.

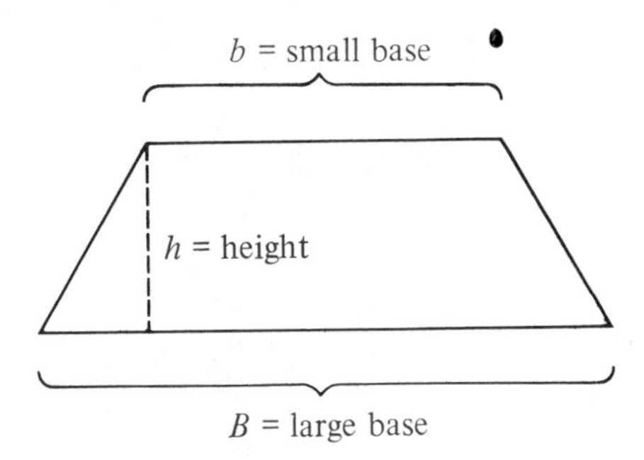

Since multiplication is both commutative and associative, we can rewrite the formula as follows:

$$A = \frac{h}{2}(b + B)$$

Multiplying both sides by $\frac{2}{h}$, we have

$$\frac{\mathbf{2}}{\mathbf{h}}A = \frac{\mathbf{2}}{\mathbf{h}} \cdot \frac{h}{2}(b + B)$$

$$\frac{2A}{h} = b + B$$

Now, adding $-b$ to both sides completes the problem:

$$\frac{2A}{h} + (\mathbf{-b}) = b + (\mathbf{-b}) + B$$

$$\frac{2A}{h} - b = B$$

or

$$B = \frac{2A}{h} - b$$

▲

There are a number of other ways to solve for B. This is only one of them. The problem is complete when the variable we are solving for appears alone on only one side of the equal sign.

Problem Set 2.5

Use the formula $P = 2l + 2w$ to find the length l of a rectangular lot if

1. the width w is 50 feet and the perimeter P is 300 feet.
2. the width w is 75 feet and the perimeter P is 300 feet.

Use the formula $2x + 3y = 6$ to find y if

3. x is 3
4. x is -2
5. x is 0
6. x is -3

The circumference of a circle is given by the formula $C = 2\pi r$. Find r if

7. the circumference C is 44 meters and π is $\frac{22}{7}$.
8. the circumference C is 176 meters and π is $\frac{22}{7}$.
9. the circumference is 9.42 inches and π is 3.14.
10. the circumference is 12.56 inches and π is 3.14.

Use the formula $2x - 5y = 20$ to find x if

11. y is 2
12. y is -4
13. y is 0
14. y is -6

The volume of a cylinder is given by the formula $V = \pi r^2 h$. Find the height h if

15. the volume V is 42 cubic feet, the radius is $\frac{7}{22}$ feet, and π is $\frac{22}{7}$.
16. the volume V is 84 cubic inches, the radius is $\frac{7}{11}$ inches, and π is $\frac{22}{7}$.
17. the volume is 6.28 cubic centimeters, the radius is 3 centimeters, and π is 3.14.
18. the volume is 12.56 cubic centimeters, the radius is 2 centimeters, and π is 3.14.

Use the equation $y = 2x - 1$ to find x when

19. y is 7
20. y is 9
21. y is 3
22. y is -1

Solve each of the following formulas for the indicated variable.

23. $A = lw$ for l
24. $A = lw$ for w
25. $d = rt$ for r
26. $d = rt$ for t
27. $V = lwh$ for h
28. $V = lwh$ for l
29. $V = IR$ for R
30. $V = IR$ for I
31. $PV = nRT$ for P
32. $PV = nRT$ for T

33. $V = \pi r^2 h$ for h
34. $V = \frac{1}{3}\pi r^2 h$ for h
35. $c^2 = a^2 + b^2$ for a^2
36. $c^2 = a^2 + b^2$ for b^2
37. $P = a + b + c$ for a
38. $P = a + b + c$ for b
39. $x - 3y = -1$ for x
40. $x + 3y = 2$ for x
41. $-3x + y = 6$ for y
42. $2x + y = -17$ for y
43. $2x + 3y = 6$ for y
44. $4x + 5y = 20$ for y
45. $6x + 3y = 12$ for y
46. $3x + 6y = 12$ for y
47. $5x - 2y = 3$ for y
48. $7x - 3y = 5$ for y
49. $P = 2l + 2w$ for w
50. $P = 2l + 2w$ for l
51. $F = \frac{9}{5}C + 32$ for C
52. $A = a + (n - 1)d$ for d
53. $h = vt + 16t^2$ for v
54. $h = vt - 16t^2$ for v
55. $A = \pi r^2 + 2\pi rh$ for h
56. $A = 2\pi r^2 + 2\pi rh$ for h
57. $A = \frac{1}{2}(b + B)h$ for b
58. $A = a + (n - 1)d$ for n
59. $3x - 4y - 12 = 0$ for y
60. $2x - 6y - 12 = 0$ for y
61. $-2x - 4y + 14 = 0$ for x
62. $-3x - 6y + 15 = 0$ for x

63. The formula $C = \frac{5}{9}(F - 32)$ gives the relationship between temperatures measured in degrees Fahrenheit and the corresponding temperature in degrees Celsius. Find the temperature in degrees Celsius by letting $F = 77$ in the formula.
64. If the temperature is 86° Fahrenheit, what is it in degrees Celsius?
65. The surface area A of a cylinder that is closed at the top and bottom is given by the formula $A = 2\pi r^2 + 2\pi rh$, where r is the radius and h is the height of the cylinder. Find the surface area of a can of corn if the radius is 4 centimeters and the height is 11 centimeters. (Use 3.14 for π.) The units on your answer will be square centimeters.
66. Find the surface area of a one-pound can of coffee if its radius is 5 centimeters and its height is 13 centimeters. (Use 3.14 for π.)
67. Find the volume of the can, described in Problem 65, by using the formula $V = \pi r^2 h$.
68. Find the volume of the can described in Problem 66.

Review Problems The problems that follow review material we covered in Chapter 1 and in Section 2.3. Reviewing these problems will help you understand the next section.

Write an equivalent expression in symbols and then simplify it.

69. Twice the sum of 6 and 4.
70. Twice the difference of 7 and 3.
71. The sum of twice 6 and 4.
72. The difference of twice 7 and 3.
73. What number is 12% of 500?
74. What number is 9% of 800?
75. What number is 10% of 2,000?
76. What number is 8% of 3,000?
77. 160 is 8% of what number?
78. 320 is 8% of what number?
79. 220 is 11% of what number?
80. 330 is 11% of what number?

Section 2.6 Word Problems

It seems that the major difference between those people who are good at working word problems and those who are not is confidence. The people with confidence know that no matter how long it takes them, they will eventually be able to solve the problem. Those without confidence begin by saying to themselves, "I'll never be able to work this problem." Are you like that? If you are, what you need to do is put your old ideas about you and word problems aside for awhile and make a decision to be successful. Sometimes that's all it takes. Instead of telling yourself that you can't do word problems, that you don't like them, or that they're not good for anything anyway, decide to do whatever it takes to master them. I think you'll find your attitude toward word problems will change once you've experienced success solving them. Let's begin our study of word problems by listing the steps we will use in the examples that follow.

To Solve a Word Problem

Step 1: Let x (or any letter you choose) represent the unknown quantity. It is usually the quantity asked for in the problem.
Step 2: Write an equation, in x, that describes the situation.
Step 3: Solve the equation you wrote in Step 2.
Step 4: Check your solution in the original problem.

Step 2 is the most difficult part. It may be necessary to read each problem a number of times before you can do this step correctly. You may make some mistakes along the way. There will be times when your first attempt at Step 2 results in the wrong equation. Mistakes are part of the process of learning to do things correctly. Many times the correct equation will become obvious after you have written an equation that is partially wrong. In any case, it is better to write an equation that is partially wrong and be actively involved with the problem than to write nothing at all. Word problems, like other problems in algebra, are not always solved correctly the first time.

Here are some common English words and phrases and their mathematical translation:

English	Algebra
the difference of a and b	$a - b$
the product of a and b	$a \cdot b$
the quotient of a and b	$\frac{a}{b}$
of	$\cdot$ (multiply)
is	$=$ (equals)
a number	x
4 more than x	$x + 4$
4 times x	$4x$
4 less than x	$x - 4$

Look over the following examples and try to solve some on your own. The examples are similar to the problems in the problem set.

Number Problems

▼ **Example 1** The sum of twice a number and three is seven. Find the number.

Solution

Step 1: Let x = the number asked for.

Step 2: "The sum of twice a number and three" translates to $2x + 3$; "is" translates to =; and "seven" is 7. An equation that describes the situation, then, is

$$2x + 3 = 7$$

Step 3: Solve the equation:

$$\begin{aligned} 2x + 3 &= 7 \\ 2x + 3 + (\mathbf{-3}) &= 7 + (\mathbf{-3}) \\ 2x &= 4 \\ \frac{\mathbf{1}}{\mathbf{2}}(2x) &= \frac{\mathbf{1}}{\mathbf{2}}(4) \\ x &= 2 \end{aligned}$$

Step 4: Check the solution in the original problem.

The sum of twice two and three is seven. ▲

▼ **Example 2** If twice the difference of a number and three were decreased by five, the result would be three. Find the number.

Solution

Step 1: Let x = the number asked for.

Step 2: "Twice the difference of a number and three" translates to $2(x - 3)$; "is decreased by five" translates to -5 (subtract 5); "the result is" translates to =; and "three" is 3. An equation that describes the situation is

$$2(x - 3) - 5 = 3$$

Step 3: Solve the equation:

$$\begin{aligned} 2(x - 3) - 5 &= 3 \\ 2x - 6 - 5 &= 3 \\ 2x - 11 &= 3 \\ 2x - 11 + \mathbf{11} &= 3 + \mathbf{11} \\ 2x &= 14 \\ x &= 7 \end{aligned}$$

Step 4: Try it and see. ▲

Age Problems

▼ **Example 3** Bill is 6 years older than Tom. Three years ago Bill's age was four times Tom's age. Find the age of each boy now.

Solution It is sometimes very helpful to make a table for age problems. The people are arranged in rows and the times are arranged in columns. For example:

	Three years ago	Now
Bill		
Tom		

Proceeding as usual we have:

Step 1: Let x = Tom's age now. That makes Bill $x + 6$ years old now.

Step 2: If Tom is x years old now, three years ago he was $x - 3$ years old. If Bill is $x + 6$ years old now, three years ago he was $x + 6 - 3 = x + 3$ years old.

Using the information from Steps 1 and 2, we fill in the table as follows:

	Three years ago	Now
Bill	$x + 3$	$x + 6$
Tom	$x - 3$	x

Reading the problem again, we see that three years ago Bill's age was four times Tom's age. Writing this as an equation, we have

$$\begin{aligned} \text{Bill's age} &= 4(\text{Tom's age}) \\ x + 3 &= 4(x - 3) \end{aligned}$$

Step 3: Solve the equation:

$$\begin{aligned} x + 3 &= 4(x - 3) \\ x + 3 &= 4x - 12 \\ x + (\boldsymbol{-x}) + 3 &= 4x + (\boldsymbol{-x}) + 12 \\ 3 &= 3x - 12 \\ 3 + \mathbf{12} &= 3x - 12 + \mathbf{12} \\ 15 &= 3x \\ x &= 5 \end{aligned}$$

Tom is 5 years old. Bill is 11 years old. ▲

Geometry Problems

▼ **Example 4** The length of a rectangle is 5 inches more than twice the width. The perimeter is 34 inches. Find the length and width.

Solution

Step 1: Let x = the width of the rectangle.

Step 2: The length is 5 more than twice the width, so it must be $2x + 5$. It is a good idea when working problems that involve geometric figures (circles, squares, rectangles, triangles, etc.) to visualize the problem by drawing a picture. A picture that describes the information given in this problem looks like this:

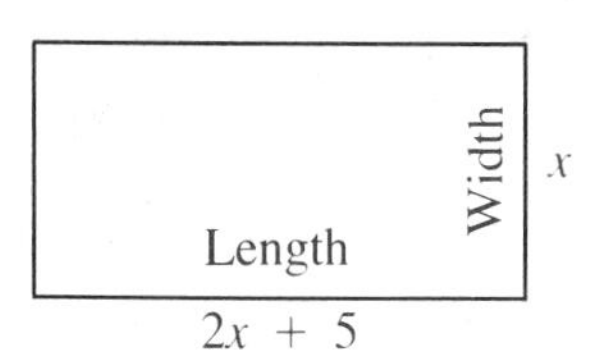

Since the perimeter of a rectangle is the total length of all the sides, we can find it by taking twice the length and twice the width. The equation that describes the situation in this problem is:

$$\begin{array}{ccccc} \text{twice length} & + & \text{twice width} & = & \text{perimeter} \\ 2(2x + 5) & + & 2x & = & 34 \end{array}$$

Step 3: Solve the equation:

$$\begin{aligned} 2(2x + 5) + 2x &= 34 & & \\ 4x + 10 + 2x &= 34 & & \text{Distributive property} \\ 6x + 10 &= 34 & & \text{Add } 4x \text{ and } 2x \\ 6x &= 24 & & \text{Add } -10 \text{ to each side} \\ x &= 4 & & \text{Divide each side by } 6 \end{aligned}$$

The width is 4 inches. The length is $2(4) + 5 = 13$ inches.

Step 4: If the length is 13 and the width is 4, then the perimeter must be $2(13 + 4) = 2(17) = 34$, which checks with the original problem. ▲

Coin Problems

▼ **Example 5** Jennifer has \$2.45 in dimes and nickels. If she has 8 more dimes than nickels, how many of each coin does she have?

Solution If we let x = the number of nickels, then $x + 8$ = the number of dimes. Since the value of each nickel is 5 cents, the amount of money in nickels is $5x$. Similarly, since each dime is worth 10 cents, the amount of money in dimes is $10(x + 8)$. Here is a table that summarizes the information we have so far:

	Nickels	Dimes
Number	x	$x + 8$
Value (in cents)	$5x$	$10(x + 8)$

Since the total value of all the coins in 245 cents, the equation that describes this situation is

$$\underbrace{\text{Amount of money in nickels}}_{5x} + \underbrace{\text{Amount of money in dimes}}_{10(x+8)} = \underbrace{\text{Total amount of money}}_{245}$$

To solve the equation, we apply the distributive property first.

$$
\begin{aligned}
5x + 10x + 80 &= 245 && \text{Distributive property} \\
15x + 80 &= 245 && \text{Add } 5x \text{ and } 10x \\
15x &= 165 && \text{Add } -80 \text{ to each side} \\
x &= 11 && \text{Divide each side by 15}
\end{aligned}
$$

The number of nickels is $x = 11$.
The number of dimes is $x + 8 = 11 + 8 = 19$.
To check our results:

11 nickels are worth $5(11)$	=	55 cents
19 dimes are worth $10(19)$	=	190 cents
The total value is		245 cents ($2.45). ▲

Interest Problems

▼ **Example 6** Suppose you invest a certain amount of money in an account that earns 8% in annual interest. At the same time, you invest $2,000 more than that in an account that pays 9% in annual interest. If the total interest from both accounts at the end of the year is $690, how much is invested at 8%?

Solution Let x = the amount of money invested at 8%. From this, $x + 2000$ = the amount of money invested at 9%. The interest earned on x dollars invested at 8% is $.08x$. The interest earned on $x + 2000$ dollars invested at 9% is $.09(x + 2000)$.

Here is a table that summarizes this information:

	Dollars invested at 8%	Dollars invested at 9%
Number of	x	$x + 2000$
Interest on	$.08x$	$.09(x + 2000)$

Since the total amount of interest earned from both accounts is $690, the equation that describes the situation is

$$\underbrace{\text{Interest earned at } 8\%}_{.08x} + \underbrace{\text{Interest earned at } 9\%}_{.09(x + 2000)} = \underbrace{\text{Total interest earned}}_{690}$$

$.08x + .09x + 180 = 690$	Distributive property
$.17x + 180 = 690$	Add $.08x$ and $.09x$
$.17x = 510$	Add -180 to each side
$x = 3000$	Divide each side by .17

The amount of money invested at 8% is $3,000, while the amount of money invested at 9% is $x + 2000 = 3000 + 2000 =$ $5,000. We check our results as follows:

The interest at 8% is 8% of 3000	= .08(3000) =	$240
The interest at 9% is 9% of 5000	= .09(5000) =	$450
The total interest is		$690 ▲

When you begin working the problems in the problem set that follows, there are a couple of things to remember. The first is that you may have to read the problems over a number of times before you begin to see how to solve them. The second thing to remember is that word problems are not always solved correctly the first time you try them. Sometimes it takes a couple of attempts and some wrong answers before you can set up and solve problems correctly.

Problem Set 2.6

Solve the following word problems. Follow the four steps given in the text and be sure to show the equation you use.

Number Problems

1. The sum of a number and five is thirteen. Find the number.
2. The difference of ten and a number is negative eight. Find the number.
3. The sum of twice a number and four is fourteen. Find the number.
4. The difference of four times a number and eight is sixteen. Find the number.
5. Five times the sum of a number and seven is thirty. Find the number.
6. Five times the difference of twice a number and six is negative twenty. Find the number.
7. One number is two more than another. Their sum is eight. Find both numbers.
8. One number is three less than another. Their sum is fifteen. Find the numbers.
9. One number is four less than three times another. If their sum is increased by five, the result is twenty-five. Find the numbers.

10. One number is five more than twice another. If their sum is decreased by ten, the result is twenty-two. Find the numbers.

Age Problems

11. Fred is 4 years older than Barney. Five years ago the sum of their ages was 48. How old are they now? (Begin by filling in the table below.)

	Five years ago	Now
Fred		
Barney		x

12. Tim is 5 years older than JoAnn. Six years from now the sum of their ages will be 79. How old are they now?

	Now	Six years from now
Tim		
JoAnn	x	

13. Jack is twice as old as Lacy. In three years the sum of their ages will be 54. How old are they now?

14. John is four times as old as Martha. Five years ago the sum of their ages was 50. How old are they now?

15. Pat is 20 years older than his son Patrick. In two years Pat will be twice as old as Patrick. How old are they now?

16. Diane is 23 years older than her daughter Amy. In 6 years Diane will be twice as old as Amy. How old are they now?

Geometry Problems

17. The length of a rectangle is 5 inches more than the width. The perimeter is 34 inches. Find the length and width.

18. The width of a rectangle is 3 feet less than the length. The perimeter is 10 feet. Find the length and width.

19. The perimeter of a square is 48 meters. Find the length of one side.

20. One side of a triangle is twice the smallest side. The third side is 3 feet more than the shortest side. The perimeter is 19 feet. Find all 3 sides.

21. The length of a rectangle is 3 inches less than twice the width. The perimeter is 54 inches. Find the length and width.

22. The length of a rectangle is 4 feet less than 3 times the width. The sum of the length and width is 14 more than the width. Find the width.

Coin Problems

23. Sue has $2.10 in dimes and nickels. If she has 9 more dimes than nickels, how many of each coin does she have? (Completing the table below may help you get started.)

	Nickels	Dimes
Number	x	
Value (in cents)		

24. Mike has \$1.55 in dimes and nickels. If he has 7 more nickels than dimes, how many of each coin does he have?

	Nickels	Dimes
Number		x
Value (in cents)		

25. Suppose you have \$9.00 in dimes and quarters. How many of each coin do you have if you have twice as many quarters as dimes?
26. A collection of dimes and quarters has a total value of \$2.20. If there are 3 times as many dimes as quarters, how many of each coin is in the collection?
27. Katie has a collection of nickels, dimes, and quarters with a total value of \$4.35. There are 3 more dimes than nickels and 5 more quarters than nickels. How many of each coin is in her collection? (Hint: Let x = the number of nickels.)
28. Mary Jo has \$3.90 worth of nickels, dimes, and quarters. The number of nickels is 3 more than the number of dimes. The number of quarters is 7 more than the number of dimes. How many of each coin does she have? (Hint: Let x = the number of dimes.)

Interest Problems

29. Suppose you invest money in two accounts. One of the accounts pays 8% annual interest, while the other pays 9% annual interest. If you have \$2,000 more invested at 9% than you have invested at 8%, how much do you have invested in each account if the total amount of interest you earn in a year is \$860? (Begin by completing the following table.)

	Dollars invested at 8%	Dollars invested at 9%
Number of	x	
Interest on		

30. Suppose you invest a certain amount of money in an account that pays 11% interest annually, and \$4,000 more than that in an account that pays 12% annually. How much money do you have in each account if the total interest for a year is \$940?

	Dollars invested at 11%	Dollars invested at 12%
Number of	x	
Interest on		

31. Tyler has two savings accounts that his grandparents opened for him. The two accounts pay 10% and 12% in annual interest; there is \$500 more in the account that pays 12% than there is in the other account. If the total interest for a year is \$214, how much money does he have in each account?
32. Travis has a savings account that his parents opened for him. It pays 6% annual interest. His uncle also opened an account for him, but it pays 8% annual

interest. If there is \$800 more in the account that pays 6%, and the total interest from both accounts is \$104, how much money is in each of the accounts?

33. A stock broker has money in three accounts. The interest rates on the three accounts are 8%, 9%, and 10%. If she has twice as much money invested at 9% as she has invested at 8%, and three times as much at 10% as she has at 8%, and the total interest for the year is \$280, how much is invested at each rate? (Hint: Let $x =$ the amount invested at 8%.)

34. An accountant has money in three accounts that pay 9%, 10%, and 11% in annual interest. He has twice as much invested at 9% as he does at 10%, and three times as much invested at 11% as he does at 10%. If the total interest from the three accounts is \$610 for the year, how much is invested at each rate? (Hint: Let $x =$ the amount invested at 10%.)

Review Problems The following problems review material we covered in Sections 1.1 and 1.2. Reviewing these problems will help you understand the material in the next section.

Write an equivalent statement in English. [1.1]

35. $4 < 10$
36. $4 \leq 10$
37. $9 \geq -5$
38. $x - 2 > 4$

Place the symbol < or the symbol > between the quantities in each expression. [1.2]

39. $12 \quad 20$
40. $-12 \quad 20$
41. $-8 \quad -6$
42. $-10 \quad -20$

Simplify. [1.2]

43. $|8 - 3| - |5 - 2|$
44. $|9 - 2| - |10 - 8|$
45. $15 - |9 - 3(7 - 5)|$
46. $10 - |7 - 2(5 - 3)|$

Section 2.7 Linear Inequalities

Linear inequalities are solved by a method similar to the one used in solving linear equations. The only real differences between the methods are in the multiplication property for inequalities and in graphing the solution set.

An inequality differs from an equation only with respect to the comparison symbol between the two quantities being compared. In place of the equal sign, we use < (less than), ≤ (less than or equal to), > (greater than), or ≥ (greater than or equal to). The addition property for inequalities is almost identical to the addition property for equality.

It makes no difference which inequality symbol we use to state the property. Adding the same amount to both sides always produces an inequality

Addition Property for Inequalities

For any three algebraic expressions A, B, and C,

$$\text{if} \quad A < B$$
$$\text{then } A + C < B + C$$

In words: Adding the same quantity to both sides of an inequality will not change the solution set.

equivalent to the original inequality. Also, since subtraction can be thought of as addition of the opposite, this property holds for subtraction as well as addition.

▼ **Example 1** Solve the inequality: $x + 5 < 7$.

Solution To isolate x, we add -5 to both sides of the inequality:

$$x + 5 < 7$$
$$x + 5 + (\mathbf{-5}) < 7 + (\mathbf{-5}) \qquad \text{Addition property for inequalities}$$
$$x < 2$$

▲

We can go one step further here and graph the solution set. The solution set is all real numbers less than 2. To graph this set, we simply draw a straight line and label the center 0 (zero) for reference. Then, we label the 2 on the right side of zero and extend an arrow beginning at 2 and pointing to the left. We use an open circle at 2, since it is not included in the solution set. Here is the graph:

▼ **Example 2** Solve: $x - 6 \leq -3$.

Solution Adding 6 to each side will isolate x on the left side:

$$x - 6 \leq -3$$
$$x - 6 + \mathbf{6} \leq -3 + \mathbf{6} \qquad \text{Add } \mathbf{6} \text{ to both sides}$$
$$x \leq 3$$

The graph of the solution set is:

▲

Notice that the 3 is darkened because 3 is included in the solution set. We will always use open circles on the graph of solution sets with $<$ or $>$ and closed (darkened) circles on the graphs of solution sets with $\leq$ or $\geq$.

To see the idea behind the multiplication property of inequalities, we will consider four true inequality statements and explore what happens when we multiply both sides by a positive number and then what happens when we multiply by a negative number.

Consider the following three true statements:

$$3 < 5 \qquad -3 < 5 \qquad -5 < -3$$

Now multiply both sides by the positive number 4:

$$\begin{array}{ccc} 4(3) < 4(5) & 4(-3) < 4(5) & 4(-5) < 4(-3) \\ 12 < 20 & -12 < 20 & -20 < -12 \end{array}$$

In each case, the inequality symbol in the result points in the same direction it did in the original inequality. We say the "sense" of the inequality doesn't change when we multiply both sides by a positive quantity.

Notice what happens when we go through the same process but multiply both sides by -4 instead of 4:

$$\begin{array}{ccc} 3 < 5 & -3 < 5 & -5 < -3 \\ \downarrow & \downarrow & \downarrow \\ -4(3) > -4(5) & -4(-3) > -4(5) & -4(-5) > -4(-3) \\ -12 > -20 & 12 > -20 & 20 > 12 \end{array}$$

In each case, we have to change the direction in which the inequality symbol points to keep each statement true. Multiplying both sides of an inequality by a negative quantity *always* reverses the sense of the inequality. Our results are summarized in the multiplication property for inequalities.

Multiplication Property for Inequalities

For any three algebraic expressions A, B, and C,

$$\begin{array}{lll} \text{if} & A < B & \\ \text{then} & AC < BC & \text{when } C \text{ is positive} \\ \text{and} & AC > BC & \text{when } C \text{ is negative} \end{array}$$

In words: Multiplying both sides of an inequality by a positive number does not change the solution set. When multiplying both sides of an inequality by a negative number, it is necessary to reverse the inequality symbol in order to produce an equivalent inequality.

We can multiply both sides of an inequality by any nonzero number we choose. If that number happens to be negative, we must also reverse the sense of the inequality.

Note Since division is defined in terms of multiplication, this property is also true for division. We can divide both sides of an inequality by any number we choose. If that number happens to be negative, we must also reverse the direction of the inequality symbol.

▼ **Example 3** Solve $3a < 15$ and graph the solution.

Solution We begin by multiplying each side by $\frac{1}{3}$. Since $\frac{1}{3}$ is a positive number, we do not reverse the direction of the inequality symbol:

$$3a < 15$$

$$\frac{1}{3}(3a) < \frac{1}{3}(15) \quad \text{Multiply each side by } \frac{1}{3}$$

$$a < 5$$

0 5

▲

▼ **Example 4** Solve $-3a \leq 18$ and graph the solution.

Solution We begin by multiplying both sides by $-\frac{1}{3}$. Since $-\frac{1}{3}$ is a negative number, we must reverse the direction of the inequality symbol at the same time that we multiply by $-\frac{1}{3}$:

$$-3a \leq 18$$

$$-\frac{1}{3}(-3a) \geq -\frac{1}{3}(18) \quad \text{Multiply both sides by } -\frac{1}{3} \text{ and reverse the direction of the inequality symbol}$$

$$a \geq -6$$

−6 0

▲

▼ **Example 5** Solve $-\frac{x}{4} > 2$ and graph the solution.

Solution To isolate x, we multiply each side by -4. Since -4 is a negative number, we must also reverse the direction of the inequality symbol:

$$-\frac{x}{4} > 2$$

$$-4\left(-\frac{x}{4}\right) < -4(2) \quad \text{Multiply each side by } -4 \text{ and reverse the direction of the inequality symbol}$$

$$x < -8$$

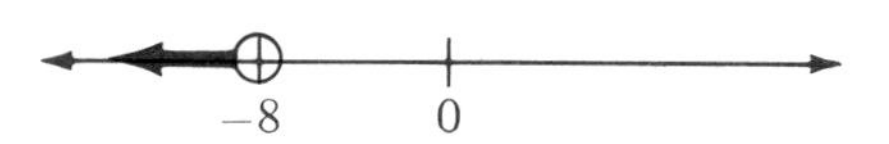

▲

To solve more complicated inequalities we use the following four steps.

To Solve a Linear Inequality in One Variable

Step 1: Use the distributive property, if necessary, to separate terms. Simplify both sides as much as possible.
Step 2: Use the addition property of inequalities to get all variable terms on one side and all constant terms on the other.
Step 3: Use the multiplication property of inequalities to get x on just one side.
Step 4: Graph the solution set.

▼ **Example 6** Use these four steps to solve $3(x - 4) \geq -2$.

Solution To begin, we use the distributive property to multiply 3 and $x - 4$.

$$3x - 12 \geq -2 \quad \text{Distributive property}$$
$$3x - 12 + \mathbf{12} \geq -2 + \mathbf{12} \quad \text{Add } \mathbf{12} \text{ to both sides}$$
$$3x \geq 10$$
$$\frac{1}{3}(3x) \geq \frac{1}{3}(10) \quad \text{Multiply both sides by } \tfrac{1}{3}$$
$$x \geq \frac{10}{3}$$

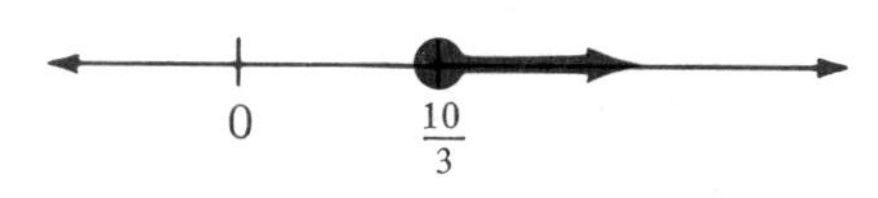

▲

▼ **Example 7** Solve and graph: $2(1 - 3x) + 4 < 4x - 14$.

Solution

$$2 - 6x + 4 < 4x - 14 \quad \text{Distributive property}$$
$$-6x + 6 < 4x - 14 \quad \text{Simplify}$$
$$-6x + 6 \mathbf{- 6} < 4x - 14 \mathbf{- 6} \quad \text{Add } \mathbf{-6} \text{ to both sides}$$
$$-6x < 4x - 20$$

$$-6x - \mathbf{4x} < 4x - \mathbf{4x} - 20 \qquad \text{Add } \mathbf{-4x} \text{ to both sides}$$

$$-10x < -20$$

$$-\frac{\mathbf{1}}{\mathbf{10}}(-10x) \overset{\downarrow}{>} -\frac{\mathbf{1}}{\mathbf{10}}(-20) \qquad \text{Multiplying by } \mathbf{-\tfrac{1}{10}}\text{, reverse the sense of the inequality}$$

$$x > 2$$

▲

▼ **Example 8** Solve $2x - 3y < 6$ for y.

Solution We can solve this formula for y by first adding $-2x$ to each side and then multiplying each side by $-\frac{1}{3}$. When we multiply by $-\frac{1}{3}$ we must reverse the direction of the inequality symbol. Since this is a formula, we will not graph the solution.

$$2x - 3y < 6 \qquad \text{Original formula}$$

$$2x + (\mathbf{-2x}) - 3y < (\mathbf{-2x}) + 6 \qquad \text{Add } \mathbf{-2x} \text{ to each side}$$

$$-3y < -2x + 6$$

$$-\frac{\mathbf{1}}{\mathbf{3}}(-3y) \overset{\downarrow}{>} -\frac{\mathbf{1}}{\mathbf{3}}(-2x + 6) \qquad \text{Multiply each side by } \mathbf{-\tfrac{1}{3}}$$

$$y > \frac{2}{3}x - 2 \qquad \text{Distributive property}$$

▲

When working word problems that involve inequalities, the phrases "at least" and "at most" translate as follows:

In words	*In symbols*
x is at least 30	$x \geq 30$
x is at most 20	$x \leq 20$

When we ask for consecutive integers, we mean integers that are next to each other on the number line, like 5 and 6, or 13 and 14, or -4 and -3. In the dictionary, consecutive is defined as following one another in uninterrupted order. If we ask for consecutive *odd* integers, then we mean odd integers that follow one another on the number line. For example, 3 and 5, 11 and 13, and -9 and -7 are consecutive odd integers. As you can see, to get from one odd integer to the next consecutive odd integer we add 2.

▼ **Example 9** The sum of two consecutive odd integers is at most 28. What are the possibilities for the first of the two integers?

Solution When we use the phrase "their sum is at most 28," we mean that their sum is less than or equal to 28. If we let x = the first of the two consecutive odd integers, then $x + 2$ is the next consecutive one. The equation that describes this situation is:

$$x + (x + 2) \leq 28$$
$$2x + 2 \leq 28 \quad \text{Simplify the left side}$$
$$2x \leq 26 \quad \text{Add } -2 \text{ to each side}$$
$$x \leq 13 \quad \text{Multiply each side by } \tfrac{1}{2}$$

The first of the two integers must be an integer that is less than or equal to 13. The second of the two integers will be two more than whatever the first one is. ▲

Problem Set 2.7

Solve the following inequalities using the addition property of inequalities. Graph each solution set.

1. $x - 5 < 7$
2. $x + 3 < -5$
3. $a - 4 \leq 8$
4. $a + 3 \leq 10$
5. $x - 5 > 8$
6. $x - 2 > 11$
7. $y + 6 \geq 10$
8. $y + 3 \geq 12$
9. $2 < x - 7$
10. $3 < x + 8$

Solve the following inequalities using the multiplication property of inequalities. If you multiply both sides by a negative number, be sure to reverse the direction of the inequality symbol. Graph the solution set.

11. $3x < 6$
12. $2x < 14$
13. $5a \leq 25$
14. $4a \leq 16$
15. $\frac{x}{3} > 5$
16. $\frac{x}{7} > 1$
17. $-2x > 6$
18. $-3x \geq 9$
19. $-3x \geq -18$
20. $-8x \geq -24$
21. $-\frac{x}{5} \leq 10$
22. $-\frac{x}{9} \geq -1$
23. $-\frac{2}{3}y > 4$
24. $-\frac{3}{4}y > 6$

Solve the following inequalities. Graph the solution set in each case.

25. $2x - 3 < 9$
26. $3x - 4 < 17$
27. $-3y - 5 \leq 10$
28. $-2y - 6 \leq 8$
29. $-4x + 1 > -11$
30. $-6x - 1 > 17$
31. $\frac{2}{3}x - 5 \leq 7$
32. $\frac{3}{4}x - 8 \leq 1$

33. $-\frac{2}{5}a - 3 > 5$
34. $-\frac{4}{5}a - 2 > 10$
35. $5 - \frac{3}{5}y > -10$
36. $4 - \frac{5}{6}y > -11$
37. $3(a + 1) \le 12$
38. $4(a - 2) \le 4$
39. $2(5 - 2x) < -20$
40. $7(8 - 2x) > 28$
41. $3x - 5 > 8x$
42. $8x - 4 > 6x$
43. $2y - 3 \le 5y + 3$
44. $7y + 8 \le 11y - 7$
45. $-2x + 3 < -10x - 5$
46. $-6x - 2 < -2x + 10$
47. $3(m - 2) - 4 \ge 7m + 14$
48. $2(3m - 1) + 5 \ge 8m - 7$
49. $3 - 4(x - 2) \le -5x + 6$
50. $8 - 6(x - 3) \le -4x + 12$

Solve each of the following formulas for y.

51. $3x + 2y < 6$
52. $-3x + 2y < 6$
53. $2x - 5y > 10$
54. $-2x - 5y > 5$
55. $-3x + 7y \le 21$
56. $-7x + 3y \le 21$
57. $2x - 4y \ge -4$
58. $4x - 2y \ge -8$

Number Problems

59. The sum of twice a number and six is less than ten. Find all solutions.
60. Twice the difference of a number and three is greater than or equal to the number increased by five. Find all solutions.
61. The product of a number and four is greater than the number minus eight. Find the solution set.
62. The quotient of a number and five is less than the sum of seven and two. Find the solution set.
63. Twice the sum of a number and five is less than or equal to twelve. Find all solutions.
64. Three times the difference of a number and four is greater than twice the number. Find all solutions.
65. The difference of three times a number and five is less than the sum of the number and seven. Find all solutions.
66. If twice a number is added to three, the result is less than the sum of three times the number and two. Find all solutions.

Geometry Problems

67. The length of a rectangle is three times the width. If the perimeter is to be at least 48 meters, what are the possible values for the width? (If the perimeter is at least 48 meters, then it is greater than or equal to 48.)
68. The length of a rectangle is 3 more than twice the width. If the perimeter is to be at least 51 meters, what are the possible values for the width? (If the perimeter is at least 51 meters, then it is greater than or equal to 51 meters.)
69. The numerical values of the three sides of a triangle are given by three consecutive even integers. If the perimeter is greater than 24 inches, what are the possibilities for the shortest side?
70. The numerical values of the three sides of a triangle are given by three consecutive odd integers. If the perimeter is greater than 27 inches, what are the possibilities for the shortest side?

Interest Problems

71. JoAnn has money invested in an account that earns 12% as long as her balance in the account never drops below \$3,000. If she keeps at least \$3,000 in the account, what can she expect to earn in interest each year?

72. Tim has an account that requires a minimum balance of \$5,000 in order to earn 10% interest per year. If the balance in this account never drops below the minimum \$5,000, how much interest can Tim expect to earn in a year?

73. Diane has invested money in an account that pays 8% interest on the first \$2,000 and 10% interest on any amount above \$2,000. If she keeps at least \$5,000 in the account for a year, how much interest can she expect to earn?

74. Each month Colleen invests money in an account that pays 9% interest on the first \$1,000 and 11% interest on any amount above \$1,000. If she keeps at least \$3,000 in the account for a year, how much interest can she expect to earn?

Review Problems The problems below review material we covered in Sections 1.2 and 1.5.

For the set $\{-3, -\frac{5}{3}, 0, 2, \pi, \frac{15}{2}\}$ list all the elements that are in the following sets.

75. Whole numbers

76. Integers

77. Rational numbers

78. Irrational numbers

Match each expression on the left with one or more of the properties on the right. [1.5]

79. $x + 4 = 4 + x$

80. $2(3x) = (2 \cdot 3)x$

81. $5(x - 3) = 5x - 15$

82. $x + (y + 4) = (x + y) + 4$

83. $x + (y + 4) = (x + 4) + y$

84. $7 \cdot 5 = 5 \cdot 7$

A. Distributive property

B. Commutative property of addition

C. Associative property of addition

D. Commutative property of multiplication

E. Associative property of multiplication

Chapter 2 Summary and Review

Examples

1. The terms $2x$, $5x$, and $-7x$ are all similar since their variable parts are the same.

SIMILAR TERMS [2.1]

A *term* is a number or a number and one or more variables multiplied together. *Similar terms* are terms with the same variable part.

SIMPLIFYING EXPRESSIONS [2.1]

In this chapter we simplified expressions that contained variables by using the distributive property to combine similar terms.

2. Simplify $3x + 4x$.

$$3x + 4x = (3 + 4)x = 7x$$

SOLUTION SET [2.2]

The *solution set* for an equation (or inequality) is all the numbers that, when used in place of the variable, make the equation a true statement.

3. The solution set for the equation $x + 2 = 5$ is $\{3\}$ because when x is 3 the equation is $3 + 2 = 5$ or $5 = 5$.

EQUIVALENT EQUATIONS [2.2]

Two equations are called *equivalent* if they have the same solution set.

4. The equations $a - 4 = 3$ and $a - 2 = 5$ are equivalent since both have solution set $\{7\}$.

ADDITION PROPERTY OF EQUALITY [2.2]

When the same quantity is added to both sides of an equation, the solution set for the equation is unchanged. Adding the same amount to both sides of an equation produces an equivalent equation.

5. Solve $x - 5 = 12$.

$$x - 5 + \mathbf{5} = 12 + \mathbf{5}$$
$$x + 0 = 17$$
$$x = 17$$

MULTIPLICATION PROPERTY OF EQUALITY [2.3]

If both sides of an equation are multiplied by the same nonzero number, the solution set is unchanged. Multiplying both sides of an equation by a nonzero quantity produces an equivalent equation.

6. Solve $3x = 18$.

$$\tfrac{1}{3}(3x) = \tfrac{1}{3}(18)$$
$$x = 6$$

TO SOLVE A LINEAR EQUATION [2.4]

Step 1: Use the distributive property, if necessary, to separate terms. Simplify both sides as much as possible.

Step 2: Use the addition property of equality to get all variable terms on one side and all constant terms on the other.

Step 3: Use the multiplication property of equality to get x by itself.

Step 4: Check your solution, if necessary.

7. Solve $2(x + 3) = 10$.

$$2x + 6 = 10$$
$$2x + 6 + (\mathbf{-6}) = 10 + (\mathbf{-6})$$
$$2x = 4$$
$$\tfrac{1}{2}(2x) = \tfrac{1}{2}(4)$$
$$x = 2$$

FORMULAS [2.5]

A formula is an equation with more than one variable. To solve a formula for one of its variables we use the addition and multiplication properties of equality to move everything except the variable in question to one side of the equal sign so the variable in question is alone on the other side.

8. Solving $P = 2l + 2w$ for l, we have

$$P - 2w = 2l$$
$$\frac{P - 2w}{2} = l$$

TO SOLVE A WORD PROBLEM [2.6]

Step 1: Let x represent the unknown quantity.
Step 2: Using x, write an equation that describes the situation.
Step 3: Solve the equation.
Step 4: Check your solution with the original problem.

9. The sum of twice a number and three is seven. Find the number.
Let x = the number:

$$2x + 3 = 7$$
$$x = 2$$

The sum of twice 2 and 3 is 7.

ADDITION PROPERTY OF INEQUALITY [2.7]

Adding the same quantity to both sides of an inequality produces an equivalent inequality, one with the same solution set.

10. Solve $x + 5 < 7$.

$$x + 5 + \mathbf{(-5)} < 7 + \mathbf{(-5)}$$
$$x < 2$$

MULTIPLICATION PROPERTY OF INEQUALITY [2.7]

Multiplying both sides of an inequality by a positive number never changes the solution set. If both sides are multiplied by a negative number, the sense of the inequality must be reversed to produce an equivalent inequality.

11. Solve $-3a \le 18$.

$$-\tfrac{1}{3}(-3a) \ge -\tfrac{1}{3}(18)$$
$$a \ge -6$$

TO SOLVE A LINEAR INEQUALITY [2.7]

Step 1: Use the distributive property, if necessary, to separate terms. Simplify both sides as much as possible.
Step 2: Use the addition property of inequality to get all variable terms on one side and all constant terms on the other.
Step 3: Use the multiplication property of inequality to get just one x. (Remember to reverse the direction of the inequality symbol if you multiply both sides by a negative number.)
Step 4: Graph the solution set.

12. Solve $3(x - 4) \ge -2$.

$$3x - 12 \ge -2$$
$$3x - 12 + \mathbf{12} \ge -2 + \mathbf{12}$$
$$3x \ge 10$$
$$\tfrac{1}{3}(3x) \ge \tfrac{1}{3}(10)$$
$$x \ge \tfrac{10}{3}$$

COMMON MISTAKES

1. Trying to subtract away coefficients (the number in front of variables) when solving equations. For example:

$$\begin{aligned} 4x &= 12 \\ 4x - 4 &= 12 - 4 \\ x &= 8 \leftarrow \text{Mistake} \end{aligned}$$

It is not incorrect to add (-4) to both sides, it's just that $4x - 4$ is not equal to x. Both sides should be multiplied by $\frac{1}{4}$ to solve for x.

2. Forgetting to reverse the direction of the inequality symbol when multiplying both sides of an inequality by a negative number. For instance:

$$\begin{aligned} -3x &< 12 \\ -\tfrac{1}{3}(-3x) &< -\tfrac{1}{3}(12) \quad \leftarrow \text{Mistake} \\ x &< -4 \end{aligned}$$

It is not incorrect to multiply both sides by $-\frac{1}{3}$. But if we do, we must also reverse the sense of the inequality.

Chapter 2 Test

Simplify each of the following expressions. [2.1]

1. $3x + 2 - 7x + 3$
2. $4a - 5 - a + 1$
3. $7 - 3(y + 5) - 4$
4. $8(2x + 1) - 5(x - 4)$
5. Find the value of $2x - 3 - 7x$ when $x = -5$. [2.2]
6. Find the value of $x^2 + 2xy + y^2$ when $x = 2$ and $y = 3$. [2.1]

Solve the following equations. [2.2, 2.3, 2.4]

7. $2x - 5 = 7$
8. $2y + 4 = 5y$
9. $5x - 1 = 2x + 5$
10. $4(a + 3) = -6$
11. $-5(2x + 1) - 6 = 19$
12. $4 - 3(3a - 5) = 1$
13. $2(t - 4) + 3(t + 5) = 2t - 2$
14. $2x - 4(5x + 1) = 3x + 17$

15. What number is 15% of 38? [2.3]
16. 240 is 12% of what number? [2.3]
17. If $2x - 3y = 12$, find x when $y = -2$. [2.5]
18. The formula for the volume of a cone is $V = \frac{1}{3}\pi r^2 h$. Find h if $V = 88$ cubic inches, $\pi = \frac{22}{7}$, and $r = 3$ inches. [2.5]
19. Solve $2x + 5y = 20$ for y. [2.5]
20. Solve $h = x + vt + 16t^2$ for v. [2.5]

Solve each word problem. [2.6]

21. Dave is twice as old as Rick. Ten years ago the sum of their ages was 40. How old are they now?

22. A rectangle is twice as long as it is wide. The perimeter is 60 inches. What are the length and width?

23. A man has a collection of dimes and quarters with a total value of $3.50. If he has 7 more dimes than quarters, how many of each coin does he have?

24. A woman has money in two accounts. One account pays 7% annual interest, while the other pays 9% annual interest. If she has $600 more invested at 9% than she does at 7%, and her total interest for a year is $182, how much does she have in each account?

Solve each inequality and graph the solution. [2.7]

25. $2x + 3 < 5$

26. $-5a > 20$

27. $4 - 2x \geq 10$

28. $4 - 5(m + 1) \leq 9$

3 Graphing and Linear Systems

In the last chapter we spent most of our time developing and using the method of solving linear equations in one variable. The equations we worked with had the form $ax + b = 0$. (If they didn't have this form, they could be put into this form.) In this chapter, we will expand our work with equations to include linear equations in *two* variables. We are going to include another variable so that most of the equations in this chapter will have the form $ax + by = c$, where a, b, and c are constants and x and y are variables. We will also extend the technique of graphing to include points associated with two number lines instead of just one. The background material needed for this chapter is in Chapter 2. You need to know how to solve a linear equation in one variable.

Section 3.1 Solutions to Linear Equations in Two Variables

If we solve the equation $3x - 2 = 10$, the solution is $x = 4$. If we were to graph this solution, we would simply draw the real number line and place a dot at the point whose coordinate is 4. The relationship between linear equations in one variable, their solutions, and the graphs of those solutions looks like this:

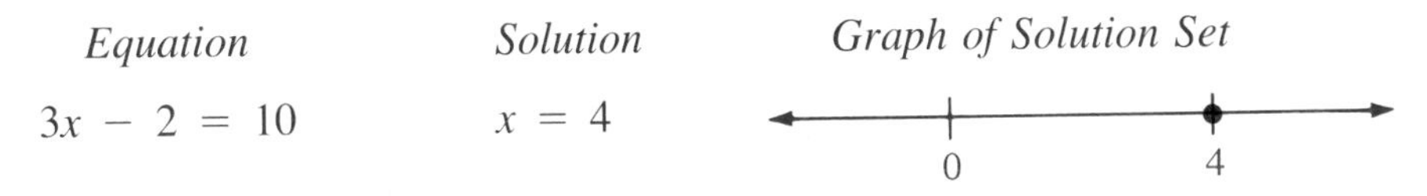

$$x + 5 = 7 \qquad x = 2$$

$$2x = -6 \qquad x = -3$$

When the equation has one variable, the solution is a single number whose graph is a point on a line.

Now consider the equation $2x + y = 3$. The first thing we notice is that there are two variables instead of one. Therefore, a solution to the equation $2x + y = 3$ will not be a single number but a pair of numbers, one for x and one for y, that make the equation a true statement. One pair of numbers that works is $x = 2$, $y = -1$, because when we substitute them for x and y in the equation, we get a true statement. That is:

$$\begin{aligned} 2(2) + (-1) &\stackrel{?}{=} 3 \\ 4 - 1 &= 3 \\ 3 &= 3 \qquad \text{A true statement} \end{aligned}$$

The pair of numbers $x = 2$, $y = -1$ is written as $(2, -1)$. When written in this form, $(2, -1)$ is called an *ordered pair* because it is a pair of numbers written in a specific order. The first number is always associated with the variable x and the second number is always associated with the variable y. We call the first number in the ordered pair the *x-coordinate* (or x component) and the second number the *y-coordinate* (or y component) of the ordered pair.

Let's look back to the equation $2x + y = 3$. The ordered pair $(2, -1)$ is not the only solution. Another solution is $(0, 3)$, because when we substitute 0 for x and 3 for y we get

$$\begin{aligned} 2(0) + 3 &\stackrel{?}{=} 3 \\ 0 + 3 &= 3 \\ 3 &= 3 \qquad \text{A true statement} \end{aligned}$$

Still another solution is the ordered pair $(5, -7)$, because

$$\begin{aligned} 2(5) + (-7) &\stackrel{?}{=} 3 \\ 10 - 7 &= 3 \\ 3 &= 3 \qquad \text{A true statement} \end{aligned}$$

As a matter of fact, for any number we want to use for x there is another number we can use for y that will make the equation a true statement. There is an infinite number of ordered pairs that satisfy (are solutions to) the equation $2x + y = 3$; we have listed just a few of them.

▼ **Example 1** Given the equation $2x + 3y = 6$, complete the following ordered pairs so they will be solutions to the equation: (0,), (, 1), (3,).

Solution To complete the ordered pair (0,), we substitute 0 for x in the equation and then solve for y:

$$\begin{aligned} 2(0) + 3y &= 6 \\ 3y &= 6 \\ y &= 2 \end{aligned}$$

The ordered pair is (0, 2).

To complete the ordered pair (, 1), we substitute 1 for y in the equation and solve for x:

$$\begin{aligned} 2x + 3(1) &= 6 \\ 2x + 3 &= 6 \\ 2x &= 3 \\ x &= \frac{3}{2} \end{aligned}$$

The ordered pair is $(\frac{3}{2}, 1)$.

To complete the ordered pair (3,), we substitute 3 for x in the equation and solve for y:

$$\begin{aligned} 2(3) + 3y &= 6 \\ 6 + 3y &= 6 \\ 3y &= 0 \\ y &= 0 \end{aligned}$$

The ordered pair is (3, 0). ▲

▼ **Example 2** Complete the following table for equation $2x - 5y = 20$.

x	y
0	
	2
	0
−5	

Solution Filling in the table is equivalent to completing the following ordered pairs: (0,), (, 2), (, 0), (−5,). So we proceed as in Example 1:

When $x = 0$, we have

$$\begin{aligned} 2(0) - 5y &= 20 \\ 0 - 5y &= 20 \\ -5y &= 20 \\ y &= -4 \end{aligned}$$

When $y = 2$, we have

$$\begin{aligned} 2x - 5(2) &= 20 \\ 2x - 10 &= 20 \\ 2x &= 30 \\ x &= 15 \end{aligned}$$

When $y = 0$, we have

$$\begin{aligned} 2x - 5(0) &= 20 \\ 2x - 0 &= 20 \\ 2x &= 20 \\ x &= 10 \end{aligned}$$

When $x = -5$, we have

$$\begin{aligned} 2(-5) - 5y &= 20 \\ -10 - 5y &= 20 \\ -5y &= 30 \\ y &= -6 \end{aligned}$$

The completed table looks like this:

x	y
0	-4
15	2
10	0
-5	-6

which is equivalent to the ordered pairs $(0, -4)$, $(15, 2)$, $(10, 0)$, and $(-5, -6)$. ▲

▼ **Example 3** Complete the following table for equation $y = 2x - 1$.

x	y
0	
5	
	7
	3

Solution When $x = 0$, we have

$$\begin{aligned} y &= 2(0) - 1 \\ y &= 0 - 1 \\ y &= -1 \end{aligned}$$

When $x = 5$, we have

$$\begin{aligned} y &= 2(5) - 1 \\ y &= 10 - 1 \\ y &= 9 \end{aligned}$$

When $y = 7$, we have

$$\begin{aligned} 7 &= 2x - 1 \\ 8 &= 2x \\ 4 &= x \end{aligned}$$

When $y = 3$, we have

$$\begin{aligned} 3 &= 2x - 1 \\ 4 &= 2x \\ 2 &= x \end{aligned}$$

The completed table is:

x	y
0	-1
5	9
4	7
2	3

which means the ordered pairs $(0, -1)$, $(5, 9)$, $(4, 7)$, and $(2, 3)$ are among the solutions to the equation $y = 2x - 1$. ▲

▼ **Example 4** Which of the ordered pairs $(2, 3)$, $(1, 5)$, and $(-2, -4)$ are solutions to the equation $y = 3x + 2$?

Solution If an ordered pair is a solution to the equation, then it must satisfy the equation. That is, when the coordinates are used in place of the variables in the equation, the equation becomes a true statement.

Try $(2, 3)$ in $y = 3x + 2$:

$$3 \stackrel{?}{=} 3(2) + 2$$
$$3 = 6 + 2$$
$$3 = 8 \qquad \text{A false statement}$$

Try $(1, 5)$ in $y = 3x + 2$:

$$5 \stackrel{?}{=} 3(1) + 2$$
$$5 = 3 + 2$$
$$5 = 5 \qquad \text{A true statement}$$

Try $(-2, -4)$ in $y = 3x + 2$:

$$-4 \stackrel{?}{=} 3(-2) + 2$$
$$-4 = -6 + 2$$
$$-4 = -4 \qquad \text{A true statement}$$

The ordered pairs $(1, 5)$ and $(-2, -4)$ are solutions to the equation $y = 3x + 2$; $(2, 3)$ is not. ▲

Problem Set 3.1

For each equation, complete the given ordered pairs.

1. $2x + y = 6$ $\quad (0, \ ), (3, \ ), (\ , -6)$
2. $3x - y = 5$ $\quad (0, \ ), (1, \ ), (\ , 5)$
3. $3x + 4y = 12$ $\quad (0, \ ), (\ , 0), (-4, \ )$
4. $5x - 5y = 20$ $\quad (0, \ ), (\ , -2), (1, \ )$
5. $y = 4x - 3$ $\quad (1, \ ), (\ , 0), (5, \ )$
6. $y = 3x - 5$ $\quad (\ , 13), (0, \ ), (-2, \ )$
7. $y = 7x - 1$ $\quad (2, \ ), (\ , 6), (0, \ )$
8. $y = 8x + 2$ $\quad (3, \ ), (\ , 0), (\ , -6)$
9. $x = -5$ $\quad (\ , 4), (\ , -3), (\ , 0)$
10. $y = 2$ $\quad (5, \ ), (-8, \ ), (\frac{1}{2}, \ )$

For each of the following equations, complete the given table.

11. $y = 3x$

x	y
1	
-3	
	12
	18

12. $y = -2x$

x	y
-4	
0	
	10
	12

13. $y = 4x$

x	y
0	
	-2
-3	
	12

14. $y = -5x$

x	y
3	
	0
-2	
	-20

15. $x + y = 5$

x	y
2	
3	
	0
	-4

16. $x - y = 8$

x	y
0	
4	
	-3
	-2

17. $2x - y = 4$

x	y
	0
	2
1	
-3	

18. $3x - y = 9$

x	y
	0
	-9
5	
-4	

19. $y = 6x - 1$

x	y
0	
-1	
-3	
	8

20. $y = 5x + 7$

x	y
0	
-2	
-4	
	-8

For the following equations, tell which of the given ordered pairs are solutions.

21. $2x - 5y = 10$ $\quad (2, 3), (0, -2), (\frac{5}{2}, 1)$

22. $3x + 7y = 21$ $\quad (0, 3), (7, 0), (1, 2)$

23. $y = 7x - 2$ $(1, 5), (0, -2), (-2, -16)$
24. $y = 8x - 3$ $(0, 3), (5, 16), (1, 5)$
25. $y = 6x$ $(1, 6), (-2, -12), (0, 0)$
26. $y = -4x$ $(0, 0), (2, 4), (-3, 12)$
27. $x + y = 0$ $(1, 1), (2, -2), (3, 3)$
28. $x - y = 1$ $(0, 1), (0, -1), (1, 2)$
29. $x = 3$ $(3, 0), (3, -3), (5, 3)$
30. $y = -4$ $(3, -4), (-4, 4), (0, -4)$

31. If the perimeter of a rectangle is 30 inches, then the relationship between the length l and the width w is given by the equation

$$2l + 2w = 30$$

What is the length when the width is 3 inches?

32. The relationship between the perimeter P of a square and the length of its side s is given by the formula $P = 4s$. If each side of a square is 5 inches, what is the perimeter? If the perimeter of a square is 28 inches, how long is a side?
33. If every ordered pair that satisfies an equation has a y-coordinate that is twice as large as its x-coordinate, then the equation is $y = 2x$ and every ordered pair has the form $(x, 2x)$. Write the ordered pair that has 5 as its x-coordinate.
34. If every ordered pair that satisfies an equation has a y-coordinate that is three times as large as its x-coordinate, then the equation is $y = 3x$. What is the form of each ordered pair? (*Hint:* Fill in the second coordinate in $(x, \quad)$; see Problem 33.) If x is 4, what is y?
35. For the equation $y = |x|$, y is equal to the absolute value of x. Complete the following ordered pairs so that they are solutions to $y = |x|$.

$$(3, \quad) \quad (-3, \quad) \quad (5, \quad) \quad (-5, \quad)$$

36. If $y = |x|$ and y is 4, there are two possible x values: either 4 or -4, because both the ordered pairs $(4, 4)$ and $(-4, 4)$ satisfy the equation. What two ordered pairs that are solutions to $y = |x|$ have a y-coordinate of 6?

Review Problems The following problems review material we covered in Section 2.5.

37. Find y when x is 4 in the formula $3x + 2y = 6$.
38. Find y when x is 0 in the formula $3x + 2y = 6$.
39. Find y when x is 0 in $y = -\frac{1}{3}x + 2$.
40. Find y when x is 3 in $y = -\frac{1}{3}x + 2$.
41. Find y when x is 2 in $y = \frac{3}{2}x - 3$.
42. Find y when x is 4 in $y = \frac{3}{2}x - 3$.
43. Solve $5x + y = 4$ for y.
44. Solve $-3x + y = 5$ for y.
45. Solve $3x - 2y = 6$ for y.
46. Solve $2x - 3y = 6$ for y.

Section 3.2 Graphing Ordered Pairs and Straight Lines

To graph an ordered pair such as (3, 4), we need two real number lines—one associated with the first coordinate and one associated with the second coordinate. A rectangular (or Cartesian) coordinate system looks like this:

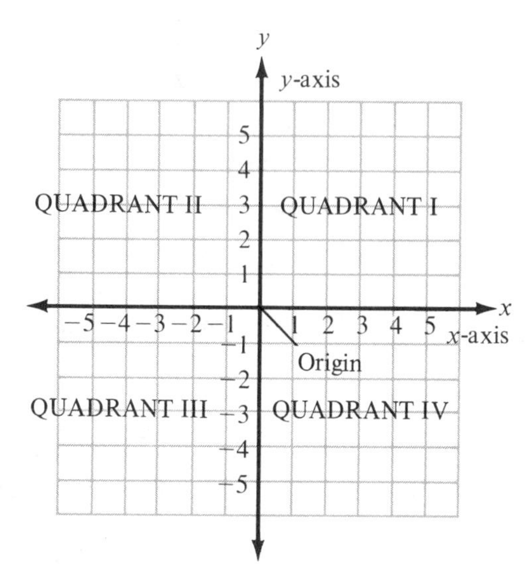

Figure 1

The rectangular coordinate system is built from two number lines oriented perpendicular to each other. The horizontal number line is exactly the same as our real number line and is called the x-axis. The vertical number line is also the same as our real number line with the positive direction up and the negative direction down. It is called the y-axis. The point where the two axes intersect is called the origin. As you can see from the diagram, the axes divide the plane into four quadrants, which are numbered I through IV in a counterclockwise direction.

Graphing Ordered Pairs

To graph the ordered pair (a, b), we start at the origin and move a units forward or back (forward if a is positive and back if a is negative). Then, we move b units up or down (up if b is positive, down if b is negative). The point where we end up is the graph of the ordered pair (a, b).

▼ **Example 1** Graph the ordered pairs (3, 4), (3, −4), (−3, 4), and (−3, −4).

Solution

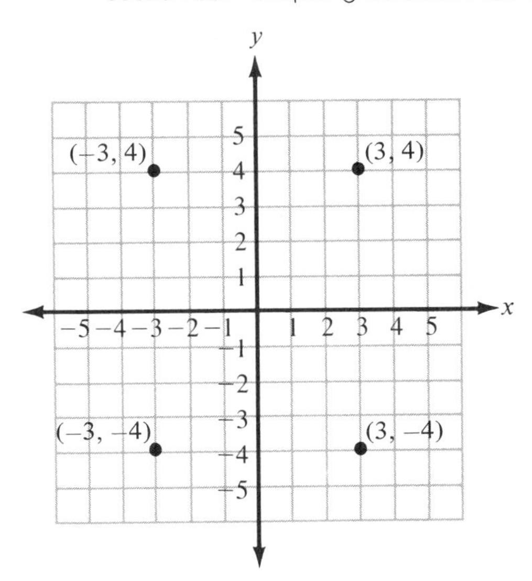

Figure 2

We can see that when we graph ordered pairs the x-coordinate corresponds to movement parallel to the x-axis (horizontal) and the y-coordinate corresponds to movement parallel to the y-axis (vertical). ▲

▼ **Example 2** Graph the following ordered pairs: $(-1, 3)$, $(2, 5)$, $(0, 0)$, $(0, -3)$, $(4, 0)$.

Solution

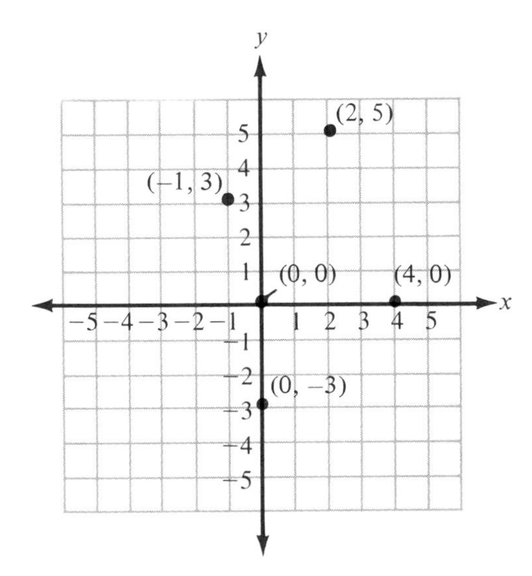

Figure 3 ▲

Graphing Straight Lines

To graph the solution set for an equation in two variables, we simply graph all the points whose coordinates satisfy the equation.

▼ **Example 3** Graph the solution set for $x + y = 5$.

Solution We know from the last section that there is an infinite number of ordered pairs that are solutions to the equation $x + y = 5$. We can't possibly list them all. What we can do is list a few of them and see if there is any pattern to their graphs.

Some ordered pairs that are solutions to $x + y = 5$ are (0, 5), (2, 3), (3, 2), (5, 0). The graph of each is shown in Figure 4.

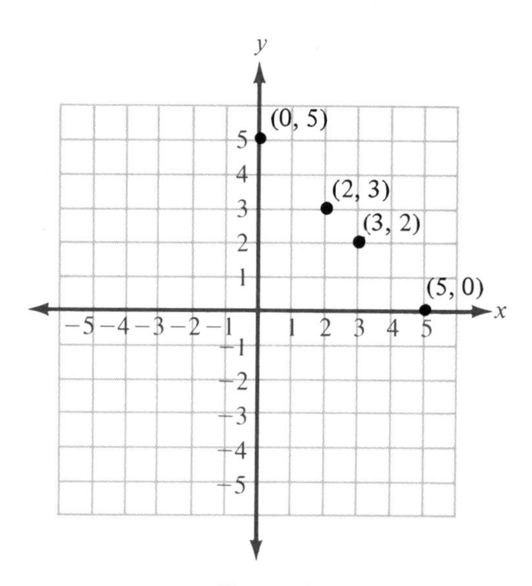

Figure 4

Now, by passing a straight line through these points we can graph the solution set for the equation $x + y = 5$. Linear equations in two variables always have graphs that are straight lines. The graph of the solution set for $x + y = 5$ is shown in Figure 5.

Every ordered pair that satisfies $x + y = 5$ has its graph on the line and any point on the line has coordinates that satisfy the equation. That is, there is a one-to-one correspondence between points on the line and solutions to the equation.

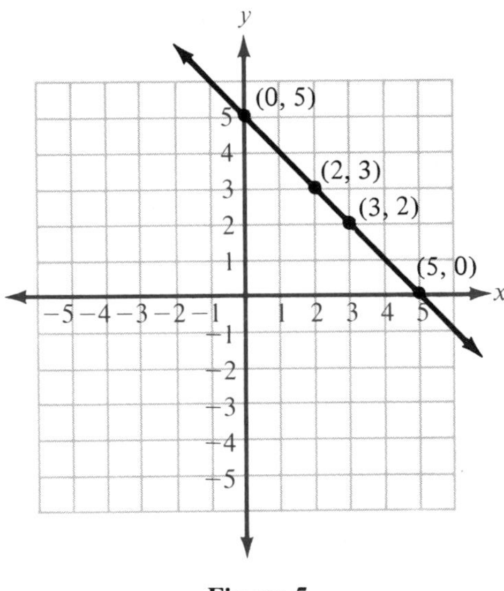

Figure 5 ▲

Here are some steps to follow when graphing straight lines.

To Graph a Straight Line

Step 1: Find any three ordered pairs that satisfy the equation. This can be done by using a convenient number for one variable and solving for the other variable.

Step 2: Graph the three ordered pairs found in Step 1. Actually, we only need two points to graph a straight line. The third point serves as a check. If all three points do not line up, there is a mistake in our work.

Step 3: Draw a straight line through the three points graphed in Step 2.

▼ **Example 4** Graph the equation $y = 3x - 1$.

Solution We first find three solutions to the equation by choosing convenient numbers for one variable and solving for the other variable.

Let $x = 0$: $\quad y = 3(0) - 1$
$y = 0 - 1$
$y = -1$ $\qquad$ $(0, -1)$ is one solution

Let $x = 2$: $\quad y = 3(2) - 1$
$y = 6 - 1$
$y = 5$ $\qquad$ $(2, 5)$ is another solution

$$\text{Let } x = -1: \quad y = 3(-1) - 1$$
$$y = -3 - 1$$
$$y = -4 \qquad (-1, -4) \text{ is a third solution}$$

Next we graph the ordered pairs $(0, -1)$, $(2, 5)$, $(-1, -4)$ and draw a straight line through them.

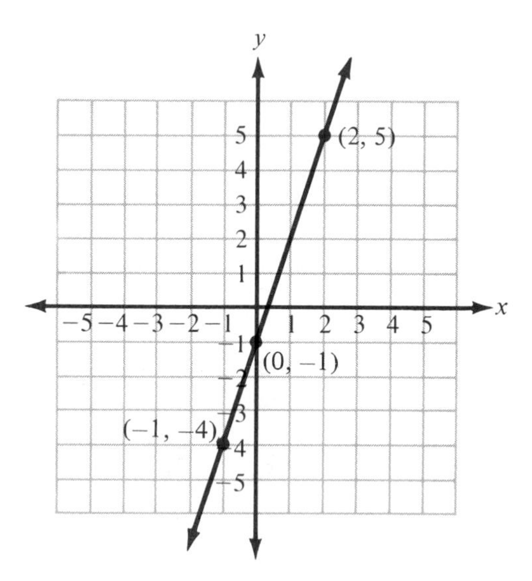

Figure 6

The line we have drawn is the graph of $y = 3x - 1$. ▲

▼ **Example 5** Graph the equation $y = -\frac{1}{3}x + 2$.

Solution We need to find three ordered pairs that satisfy the equation. To do so, we can let x equal any numbers we choose and find corresponding values of y. But, since every value of x we substitute into the equation is going to be multiplied by $-\frac{1}{3}$, let's use numbers for x that are divisible by 3, like -3, 0, and 3. That way, when we multiply them by $-\frac{1}{3}$, the result will be an integer.

$$\text{Let } x = -3; \quad y = -\frac{1}{3}(-3) + 2$$
$$y = 1 + 2$$
$$y = 3 \qquad (-3, 3) \text{ is one solution}$$

$$\text{Let } x = 0; \quad y = -\frac{1}{3}(0) + 2$$
$$y = 0 + 2$$
$$y = 2 \qquad (0, 2) \text{ is another solution}$$

Let $x = 3$; $\quad y = -\frac{1}{3}(3) + 2$

$y = -1 + 2$

$y = 1 \qquad$ (3, 1) is a third solution

Graphing the ordered pairs $(-3, 3)$, $(0, 2)$, and $(3, 1)$ and drawing a straight line through their graphs, we have the graph of the equation $y = -\frac{1}{3}x + 2$.

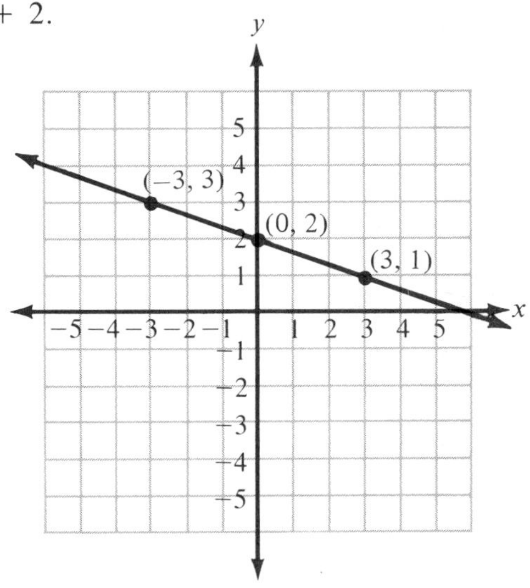

Figure 7 ▲

Note In Example 5, the values of x we used, -3, 0, and 3, are referred to as convenient values of x because they are easier to work with than some other numbers. For instance, if we let $x = 2$ in our original equation, we would have to add $-\frac{2}{3}$ and 2 to find the corresponding value of y. Not only would the arithmetic be more difficult, but the ordered pair we would obtain would have a fraction for its y-coordinate, making it more difficult to graph accurately also.

▼ **Example 6** Graph the solution set for $3x - 2y = 6$.

Solution It will be easier to find convenient values of x to use in the equation if we first solve the equation for y. To do so, we add $-3x$ to each side and then we multiply each side by $-\frac{1}{2}$.

$3x - 2y = 6$	Original equation
$-2y = -3x + 6$	Add $-3x$ to each side
$-\frac{1}{2}(-2y) = -\frac{1}{2}(-3x + 6)$	Multiply each side by $-\frac{1}{2}$
$y = \frac{3}{2}x - 3$	Simplify each side

Now, since each value of x will be multiplied by $-\frac{3}{2}$, it will be to our advantage to choose values of x that are divisible by 2. That way, we will obtain values of y that do not contain fractions. This time, let's use 0, 2, and 4 for x.

When $x = 0$; $\quad y = \dfrac{3}{2}(0) - 3$

$y = 0 - 3$

$y = -3 \qquad$ $(0, -3)$ is one solution

When $x = 2$; $\quad y = \dfrac{3}{2}(2) - 3$

$y = 3 - 3$

$y = 0 \qquad$ $(2, 0)$ is a second solution

When $x = 4$; $\quad y = \dfrac{3}{2}(4) - 3$

$y = 6 - 3$

$y = 3 \qquad$ $(4, 3)$ is a third solution

Graphing the ordered pairs $(0, -3)$, $(2, 0)$, and $(4, 3)$, and drawing a line through them, we have the graph shown in Figure 8.

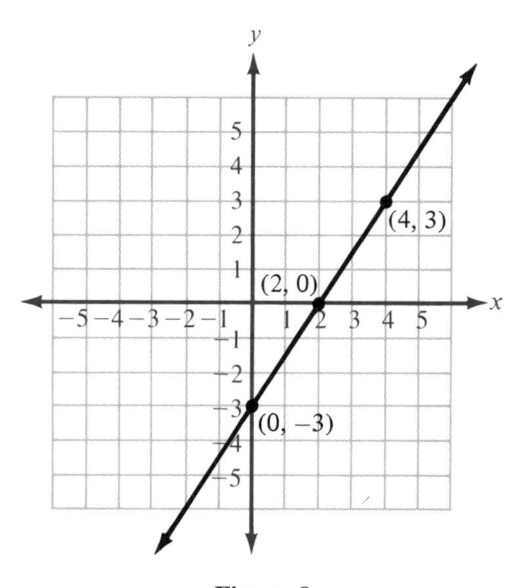

Figure 8

▲

Note After reading through Example 6, many students ask why we didn't use -2 for x when we were finding ordered pairs that were solutions to the original equation. The answer is, we could have. If we were to let $x = -2$, the corresponding value of y would have been -6. As you can see by looking at the graph in Figure 8, the ordered pair $(-2, -6)$ is on the graph.

▼ **Example 7** Graph the lines $x = 2$ and $y = -3$ on the same coordinate system.

Solution The line $x = 2$ is the set of all points whose x-coordinate is 2. When the variable y does not appear in the equation, it means that y can be any number. Therefore, the points $(2, 0)$, $(2, 3)$, $(2, -4)$, and $(2, 2)$ all satisfy the equation $x = 2$, simply because their x-coordinate is 2. The graph of $x = 2$, along with the points we named above, is given in Figure 9.

The line $y = -3$ is the set of all points whose y-coordinate is -3. The points $(0, -3)$, $(4, -3)$, and $(-4, -3)$ all satisfy the equation $y = -3$, because their y-coordinates are all -3. The graph of the line $y = -3$ is also shown in Figure 9.

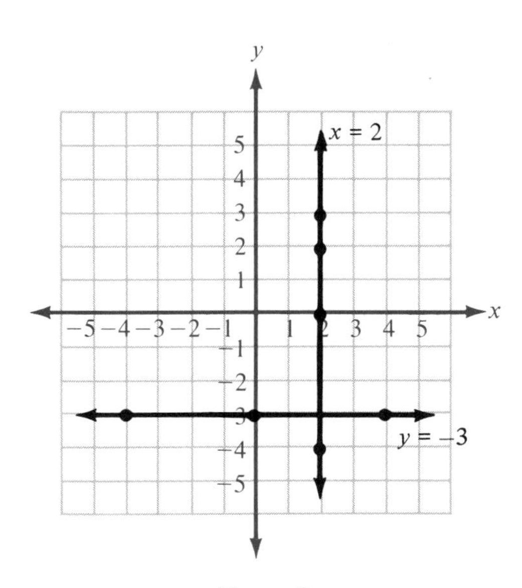

Figure 9 ▲

Problem Set 3.2

Graph the following ordered pairs.

1. $(3, 2)$
2. $(3, -2)$
3. $(-3, 2)$
4. $(-3, -2)$
5. $(5, 1)$
6. $(5, -1)$
7. $(1, 5)$
8. $(1, -5)$
9. $(-1, 5)$
10. $(-1, -5)$

11. $(2, \frac{1}{2})$
12. $(3, \frac{3}{2})$
13. $(-4, -\frac{5}{2})$
14. $(-5, -\frac{3}{2})$
15. $(3, 0)$
16. $(-2, 0)$
17. $(0, 5)$
18. $(0, 0)$

For the following equations, complete the given ordered pairs and use the results to graph the solution set for the equation.

19. $x + y = 3$ $(0,\ ), (2,\ ), (\ , -1)$
20. $x - y = 4$ $(1,\ ), (-1,\ ), (\ , 0)$
21. $y = 2x$ $(0,\ ), (-2,\ ), (2,\ )$
22. $y = \frac{1}{2}x$ $(0,\ ), (-2,\ ), (2,\ )$
23. $y = \frac{1}{3}x$ $(-3,\ ), (0,\ ), (3,\ )$
24. $y = 3x$ $(-2,\ ), (0,\ ), (2,\ )$
25. $y = 2x + 1$ $(0,\ ), (-1,\ ), (1,\ )$
26. $y = -2x + 1$ $(0,\ ), (-1,\ ), (1,\ )$
27. $y = 4x - 3$ $(0,\ ), (\ , 1), (\ , 5)$
28. $y = 4x + 3$ $(\ , 3), (-1,\ ), (-2,\ )$
29. $y = \frac{1}{2}x + 3$ $(-2,\ ), (0,\ ), (2,\ )$
30. $y = \frac{1}{2}x - 3$ $(-2,\ ), (0,\ ), (2,\ )$
31. $y = -\frac{2}{3}x + 1$ $(-3,\ ), (0,\ ), (3,\ )$
32. $y = -\frac{2}{3}x - 1$ $(-3,\ ), (0,\ ), (3,\ )$

Solve each equation for y. Then, complete the given ordered pairs and use them to draw the graph.

33. $2x + y = 3$ $(-1,\ ), (0,\ ), (1,\ )$
34. $3x + y = 2$ $(-1,\ ), (0,\ ), (1,\ )$
35. $3x + 2y = 6$ $(0,\ ), (2,\ ), (4,\ )$
36. $2x + 3y = 6$ $(0,\ ), (3,\ ), (6,\ )$
37. $-x + 2y = 6$ $(-2,\ ), (0,\ ), (2,\ )$
38. $-x + 3y = 6$ $(-3,\ ), (0,\ ), (3,\ )$

Find three solutions to each of the following equations and then graph the solution set.

39. $y = -\frac{1}{2}x$
40. $y = -2x$
41. $y = 3x - 1$
42. $y = -3x - 1$
43. $-2x + y = 1$
44. $-3x + y = 1$
45. $3x + 4y = 8$
46. $3x - 4y = 8$
47. $x = -2$
48. $y = 3$
49. $y = 2$
50. $x = -3$

51. If the perimeter of a rectangle is 10 inches, then the equation that describes the relationship between the length l and width w is

$$2l + 2w = 10$$

To graph this equation, we use a coordinate system in which the horizontal axis is labeled l and the vertical axis is labeled w.

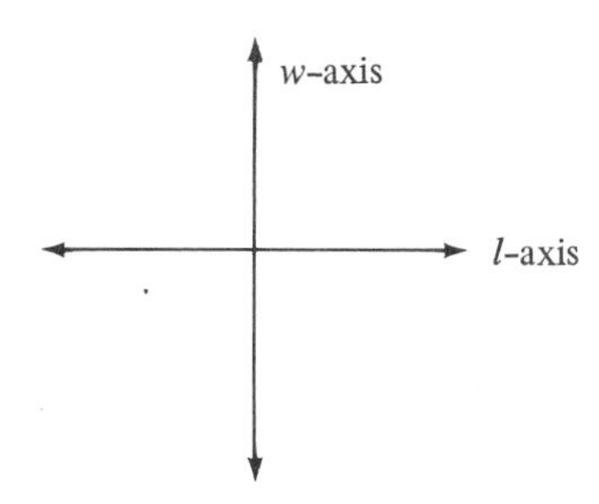

The ordered pairs that satisfy the equation look like (l, w). Complete the ordered pairs $(1,\ )$, $(2,\ )$, and $(3,\ )$ for the equation $2l + 2w = 10$ and use the results to graph the equation. Compare your answer with Example 3 in this section.

52. The perimeter of a rectangle is 6 inches. Graph the equation that describes the relationship between the length l and width w.

53. The solutions to the equation $y = x$ are ordered pairs of the form (x, x). The x- and y-coordinates are equal. Graph the line $y = x$.

54. The solutions to a certain equation are ordered pairs of the form $(x, 2x)$. Find the equation and its graph.

55. The y-coordinate of each ordered pair that satisfies the equation $y = |x|$ will never be negative. Each value of y is the absolute value of the corresponding value of x. Fill in the following ordered pairs so that they are solutions to the equation $y = |x|$. Then graph them and connect their graphs in a way that makes the most sense to you.

$(-3,\ )\quad(-2,\ )\quad(-1,\ )\quad(0,\ )\quad(1,\ )\quad(2,\ )\quad(3,\ )$

56. Fill in the following ordered pairs so that they are solutions to the equation $y = |x - 2|$. Graph each point and then connect the graphs in a way that makes the most sense to you.

$(-1,\ )\quad(0,\ )\quad(1,\ )\quad(2,\ )\quad(3,\ )\quad(4,\ )\quad(5,\ )$

57. Graph the lines $y = x + 1$ and $y = x - 3$ on the same coordinate system. Can you tell from looking at these first two graphs where the graph of $y = x + 3$ would be?

58. Graph the lines $y = 2x + 2$ and $y = 2x - 1$ on the same coordinate system. Use the similarities between these two graphs to graph the line $y = 2x - 4$.

Review Problems The problems below review material we covered in Section 2.4.

Solve each equation.

59. $3(x - 2) = 9$

60. $-4(x - 3) = -16$

61. $2(3x - 1) + 4 = -10$

62. $-5(2x + 3) - 10 = 15$

63. $6 - 2(4x - 7) = -4$

64. $5 - 3(2 - 3x) = 8$

Section 3.3 More on Graphing: Intercepts and Slope

There are three special quantities associated with straight lines and their graphs. One is the slope of the line. The other two are the x- and y-intercepts of the graph.

Intercepts

DEFINITION The *x-intercept* of a straight line is the x-coordinate of the point where the graph crosses the x-axis. The *y-intercept* is defined similarly. It is the y-coordinate of the point where the graph crosses the y-axis.

If the x-intercept is a, then the point $(a, 0)$ lies on the graph. (This is true because any point on the x-axis has a y-coordinate of 0.)

If the y-intercept is b, then the point $(0, b)$ lies on the graph. (This is true because any point on the y-axis has an x-coordinate of 0.)

Graphically, the relationship is shown in Figure 1.

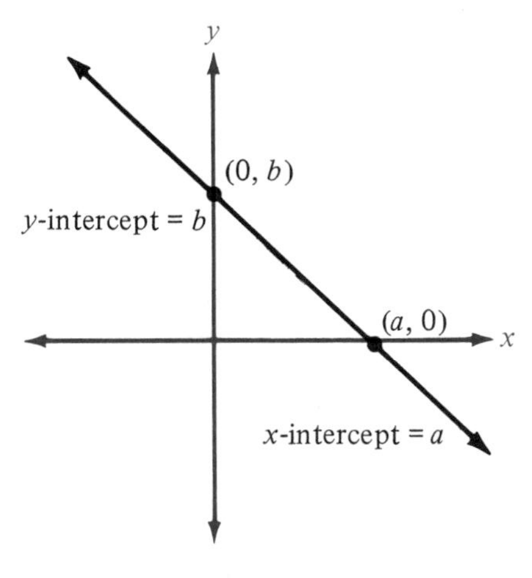

Figure 1

▼ **Example 1** Find the x- and y-intercepts for $3x - 2y = 6$ and then use them to draw the graph.

Solution To find where the graph crosses the x-axis, we let $y = 0$. (The y-coordinate of any point on the x-axis is 0.)

***x*-intercept:**

When $y = 0$

the equation $3x - 2y = 6$

becomes $3x - 2(0) = 6$

$$3x - 0 = 6$$

$$3x = 6$$

$$x = 2 \qquad \text{Multiply each side by } \tfrac{1}{3}$$

The graph crosses the x-axis at (2, 0), which means the x-intercept is 2.

y-intercept:

When $x = 0$

the equation $3x - 2y = 6$

becomes $3(0) - 2y = 6$

$$0 - 2y = 6$$

$$-2y = 6$$

$$y = -3 \quad \text{Multiply each side by } -\tfrac{1}{2}$$

The graph crosses the y-axis at $(0, -3)$, which means the y-intercept is -3.

Plotting the x- and y-intercepts and then drawing a line through them, we have the graph of $3x - 2y = 6$, as shown in Figure 2.

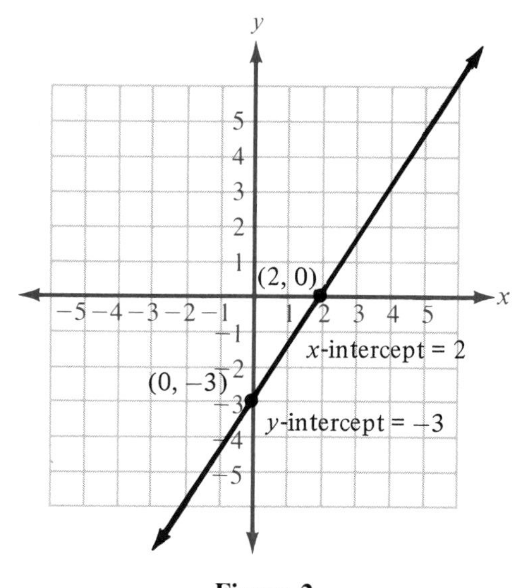

Figure 2

Note that this is the same graph we obtained in Example 6 of Section 3.2 when we graphed $3x - 2y = 6$ by first solving for y and then finding three ordered pairs that satisfied the resulting equation. ▲

▼ **Example 2** Graph $-x + 2y = 4$ by finding the intercepts and using them to draw the graph.

Solution Again, we find the x-intercept by letting $y = 0$ in the equation and solving for x. Similarly, we find the y-intercept by letting $x = 0$ and solving for y.

x-intercept:

When $\qquad y = 0$

the equation $\qquad -x + 2y = 4$

becomes

$$\begin{aligned} -x + 2(0) &= 4 \\ -x + 0 &= 4 \\ -x &= 4 \\ x &= -4 \qquad \text{Multiply each side by } -1 \end{aligned}$$

The x-intercept is -4, indicating that the point $(-4, 0)$ is on the graph of $-x + 2y = 4$.

y-intercept:

When $\qquad x = 0$

the equation $\qquad -x + 2y = 4$

becomes

$$\begin{aligned} -0 + 2y &= 4 \\ 2y &= 4 \\ y &= 2 \qquad \text{Multiply each side by } \tfrac{1}{2} \end{aligned}$$

The y-intercept is 2, indicating that the point $(0, 2)$ is on the graph of $-x + 2y = 4$.

Plotting the intercepts and drawing a line through them, we have the graph of $-x + 2y = 4$, as shown in Figure 3.

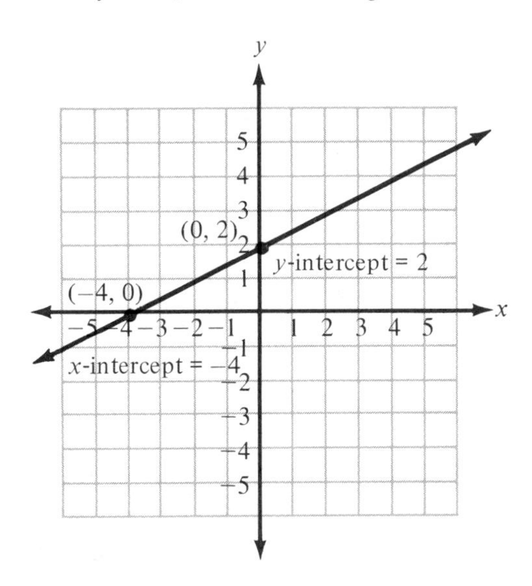

Figure 3

▲

Graphing a line by finding the intercepts, as we have done in Examples 1 and 2, is an easy method of graphing if the equation has the form $ax + by = c$ and both the numbers a and b divide the number c evenly.

In our next example we use the intercepts to graph a line in which y is given in terms of x.

▼ **Example 3** Use the intercepts for $y = -\frac{1}{3}x + 2$ to draw its graph.

Solution We graphed this line previously in Example 5 of Section 3.2 by substituting 3 different values of x into the equation and solving for y. This time we will graph the line by finding the intercepts.

***x*-intercept:**

When	$y = 0$	
the equation	$y = -\frac{1}{3}x + 2$	
becomes	$0 = -\frac{1}{3}x + 2$	
	$-2 = -\frac{1}{3}x$	Add -2 to each side
	$6 = x$	Multiply each side by -3

The x-intercept is 6, which means the graph passes through the point (6, 0).

***y*-intercept:**

When	$x = 0$
the equation	$y = -\frac{1}{3}x + 2$
becomes	$y = -\frac{1}{3}(0) + 2$
	$y = 2$

The y-intercept is 2, which means the graph passes through the point (0, 2).

The graph of $y = -\frac{1}{3}x + 2$ is shown in Figure 4. Compare this graph, and the method used to obtain it, with Example 5 in Section 3.2.

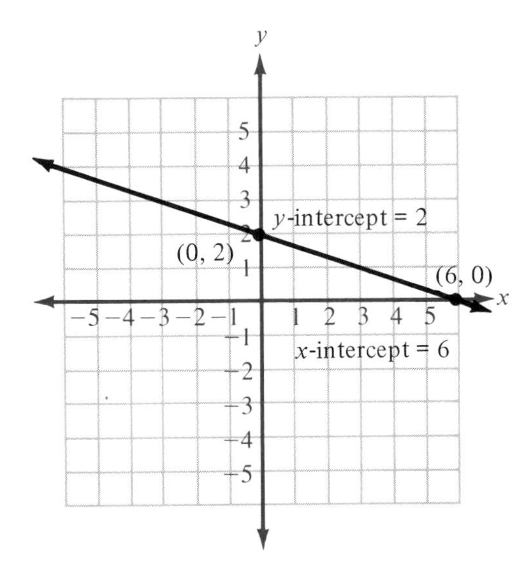

Figure 4 ▲

The Slope of a Line

The idea behind developing what is called the slope of a line is to find a number we can associate with the steepness of the graph of a straight line.

Suppose we know the coordinates of two points on a straight line. Since we are trying to develop a general formula for the slope of a straight line, we will use general points—call the two points (x_1, y_1) and (x_2, y_2). They represent the coordinates of any two different points on our straight line. We define the *slope* of our line to be the ratio of the vertical change to the horizontal change as we move from point (x_1, y_1) to point (x_2, y_2) on the line. (See Figure 5.)

Note The 2 in x_2 is called a *subscript.* It is notation that allows us to distinguish between the variables x_1 and x_2, while still showing that they are both x-coordinates.

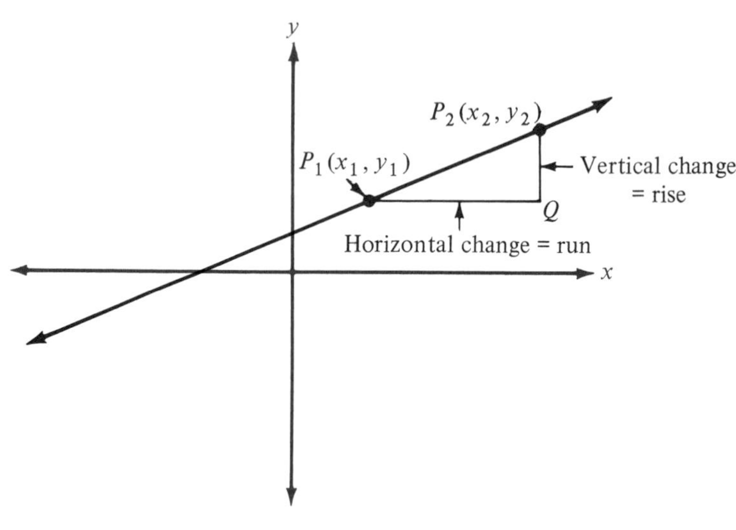

Figure 5

We call the vertical change the *rise* in the graph and the horizontal change the *run* in the graph. The slope, then, is

$$\text{Slope} = \frac{\text{vertical change}}{\text{horizontal change}} = \frac{\text{rise}}{\text{run}}$$

We would like to have a numerical value to associate with the rise in the graph and a numerical value to associate with the run in the graph. A quick study of Figure 5 shows that the coordinates of point Q must be (x_2, y_1), since

Q is directly below point P_2 and right across from point P_1. We can draw our diagram again in the manner shown in Figure 6. It is apparent from this graph that the rise can be expressed as $(y_2 - y_1)$ and the run as $(x_2 - x_1)$. We usually denote the slope of a line by the letter m. Here is the complete definition of slope, along with a diagram (Figure 6) that illustrates the definition.

DEFINITION If points (x_1, y_1) and (x_2, y_2) are any two different points, then the slope of the line on which they lie is:

$$\text{Slope} = m = \frac{\text{rise}}{\text{run}} = \frac{y_2 - y_1}{x_2 - x_1}$$

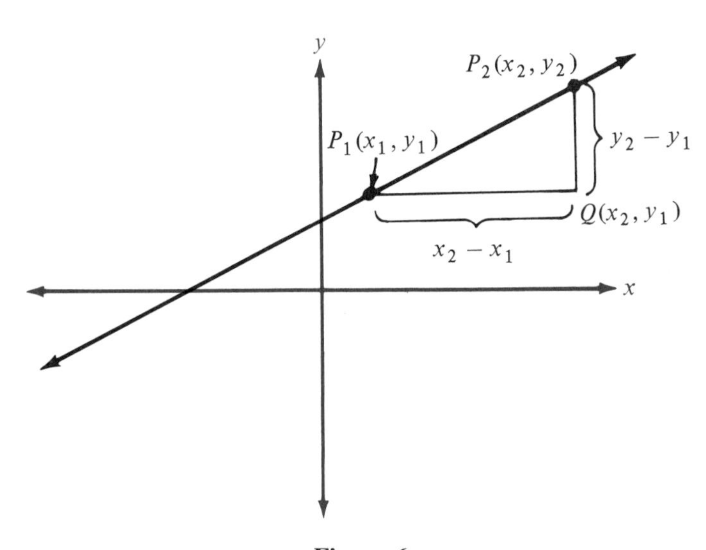

Figure 6

This definition of the slope of a line does just what we want it to do. If the line rises going from left to right, the slope will be positive. If the line falls from left to right, the slope will be negative. Also, the steeper the line, the larger numerical value the slope will have.

▼ **Example 4** Find the slope of the line between the points (1, 2) and (3, 5).

Solution We can let

$$(x_1, y_1) = (1, 2)$$

and

$$(x_2, y_2) = (3, 5)$$

then

$$m = \frac{y_2 - y_1}{x_2 - x_1} = \frac{5 - 2}{3 - 1} = \frac{3}{2}$$

The slope is $\frac{3}{2}$. For every vertical change of 3 units, there will be a corresponding horizontal change of 2 units. (See Figure 7.)

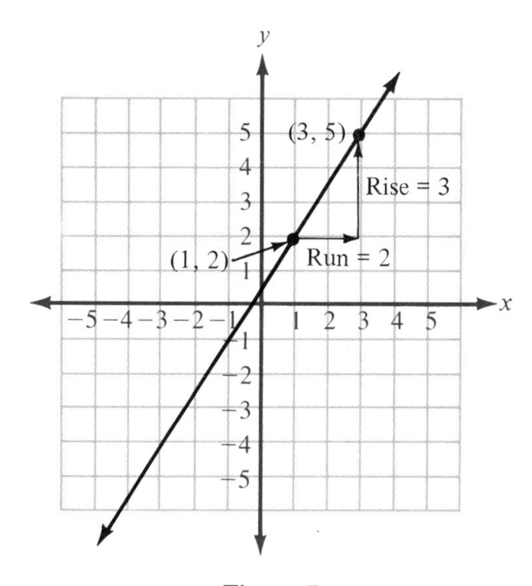

Figure 7 ▲

▼ **Example 5** Find the slope of the line through $(-2, 1)$ and $(5, -4)$.

Solution It makes no difference which ordered pair we call (x_1, y_1) and which we call (x_2, y_2).

$$\text{Slope} = m = \frac{y_2 - y_1}{x_2 - x_1} = \frac{-4 - 1}{5 - (-2)} = \frac{-5}{7}$$

The slope is $-\frac{5}{7}$. Every vertical change of -5 units (down 5 units) is accompanied by a horizontal change of 7 units (to the right 7 units). (See Figure 8.)

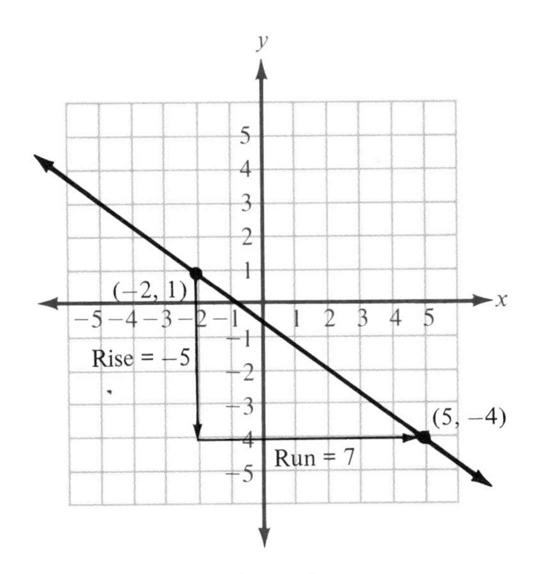

Figure 8 ▲

Problem Set 3.3

Find the x- and y-intercepts for the following equations. Then, use the intercepts to graph each equation.

1. $2x + y = 4$
2. $2x + y = 2$
3. $-x + y = 3$
4. $-x + y = 4$
5. $-x + 2y = 2$
6. $-x + 2y = 4$
7. $5x + 2y = 10$
8. $2x + 5y = 10$
9. $4x - 2y = 8$
10. $2x - 4y = 8$
11. $-4x + 5y = 20$
12. $-5x + 4y = 20$
13. $y = 2x - 6$
14. $y = 2x + 6$
15. $y = 2x + 2$
16. $y = -2x + 2$
17. $y = 2x - 1$
18. $y = -2x - 1$
19. $y = \frac{1}{2}x + 3$
20. $y = \frac{1}{2}x - 3$
21. $y = -\frac{1}{3}x - 2$
22. $y = -\frac{1}{3}x + 2$

Find the slope of the line through the following pairs of points. Then, plot each pair of points, draw a line through them, and indicate the rise and run in the graph in the same manner shown in Examples 4 and 5.

23. $(2, 1), (4, 4)$
24. $(3, 1), (5, 4)$
25. $(1, 4), (5, 2)$
26. $(1, 3), (5, 2)$
27. $(1, -3), (4, 2)$
28. $(2, -3), (5, 2)$
29. $(-3, -2), (1, 3)$
30. $(-3, -1), (1, 4)$
31. $(-3, 2), (3, -2)$
32. $(-3, 3), (3, -1)$
33. $(2, -5), (3, -2)$
34. $(2, -4), (3, -1)$

35. Graph the line that has an x-intercept of 3 and a y-intercept of -2. What is the slope of this line?
36. Graph the line that has an x-intercept of 2 and a y-intercept of -3. What is the slope of this line?
37. Graph the line with x-intercept 4 and y-intercept 2. What is the slope of this line?
38. Graph the line with x-intercept -4 and y-intercept -2. What is the slope of this line?
39. The vertical line $x = 3$ has only one intercept. Graph $x = 3$ and name its intercept. (Remember, ordered pairs (x, y) that are solutions to the equation $x = 3$ are ordered pairs with an x-coordinate of 3 and any y-coordinate.)
40. Graph the vertical line $x = -2$. Then, name its intercept.
41. The horizontal line $y = 4$ has only one intercept. Graph $y = 4$ and name its intercept. (Ordered pairs (x, y) that are solutions to the equation $y = 4$ are ordered pairs with a y-coordinate of 4 and any x-coordinate.)

42. Graph the horizontal line $y = -3$. Then, name its intercept.

Review Problems The problems below review material we covered in Section 1.6.

Simplify each expression.

43. $(-7)^2$
44. 2^3
45. $5 - 2 \cdot 6$
46. $7 - 4 \cdot 5$
47. $6 - 3(8 - 2)$
48. $7 - 2(3 - 5)$
49. $2(3)^2 - 4(3)^2$
50. $7(2)^3 - 3(3)^3$
51. $5 - 2[3 - 4(-2)]$
52. $6 - 3[5 - 2(-3)]$

Section 3.4 The Slope-Intercept Form of the Equation of a Line

In this section we will use the slope and y-intercept of a line to find the equation of the line. But, before we do so, let's look at how we can use the slope and y-intercept to graph a line.

▼ **Example 1** Graph the line with slope $\frac{3}{2}$ and y-intercept 1.

Solution Since the y-intercept is 1, we know that one point on the line is $(0, 1)$. So, we begin by plotting the point $(0, 1)$, as shown in Figure 1.

There are many lines that pass through the point shown in Figure 1, but only one of those lines has a slope of $\frac{3}{2}$. The slope, $\frac{3}{2}$, can be thought of as the rise in the graph divided by the run in the graph. Therefore, if we start at the point $(0, 1)$ and move 3 units up (that's a rise of 3) and then

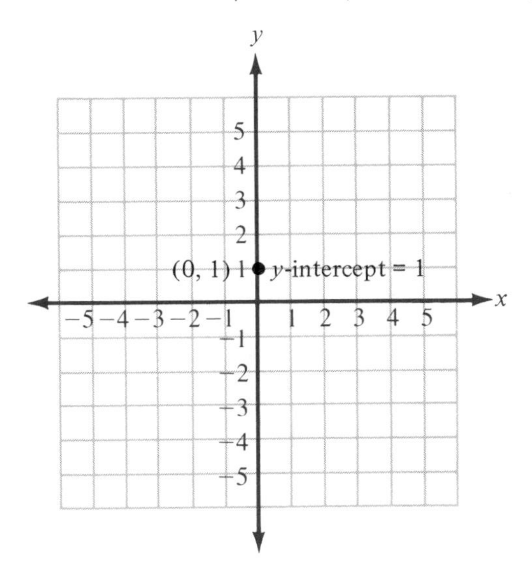

Figure 1

2 units to the right (a run of 2) we will be at another point on the graph. Figure 2 shows that the point we reach by doing so is the point (2, 4).

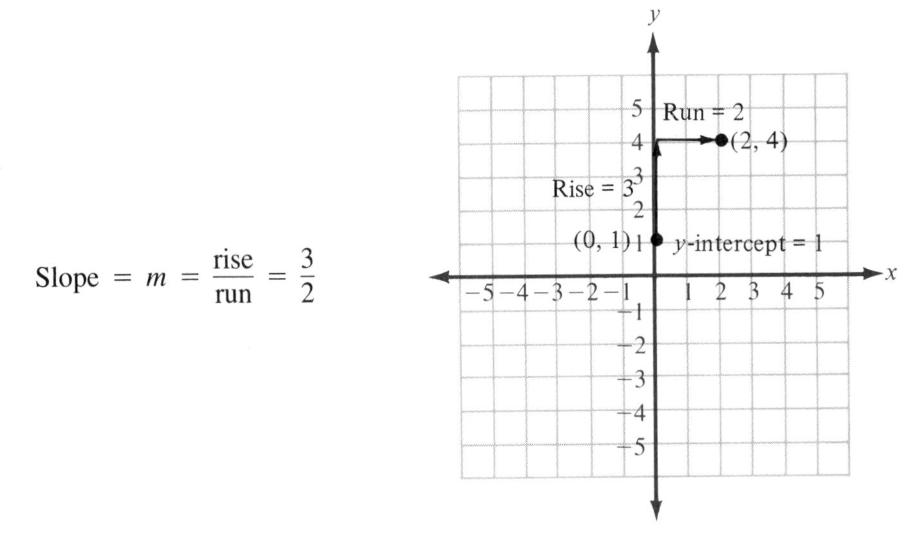

Figure 2

To graph the line with slope $\frac{3}{2}$ and y-intercept 1, we simply draw a line through the two points in Figure 2 to obtain the graph shown in Figure 3.

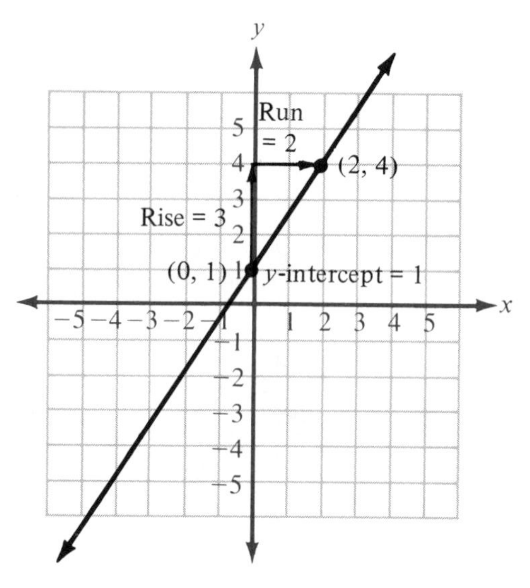

Figure 3 ▲

Next, we want to find the equation of the line we graphed in Example 1. Suppose the point (x, y) is any other point on the line shown in Figure 3. Applying our slope formula to the points (x, y) and $(0, 1)$ and setting the result equal to the slope $\frac{3}{2}$, we have

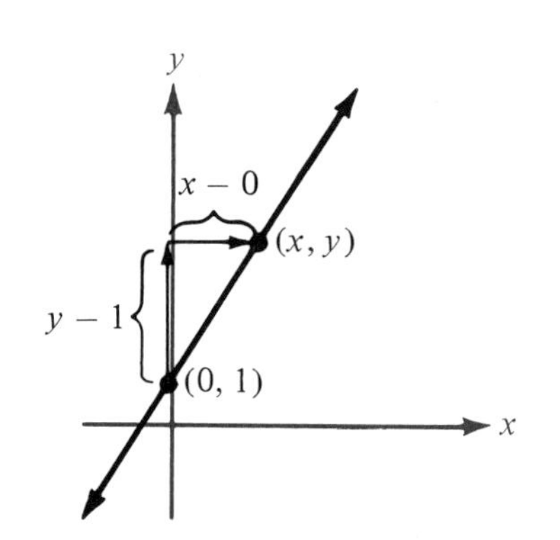

$$\frac{y-1}{x-0} = \frac{3}{2} \qquad \text{Slope} = \frac{\text{vertical change}}{\text{horizontal change}}$$

$$\frac{y-1}{x} = \frac{3}{2} \qquad x - 0 = x$$

$$y - 1 = \frac{3}{2}x \qquad \text{Multiply each side by } x$$

$$y = \frac{3}{2}x + 1 \qquad \text{Add 1 to each side}$$

What is interesting and useful about the equation we just found is that the number in front of x is the slope of the line and the constant term is the y-intercept. It is no coincidence that it turned out this way. Whenever an equation has the form $y = mx + b$, the graph is always a straight line with slope m and y-intercept b. To see that this is true in general, suppose we want the equation of a line with slope m and y-intercept b. Since the y-intercept is b, then the point $(0, b)$ is on the line. If (x, y) is any other point on the line, then we apply our slope formula to get

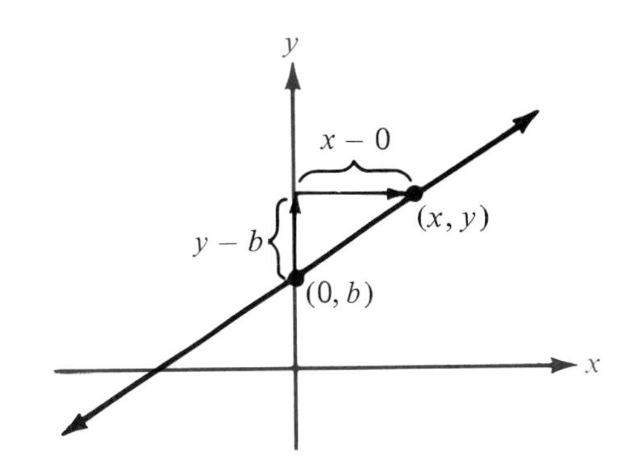

$\dfrac{y - b}{x - 0} = m$	Slope $= \dfrac{\text{vertical change}}{\text{horizontal change}}$
$\dfrac{y - b}{x} = m$	$x - 0 = x$
$y - b = mx$	Multiply each side by x
$y = mx + b$	Add b to each side

Here is a summary of what we have just found.

Slope-Intercept Form of the Equation of a Line

The equation of the line with slope m and y-intercept b is always given by

$$y = mx + b$$

▼ **Example 2** Find the equation of the line with slope $-\frac{4}{3}$ and y-intercept 5. Then, graph the line.

Solution Substituting $m = -\frac{4}{3}$ and $b = 5$ into the equation $y = mx + b$, we have

$$y = -\frac{4}{3}x + 5$$

Finding the equation from the slope and y-intercept is just that easy. If the slope is m and the y-intercept is b, then the equation is always $y = mx + b$.

Since the y-intercept is 5, the graph goes through the point $(0, 5)$. To find a second point on the graph, we start at $(0, 5)$ and move 4 units down (that's a rise of -4) and 3 units to the right (a run of 3). The point we reach is $(3, 1)$. Drawing a line that passes through $(0, 5)$ and $(3, 1)$, we have the graph of our equation. (Note that we could also let the rise $= 4$ and the run $= -3$ and obtain the same graph.) The graph is shown in Figure 4.

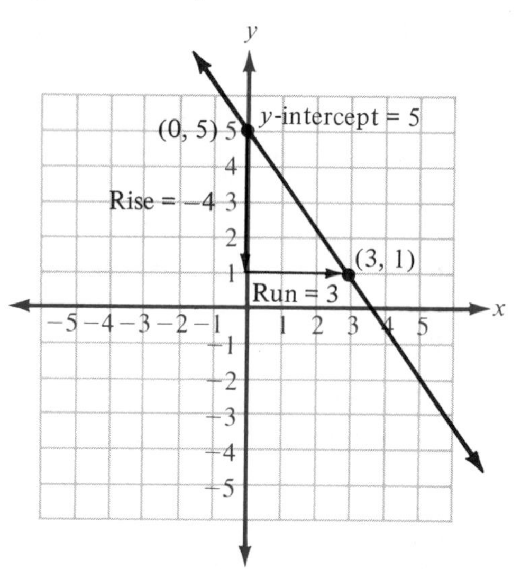

Figure 4

▲

▼ **Example 3** Find the slope and y-intercept for the equation $-2x + y = -4$. Then, use them to draw the graph.

Solution To identify the slope and y-intercept from the equation, the equation must be in the form $y = mx + b$ (slope-intercept form). To write our equation in this form, we must solve the equation for y. To do so, we simply add $2x$ to each side of the equation.

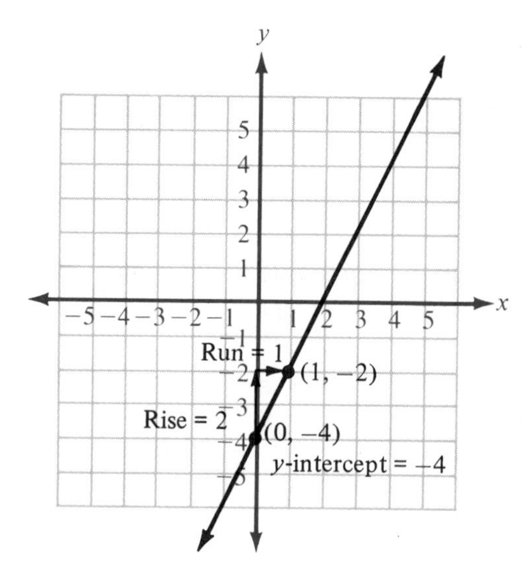

Figure 5

$$-2x + y = -4 \qquad \text{Original equation}$$
$$y = 2x - 4 \qquad \text{Add } 2x \text{ to each side}$$

The equation is now in slope-intercept form, so the slope must be 2 and the y-intercept -4. The graph, therefore, crosses the y-axis at $(0, -4)$. Since the slope is 2, we can let the rise $= 2$ and the run $= 1$ and find a second point on the graph. The graph is shown in Figure 5. ▲

▼ **Example 4** Find the slope and y-intercept for $3x - 2y = 6$.

Solution To find the slope and y-intercept from the equation, we must write the equation in the form $y = mx + b$. This means we must solve the equation $3x - 2y = 6$ for y.

$$3x - 2y = 6 \qquad \text{Original equation}$$
$$-2y = -3x + 6 \qquad \text{Add } -3x \text{ to each side}$$
$$-\frac{1}{2}(-2y) = -\frac{1}{2}(-3x + 6) \qquad \text{Multiply each side by } -\tfrac{1}{2}$$
$$y = \frac{3}{2}x - 3 \qquad \text{Simplify each side}$$

Now that the equation is written in slope-intercept form, we can identify the slope as $\frac{3}{2}$ and the y-intercept as -3. The graph is shown in Figure 6.

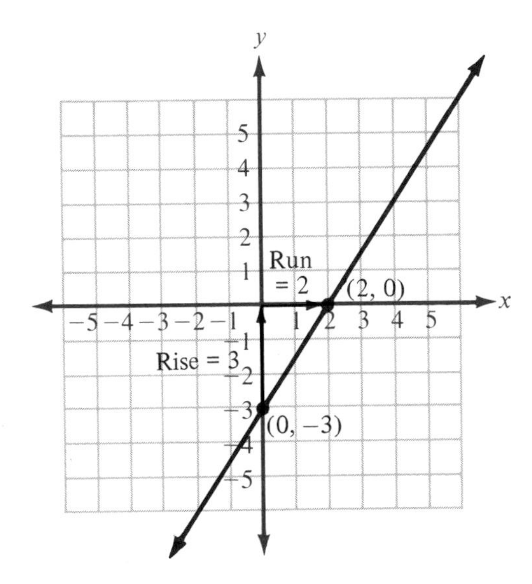

Figure 6

▲

▼ **Example 5** Find the equation of the line with slope 2 that goes through the point $(-3, -5)$.

Solution To find the equation of a line using the slope-intercept form of a line, we need the slope m and the y-intercept b. Since the slope we are given in this problem is $m = 2$, we can partially fill in the equation $y = mx + b$ as follows:

$$y = 2x + b$$

The only thing left to find is the y-intercept b. To do so, we substitute the coordinates of our given point $(-3, -5)$ into the equation for x and y. (It makes sense to do this since the line we are interested in passes through the point $(-3, -5)$, meaning that its coordinates must satisfy the equation of that line.)

Substituting	$x = -3$ and $y = -5$	
into the equation	$y = 2x + b$	
gives us	$-5 = 2(-3) + b$	
	$-5 = -6 + b$	Simplify the right side
	$1 = b$	Add 6 to each side

As you can see, the y-intercept is 1. The complete equation is

$$y = 2x + 1$$

The graph is shown in Figure 7.

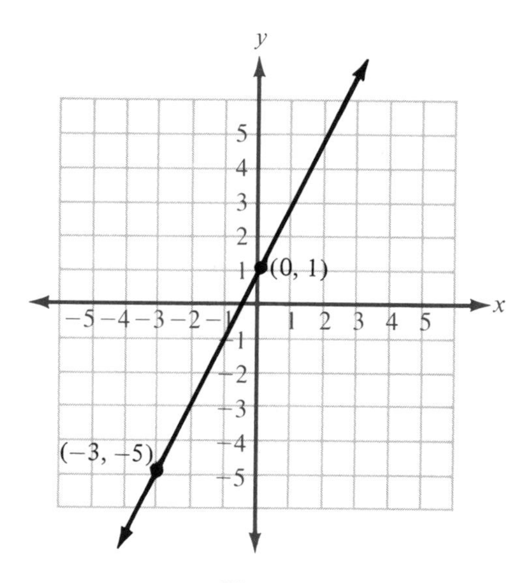

Figure 7 ▲

▼ **Example 6** Find the equation of the line that contains the points (1, 2) and (−2, 5).

Solution As was the case in Example 5, to find the equation of this line we need its slope and y-intercept. To find the slope, we use our slope formula

$$m = \frac{y_2 - y_1}{x_2 - x_1} = \frac{5 - 2}{-2 - 1} = \frac{3}{-3} = -1$$

Knowing the slope, we fill in part of our equation:

$$y = -1x + b$$
$$y = -x + b$$

We can substitute the coordinates of either of our two points into this last equation and find the y-intercept b. Let's use the coordinates of (1, 2).

When	$x = 1$ and $y = 2$	
the equation	$y = -x + b$	
becomes	$2 = -1 + b$	
	$3 = b$	Add 1 to each side

The complete equation is $y = -x + 3$. See Figure 8.

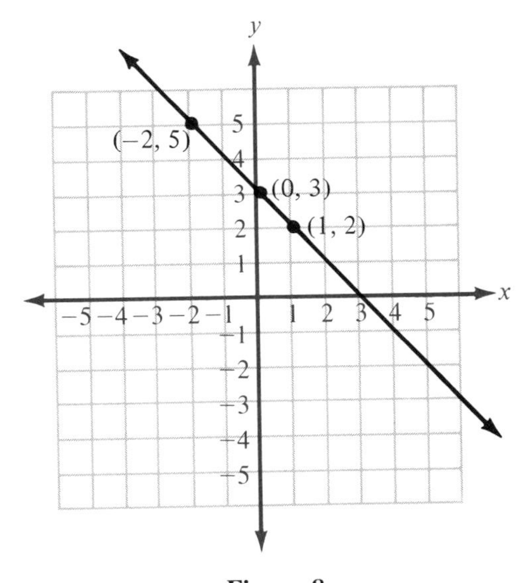

Figure 8 ▲

As you look back over the last three sections of this chapter, you will see that we have shown three different methods of graphing lines. To review, here is a short summary of each of the three methods.

Methods of Graphing Lines

1. Substitute convenient values of x into the equation and find the corresponding values of y. We used this method first for equations like $y = 2x - 3$. To use this method for equations that looked like $2x - 3y = 6$, we first solved them for y.
2. Find the x- and y-intercepts. This method works best for equations of the form $3x + 2y = 6$ where the numbers in front of x and y divide the constant term evenly.
3. Find the slope and y-intercept. This method works best when the equation has the form $y = mx + b$ and b is an integer.

Problem Set 3.4

In each of the following problems, graph the line with the given slope and y-intercept.

1. $m = \frac{2}{3}, b = 1$
2. $m = \frac{3}{4}, b = -2$
3. $m = \frac{3}{2}, b = -1$
4. $m = \frac{4}{3}, b = 2$
5. $m = -\frac{2}{5}, b = 3$
6. $m = -\frac{3}{5}, b = 4$
7. $m = 2, b = -4$
8. $m = -2, b = 4$
9. $m = -3, b = 2$
10. $m = 3, b = -2$

Find the slope and y-intercept for each of the following equations by writing them in the form $y = mx + b$. Then, graph each equation.

11. $-2x + y = 4$
12. $-2x + y = 2$
13. $3x + y = 3$
14. $3x + y = 6$
15. $3x + 2y = 6$
16. $2x + 3y = 6$
17. $4x - 5y = 20$
18. $2x - 5y = 10$
19. $-2x - 5y = 10$
20. $-4x + 5y = 20$

For each problem below, the slope and one point on a line are given. In each case, find the equation of that line. (Write the equation of each line in slope-intercept form.)

21. $(-2, -5), m = 2$
22. $(-1, -5), m = 2$
23. $(-4, 1), m = -\frac{1}{2}$
24. $(-2, 1), m = -\frac{1}{2}$
25. $(2, -3), m = \frac{3}{2}$
26. $(3, -4), m = \frac{4}{3}$
27. $(-1, 4), m = -3$
28. $(-2, 5), m = -3$
29. $(2, 4), m = 1$
30. $(4, 2), m = -1$

Find the equation of the line that passes through each pair of points. Write your answers in slope-intercept form.

31. $(-2, -4), (1, -1)$
32. $(2, 4), (-3, -1)$
33. $(-1, -5), (2, 1)$
34. $(-1, 6), (1, 2)$
35. $(-3, -2), (3, 6)$
36. $(-3, 6), (3, -2)$
37. $(-3, -1), (3, -5)$
38. $(-3, -5), (3, 1)$
39. $(-2, 1), (1, -2)$
40. $(1, -2), (2, 1)$

41. Find the equation of the line with x-intercept 3 and y-intercept 2.
42. Find the equation of the line with x-intercept 2 and y-intercept 3.
43. Find the equation of the line with x-intercept -2 and y-intercept -5.
44. Find the equation of the line with x-intercept -3 and y-intercept -5.
45. The equation of the vertical line that passes through the points $(3, -2)$ and $(3, 4)$ is either $x = 3$ or $y = 3$. Which one is it?
46. Find the equation of the vertical line that passes through the points $(2, 5)$ and $(2, -3)$.
47. The equation of the horizontal line that passes through the points $(2, 3)$ and $(-1, 3)$ is either $x = 3$ or $y = 3$. Which one is it?
48. Find the equation of the horizontal line that passes through the points $(-4, -2)$ and $(3, -2)$.

Review Problems The problems that follow review material we covered in Section 2.3.

Translate each of the following percent problems into an equation and then solve the equation.

49. What number is 25% of 300?
50. What number is 40% of 250?
51. 25 is what percent of 125?
52. 30 is what percent of 120?
53. 60 is 15% of what number?
54. 75 is 30% of what number?
55. A savings account pays 12% interest each year. If \$2,000 is deposited in the account at the beginning of the year, how much money will be in the account at the end of the year?
56. Suppose you deposit \$3,000 in an account that pays 9% per year. How much money will you have in the account at the end of a year?

Section 3.5 Solving Linear Systems by Graphing

Two linear equations considered at the same time make up what is called a *system* of linear equations. Both equations contain two variables and, of course, have graphs that are straight lines. The following are systems of linear equations:

$$\begin{aligned} x + y &= 3 \\ 3x + 4y &= 2 \end{aligned} \qquad \begin{aligned} y &= 2x + 1 \\ y &= 3x + 2 \end{aligned} \qquad \begin{aligned} 2x - y &= 1 \\ 3x - 2y &= 6 \end{aligned}$$

The solution set for a system of linear equations is all ordered pairs that are solutions to both equations. Since each linear equation has a graph that is a straight line, we can expect the intersection of the graphs to be a point whose coordinates are solutions to the system. That is, if we graph both equations on the same coordinate system, we can read the coordinates of the point of intersection and have the solution to our system. Here is an example.

▼ **Example 1** Solve the following system by graphing:

$$x + y = 4$$
$$x - y = -2$$

Solution On the same set of coordinate axes we graph each equation separately. Figure 1 shows both graphs, without showing the work necessary to solve them. We can see from the graphs that they intersect at the point (1, 3). The point (1, 3) must therefore be the solution to our system, since it is the only ordered pair whose graph lies on both lines. Its coordinates satisfy both equations.

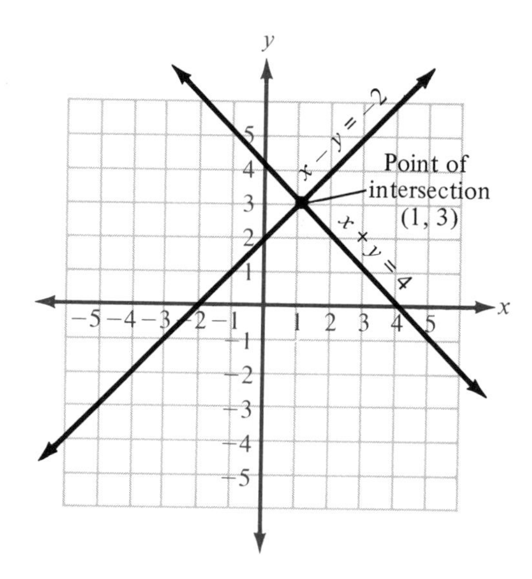

Figure 1

We can check our results by substituting the coordinates $x = 1$, $y = 3$ into both equations to see if they work.

When	$x = 1$	When	$x = 1$
and	$y = 3$	and	$y = 3$
the equation	$x + y = 4$	the equation	$x - y = -2$
becomes	$1 + 3 \stackrel{?}{=} 4$	becomes	$1 - 3 \stackrel{?}{=} -2$
or	$4 = 4$	or	$-2 = -2$

The point (1, 3) satisfies both equations. ▲

Here are some steps to follow in solving linear systems by graphing.

To Solve a Linear System by Graphing

Step 1: Graph the first equation.
Step 2: Graph the second equation on the same set of axes used for the first equation.
Step 3: Read the coordinates of the point of intersection of the two graphs. The ordered pair is the solution to the system.
Step 4: Check the solution in both equations, if necessary.

▼ **Example 2** Solve the following system by graphing:

$$x + 2y = 8$$
$$2x - 3y = 2$$

Solution We graph each equation on the same set of coordinate axes and get

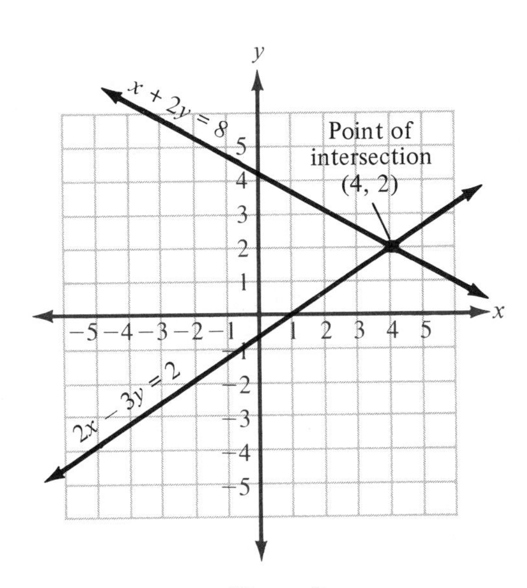

Figure 2

From the graph, we can see the solution for our system is (4, 2). We check this solution as follows:

When	$x = 4$	When	$x = 4$
and	$y = 2$	and	$y = 2$
the equation	$x + 2y = 8$	the equation	$2x - 3y = 2$
becomes	$4 + 2(2) \stackrel{?}{=} 8$	becomes	$2(4) - 3(2) \stackrel{?}{=} 2$
	$4 + 4 = 8$		$8 - 6 = 2$
	$8 = 8$		$2 = 2$

The point (4, 2) satisfies both equations and, therefore, must be the solution to our system. ▲

▼ **Example 3** Solve this system by graphing:

$$y = 2x - 3$$
$$x = 3$$

Solution Graphing both equations on the same set of axes, we have Figure 3.

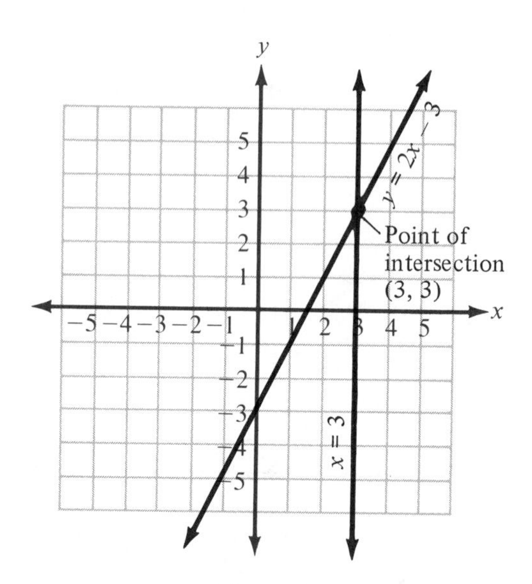

Figure 3

The solution to the system is the point (3, 3). ▲

▼ **Example 4** Solve by graphing:

$$y = x - 2$$
$$y = x + 1$$

Solution Graphing both equations, we have the following:

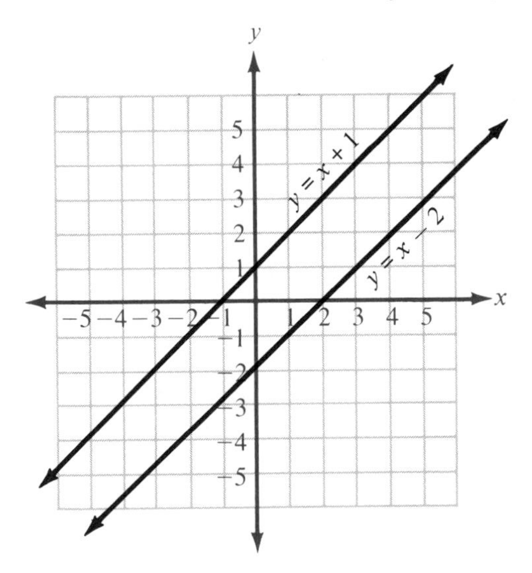

Figure 4

We can see in Figure 4 that the lines are parallel and therefore do not intersect. Our system has no ordered pair as a solution since there is no ordered pair that satisfies both equations. We say the solution set is the empty set and write $\emptyset$. ▲

Example 4 is one example of two special cases associated with linear systems. The other special case happens when the two graphs coincide. Here is an example:

▼ **Example 5** Graph the system:

$$\begin{aligned} 2x + y &= 4 \\ 4x + 2y &= 8 \end{aligned}$$

Solution Both graphs are shown in Figure 5.

The two graphs coincide. The reason becomes apparent when we multiply both sides of the first equation by 2:

$$\begin{aligned} 2x + y &= 4 \\ 2(2x + y) &= 2(4) && \text{Multiply both sides by 2} \\ 4x + 2y &= 8 \end{aligned}$$

The equations have the same solution set. Any ordered pair that is a solution to one is a solution to the system. The system has an infinite number of solutions. (Any point on the line is a solution to the system.)

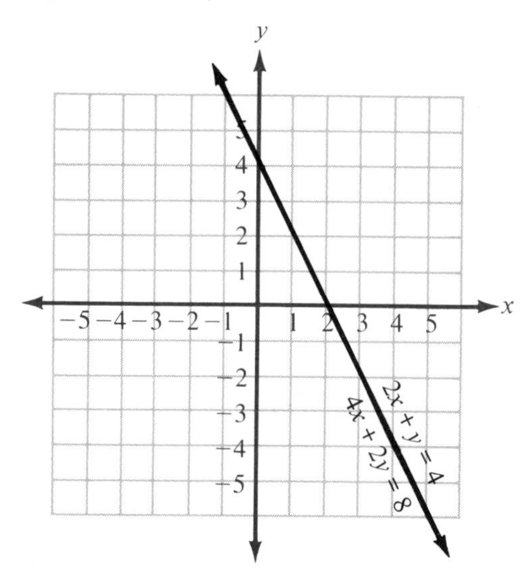

Figure 5

The two special cases illustrated in the last two examples do not happen often. Usually a system has a single ordered pair as a solution. Solving a system of linear equations by graphing is useful only when the ordered pair in the solution set has integers for coordinates. There are two other solution methods that work well in all cases. We will develop the other two methods in the next two sections.

Problem Set 3.5

Solve the following systems of linear equations by graphing.

1. $x + y = 3$
$x - y = 1$

2. $x + y = 2$
$x - y = 4$

3. $x + y = 1$
$-x + y = 3$

4. $x + y = 1$
$x - y = -5$

5. $x + y = 8$
$-x + y = 2$

6. $x + y = 6$
$-x + y = -2$

7. $3x - 2y = 6$
$x - y = 1$

8. $5x - 2y = 10$
$x - y = -1$

9. $6x - 2y = 12$
$3x + y = -6$

10. $4x - 2y = 8$
$2x + y = -4$

11. $4x + y = 4$
$3x - y = 3$

12. $5x - y = 10$
$2x + y = 4$

13. $x + 2y = 0$
$2x - y = 0$

14. $3x + y = 0$
$5x - y = 0$

15. $3x - 5y = 15$
$-2x + y = 4$

16. $2x - 4y = 8$
$2x - y = -1$

17. $y = 2x + 1$
$y = -2x - 3$

18. $y = 3x - 4$
$y = -2x + 1$

19. $x + 3y = 3$
$y = x + 5$

20. $2x + y = -2$
$y = x + 4$

21. $x + y = 2$
$x = -3$

22. $x + y = 6$
$y = 2$

23. $x = -4$
$y = 6$

24. $x = 5$
$y = -1$

25. $x + y = 4$
$2x + 2y = -6$

26. $x - y = 3$
$2x - 2y = 6$

27. $4x - 2y = 8$
$2x - y = 4$

28. $3x - 6y = 6$
$x - 2y = 4$

Review Problems The following problems review material we covered in Section 2.1.

Find the value of each expression when x is -3.

29. $2x - 9$
30. $-4x + 3$
31. $9 - 6x$
32. $7 - 5x$
33. $4(3x + 2) + 1$
34. $3(6x + 5) + 10$
35. $2x^2 + 3x + 4$
36. $4x^2 + 3x + 2$

Section 3.6 The Elimination Method

The addition property states that if equal quantities are added to both sides of an equation the solution set is unchanged. In the past we have used this property to help solve equations in one variable. We will now use it to solve systems of linear equations. Here is another way to state the addition property of equality.

Let A, B, C, and D represent algebraic expressions.

$$\begin{array}{ll} \text{If} & A = B \\ \text{and} & C = D \\ \hline \text{then} & A + C = B + D \end{array}$$

Since C and D are equal (that is, they represent the same number), what we have done is added the same amount to both sides of the equation $A = B$. Let's see how we can use this form of the addition property of equality to solve a system of linear equations.

▼ **Example 1** Solve the system:

$$x + y = 4$$
$$x - y = 2$$

Solution The system is written in the form of the addition property of equality, as described in this section. It looks like this:

$$\begin{aligned} A &= B \\ C &= D \end{aligned}$$

where A is $x + y$, B is 4, C is $x - y$, and D is 2.

We use the addition property of equality to add the left sides together and the right sides together:

$$\begin{array}{r} x + y = 4 \\ x - y = 2 \\ \hline 2x + 0 = 6 \end{array}$$

We now solve the resulting equation for x:

$$\begin{aligned} 2x + 0 &= 6 \\ 2x &= 6 \\ x &= 3 \end{aligned}$$

The value we get for x is the value of the x-coordinate of the point of intersection of the two lines $x + y = 4$ and $x - y = 2$. To find the y-coordinate we simply substitute $x = 3$ into either of the two original equations and get:

$$\begin{aligned} 3 + y &= 4 \\ y &= 1 \end{aligned}$$

The solution to our system is the ordered pair (3, 1). It satisfies both equations.

When	$x = 3$	When	$x = 3$
and	$y = 1$	and	$y = 1$
the equation	$x + y = 4$	the equation	$x - y = 2$
becomes	$3 + 1 \stackrel{?}{=} 4$	becomes	$3 - 1 \stackrel{?}{=} 2$
or	$4 = 4$	or	$2 = 2$ ▲

The most important part of this method of solving linear systems is eliminating one of the variables when we add the left and right sides together. In our first example, the equations were written so that the y variable was eliminated when we added the left and right sides together. If the equations are not set up this way to begin with, we have to work on one or both of them separately before we can add them together to eliminate one variable.

▼ **Example 2** Solve the following system:

$$\begin{aligned} x + 2y &= 4 \\ x - y &= -5 \end{aligned}$$

Solution Notice that if we were to add the equations together as they are, the resulting equation would have terms in both x and y. Let's eliminate the variable x by multiplying both sides of the second equation by -1 before we add the equations together. (As you will see, we can choose to eliminate either the x or the y variable.) Multiplying both sides of the second equation by -1 will not change its solution, so we do not need to be concerned that we have altered the system.

$$\begin{array}{rcl} x + 2y = 4 & \xrightarrow{\text{no change}} & x + 2y = 4 \\ x - y = -5 & \xrightarrow[\text{multiply by } -1]{} & -x + y = 5 \\ \hline & & 0 + 3y = 9 \\ & & 3y = 9 \\ & & y = 3 \end{array}$$

Add left and right sides to get ($0 + 3y = 9$)

y-coordinate of the point of intersection ($y = 3$)

Substituting $y = 3$ into either of the two original equations, we get $x = -2$. The solution to the system is $(-2, 3)$. It satisfies both equations. ▲

▼ **Example 3** Solve the system:

$$\begin{aligned} 2x - y &= 6 \\ x + 3y &= 3 \end{aligned}$$

Solution Let's eliminate the y variable from the two equations. We can do this by multiplying the first equation by 3 and leaving the second equation unchanged:

$$\begin{array}{rcl} 2x - y = 6 & \xrightarrow{\text{3 times both sides}} & 6x - 3y = 18 \\ x + 3y = 3 & \xrightarrow[\text{no change}]{} & x + 3y = 3 \end{array}$$

The important thing about our system now is that the coefficients (the numbers in front) of the y variables are opposites. When we add the terms on each side of the equal sign, then the terms in y will add to zero and be eliminated:

$$\begin{array}{rl} 6x - 3y &= 18 \\ x + 3y &= 3 \\ \hline 7x \quad &= 21 \end{array} \qquad \text{Add corresponding terms}$$

which gives us $x = 3$. Using this value of x in the second equation of our original system, $x + 3y = 3$, we have

$$\begin{aligned} 3 + 3y &= 3 \\ 3y &= 0 \\ y &= 0 \end{aligned}$$

We could substitute $x = 3$ into any of the equations with both x and y variables and also get $y = 0$. The solution to our system is the ordered pair $(3, 0)$. ▲

▼ **Example 4** Solve the system:

$$\begin{aligned} 2x + 3y &= -1 \\ 3x + 5y &= -2 \end{aligned}$$

Solution Let's eliminate x from the two equations. If we multiply the first equation by 3 and the second by -2, the coefficients of x will be 6 and -6, respectively. The x terms in the two equations will then add to zero:

$$\begin{aligned} 2x + 3y &= -1 \xrightarrow{\text{multiply by } 3} & 6x + 9y &= -3 \\ 3x + 5y &= -2 \xrightarrow[\text{multiply by } -2]{} & -6x - 10y &= 4 \end{aligned}$$

We now add the left and right sides of our new system together:

$$\begin{aligned} 6x + 9y &= -3 \\ -6x - 10y &= 4 \\ \hline -y &= 1 \\ y &= -1 \end{aligned}$$

Substituting $y = -1$ into the first equation in our original system, we have:

$$\begin{aligned} 2x + 3(-1) &= -1 \\ 2x - 3 &= -1 \\ 2x &= 2 \\ x &= 1 \end{aligned}$$

The solution to our system is $(1, -1)$. It is the only ordered pair that satisfies both equations. ▲

▼ **Example 5** Solve the system:

$$\begin{aligned} 3x + 5y &= -7 \\ 5x + 4y &= 10 \end{aligned}$$

Solution Let's eliminate y by multiplying the first equation by -4 and the second equation by 5:

$$\begin{aligned} 3x + 5y &= -7 \xrightarrow{\text{multiply by } -4} & -12x - 20y &= 28 \\ 5x + 4y &= 10 \xrightarrow[\text{multiply by } 5]{} & 25x + 20y &= 50 \\ & & \hline 13x \quad &= 78 \\ & & x &= 6 \end{aligned}$$

Substitute $x = 6$ into any equation with both x and y, and the result will be $y = -5$. The solution is therefore $(6, -5)$. ▲

▼ **Example 6** Solve the system:

$$\begin{aligned} 2x - y &= 2 \\ 4x - 2y &= 8 \end{aligned}$$

Solution Let us choose to eliminate y from the system. We can do this by multiplying the first equation by -2 and leaving the second equation unchanged:

$$\begin{aligned} 2x - y = 2 &\xrightarrow{\text{multiply by } -2} -4x + 2y = -4 \\ 4x - 2y = 8 &\xrightarrow[\text{no change}]{} \quad 4x - 2y = 8 \end{aligned}$$

If we add both sides of the resulting system, we have

$$\begin{array}{r} -4x + 2y = -4 \\ 4x - 2y = 8 \\ \hline 0 + 0 = 4 \\ \text{or} \quad 0 = 4 \end{array} \qquad \text{A false statement}$$

Both variables have been eliminated and we end up with the false statement $0 = 4$. We have tried to solve a system that consists of two parallel lines. There is no solution and that is the reason we end up with a false statement. ▲

If, on the other hand, we had eliminated both variables and ended up with a true statement, it would mean that the lines coincided and every ordered pair whose graph was on the line would be a solution to the system.

Both variables eliminated and the resulting statement false	$\leftrightarrow$	The lines are parallel and there is no solution to the system
Both variables eliminated and the resulting statement true	$\leftrightarrow$	The lines coincide and there is an infinite number of solutions to the system

The main idea in solving a system of linear equations by the elimination method is to use the multiplication property of equality on one or both of the original equations, if necessary, to make the coefficients of either variable opposites. Here are some steps to follow when solving a system of linear equations by the elimination method.

To Solve a System of Linear Equations by the Elimination Method

Step 1: Decide which variable to eliminate. (In some cases one variable will be easier to eliminate than the other. With some practice you will notice which one it is.)

Step 2: Use the multiplication property of equality on each equation separately to make the coefficients of the variable that is to be eliminated opposites.

Step 3: Add the respective left and right sides of the system together.

Step 4: Solve for the variable remaining.

Step 5: Substitute the value of the variable from Step 4 into an equation containing both variables and solve for the other variable.

Step 6: Check your solution in both equations, if necessary.

Problem Set 3.6

Solve the following systems of linear equations by elimination.

1. $x + y = 3$
$x - y = 1$

2. $x + y = -2$
$x - y = 6$

3. $x + y = 10$
$-x + y = 4$

4. $x - y = 1$
$-x - y = -7$

5. $x - y = 7$
$-x - y = 3$

6. $x - y = 4$
$2x + y = 8$

7. $x + y = -1$
$3x - y = -3$

8. $2x - y = -2$
$-2x - y = 2$

9. $3x + 2y = 1$
$-3x - 2y = -1$

10. $-2x - 4y = 1$
$2x + 4y = -1$

Solve each of the following systems by eliminating the y variables.

11. $3x - y = 4$
$2x + 2y = 24$

12. $2x + y = 3$
$3x + 2y = 1$

13. $5x - 3y = 1$
$x - y = -1$

14. $4x - y = 13$
$2x + 4y = 2$

15. $11x - 4y = 11$
$5x + y = 5$

16. $3x - y = 7$
$10x - 5y = 25$

Solve each of the following systems by eliminating the x variable.

17. $3x - 5y = 7$
$-x + y = -1$

18. $4x + 2y = 32$
$x + y = -2$

19. $-x - 8y = 3$
$-2x + y = -11$

20. $-x + 10y = 3$
$-5x + 11y = -24$

21. $-3x - y = 7$
$6x + 7y = 11$

22. $-5x + 2y = -6$
$10x + 7y = 34$

Solve each of the following systems of linear equations by the elimination method.

23. $6x - y = -8$
$2x + y = -16$

24. $5x - 3y = -3$
$3x + 3y = -21$

25. $x + 3y = 9$
$2x - y = 4$

26. $x + 2y = 0$
$2x - y = 0$

27. $x - 3y = -9$
$4x + 3y = 39$

28. $3x + y = 7$
$4x - 5y = 3$

29. $2x + y = 5$
$5x + 3y = 11$

30. $5x + 2y = 11$
$7x + y = 10$

31. $4x + 3y = 14$
$9x - 2y = 14$

32. $7x - 6y = 13$
$6x - 5y = 11$

33. $3x + 2y = -1$
$6x + 4y = 0$

34. $8x - 2y = 2$
$4x - y = 2$

35. $11x + 6y = 17$
$5x - 4y = 1$

36. $3x - 8y = 7$
$10x - 5y = 45$

37. $3x + y = 2$
$-6x - 2y = -1$

38. $2x - 3y = 4$
$-4x + 6y = -8$

39. For some systems of equations it is necessary to apply the addition property of equality to each equation to line up the x variables and y variables before trying to eliminate a variable. Solve the following system by first writing each equation so that the variable terms with x in them come first, the variable terms with y in them come second, and the constant terms are on the right side of each equation.

$$\begin{aligned} 4x - 5y &= 17 - 2x \\ 5y &= 3x + 4 \end{aligned}$$

40. Solve the following system by first writing each equation so that the variable terms with x in them come first, the variable terms with y in them come second, and the constant terms are on the right side of each equation.

$$\begin{aligned} 3x - 6y &= -20 + 7x \\ 4x &= 3y - 34 \end{aligned}$$

41. Multiply both sides of the second equation in the following system by 100 and then solve as usual.

$$\begin{aligned} x + y &= 22 \\ 0.05x + 0.10y &= 1.70 \end{aligned}$$

42. Multiply both sides of the second equation in the following system by 100 and then solve as usual.

$$\begin{aligned} x + y &= 15{,}000 \\ 0.06x + 0.07y &= 980 \end{aligned}$$

Review Problems The problems below review material we covered in Section 2.7.

Solve each inequality.

43. $x - 3 < 2$

44. $x + 4 \leq 6$

45. $-3x \geq 12$

46. $-2x > 10$

47. $-\frac{x}{3} \leq -1$

48. $-\frac{x}{5} < -2$

49. $-4x + 1 < 17$

50. $-3x + 2 \leq -7$

Section 3.7 The Substitution Method

There is a third method of solving systems of equations. It is the substitution method, and like the elimination method, it can be used on any system of linear equations. Some systems, however, lend themselves more to the substitution method than others.

▼ **Example 1** Solve the following system:

$$\begin{aligned} x + y &= 2 \\ y &= 2x - 1 \end{aligned}$$

Solution If we were to solve this system by the methods used in the last section, we would have to rearrange the terms of the second equation so that similar terms would be in the same column. There is no need to do this, however, since the second equation tells us that y is $2x - 1$. We can replace the y variable in the first equation with the expression $2x - 1$ from the second equation. That is, we *substitute* $2x - 1$ from the second equation for y in the first equation. Here is what it looks like:

$$\begin{aligned} x + y &= 2 \\ y &= 2x - 1 \end{aligned}$$

Substituting $2x - 1$ for y in the first equation, we have

$$x + (2x - 1) = 2$$

The equation we end up with contains only the variable x. The y variable has been eliminated by substitution.

Solving the resulting equation, we have

$$\begin{aligned} x + (2x - 1) &= 2 \\ 3x - 1 &= 2 \\ 3x &= 3 \\ x &= 1 \end{aligned}$$

This is the x-coordinate of the solution to our system. To find the y-coordinate, we substitute $x = 1$ into the second equation of our system. (We could substitute $x = 1$ into the first equation also and have the same result.)

$$y = 2(1) - 1$$
$$y = 2 - 1$$
$$y = 1$$

The solution to our system is the ordered pair (1, 1). It satisfies both of the original equations. ▲

The key to this method of solving a system of equations is again elimination of one of the variables. With this method, we eliminate the variable by substitution rather than addition.

▼ **Example 2** Solve the following system by the substitution method:

$$2x - 3y = 12$$
$$y = 2x - 8$$

Solution Again, the second equation says y is $2x - 8$. Since we are looking for the ordered pair that satisfies both equations, the y in the first equation must also be $2x - 8$. Substituting $2x - 8$ from the second equation for y in the first equation, we have

$$2x - 3(2x - 8) = 12$$
$$2x - 6x + 24 = 12$$
$$-4x + 24 = 12$$
$$-4x = -12$$
$$x = 3$$

To find the y-coordinate of our solution, we substitute $x = 3$ into the second equation in the original system.

When	$x = 3$
the equation	$y = 2x - 8$
becomes	$y = 2(3) - 8$
	$y = 6 - 8$
	$y = -2$

The solution to our system is $(3, -2)$. ▲

▼ **Example 3** Solve the following system by solving the first equation for x and then using the substitution method:

$$x - 3y = -1$$
$$2x - 3y = 4$$

Solution We solve the first equation for x by adding $3y$ to both sides to get

$$x = 3y - 1$$

Using this value of x in the second equation, we have

$$\begin{aligned} 2(3y - 1) - 3y &= 4 \\ 6y - 2 - 3y &= 4 \\ 3y - 2 &= 4 \\ 3y &= 6 \\ y &= 2 \end{aligned}$$

When $\quad y = 2$

the equation $\quad x = 3y - 1$

becomes $\quad x = 3(2) - 1$

$$\begin{aligned} x &= 6 - 1 \\ x &= 5 \end{aligned}$$

The solution to our system is (5, 2). ▲

Here are the steps to use in solving a system of equations by the substitution method:

To Solve a System of Equations by the Substitution Method

Step 1: Solve either one of the equations for x or y. (This step is not necessary if one of the equations is already in the correct form, as in Examples 1 and 2.)

Step 2: Substitute the expression for the variable obtained in Step 1 into the other equation and solve it.

Step 3: Substitute the solution from Step 2 into any equation in the system that contains both variables and solve it.

Step 4: Check your results, if necessary.

▼ **Example 4** Solve by substitution:

$$\begin{aligned} -2x + 4y &= 14 \\ -3x + y &= 6 \end{aligned}$$

Solution We can solve either equation for either variable. If we look at the system closely, it becomes apparent that solving the second equation for y is the easiest way to go. If we add $3x$ to both sides of the second equation, we have

$$y = 3x + 6$$

Substituting the expression $3x + 6$ back into the first equation in place of y yields the following results:

$$\begin{aligned} -2x + 4(3x + 6) &= 14 \\ -2x + 12x + 24 &= 14 \\ 10x + 24 &= 14 \\ 10x &= -10 \\ x &= -1 \end{aligned}$$

Substituting $x = -1$ into the equation $y = 3x + 6$ leaves us with

$$\begin{aligned} y &= 3(-1) + 6 \\ y &= -3 + 6 \\ y &= 3 \end{aligned}$$

The solution to our system is $(-1, 3)$. ▲

▼ **Example 5** Solve by substitution:

$$\begin{aligned} 4x + 2y &= 8 \\ y &= -2x + 4 \end{aligned}$$

Solution Substituting the expression $-2x + 4$ for y from the second equation into the first equation, we have

$$\begin{aligned} 4x + 2(-2x + 4) &= 8 \\ 4x - 4x + 8 &= 8 \\ 8 &= 8 \qquad \text{A true statement} \end{aligned}$$

Both variables have been eliminated and we are left with a true statement. Recall from the last section that a true statement in this situation tells us the lines coincide. That is, the equations $4x + 2y = 8$ and $y = -2x + 4$ have exactly the same graph. Any point on that graph has coordinates that satisfy both equations and is a solution to the system. ▲

Problem Set 3.7

Solve the following systems by substitution. Substitute the expression in the second equation into the first equation and solve.

1. $x + y = 11$
$y = 2x - 1$

2. $x - y = -3$
$y = 3x + 5$

3. $x + y = 20$
$y = 5x + 2$

4. $3x - y = -1$
$y = -2x + 6$

5. $-2x + y = 0$
$y = -3x - 5$

6. $4x - y = 6$
$y = -x + 4$

7. $3x - 2y = -2$
 $y = -x + 6$

8. $2x - 3y = 17$
 $y = -x + 6$

9. $5x - 4y = -16$
 $y = 4x + 4$

10. $6x + 2y = 18$
 $y = 2x - 6$

11. $5x + 4y = 7$
 $y = -3x$

12. $10x + 2y = -6$
 $y = -5x$

Solve the following systems by solving the second equation for x and then using the substitution method.

13. $x + 3y = 4$
 $x - 2y = -1$

14. $x - y = 5$
 $x + 2y = -1$

15. $2x + y = 1$
 $x - 5y = 17$

16. $2x - 2y = 2$
 $x - 3y = -7$

17. $3x + 5y = 14$
 $x - 5y = 18$

18. $2x - 4y = 2$
 $x + 2y = 5$

19. $5x + 3y = 0$
 $x - 3y = -18$

20. $x - 3y = -5$
 $x - 2y = 0$

21. $-3x - 9y = 7$
 $x + 3y = 2$

22. $2x + 6y = 18$
 $x + 3y = 9$

Solve the following systems using the substitution method.

23. $5x - 8y = 7$
 $y = 2x - 5$

24. $3x + 4y = 10$
 $y = 8x - 15$

25. $7x - y = 24$
 $x = 2y + 9$

26. $3x - y = -8$
 $x = 6y + 3$

27. $-3x + 2y = 6$
 $y = 3x$

28. $-2x - y = -3$
 $y = -3x$

29. $x - 6y = -20$
 $x = y$

30. $2x - 4y = 0$
 $x = y$

31. $2x - y = 12$
 $3x + 3y = 9$

32. $2x - y = -4$
 $4x - y = -2$

33. $x - y = 8$
 $x + 3y = 48$

34. $-3x + y = -2$
 $2x + y = -17$

35. $x - y = 2$
 $-4x + 4y = -8$

36. $2x - y = 5$
 $-4x + 2y = -10$

Review Problems The problems that follow review material we covered in Sections 2.3 and 2.6. Reviewing these problems will help you understand the next section.

37. The sum of twice a number and 4 is 12. Find the number.
38. Twice the sum of a number and 5 is 48. Find the number.
39. A rectangle is 3 times as long as it is wide. If the perimeter is 24 meters, find the length and width.

40. The perimeter of a square is 48 inches. Find the length of its side.

41. A collection of coins consists of nickels and dimes and is worth \$2.10. If there are 3 more dimes than nickels, how many of each coin are in the collection?

42. Mary has \$5.75 in nickels and quarters. If she has 5 more quarters than nickels, how many of each coin does she have?

43. What number is 8% of 6,000?

44. 540 is 9% of what number?

45. A man invests twice as much money at 10% annual interest as he does at 8% annual interest. If his total interest for the year is \$224, how much does he have invested at each rate?

46. A woman invests three times as much money at 8% annual interest as she does at 6% annual interest. If the total interest for the year is \$66, how much does she have invested at each rate?

Section 3.8 Word Problems

I have often heard students remark about the word problems in beginning algebra: "What does this have to do with real life?" Most of the word problems we will encounter don't have much to do with "real life." We are actually just practicing. Ultimately, all problems requiring the use of algebra are word problems. That is, they are stated in words first, then translated to symbols. The problems are then solved by some system of mathematics, like algebra. Most real applications involve calculus or higher levels of mathematics. So, if the problems we solve are upsetting or frustrating to you, then you are probably taking them too seriously.

The word problems in this section have two unknown quantities. We will write two equations in two variables (each of which represents one of the unknown quantities), which of course is a system of equations. We then solve the system by one of the methods developed in the previous sections of this chapter. Here are the steps to follow in solving these word problems.

To Solve Word Problems Involving a System of Equations

Step 1: Read the problem carefully (and more than once) and decide what is needed. Let x represent one of the unknown quantities and y the other.

Step 2: Write two equations using the two variables from Step 1 that together describe the situation. (This is the step that tends to make people anxious. Read the problem as many times as it takes to understand the situation clearly.)

Step 3: Solve the system obtained from Step 2.

Step 4: Check the values of x and y from Step 3 to see that they make sense in the original problem.

Remember, the more problems you work, the more problems you will be able to work. If you have trouble getting started on the problem set, come back to the examples and work through them yourself. The examples are similar to the problems found in the problem set.

Number Problems

▼ **Example 1** One number is two more than five times another number. Their sum is 20. Find the two numbers.

Solution

Step 1: Let x represent one of the numbers and y represent the other.

Step 2: "One number is two more than five times another" translates to

$$y = 5x + 2$$

"Their sum is 20" translates to

$$x + y = 20$$

The system that describes the situation must be

$$\begin{aligned} x + y &= 20 \\ y &= 5x + 2 \end{aligned}$$

Step 3: We can solve this system by substituting the expression $5x + 2$ in the second equation for y in the first equation:

$$\begin{aligned} x + 5x + 2 &= 20 \\ 6x + 2 &= 20 \\ 6x &= 18 \\ x &= 3 \end{aligned}$$

Using $x = 3$ in either of the first two equations and then solving for y, we get $y = 17$.

Step 4: The number 17 is 2 more than 5 times 3, and the sum of 17 and 3 is 20.

So 17 and 3 are the numbers we are looking for. ▲

Interest Problems

▼ **Example 2** Mr. Smith had \$15,000 to invest. He invested part at 6% and the rest at 7%. If he earns \$980 in interest, how much did he invest at each rate?

Solution

Step 1: Let x = the amount invested at 6% and y = the amount invested at 7%.

Step 2: Since Mr. Smith invested a total of \$15,000, we have

$$x + y = 15{,}000$$

The interest he earns comes from 6% of the amount invested at 6% (x) and 7% of the amount invested at 7% (y). To find 6% of x, we simply multiply x by 0.06, which gives us $0.06x$ (6% is equivalent to 0.06). To find 7% of y, we multiply 0.07 times y and get $0.07y$.

$$\begin{array}{ccccc} \text{Interest} & + & \text{interest} & = & \text{total} \\ \text{at } 6\% & & \text{at } 7\% & & \text{interest} \\ 0.06x & + & 0.07y & = & 980 \end{array}$$

The system is

$$\begin{aligned} x + \quad y &= 15{,}000 \qquad \text{The number of dollars} \\ 0.06x + 0.07y &= 980 \qquad \text{The value of the dollars} \end{aligned}$$

Step 3: We multiply the first equation by -6 and the second by 100 to eliminate x:

$$\begin{array}{rcl} x + \quad y = 15{,}000 & \xrightarrow{\text{multiply by } -6} & -6x - 6y = -90{,}000 \\ 0.06x + 0.07y = 980 & \xrightarrow[\text{multiply by } 100]{} & 6x + 7y = 98{,}000 \\ \hline & & y = 8{,}000 \end{array}$$

Substituting $y = 8000$ into the first equation and solving for x, we get $x = 7000$.

Step 4: The sum of 8000 and 7000 is 15,000. Six percent of 7000 is $(0.06)(7000) = 420$. Seven percent of 8000 is $(0.07)(8000) = 560$. The total interest then is \$420 + \$560 = \$980.

He invested \$7000 at 6% and \$8000 at 7%. ▲

Coin Problems

▼ **Example 3** John has \$1.70 all in dimes and nickels. He has a total of 22 coins. How many of each kind does he have?

Solution

Step 1: Let x = the number of nickels and y = the number of dimes.

Step 2: The total number of coins is 22, so

$$x + y = 22$$

The total amount of money he has is 1.70, which comes from nickels and dimes:

$$\begin{array}{ccccc} \text{Amount of money in nickels} & + & \text{amount of money in dimes} & = & \text{total amount of money} \\ 0.05x & + & 0.10y & = & 1.70 \end{array}$$

The system that represents the situation is

$$\begin{aligned} x + y &= 22 \qquad \text{The number of coins} \\ 0.05x + 0.10y &= 1.70 \qquad \text{The value of the coins} \end{aligned}$$

Step 3: We multiply the first equation by -5 and the second by 100 to eliminate the variable x:

$$\begin{aligned} x + y &= 22 \xrightarrow{\text{multiply by } -5} & -5x - 5y &= -110 \\ 0.05x + 0.10y &= 1.70 \xrightarrow[\text{multiply by } 100]{} & 5x + 10y &= 170 \\ & & 5y &= 60 \\ & & y &= 12 \end{aligned}$$

Substituting $y = 12$ into our first equation, we get $x = 10$.

Step 4: Twelve dimes and 10 nickels total 22 coins with a total value of $0.10(12) + 0.05(10) = 1.20 + 0.50 = \$1.70$.

John has 12 dimes and 10 nickels. ▲

Mixture Problems

▼ **Example 4** How much 20% alcohol solution and 50% alcohol solution must be mixed to get 12 gallons of 30% alcohol solution?

Solution To solve this problem, we must first understand that a 20% alcohol solution is 20% alcohol and 80% water.

Step 1: Let $x =$ the number of gallons of 20% alcohol solution needed, and $y =$ the number of gallons of 50% alcohol solution needed.

Step 2: Since the total number of gallons we will end up with is 12, and this 12 gallons must come from the two solutions we are mixing, our first equation is

$$x + y = 12$$

To obtain our second equation, we look at the amount of alcohol in our two original solutions and our final solution. The amount of alcohol in the x gallons of 20% solution is $0.20x$, while the amount of alcohol in y gallons of 50% solution is $0.50y$. The amount of alcohol in the 12 gallons of 30% solution is $0.30(12)$. Since the amount of alcohol we start with must equal the amount of alcohol we end up with, our second equation is

$$0.20x + 0.50y = 0.30(12)$$

The information we have so far can also be summarized with a table. Sometimes by looking at a table like the one that follows, it is easier to see where the equations come from.

	20% Solution	50% Solution	Final Solution
Number of Gallons	x	y	12
Gallons of Alcohol	$0.20x$	$0.50y$	$0.30(12)$

Step 3: Our system of equations is

$$\begin{aligned} x + \quad y &= 12 \\ 0.20x + 0.50y &= 0.30(12) \end{aligned}$$

We can solve this system by substitution. Solving the first equation for y and substituting the result into the second equation, we have

$$0.20x + 0.50(12 - x) = 0.30(12)$$

Multiplying each side by 10 gives us an equivalent equation that is a little easier to work with:

$$\begin{aligned} 2x + 5(12 - x) &= 3(12) \\ 2x + 60 - 5x &= 36 \\ -3x + 60 &= 36 \\ -3x &= -24 \\ x &= 8 \end{aligned}$$

If x is 8, then y must be 4 because $x + y = 12$.

Step 4: It takes 8 gallons of 20% alcohol solution and 4 gallons of 50% alcohol solution to produce 12 gallons of 30% alcohol solution. ▲

Problem Set 3.8

Solve the following word problems. Be sure to show the equations used.

Number Problems

1. Two numbers have a sum of 25. One number is five more than the other. Find the numbers.

2. The difference of two numbers is 6. Their sum is 30. Find the two numbers.

3. The sum of two numbers is 15. One number is four times the other. Find the numbers.

4. The difference of two positive numbers is 28. One number is three times the other. Find the two numbers.
5. Two positive numbers have a difference of 5. The larger number is one more than twice the smaller. Find the two numbers.
6. One number is two more than three times another. Their sum is 26. Find the two numbers.
7. One number is five more than four times another. Their sum is 35. Find the two numbers.
8. The difference of two positive numbers is 8. The smaller is twice the larger decreased by 17. Find the two numbers.

Interest Problems

9. Mr. Wilson invested money in two accounts. His total investment was $20,000. If one account pays 6% in interest and the other pays 8% in interest, how much does he have in each account if he earned a total of $1380 in interest in one year?
10. A total of $11,000 was invested. Part of the $11,000 was invested at 4% and the rest was invested at 7%. If the investment earns $680 per year, how much was invested at each rate?
11. A woman invested four times as much at 5% as she did at 6%. The total amount of interest she earns in one year from both accounts is $520. How much did she invest at each rate?
12. Ms. Hagan invested twice as much money in an account that pays 7% interest as she did in an account that pays 6% in interest. Her total investment pays her $1000 a year in interest. How much did she invest at each amount?

Coin Problems

13. Ron has 14 coins with a total value of $2.30. The coins are nickels and quarters. How many of each coin does he have?
14. Diane has $0.95 in dimes and nickels. She has a total of 11 coins. How many of each kind does she have?
15. Suppose Tom has 21 coins totaling $3.45. If he has only dimes and quarters, how many of each type does he have?
16. A coin collector has 31 dimes and nickels with a total face value of $2.40. (They are actually worth a lot more.) How many of each coin does she have?

Mixture Problems

17. How many liters of 50% alcohol solution and 20% alcohol solution must be mixed to obtain 18 liters of 30% alcohol solution?
18. How many liters of 10% alcohol solution and 5% alcohol solution must be mixed to obtain 40 liters of 8% alcohol solution?
19. A mixture of 8% disinfectant solution is to be made from 10% and 7% disinfectant solutions. How much of each solution should be used if 30 gallons of 8% solution are needed?
20. How much 50% antifreeze solution and 40% antifreeze solution should be combined to give 50 gallons of 46% antifreeze solution?

Review Problems The following problems review material we covered in Section 2.7.

Solve each inequality.

21. $2x - 6 > 5x + 6$

22. $4x - 3 \geq 8x + 1$

23. $3(2x + 4) \leq -6$

24. $5(3x - 2) < 20$

25. $4(2 - x) \geq 12$

26. $6(1 - x) > -12$

27. $6 - 2(x + 3) < -2$

28. $7 - 3(x - 5) \leq -2$

Chapter 3 Summary and Review

LINEAR EQUATIONS IN TWO VARIABLES [3.1, 3.2]

A linear equation in two variables is any equation that can be put in the form $ax + by = c$. The graph of every linear equation is a straight line.

INTERCEPTS [3.3]

The x-intercept of an equation is the x-coordinate of the point where the graph crosses the x-axis. The y-intercept is the y-coordinate of the point where the graph crosses the y-axis. We find the y-intercept by substituting $x = 0$ into the equation and solving for y. The x-intercept is found by letting $y = 0$ and solving for x.

THE SLOPE OF A LINE [3.3]

The slope of the line containing points (x_1, y_1) and (x_2, y_2) is given by

$$\text{Slope} = m = \frac{\text{rise}}{\text{run}} = \frac{y_2 - y_1}{x_2 - x_1}$$

Horizontal lines have 0 slope, and vertical lines have no slope.

THE SLOPE-INTERCEPT FORM OF A STRAIGHT LINE [3.4]

The equation of a line with slope m and y-intercept b is given by

$$y = mx + b$$

Examples

1. The equation $3x + 2y = 6$ is an example of a linear equation in two variables.

2. To find the x-intercept for $3x + 2y = 6$ we let $y = 0$ and get

$$3x = 6$$
$$x = 2$$

In this case the x-intercept is 2, and the graph crosses the x-axis at $(2, 0)$.

3. The slope of the line through $(6, 9)$ and $(1, -1)$ is

$$m = \frac{9 - (-1)}{6 - 1} = \frac{10}{5} = 2$$

4. The equation of the line with slope 5 and y-intercept 3 is

$$y = 5x + 3$$

5. The solution to the system

$$\begin{aligned} x + 2y &= 4 \\ x - y &= 1 \end{aligned}$$

is the ordered pair (2, 1). It is the only ordered pair that satisfies both equations.

6. Solving the system in Example 5 by graphing looks like

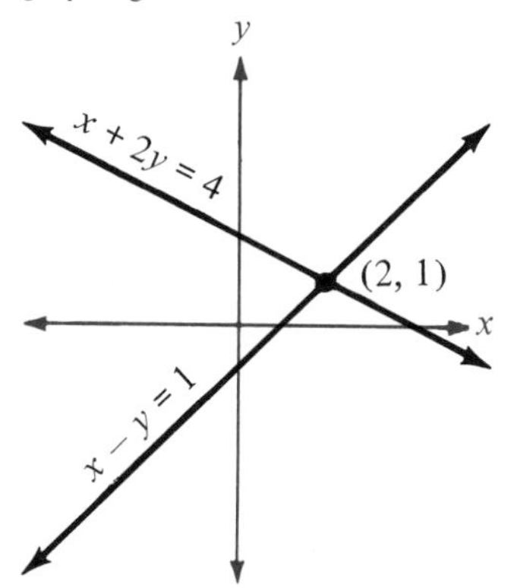

7. We can eliminate the y variable from the system in Example 5 by multiplying both sides of the second equation by 2 and adding the result to the first equation:

$$\begin{aligned} x + 2y &= 4 & \qquad x + 2y &= 4 \\ x - y &= 1 \xrightarrow{\text{multiply by 2}} & 2x - 2y &= 2 \\ & & 3x &= 6 \\ & & x &= 2 \end{aligned}$$

Substituting $x = 2$ into either of the original two equations gives $y = 1$. The solution is (2, 1).

DEFINITIONS [3.5]

1. A *system of linear equations,* as the term is used in this book, is two linear equations that each contain the same two variables.
2. The *solution* set for a system of equations is the set of all ordered pairs that satisfy *both* equations. The solution set to a system of linear equations will contain:

   Case I One ordered pair when the graphs of the two equations intersect at only one point (this is the most common situation)
   Case II No ordered pairs when the graphs of the two equations are parallel lines
   Case III An infinite number of ordered pairs when the graphs of the two equations coincide (are the same line)

TO SOLVE A SYSTEM BY GRAPHING [3.5]

Step 1: Graph the first equation.
Step 2: Graph the second equation on the same set of axes.
Step 3: The solution to the system consists of the coordinates of the point where the graphs cross each other (the coordinates of the point of intersection).
Step 4: Check the solution to see that it satisfies *both* equations, if necessary.

TO SOLVE A SYSTEM BY THE ELIMINATION METHOD [3.6]

Step 1: Look the system over to decide which variable will be easier to eliminate.
Step 2: Use the multiplication property of equality on each equation separately to insure that the coefficients of the variable to be eliminated are opposites.
Step 3: Add the left and right sides of the system produced in Step 2 and solve the resulting equation.
Step 4: Substitute the solution from Step 3 back into any equation with both x and y variables and solve.
Step 5: Check your solutions in both equations, if necessary.

TO SOLVE A SYSTEM BY THE SUBSTITUTION METHOD [3.7]

Step 1: Solve either of the equations for one of the variables (this step is not necessary if one of the equations has the correct form already).

Step 2: Substitute the results of Step 1 into the other equation and solve.

Step 3: Substitute the results of Step 2 into an equation with both x and y variables and solve. (The equation produced in Step 1 is usually a good one to use.)

Step 4: Check your solution, if necessary.

8. We can apply the substitution method to the system in Example 5 by first solving the second equation for x to get

$$x = y + 1$$

Substituting this expression of x into the first equation, we have

$$\begin{aligned} y + 1 + 2y &= 4 \\ 3y + 1 &= 4 \\ 3y &= 3 \\ y &= 1 \end{aligned}$$

Using $y = 1$ in either of the original equations gives $x = 2$.

SPECIAL CASES [3.5, 3.6, 3.7]

In some cases, using the elimination or substitution method eliminates both variables. The situation is interpreted as follows:

1. If the resulting statement is *false,* then the lines are parallel and there is no solution to the system.
2. If the resulting statement is *true,* then the equations represent the same line (the lines coincide). In this case, any ordered pair that satisfies either equation is a solution to the system.

COMMON MISTAKES

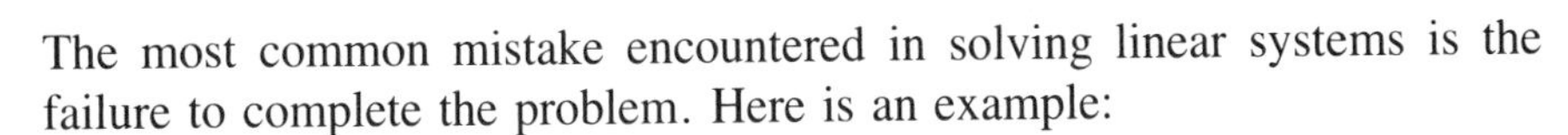

The most common mistake encountered in solving linear systems is the failure to complete the problem. Here is an example:

$$\begin{array}{r} x + y = 8 \\ x - y = 4 \\ \hline 2x \quad = 12 \\ x = 6 \end{array}$$

This is only half the solution. To find the other half, we must substitute the 6 back into one of the original equations and then solve for y.

Remember, solutions to systems of linear equations always consist of ordered pairs. We need an x-coordinate and a y-coordinate; $x = 6$ can never be a solution to a system of linear equations.

Chapter 3 Test

1. Fill in the following ordered pairs for the equation $2x - 5y = 10$. [3.1]

   $(0, \ )$ $(\ , 0)$ $(10, \ )$ $(\ , -3)$

2. Which of the following ordered pairs are solutions to $y = 4x - 3$? [3.1]

   $(2, 5)$ $(0, -3)$ $(3, 0)$ $(-2, 11)$

Graph each line. [3.2]

3. $y = 3x - 2$
4. $x = -2$

Give the slope, x-intercept, and y-intercept of the following straight lines and sketch their graphs. [3.3, 3.4]

5. $3x - 2y = 6$
6. $2x - y = 5$
7. Find the equation of the line with a slope of 4 and y-intercept 8. [3.4]
8. Line l passes through the points $(-3, 1)$ and $(-2, 4)$. Find the equation of line l. [3.4]

Solve the system by graphing. [3.5]

9. $x + 2y = 5$
   $y = 2x$

Solve each system by the elimination method. [3.6]

10. $x - y = 1$
    $2x + y = -10$
11. $2x + y = 7$
    $3x + y = 12$
12. $7x + 8y = -2$
    $3x - 2y = 10$
13. $6x - 10y = 6$
    $9x - 15y = 9$

Solve each system by the substitution method. [3.7]

14. $3x + 2y = 20$
    $y = 2x + 3$
15. $3x - 6y = -6$
    $x = y + 1$
16. $7x - 2y = -4$
    $-3x + y = 3$

Solve the following word problems. In each case, be sure to show the system of equations that describes the situation. [3.8]

17. The sum of two numbers is 12. Their difference is 2. Find the numbers.
18. The sum of two numbers is 15. One number is six more than twice the other. Find the two numbers.
19. Dr. Stork has $10,000 to invest. He would like to earn $980 per year in interest. How much should he invest at 9% if the rest is to be invested at 11%?
20. Diane has 12 coins that total $1.60. If the coins are all nickels and quarters, how many of each type does she have?

In this chapter we take a more detailed look at exponents. We begin by reviewing the definition for positive integer exponents. The rest of the chapter deals with developing properties associated with exponents.

To perform successfully in this chapter you should understand the following concepts. The concepts that are not familiar should be reviewed.

1. Subtraction is addition of the opposite.
2. Division is multiplication by the reciprocal.
3. Multiplication by positive integers is repeated addition.
4. Addition and multiplication are commutative operations.
5. Multiplication distributes over addition.

Section 4.1 Multiplication with Exponents

Recall that an *exponent* is a number written just above and to the right of another number, which is called the *base*. In the expression 5^2, for example, the exponent is 2 and the base is 5. The expression 5^2 is read "5 to the second power," or "5 squared." The meaning of the expression is

$$5^2 = 5 \cdot 5 = 25$$

In the expression 5^3, the exponent is 3 and the base is 5. The expression 5^3 is read "5 to the third power," or "5 cubed." The meaning of the expression is

$$5^3 = 5 \cdot 5 \cdot 5 = 125$$

Here are some further examples:

▼ Examples

1. $4^3 = 4 \cdot 4 \cdot 4 = 16 \cdot 4 = 64$ — exponent 3, base 4
2. $-3^4 = -3 \cdot 3 \cdot 3 \cdot 3 = -81$ — exponent 4, base 3
3. $(-2)^5 = (-2)(-2)(-2)(-2)(-2) = -32$ — exponent 5, base -2
4. $\left(-\frac{3}{4}\right)^2 = \left(-\frac{3}{4}\right)\left(-\frac{3}{4}\right) = \frac{9}{16}$ — exponent 2, base $-\frac{3}{4}$

▲

QUESTION In what way are $(-5)^2$ and -5^2 different?
ANSWER In the first case, the base is -5. In the second case, the base is 5. The answer to the first is 25. The answer to the second is -25. Can you tell why? Would there be a difference in the answers if the exponent in each case were changed to 3?

We can simplify our work with exponents by developing some properties of exponents. We want to list the things we know are true about exponents and then use these properties to simplify expressions that contain exponents.

The first property of exponents applies to products with the same base. We can use the definition of exponents, as indicating repeated multiplication, to simplify expressions like $2^4 \cdot 2^3$:

$$\begin{aligned} 2^4 \cdot 2^3 &= (2 \cdot 2 \cdot 2 \cdot 2)(2 \cdot 2 \cdot 2) \\ &= (2 \cdot 2 \cdot 2 \cdot 2 \cdot 2 \cdot 2 \cdot 2) \\ &= 2^7 \qquad \text{Notice: } 4 + 3 = 7 \end{aligned}$$

As you can see, multiplication with the same base resulted in addition of exponents. We can summarize this result with the following property.

Property 1 for Exponents

If a is any real number and r and s are integers, then

$$a^r \cdot a^s = a^{r+s}$$

In words: To multiply two expressions with the same base, add exponents and use the common base.

Here are some examples using Property 1.

▼ **Examples** Use Property 1 to simplify the following expressions. Leave your answers in terms of exponents.

5. $5^3 \cdot 5^6 = 5^{3+6} = 5^9$

6. $x^7 \cdot x^8 = x^{7+8} = x^{15}$

7. $3^4 \cdot 3^8 \cdot 3^5 = 3^{4+8+5} = 3^{17}$ ▲

Note In the preceding examples, notice that in each case the base in the original problem is the same base that appears in the answer and that it is written only once in the answer. A very common mistake that people make when they first begin to use Property 1 is to write a 2 in front of the base in the answer. For example, people making this mistake would get $2x^{15}$ or $(2x)^{15}$ as the result in Example 6. To avoid this mistake, you must be sure you understand the meaning of Property 1 exactly as it is written.

Another common type of expression involving exponents is one in which an expression containing an exponent is raised to another power. The expression $(5^3)^2$ is an example:

$$\begin{aligned}(5^3)^2 &= (5^3)(5^3)\\ &= 5^{3+3}\\ &= 5^6 \qquad \text{Notice: } 3 \cdot 2 = 6\end{aligned}$$

This result offers justification for the second property of exponents.

Property 2 for Exponents

If a is any real number and r and s are integers, then

$$(a^r)^s = a^{r \cdot s}$$

In words: A power raised to another power is the base raised to the product of the powers.

▼ **Examples** Simplify the following expressions.

8. $(4^5)^6 = 4^{5 \cdot 6} = 4^{30}$

9. $(x^3)^5 = x^{3 \cdot 5} = x^{15}$ ▲

The third property of exponents applies to expressions in which the product of two or more numbers or variables is raised to a power. Let's look at how the expression $(2x)^3$ can be simplified:

$$
\begin{aligned}
(2x)^3 &= (2x)(2x)(2x) \\
&= (2 \cdot 2 \cdot 2)(x \cdot x \cdot x) \\
&= 2^3 \cdot x^3 \\
&= 8x^3
\end{aligned}
$$

Notice: The exponent 3 distributes over the product $2x$

We can generalize this result into a third property of exponents.

Property 3 for Exponents

If a and b are any two real numbers and r is an integer, then

$$(ab)^r = a^r b^r$$

In words: The power of a product is the product of the powers.

Here are some examples using Property 3 to simplify expressions.

▼ **Examples** Simplify the following expressions.

10. $(5x)^2 = 5^2 \cdot x^2$ — Property 3
$= 25x^2$

11. $(2xy)^3 = 2^3 \cdot x^3 \cdot y^3$ — Property 3
$= 8x^3y^3$

12. $(3x^2)^3 = 3^3(x^2)^3$ — Property 3
$= 27x^6$ — Property 2

13. $\left(-\frac{1}{4}x^2y^3\right)^2 = \left(-\frac{1}{4}\right)^2(x^2)^2(y^3)^2$ — Property 3
$= \frac{1}{16}x^4y^6$ — Property 2

14. $(x^4)^3(x^2)^5 = x^{12} \cdot x^{10}$ — Property 2
$= x^{22}$ — Property 1

15. $(2y)^3(3y^2) = 2^3y^3(3y^2)$ — Property 3
$= 8 \cdot 3(y^3 \cdot y^2)$ — Commutative and associative properties
$= 24y^5$ — Property 1

16. $(2x^2y^5)^3(3x^4y)^2 = 2^3(x^2)^3(y^5)^3 \cdot 3^2(x^4)^2y^2$ — Property 3
$= 8x^6y^{15} \cdot 9x^8y^2$ — Property 2
$= (8 \cdot 9)(x^6x^8)(y^{15}y^2)$ — Commutative and associative properties
$= 72x^{14}y^{17}$ — Property 1 ▲

Scientific Notation

Many branches of science require working with very large numbers. In astronomy, for example, distances commonly are given in light-years. A light-year is the distance light travels in a year. It is approximately

$$5{,}880{,}000{,}000{,}000 \text{ miles}$$

This number is difficult to use in calculations because of the number of zeros it contains. Scientific notation provides a way of writing very large numbers in a more manageable form.

DEFINITION A number is in *scientific notation* when it is written as the product of a number between 1 and 10 and an integer power of 10. A number written in scientific notation has the form

$$n \times 10^r$$

where $1 \le n < 10$ and $r =$ an integer.

▼ **Example 17** Write 376,000 in scientific notation.

Solution We must rewrite 376,000 as the product of a number between 1 and 10 and a power of 10. To do so, we move the decimal point 5 places to the left so that it appears between the 3 and the 7. Then we multiply this number by 10^5. The number that results has the same value as our original number and is written in scientific notation.

$$376{,}000 = 3.76 \times 10^5$$

Moved 5 places

Decimal point originally here

Keeps track of the 5 places we moved the decimal point ▲

▼ **Example 18** Write 4.52×10^3 in expanded form.

Solution Since 10^3 is 1,000, we can think of this as simply a multiplication problem. That is

$$4.52 \times 10^3 = 4.52 \times 1{,}000 = 4{,}520$$

On the other hand, we can think of the exponent 3 as indicating the number of places we need to move the decimal point in order to write our number in expanded form. Since our exponent is positive 3, we move the decimal point three places to the right:

$$4.52 \times 10^3 = 4{,}520$$

▲

Problem Set 4.1

Name the base and exponent in each of the following expressions. Then use the definition of exponents as repeated multiplication to simplify.

1. 4^2
2. 6^2
3. 7^2
4. 8^2
5. 4^3
6. 10^3
7. $(-5)^2$
8. -5^2
9. -2^3
10. $(-2)^3$
11. 3^4
12. $(-3)^4$
13. $(\frac{2}{3})^2$
14. $(\frac{2}{3})^3$
15. $(\frac{1}{2})^4$
16. $(\frac{4}{5})^2$

Use Property 1 for Exponents to simplify each expression. Leave all answers in terms of exponents.

17. $x^4 \cdot x^5$
18. $x^7 \cdot x^3$
19. $7^6 \cdot 7^1$
20. $8^9 \cdot 8^1$
21. $y^{10} \cdot y^{20}$
22. $y^{30} \cdot y^{30}$
23. $2^5 \cdot 2^4 \cdot 2^3$
24. $4^2 \cdot 4^3 \cdot 4^4$
25. $x^4 \cdot x^6 \cdot x^8 \cdot x^{10}$
26. $x^{20} \cdot x^{18} \cdot x^{16} \cdot x^{14}$

Use Property 2 for Exponents to write each of the following problems with a single exponent. (Assume all variables are positive numbers.)

27. $(x^2)^5$
28. $(x^5)^2$
29. $(5^4)^3$
30. $(5^3)^4$
31. $(y^3)^3$
32. $(y^2)^2$
33. $(2^5)^{10}$
34. $(10^5)^2$
35. $(a^3)^x$
36. $(a^5)^x$
37. $(b^x)^y$
38. $(b^r)^s$

Use Property 3 for Exponents to simplify each of the following expressions.

39. $(4x)^2$
40. $(2x)^4$
41. $(2y)^5$
42. $(5y)^2$
43. $(-3x)^4$
44. $(-3x)^3$
45. $(7ab)^2$
46. $(9ab)^2$
47. $(4xyz)^3$
48. $(5xyz)^3$

Simplify the following expressions by using the properties of exponents.

49. $(2x^4)^3$
50. $(3x^5)^2$
51. $(4a^3)^2$
52. $(5a^2)^2$
53. $(x^2)^3(x^4)^2$
54. $(x^5)^2(x^3)^5$
55. $(a^3)^1(a^2)^4$
56. $(a^4)^1(a^1)^3$
57. $(2x)^3(2x)^4$
58. $(3x)^2(3x)^3$
59. $(3x^2)^3(2x)^4$
60. $(3x)^3(2x^3)^2$
61. $(4x^2y^3)^2$
62. $(9x^3y^5)^2$
63. $(2a^4b^1)^3$
64. $(3a^1b^7)^3$
65. $(3x^2)(2x^3)(5x^4)$
66. $(x^3)(2x^1)(5x^5)$
67. $(4x^2y)^3(2xy)^2$
68. $(3x^3y^2)^2(2x^4y^5)^4$

Write each number in scientific notation.

69. 43,200
70. 432,000
71. 570
72. 5,700
73. 238,000
74. 2,380,000

Write each number in expanded form.

75. 2.49×10^3

76. 2.49×10^4

77. 3.52×10^2

78. 3.52×10^5

79. 2.8×10^4

80. 2.8×10^3

81. If you are 21 years old, you have been alive for over 650,000,000 seconds. Write this last number in scientific notation.

82. The distance around the earth at the equator is more than 130,000,000 feet. Write this number in scientific notation.

83. If you earn at least \$12 an hour and work full-time for 30 years, you will make at least 7.4×10^5 dollars. Write this last number in expanded form.

84. If your pulse is 72, then in one year your heart will beat at least 3.78×10^7 times. Write this last number in expanded form.

85. If you were to use the definition of exponents as repeated multiplication to evaluate 10^2, 10^3, and 10^4, you would get $10^2 = 100$, $10^3 = 1000$, and $10^4 = 10{,}000$. Without actually multiplying, what is 10^5? What is the relationship between the exponent on each of these powers of 10 and the number of zeros in the result?

86. Use the results found in Problem 85 to evaluate 10^1 and 10^0.

87. In this section we evaluated expressions like $(2^3)^2$, but not expressions like 2^{3^2}. What is the difference in the meanings of the two expressions? Evaluate each of them.

88. Evaluate the expressions $(2^2)^3$ and 2^{2^3}. (The answers in both cases are a little larger than usual, so be careful when you multiply.)

89. Here is an example of an analogy. These kinds of questions are common on entrance exams and aptitude tests.

x^4 is to x^2 as 16 is to:

(a) 8 (b) 4 (c) 2 (d) 1

90. Here is an analogy problem that is a little more difficult than the first one:

x^2 is to x as Superman is to:

(a) Mr. White (b) Jimmy (c) Clark Kent (d) Lois Lane

Review Problems The following problems review material on subtraction with negative numbers that we covered in Section 1.4. Reviewing these problems will help you understand the material in the next section.

Subtract.

91. $4 - 7$

92. $-4 - 7$

93. $4 - (-7)$

94. $-4 - (-7)$

95. $15 - 20$

96. $15 - (-20)$

97. $-15 - (-20)$

98. $-15 - 20$

Section 4.2 Division with Exponents

In Section 4.1 we found that multiplication with the same base results in addition of exponents, that is, $a^r \cdot a^s = a^{r+s}$. Since division is the inverse operation of multiplication, we can expect division with the same base to result in subtraction of exponents.

To develop the properties for exponents under division we again apply the definition of exponents:

$$\begin{aligned}\frac{x^5}{x^3} &= \frac{x \cdot x \cdot x \cdot x \cdot x}{x \cdot x \cdot x} \\ &= \frac{x \cdot x \cdot x}{x \cdot x \cdot x}(x \cdot x) \\ &= 1(x \cdot x) \\ &= x^2 \qquad \text{Notice: } 5 - 3 = 2\end{aligned}$$

$$\begin{aligned}\frac{2^4}{2^7} &= \frac{2 \cdot 2 \cdot 2 \cdot 2}{2 \cdot 2 \cdot 2 \cdot 2 \cdot 2 \cdot 2 \cdot 2} \\ &= \frac{2 \cdot 2 \cdot 2 \cdot 2}{2 \cdot 2 \cdot 2 \cdot 2} \cdot \frac{1}{2 \cdot 2 \cdot 2} \\ &= \frac{1}{2 \cdot 2 \cdot 2} \\ &= \frac{1}{2^3} \qquad \text{Notice: } 7 - 4 = 3\end{aligned}$$

In both cases, division with the same base resulted in subtraction of the smaller exponent from the larger. The problem is deciding whether the answer is a fraction or not. The problem is resolved quite easily by the following definition.

DEFINITION If r is a positive integer, then $a^{-r} = \dfrac{1}{a^r} = \left(\dfrac{1}{a}\right)^r$. $(a \neq 0)$

The following examples illustrate how we use this definition to simplify expressions that contain negative exponents.

▼ **Examples** Write each expression with a positive exponent and then simplify.

1. $2^{-3} = \dfrac{1}{2^3} = \dfrac{1}{8}$ Notice: Negative exponents do not indicate negative numbers. They indicate reciprocals.

2. $5^{-2} = \dfrac{1}{5^2} = \dfrac{1}{25}$

3. $3x^{-6} = 3 \cdot \dfrac{1}{x^6} = \dfrac{3}{x^6}$ ▲

Now let us look back to our original problem and try to work it again with the help of a negative exponent. We know that $2^4/2^7 = 1/2^3$. Let us decide now that with division of the same base, we will always subtract the exponent in the denominator from the exponent in the numerator and see if this conflicts with what we know is true.

$\dfrac{2^4}{2^7} = 2^{4-7}$ Subtracting the bottom exponent from the top exponent

$= 2^{-3}$ Subtraction

$= \frac{1}{2^3}$ Definition of negative exponents

Subtracting the exponent in the denominator from the exponent in the numerator and then using the definition of negative exponents gives us the same result we obtained previously. We can now continue the list of properties of exponents we started in Section 4.1.

Property 4 for Exponents

If a is any real number and r and s are integers, then

$$\frac{a^r}{a^s} = a^{r-s} \quad (a \neq 0)$$

In words: To divide with the same base, subtract the exponent in the denominator from the exponent in the numerator and raise the base to the exponent that results.

The following examples show how we use Property 4 and the definition for negative exponents to simplify expressions involving division.

▼ **Examples** Simplify the following expressions.

4. $\frac{x^9}{x^6} = x^{9-6} = x^3$

5. $\frac{x^4}{x^{10}} = x^{4-10} = x^{-6} = \frac{1}{x^6}$

6. $\frac{2^{15}}{2^{20}} = 2^{15-20} = 2^{-5} = \frac{1}{2^5} = \frac{1}{32}$ ▲

Our final property of exponents is similar to Property 3 from Section 4.1, but it involves division instead of multiplication. After we have stated the property, we will give a proof of it. The proof shows why this property is true.

Property 5 for Exponents

If a and b are any two real numbers $(b \neq 0)$ and r is an integer, then

$$\left(\frac{a}{b}\right)^r = \frac{a^r}{b^r}$$

In words: A quotient raised to a power is the quotient of the powers.

PROOF

$$\left(\frac{a}{b}\right)^r = \left(a \cdot \frac{1}{b}\right)^r \quad \text{By the definition of division}$$

$$= a^r \cdot \left(\frac{1}{b}\right)^r \quad \text{By Property 3}$$

$$= a^r \cdot b^{-r} \quad \text{By the definition of negative exponents}$$

$$= a^r \cdot \frac{1}{b^r} \quad \text{By the definition of negative exponents}$$

$$= \frac{a^r}{b^r} \quad \text{By the definition of division}$$

▼ **Examples** Simplify these expressions.

7. $\left(\frac{x}{2}\right)^3 = \frac{x^3}{2^3} = \frac{x^3}{8}$

8. $\left(\frac{5}{y}\right)^2 = \frac{5^2}{y^2} = \frac{25}{y^2}$

9. $\left(\frac{2}{3}\right)^4 = \frac{2^4}{3^4} = \frac{16}{81}$ ▲

Zero and One as Exponents

We have two special exponents left to deal with before our rules for exponents are complete. We have learned to use positive and negative exponents, but we have not run across the exponents 0 and 1. To obtain an expression for x^1, we will solve a problem two different ways:

$$\left.\begin{aligned} \frac{x^3}{x^2} &= \frac{x \cdot x \cdot x}{x \cdot x} = x \\ \frac{x^3}{x^2} &= x^{3-2} = x^1 \end{aligned}\right\} \quad \text{Hence } x^1 = x$$

Stated generally, this rule says that $a^1 = a$. This seems reasonable and we will use it, since it is consistent with our property of division using the same base.

We use the same procedure to obtain an expression for x^0:

$$\left.\begin{aligned} \frac{5^2}{5^2} &= \frac{25}{25} = 1 \\ \frac{5^2}{5^2} &= 5^{2-2} = 5^0 \end{aligned}\right\} \quad \text{Hence } 5^0 = 1$$

It seems, therefore, that the best definition of x^0 is 1 for all x except $x = 0$. In the case of $x = 0$, we have 0^0, which we will not define. This definition will probably seem awkward at first. Most people would like to define x^0 as 0 when they first encounter it. Remember, the zero in this expression is an exponent, so x^0 does not mean to multiply by zero. Thus, we can make the general statement that $a^0 = 1$ for all real numbers, except $a = 0$.

Here are some examples involving the exponents 0 and 1.

▼ **Examples** Simplify the following expressions.

10. $8^0 = 1$

11. $8^1 = 8$

12. $4^0 + 4^1 = 1 + 4 = 5$

13. $(2x^2y)^0 = 1$ ▲

Here is a summary of the definitions and properties of exponents we have developed so far. For each definition or property in the list, a and b are real numbers, and r and s are integers.

Definitions		Properties	
$a^{-r} = \frac{1}{a^r} = \left(\frac{1}{a}\right)^r$	$a \neq 0$	**1.** $a^r \cdot a^s = a^{r+s}$	
$a^1 = a$		**2.** $(a^r)^s = a^{rs}$	
$a^0 = 1$	$a \neq 0$	**3.** $(ab)^r = a^r b^r$	
		4. $\frac{a^r}{a^s} = a^{r-s}$	$a \neq 0$
		5. $\left(\frac{a}{b}\right)^r = \frac{a^r}{b^r}$	$b \neq 0$

Here are some additional examples. These examples use a combination of the preceding properties and definitions.

▼ **Examples** Simplify each expression. Write all answers with positive exponents only.

14. $\frac{(5x^3)^2}{x^4} = \frac{25x^6}{x^4}$ Properties 2 and 3

$= 25x^2$ Property 4

15. $\frac{x^{-8}}{(x^2)^3} = \frac{x^{-8}}{x^6}$ Property 2

$= x^{-8-6}$ Property 4

$= x^{-14}$ Subtraction

$= \frac{1}{x^{14}}$ Definition of negative exponents

16. $\left(\dfrac{y^5}{y^3}\right)^2 = \dfrac{(y^5)^2}{(y^3)^2}$ Property 5

$= \dfrac{y^{10}}{y^6}$ Property 2

$= y^4$ Property 4

Notice in Example 16 that we could have simplified inside the parentheses first and then raised the result to the second power:

$$\left(\frac{y^5}{y^3}\right)^2 = (y^2)^2 = y^4$$

17. $(3x^5)^{-2} = \dfrac{1}{(3x^5)^2}$ Definition of negative exponents

$= \dfrac{1}{9x^{10}}$ Properties 2 and 3

18. $x^{-8} \cdot x^5 = x^{-8+5}$ Property 1

$= x^{-3}$ Addition

$= \dfrac{1}{x^3}$ Definition of negative exponents

19. $\dfrac{(a^3)^2a^{-4}}{(a^{-4})^3} = \dfrac{a^6a^{-4}}{a^{-12}}$ Property 2

$= \dfrac{a^2}{a^{-12}}$ Property 1

$= a^{14}$ Property 4 ▲

More on Scientific Notation

Now that we have completed our list of definitions and properties of exponents, we can expand the work we did previously with scientific notation.

Recall that a number is in scientific notation when it is written in the form

$$n \times 10^r$$

where $1 \leq n < 10$ and r is an integer.

Since negative exponents give us reciprocals, we can use negative exponents to write very small numbers in scientific notation. For example, the number 0.00057, when written in scientific notation, is equivalent to 5.7×10^{-4}. Here's why:

$$5.7 \times 10^{-4} = 5.7 \times \frac{1}{10^4} = 5.7 \times \frac{1}{10000} = \frac{5.7}{10000} = 0.00057$$

The table below lists some other numbers both in scientific notation and in expanded form.

Number Written the Long Way		Number Written Again in Scientific Notation
376,000	=	3.76×10^5
49,500	=	4.95×10^4
3,200	=	3.2×10^3
591	=	5.91×10^2
46	=	4.6×10^1
8	=	8×10^0
0.47	=	4.7×10^{-1}
0.093	=	9.3×10^{-2}
0.00688	=	6.88×10^{-3}
0.0002	=	2×10^{-4}
0.000098	=	9.8×10^{-5}

Notice that in each case, when the number is written in scientific notation, the decimal point in the first number is placed so that the number is between 1 and 10. The exponent on 10 in the second number keeps track of the number of places we moved the decimal point in the original number to get a number between 1 and 10:

$$376{,}000 = 3.76 \times 10^5$$

Moved 5 places

Decimal point originally here

Keeps track of the 5 places we moved the decimal point

$$0.00688 = 6.88 \times 10^{-3}$$

Moved 3 places

Keeps track of the 3 places we moved the decimal point

Problem Set 4.2

Write each of the following with positive exponents and then simplify, when possible.

1. 3^{-2} **2.** 3^{-3} **3.** 6^{-2} **4.** 2^{-6}
5. 8^{-2} **6.** 3^{-4} **7.** 5^{-3} **8.** 9^{-2}
9. $2x^{-3}$ **10.** $5x^{-1}$ **11.** $(2x)^{-3}$ **12.** $(5x)^{-1}$
13. $(5y)^{-2}$ **14.** $5y^{-2}$ **15.** 10^{-2} **16.** 10^{-3}

Use Property 4 to simplify each of the following expressions. Write all answers that contain exponents with positive exponents only.

17. $\frac{5^3}{5^1}$ **18.** $\frac{7^8}{7^6}$ **19.** $\frac{5^1}{5^3}$ **20.** $\frac{7^6}{7^8}$

21. $\frac{x^{10}}{x^4}$ **22.** $\frac{x^4}{x^{10}}$ **23.** $\frac{4^3}{4^0}$ **24.** $\frac{4^0}{4^3}$

25. $\frac{(2x)^7}{(2x)^4}$ **26.** $\frac{(2x)^4}{(2x)^7}$ **27.** $\frac{6^{11}}{6}$ **28.** $\frac{8^7}{8}$

29. $\frac{6}{6^{11}}$ **30.** $\frac{8}{8^7}$ **31.** $\frac{2^{-5}}{2^3}$ **32.** $\frac{2^{-5}}{2^{-3}}$

33. $\frac{2^5}{2^{-3}}$ **34.** $\frac{2^{-3}}{2^{-5}}$ **35.** $\frac{(3x)^{-5}}{(3x)^{-8}}$ **36.** $\frac{(2x)^{-10}}{(2x)^{-15}}$

Simplify the following expressions. Any answers that contain exponents should contain positive exponents only.

37. $(2x^{-3})^4$ **38.** $(3x^{-2})^3$ **39.** 10^0 **40.** 10^1

41. $(2a^2b)^1$ **42.** $(2a^2b)^0$ **43.** $(7y^3)^{-2}$ **44.** $(5y^4)^{-2}$

45. $x^{-3}x^{-5}$ **46.** $x^{-6} \cdot x^8$ **47.** $y^7 \cdot y^{-10}$ **48.** $y^{-4} \cdot y^{-6}$

49. $\frac{(x^2)^3}{x^4}$ **50.** $\frac{(x^5)^3}{x^{10}}$ **51.** $\frac{(a^4)^3}{(a^3)^2}$ **52.** $\frac{(a^5)^3}{(a^5)^2}$

53. $\frac{y^7}{(y^2)^8}$ **54.** $\frac{y^2}{(y^3)^4}$ **55.** $\frac{(x^0)^5}{x^3}$ **56.** $\frac{(x^5)^0}{x^3}$

57. $\frac{(x^{-2})^3}{x^{-5}}$ **58.** $\frac{(x^2)^{-3}}{x^{-5}}$ **59.** $\frac{(x^{-4})^{-2}}{x^{-10}}$ **60.** $\frac{(x^{-6})^{-5}}{x^{-20}}$

61. $\frac{(a^3)^2(a^4)^5}{(a^5)^2}$ **62.** $\frac{(a^4)^8(a^2)^5}{(a^3)^4}$ **63.** $\frac{(a^{-2})^3(a^4)^2}{(a^{-3})^{-2}}$ **64.** $\frac{(a^{-5})^{-3}(a^7)^{-1}}{(a^{-3})^5}$

65. $\frac{(x^{-7})^3(x^4)^5}{(x^3)^2(x^{-1})^8}$ **66.** $\frac{(x^{-9})^0(x^4)^{-3}}{(x^{-1})^5(x^5)^{-2}}$

Write each of the following numbers in scientific notation.

67. 0.000357 **68.** 0.00357

69. 35,700 **70.** 3,570

71. 0.0048 **72.** 0.000048

73. 25 **74.** 35

75. 0.000009 **76.** 0.0009

Write each of the following numbers in expanded form.

77. 4.23×10^{-3} **78.** 4.23×10^3

79. 5.6×10^4 **80.** 5.6×10^{-4}

81. 8×10^{-5} **82.** 8×10^5

83. 7.89×10^1

84. 7.89×10^0

85. 4.2×10^0

86. 4.2×10^1

87. Some home computers can do a calculation in 2×10^{-3} seconds. Write this number in expanded form.

88. Some of the cells in a human body have a radius of 3×10^{-5} inches. Write this number in expanded form.

89. One of the smallest ants in the world has a length of only 0.006 inches. Write this number in scientific notation.

90. Some cameras used in scientific research can take 1 picture every 0.000000167 seconds. Write this number in scientific notation.

91. The number 25×10^3 is not in scientific notation because 25 is larger than 10. Write 25×10^3 in scientific notation.

92. The number $.25 \times 10^3$ is not in scientific notation because .25 is less than 1. Write $.25 \times 10^3$ in scientific notation.

Review Problems The following problems review material we covered in Section 2.1. They will help you understand some of the material in the next section.

Simplify the following expressions.

93. $4x + 3x$

94. $9x + 7x$

95. $5a - 3a$

96. $10a - 2a$

97. $4y + 5y + y$

98. $6y - y + 2y$

Section 4.3 Operations with Monomials

We have developed all the tools necessary to perform the four basic operations on the simplest of polynomials: monomials.

DEFINITION A *monomial* is a one-term expression that is either a constant (number) or the product of a constant and one or more variables raised to whole number exponents.

The following are examples of monomials:

$$-3 \qquad 15x \qquad -23x^2y \qquad 49x^4y^2z^4 \qquad \frac{3}{4}a^2b^3$$

The numerical part of each monomial is called the *numerical coefficient,* or just *coefficient.* Monomials are also called *terms.*

Multiplication and Division of Monomials

There are two basic steps involved in the multiplication of monomials. First, we rewrite the products using the commutative and associative properties. Then, we simplify by multiplying coefficients and adding exponents of like bases.

▼ **Examples** Multiply.

1. $(-3x^2)(4x^3) = (-3 \cdot 4)(x^2 \cdot x^3)$ — Commutative and associative properties

$= -12x^5$ — Multiply coefficients, add exponents

2. $\left(\frac{4}{5}x^5 \cdot y^2\right)(10x^3 \cdot y) = \left(\frac{4}{5} \cdot 10\right)(x^5 \cdot x^3)(y^2 \cdot y)$ — Commutative and associative properties

$= 8x^8y^3$ — Multiply coefficients, add exponents ▲

You can see that in each case the work was the same—multiply coefficients and add exponents of the same base. We can expect division of monomials to proceed in a similar way. Since our properties are consistent, division of monomials will result in division of coefficients and subtraction of exponents of like bases.

▼ **Examples** Divide.

3. $\frac{15x^3}{3x^2} = \frac{15}{3} \cdot \frac{x^3}{x^2}$ — Write as separate fractions

$= 5x$ — Divide coefficients, subtract exponents

4. $\frac{39x^2y^3}{3xy^5} = \frac{39}{3} \cdot \frac{x^2}{x} \cdot \frac{y^3}{y^5}$ — Write as separate fractions

$= 13x \cdot \frac{1}{y^2}$ — Divide coefficients, subtract exponents

$= \frac{13x}{y^2}$ — Write answer as a single fraction ▲

In Example 4, the expression y^3/y^5 simplifies to $1/y^2$ because of Property 4 for exponents and the definition of negative exponents. If we were to show all the work in this simplification process, it would look like this:

$\frac{y^3}{y^5} = y^{3-5}$ — Property 4 for exponents

$= y^{-2}$ — Subtraction

$= \frac{1}{y^2}$ — Definition of negative exponents

The point of this explanation is this: Even though we may not show all the steps when simplifying an expression involving exponents, the result we obtain can still be justified using the properties of exponents. We have not introduced any new properties in Example 4; we have just not shown the details of each simplification.

▼ **Example 5** Divide.

$$\frac{25a^5b^3}{50a^2b^7} = \frac{25}{50} \cdot \frac{a^5}{a^2} \cdot \frac{b^3}{b^7} \quad \text{Write as separate fractions}$$

$$= \frac{1}{2} \cdot a^3 \cdot \frac{1}{b^4} \quad \text{Divide coefficients, subtract exponents}$$

$$= \frac{a^3}{2b^4} \quad \text{Write answer as a single fraction}$$

▲

Notice in Example 5 that dividing 25 by 50 results in $\frac{1}{2}$. This is the same result we would obtain if we reduced the fraction $\frac{25}{50}$ to lowest terms, and there is no harm in thinking of it in this way. Also, notice that the expression b^3/b^7 simplifies to $1/b^4$ by Property 4 for exponents and the definition of negative exponents, even though we have not shown the steps involved in doing so.

Multiplication and Division of Numbers Written in Scientific Notation

We multiply and divide numbers written in scientific notation using the same steps we used to multiply and divide monomials.

▼ **Example 6** Multiply: $(4 \times 10^7)(2 \times 10^{-4})$.

Solution Since multiplication is commutative and associative, we can rearrange the order of these numbers and group them as follows:

$$(4 \times 10^7)(2 \times 10^{-4}) = (4 \times 2)(10^7 \times 10^{-4})$$
$$= 8 \times 10^3$$

Notice that we add exponents, $7 + (-4) = 3$, when we multiply with the same base. ▲

▼ **Example 7** Divide: $\dfrac{9.6 \times 10^{12}}{3 \times 10^4}$.

Solution We group the numbers between 1 and 10 separately from the powers of 10 and proceed as we did in Example 6:

$$\frac{9.6 \times 10^{12}}{3 \times 10^4} = \frac{9.6}{3} \times \frac{10^{12}}{10^4}$$
$$= 3.2 \times 10^8$$

Notice that the procedure we used in both of these examples is very similar to multiplication and division of monomials, for which we multiplied or divided coefficients and added or subtracted exponents. ▲

Addition and Subtraction of Monomials

Addition and subtraction of monomials will be almost identical, since subtraction is defined as addition of the opposite. With multiplication and division of monomials, the key was rearranging the numbers and variables using the commutative and associative properties. With addition, the key is application of the distributive property. We sometimes use the phrase *combine monomials* to describe addition and subtraction of monomials.

DEFINITION Two terms (monomials) with the same variable part (same variables raised to the same powers) are called *similar* (or *like*) terms.

You can add only similar terms. This is because the distributive property (which is the key to addition of monomials) cannot be applied to terms that are not similar.

▼ **Examples** Combine the following monomials.

8. $-3x^2 + 15x^2 = (-3 + 15)x^2$ Distributive property

$= 12x^2$ Add coefficients

9. $9x^2y - 20x^2y = (9 - 20)x^2y$ Distributive property

$= -11x^2y$ Add coefficients

10. $5x^2 + 8y^2$ In this case we cannot apply the distributive property, so we cannot add the monomials ▲

Adding Fractions with the Same Denominators

A topic closely related to combining similar terms is adding fractions with the same denominators. You may recall from previous math classes that to add two fractions with the same denominator, you simply add their numerators and put the result over the common denominator:

$$\frac{3}{4} + \frac{2}{4} = \frac{3 + 2}{4} = \frac{5}{4}$$

The reason we add numerators but do not add denominators is that we must follow the distributive property. To see this, you first have to recall that $\frac{3}{4}$ can be written as $3 \cdot \frac{1}{4}$, and $\frac{2}{4}$ can be written as $2 \cdot \frac{1}{4}$ (dividing by 4 is equivalent to multiplying by $\frac{1}{4}$). Here is the addition problem again, this time showing the use of the distributive property:

$$\frac{3}{4} + \frac{2}{4} = 3 \cdot \frac{1}{4} + 2 \cdot \frac{1}{4}$$

$$= (3 + 2) \cdot \frac{1}{4} \qquad \text{Distributive property}$$

$$= 5 \cdot \frac{1}{4}$$

$$= \frac{5}{4}$$

Here are some further examples.

▼ **Examples** Combine.

11. $\frac{9}{2} + \frac{15}{2} = 9 \cdot \frac{1}{2} + 15 \cdot \frac{1}{2}$

$$= (9 + 15) \cdot \frac{1}{2} \qquad \text{Distributive property}$$

$$= 24 \cdot \frac{1}{2}$$

$$= 12$$

12. $\frac{2}{8} - \frac{7}{8} = 2 \cdot \frac{1}{8} - 7 \cdot \frac{1}{8}$

$$= (2 - 7) \cdot \frac{1}{8} \qquad \text{Distributive property}$$

$$= -5 \cdot \frac{1}{8}$$

$$= -\frac{5}{8}$$

▲

The next examples show how we simplify expressions containing monomials when more than one operation is involved.

▼ **Example 13** Simplify: $\frac{(6x^4y)(3x^7y^5)}{9x^5y^2}$.

Solution We begin by multiplying the two monomials in the numerator:

$$\frac{(6x^4y)(3x^7y^5)}{9x^5y^2} = \frac{18x^{11}y^6}{9x^5y^2} \qquad \text{Simplify numerator}$$

$$= 2x^6y^4 \qquad \text{Divide}$$

▲

▼ **Example 14** Simplify: $\dfrac{(6.8 \times 10^5)(3.9 \times 10^{-7})}{7.8 \times 10^{-4}}$.

Solution We group the numbers between 1 and 10 separately from the powers of 10:

$$\frac{(6.8)(3.9)}{7.8} \times \frac{(10^5)(10^{-7})}{10^{-4}} = 3.4 \times 10^{5+(-7)-(-4)}$$
$$= 3.4 \times 10^2$$ ▲

▼ **Example 15** Simplify: $\dfrac{14x^5}{2x^2} + \dfrac{15x^8}{3x^5}$.

Solution Simplifying each expression separately and then combining similar terms gives

$$\frac{14x^5}{2x^2} + \frac{15x^8}{3x^5} = 7x^3 + 5x^3 \qquad \text{Divide}$$
$$= 12x^3 \qquad \text{Add}$$ ▲

We end this section with a list of the rules for working with monomials.

Multiplication:	*Multiply* coefficients and *add* exponents with common bases.
Division:	*Divide* coefficients and *subtract* exponents with common bases.
Addition:	*Add* coefficients of *similar* terms.
Subtraction:	*Subtract* coefficients of *similar* terms.

Problem Set 4.3

Multiply.

1. $(3x^4)(4x^3)$
2. $(6x^5)(-2x^2)$
3. $(-2y^4)(8y^7)$
4. $(5y^{10})(2y^5)$
5. $(8x)(4x)$
6. $(7x)(5x)$
7. $(10a^3)(10a)(2a^2)$
8. $(5a^4)(10a)(10a^4)$
9. $(6ab^2)(-4a^2b)$
10. $(-5a^3b)(4ab^4)$
11. $(4x^2y)(3x^3y^3)(2xy^4)$
12. $(5x^6)(-10xy^4)(-2x^2y^6)$

Divide. Write all answers with positive exponents only.

13. $\dfrac{15x^3}{5x^2}$

14. $\dfrac{25x^5}{5x^4}$

15. $\dfrac{18y^9}{3y^{12}}$

16. $\dfrac{24y^4}{-8y^7}$

17. $\dfrac{32a^3}{64a^4}$

18. $\dfrac{25a^5}{75a^6}$

19. $\dfrac{21a^2b^3}{-7ab^5}$

20. $\dfrac{32a^5b^6}{8ab^5}$

21. $\dfrac{3x^3y^2z}{27xy^2z^3}$

22. $\dfrac{5x^5y^4z}{30x^3yz^2}$

23. $\dfrac{144x^9y^2}{-12x^{10}y^8}$

24. $\dfrac{256x^9y^2}{32x^4}$

Find each product. Write all answers in scientific notation.

25. $(3 \times 10^3)(2 \times 10^5)$

26. $(4 \times 10^8)(1 \times 10^6)$

27. $(3.5 \times 10^4)(5 \times 10^{-6})$

28. $(7.1 \times 10^5)(2 \times 10^{-8})$

29. $(5.5 \times 10^{-3})(2.2 \times 10^{-4})$

30. $(3.4 \times 10^{-2})(4.5 \times 10^{-6})$

Find each quotient. Write all answers in scientific notation.

31. $\dfrac{8.4 \times 10^5}{2 \times 10^2}$

32. $\dfrac{9.6 \times 10^{20}}{3 \times 10^6}$

33. $\dfrac{6 \times 10^8}{2 \times 10^{-2}}$

34. $\dfrac{8 \times 10^{12}}{4 \times 10^{-3}}$

35. $\dfrac{2.5 \times 10^{-6}}{5 \times 10^{-4}}$

36. $\dfrac{4.5 \times 10^{-8}}{9 \times 10^{-4}}$

Combine by adding or subtracting as indicated.

37. $3x^2 + 5x^2$

38. $4x^3 + 8x^3$

39. $8x^5 - 19x^5$

40. $75x^6 - 50x^6$

41. $2a + a - 3a$

42. $5a + a - 6a$

43. $10x^3 - 8x^3 + 2x^3$

44. $7x^5 + 8x^5 - 12x^5$

45. $20ab^2 - 19ab^2 + 30ab^2$

46. $18a^3b^2 - 20a^3b^2 + 10a^3b^2$

47. $-4abc - 9abc - abc$

48. $-7abc - abc - abc$

Combine the following using the method shown in Examples 11 and 12.

49. $\frac{3}{5} + \frac{4}{5}$

50. $\frac{3}{5} - \frac{4}{5}$

51. $\frac{9}{25} + \frac{5}{25}$

52. $\frac{3}{4} + \frac{5}{4}$

53. $\frac{3}{8} - \frac{11}{8}$

54. $\frac{1}{6} - \frac{7}{6}$

55. $\frac{4}{12} + \frac{3}{12}$

56. $\frac{6}{21} + \frac{5}{21}$

Simplify. Write all answers with positive exponents only.

57. $\dfrac{(3x^2)(8x^5)}{6x^4}$

58. $\dfrac{(7x^3)(6x^8)}{14x^5}$

59. $\dfrac{(9a^2b)(2a^3b^4)}{18a^5b^7}$

60. $\dfrac{(21a^5b)(2a^8b^4)}{14ab}$

61. $\dfrac{(4x^3y^2)(9x^4y^{10})}{(3x^5y)(2x^6y)}$

62. $\dfrac{(5x^4y^4)(10x^3y^3)}{(25xy^5)(2xy^7)}$

Simplify each expression and write all answers in scientific notation.

63. $\dfrac{(6 \times 10^8)(3 \times 10^5)}{9 \times 10^7}$

64. $\dfrac{(8 \times 10^4)(5 \times 10^{10})}{2 \times 10^7}$

65. $\dfrac{(5 \times 10^3)(4 \times 10^{-5})}{2 \times 10^{-2}}$

66. $\dfrac{(7 \times 10^6)(4 \times 10^{-4})}{1.4 \times 10^{-3}}$

67. $\dfrac{(2.8 \times 10^{-7})(3.6 \times 10^4)}{2.4 \times 10^3}$

68. $\dfrac{(5.4 \times 10^2)(3.5 \times 10^{-9})}{4.5 \times 10^6}$

Simplify.

69. $\dfrac{18x^4}{3x} + \dfrac{21x^7}{7x^4}$

70. $\dfrac{24x^{10}}{6x^4} + \dfrac{32x^7}{8x}$

71. $\dfrac{45a^6}{9a^4} - \dfrac{50a^8}{2a^6}$

72. $\dfrac{16a^9}{4a} - \dfrac{28a^{12}}{4a^4}$

73. $\dfrac{6x^7y^4}{3x^2y^2} + \dfrac{8x^5y^8}{2y^6}$

74. $\dfrac{40x^{10}y^{10}}{8x^2y^5} + \dfrac{10x^8y^8}{5y^3}$

Use your knowledge of the properties and definitions of exponents to find x in each of the following.

75. $4^x \cdot 4^5 = 4^7$

76. $\dfrac{5^x}{5^3} = 5^4$

77. $(7^3)^x = 7^{12}$

78. $\dfrac{3^x}{3^4} = 9$

79. The statement $(a + b)^2 = a^2 + b^2$ looks similar to Property 3 for exponents. However, it is not a property of exponents because almost every time we replace a and b with numbers, this expression becomes a false statement. Let $a = 4$ and $b = 5$ in the expressions $(a + b)^2$ and $a^2 + b^2$ and see what each simplifies to.

80. Show that the statement $(a - b)^2 = a^2 - b^2$ is not, in general, true by substituting 3 for a and 5 for b in each of the expressions and then simplifying each result.

81. Show that the expressions $(a + b)^2$ and $a^2 + 2ab + b^2$ are equal when $a = 3$ and $b = 4$.

82. Show that the expressions $(a - b)^2$ and $a^2 - 2ab + b^2$ are equal when $a = 7$ and $b = 5$.

Review Problems The problems that follow review material we covered in Sections 2.1 and 3.2. Reviewing the problems from Section 2.1 will help you understand some of the material in the next section.

Find the value of each expression when $x = -2$. [2.1]

83. $4x$

84. $-3x$

85. $-2x + 5$

86. $-4x - 1$

87. $x^2 + 5x + 6$

88. $x^2 - 5x + 6$

For each equation below, complete the given ordered pairs so each is a solution to the equation and then use the ordered pairs to graph the equation. [3.2]

89. $y = 2x + 2$ $(-2,\ ), (0,\ ), (2,\ )$

90. $y = 2x - 3$ $(-2,\ ), (0,\ ), (2,\ )$

91. $y = \frac{1}{3}x + 1$ $(-3,\ ), (0,\ ), (3,\ )$

92. $y = \frac{1}{2}x - 2$ $(-2,\ ), (0,\ ), (2,\ )$

Section 4.4 Addition and Subtraction of Polynomials

In this section we will extend what we learned in Section 4.3 to expressions called polynomials. We begin this section with the definition of a polynomial.

DEFINITION A *polynomial* is a finite sum of monomials (terms).

The following are examples of polynomials:

$$3x^2 + 2x + 1 \qquad 15x^2y + 21xy^2 - y^2 \qquad 3a - 2b + 4c - 5d$$

Polynomials can be further classified by the number of terms they contain. A polynomial with two terms is called a binomial. If it has three terms, it is a trinomial. As stated before, a monomial has only one term.

DEFINITION The *degree* of a polynomial in one variable is the highest power to which the variable is raised.

The following are examples of degrees:

$3x^5 + 2x^3 + 1$	A trinomial of degree 5
$2x + 1$	A binomial of degree 1
$3x^2 + 2x + 1$	A trinomial of degree 2
$3x^5$	A monomial of degree 5
-9	A monomial of degree 0

There are no new rules for adding one or more polynomials. We rely only on our previous knowledge. Here are some examples.

▼ **Example 1** Add: $(2x^2 - 5x + 3) + (4x^2 + 7x - 8)$.

Solution We use the commutative and associative properties to group similar terms together and then apply the distributive property to add:

$(2x^2 - 5x + 3) + (4x^2 + 7x - 8)$	
$= (2x^2 + 4x^2) + (-5x + 7x) + (3 - 8)$	Commutative and associative properties
$= (2 + 4)x^2 + (-5 + 7)x + (3 - 8)$	Distributive property
$= 6x^2 + 2x - 5$	Addition

The results here indicate that to add two polynomials, we add coefficients of similar terms. ▲

▼ **Example 2** Add: $x^2 + 3x + 2x + 6$.

Solution The only similar terms here are the two middle terms. We combine them as usual to get

$$x^2 + 3x + 2x + 6 = x^2 + 5x + 6$$ ▲

You will recall from Chapter 1 the definition of subtraction: $a - b = a + (-b)$. To subtract one expression from another, we simply add its opposite. The letters a and b in the definition can each represent polynomials. The opposite of a polynomial is the opposite of each of its terms. When you subtract one polynomial from another, you subtract each of its terms.

▼ **Example 3** Subtract: $(3x^2 + x + 4) - (x^2 + 2x + 3)$.

Solution To subtract $x^2 + 2x + 3$, we change the sign of each of its terms and add. If you are having trouble remembering why we do this, remember that we can think of $-(x^2 + 2x + 3)$ as $-1(x^2 + 2x + 3)$. If we distribute the -1 across $x^2 + 2x + 3$, we get $-x^2 - 2x - 3$:

$(3x^2 + x + 4) - (x^2 + 2x + 3)$	
$= 3x^2 + x + 4 - x^2 - 2x - 3$	Take the opposite of each term in the second polynomial
$= (3x^2 - x^2) + (x - 2x) + (4 - 3)$	
$= 2x^2 - x + 1$	

▲

▼ **Example 4** Subtract $-4x^2 + 5x - 7$ from $x^2 - x - 1$.

Solution The polynomial $x^2 - x - 1$ comes first, then the subtraction sign, and finally the polynomial $-4x^2 + 5x - 7$ in parentheses:

$$
\begin{aligned}
&(x^2 - x - 1) - (-4x^2 + 5x - 7) \\
&= x^2 - x - 1 + 4x^2 - 5x + 7 && \text{We take the opposite of each term in the second polynomial} \\
&= (x^2 + 4x^2) + (-x - 5x) + (-1 + 7) \\
&= 5x^2 - 6x + 6
\end{aligned}
$$

▲

There are two important points to remember when adding or subtracting polynomials. First, to add or subtract two polynomials, you always add or subtract *coefficients* of similar terms. Second, the exponents never increase in value when you are adding or subtracting similar terms.

The last topic we want to consider in this section is finding the value of a polynomial for a given value of the variable.

To find the value of the polynomial $3x^2 + 1$ when x is 5, we replace x with 5 and simplify the result:

When $x = 5$

the polynomial $3x^2 + 1$

becomes

$$
\begin{aligned}
3(5)^2 + 1 &= 3(25) + 1 \\
&= 75 + 1 \\
&= 76
\end{aligned}
$$

▼ **Example 5** Find the value of $3x^2 - 5x + 4$ when $x = -2$.

Solution When $x = -2$

the polynomial $3x^2 - 5x + 4$

becomes

$$
\begin{aligned}
3(-2)^2 - 5(-2) + 4 &= 3(4) + 10 + 4 \\
&= 12 + 10 + 4 \\
&= 26
\end{aligned}
$$

▲

Problem Set 4.4

Identify each of the following polynomials as a trinomial, binomial, or monomial, and give the degree in each case.

1. $2x^3 - 3x^2 + 1$

2. $4x^2 - 4x + 1$

3. $5 + 8a - 9a^3$

4. $6 + 12x^3 + x^4$

5. $2x - 1$

6. $4 + 7x$

7. $45x^2 - 1$

8. $3a^3 + 8$

9. $7a^2$

10. $90x$

11. -4

12. 56

Perform the following additions and subtractions.

13. $(2x^2 + 3x + 4) + (3x^2 + 2x + 5)$
14. $(x^2 + 5x + 6) + (x^2 + 3x + 4)$
15. $(3a^2 - 4a + 1) + (2a^2 - 5a + 6)$
16. $(5a^2 - 2a + 7) + (4a^2 - 3a + 2)$
17. $x^2 + 4x + 2x + 8$
18. $x^2 + 5x - 3x - 15$
19. $6x^2 - 3x - 10x + 5$
20. $10x^2 + 30x - 2x - 6$
21. $x^2 - 3x + 3x - 9$
22. $x^2 - 5x + 5x - 25$
23. $3y^2 - 5y - 6y + 10$
24. $y^2 - 18y + 2y - 12$
25. $(6x^3 - 4x^2 + 2x) + (9x^2 - 6x + 3)$
26. $(5x^3 + 2x^2 + 3x) + (2x^2 + 5x + 1)$
27. $(3x^2 - 4x - 5) + (5x^2 + 4x + 3)$
28. $(22x^3 - 12x^2 + x) - (5x^3 - 3x^2 + 5)$
29. $(a^2 - a - 1) - (-a^2 + a + 1)$
30. $(2x^2 - 7x - 8) - (6x^2 + 6x - 8) + (4x^2 - 2x + 3)$
31. $(-8x^2 + 2x + 1) + (10x^2 - 33x - 5) - (3x^2 - 5x - 8)$
32. $(7x^4 - 5x^2 - 4) - (8x^4 + 9x^2 - 3)$
33. $(4y^2 - 3y + 2) + (5y^2 + 12y - 4) - (13y^2 - 6y + 20)$
34. $(9x^3 - 8x^2 + 7x) - (2x - 3x^2 - 4x^3)$
35. Subtract $10x^2 + 23x - 50$ from $11x^2 - 10x + 13$.
36. Subtract $2x^2 - 3x + 5$ from $4x^2 - 5x + 10$.
37. Subtract $3y^2 + 7y - 15$ from $11y^2 + 11y + 11$.
38. Subtract $15y^2 - 8y - 2$ from $3y^2 - 3y + 2$.
39. Add $50x^2 - 100x - 150$ to $25x^2 - 50x + 75$.
40. Add $7x^2 - 8x + 10$ to $-8x^2 + 2x - 12$.
41. Subtract $2x + 1$ from the sum of $3x - 2$ and $11x + 5$.
42. Subtract $3x - 5$ from the sum of $5x + 2$ and $9x - 1$.
43. Find the value of the polynomial $x^2 - 2x + 1$ when x is 3.
44. Find the value of the polynomial $(x - 1)^2$ when x is 3.
45. Find the value of the polynomial $(y - 5)^2$ when y is 10.
46. Find the value of the polynomial $y^2 - 10y + 25$ when y is 10.
47. Find the value of $a^2 + 4a + 4$ when a is 2.
48. Find the value of $(a + 2)^2$ when a is 2.

Review Problems The problems below review material we covered in Sections 3.3 and 4.3. Reviewing the problems from Section 4.3 will help you understand the next section.

Multiply: [4.3]

49. $2x(5x)$
50. $2x(-2x)$
51. $3x(-5x)$
52. $-3x(-7x)$

53. $2x(3x^2)$

54. $x^2(3x)$

55. $3x^2(2x^2)$

56. $4x^2(2x^2)$

Find the x- and y-intercepts for each equation and then use the intercepts to draw the graph of the equation. [3.3]

57. $3x - 2y = 6$

58. $2x - 3y = 6$

59. $2x + y = 4$

60. $x + 2y = 4$

Find the slope of the line through each pair of points. [3.3]

61. (2, 3) (5, 7)

62. (4, 2) (7, 5)

Section 4.5 Multiplication with Polynomials

We begin our discussion of multiplication of polynomials by finding the product of a monomial and a trinomial.

▼ **Example 1** Multiply: $3x^2(2x^2 + 4x + 5)$.

Solution Applying the distributive property gives us

$$
\begin{aligned}
3x^2(2x^2 + 4x + 5) &= 3x^2(2x^2) + 3x^2(4x) + 3x^2(5) && \text{Distributive property} \\
&= 6x^4 + 12x^3 + 15x^2 && \text{Multiplication}
\end{aligned}
$$

▲

The distributive property is the key to multiplication of polynomials. We can use it to find the product of any two polynomials. There are some shortcuts we can use in certain situations, however. Let's look at an example that involves the product of two binomials.

▼ **Example 2** Multiply: $(3x - 5)(2x - 1)$.

Solution

$$
\begin{aligned}
(3x - 5)(2x - 1) &= 3x(2x - 1) - 5(2x - 1) \\
&= 3x(2x) + 3x(-1) + (-5)(2x) + (-5)(-1) \\
&= 6x^2 - 3x - 10x + 5 \\
&= 6x^2 - 13x + 5
\end{aligned}
$$

▲

If we look closely at the second and third lines of work in this example, we can see that the terms in the answer come from all possible products of terms in the first binomial with terms in the second binomial. This result is generalized as follows.

> **Rule** To multiply any two polynomials, multiply each term in the first with each term in the second.

There are two ways we can put this rule to work.

FOIL Method

If we look at the original problem in Example 2 and then at the answer, we see that the first term in the answer came from multiplying the first terms in each binomial:

$$3x \cdot 2x = 6x^2 \qquad \text{FIRST}$$

The middle term in the answer came from adding the products of the two outside terms with the two inside terms in each binomial:

$$\begin{aligned} 3x(-1) &= -3x \qquad \text{OUTSIDE} \\ -5(2x) &= \underline{-10x} \qquad \text{INSIDE} \\ &\;\; -13x \end{aligned}$$

The last term in the answer came from multiplying the two last terms:

$$-5(-1) = 5 \qquad \text{LAST}$$

To summarize the FOIL method we will multiply another two binomials.

▼ **Example 3** Multiply: $(2x + 3)(5x - 4)$.

Solution

$$(2x + 3)(5x - 4) = \underbrace{2x(5x)}_{\text{First}} + \underbrace{2x(-4)}_{\text{Outside}} + \underbrace{3(5x)}_{\text{Inside}} + \underbrace{3(-4)}_{\text{Last}}$$

$$\begin{aligned} &= 10x^2 - 8x + 15x - 12 \\ &= 10x^2 + 7x - 12 \end{aligned}$$

With practice $-8x + 15x = 7x$ can be done mentally. ▲

COLUMN Method

The FOIL method can be applied only when multiplying two binomials. To find products of polynomials with more than two terms we use what is called the COLUMN method.

The COLUMN method of multiplying two polynomials is very similar to long multiplication with whole numbers. It is just another way of finding all possible products of terms in one polynomial with terms in another polynomial.

▼ **Example 4** Multiply: $(2x + 3)(3x^2 - 2x + 1)$.

Solution

$$
\begin{array}{rl}
3x^2 - 2x + 1 & \\
2x + 3 & \\
\hline
9x^2 - 6x + 3 & \leftarrow 3(3x^2 - 2x + 1) \\
6x^3 - 4x^2 + 2x \phantom{{}+3} & \leftarrow 2x(3x^2 - 2x + 1) \\
\hline
6x^3 + 5x^2 - 4x + 3 & \leftarrow \text{Add similar terms}
\end{array}
$$

▲

It will be to your advantage to become very fast and accurate at multiplying polynomials. You should be comfortable using either method. The following examples illustrate the three types of multiplication.

▼ **Examples** Multiply.

5. $4a^2(2a^2 - 3a + 5) = 4a^2(2a^2) + 4a^2(-3a) + 4a^2(5)$
$= 8a^4 - 12a^3 + 20a^2$

6. $(x - 2)(y + 3) = x(y) + x(3) + (-2)(y) + (-2)(3)$
F O I L
$= xy + 3x - 2y - 6$

7. $(x + y)(a - b) = x(a) + x(-b) + y(a) + y(-b)$
F O I L
$= xa - xb + ya - yb$

8. $(5x - 1)(2x + 6) = 5x(2x) + 5x(6) + (-1)(2x) + (-1)(6)$
F O I L
$= 10x^2 + 30x + (-2x) + (-6)$
$= 10x^2 + 28x - 6$

9. $(3x + 2)(x^2 - 5x + 6)$

$$
\begin{array}{r}
x^2 - 5x + 6 \\
3x + 2 \\
\hline
2x^2 - 10x + 12 \\
3x^3 - 15x^2 + 18x \phantom{{}+12} \\
\hline
3x^3 - 13x^2 + 8x + 12
\end{array}
$$

▲

▼ **Example 10** The length of a rectangle is 3 more than twice the width. Write an expression for the area of the rectangle.

Solution We begin by drawing a rectangle and labeling the width with x. Since the length is 3 more than twice the width, we label the length with $2x + 3$.

$2x + 3$

x

Since the area, A, of a rectangle is the product of the length and width, we write our formula for the area of this rectangle as

$$A = x(2x + 3)$$
$$A = 2x^2 + 3x \qquad \text{Multiply}$$

▲

Suppose that a store sells x items at p dollars per item. The total amount of money obtained by selling the items is called the *revenue*. It can be found by multiplying the number of items sold, x, by the price per item, p. For example, if 100 items are sold for \$6 each, the revenue is $100(6) = \$600$. Similarly, if 500 items are sold for \$8 each, the total revenue is $500(8) = \$4{,}000$. If we denote the revenue with the letter R, then the formula that relates R, x, and p is

$$\text{Revenue} = \begin{pmatrix}\text{number of}\\ \text{items sold}\end{pmatrix}\begin{pmatrix}\text{price of}\\ \text{each item}\end{pmatrix}$$

In symbols: $R = xp$.

▼ **Example 11** A store selling diskettes for home computers knows from past experience that they can sell x diskettes each day at a price of p dollars per diskette, according to the equation $x = 800 - 100p$. Write a formula for the weekly revenue that involves only the variables R and p.

Solution From our discussion above, we know that the revenue R is given by the formula

$$R = xp$$

But, since $x = 800 - 100p$, we can substitute $800 - 100p$ for x in the revenue equation to obtain

$$R = (800 - 100p)p$$
$$R = 800p - 100p^2$$

This last formula gives the revenue, R, in terms of the price, p. ▲

Problem Set 4.5

Multiply the following by applying the distributive property.

1. $2x(3x + 1)$
2. $4x(2x - 3)$
3. $2x^2(3x^2 - 2x + 1)$
4. $5x(4x^3 - 5x^2 + x)$

5. $2ab(a^2 - ab + 1)$
6. $3a^2b(a^3 + a^2b^2 + b^3)$
7. $y^2(3y^2 + 9y + 12)$
8. $5y(2y^2 - 3y + 5)$
9. $4x^2y(2x^3y + 3x^2y^2 + 8y^3)$
10. $6xy^3(2x^2 + 5xy + 12y^2)$

Multiply the binomials below. You should do about half the problems using the FOIL method and the other half using the COLUMN method. Remember, you want to be comfortable using each method.

11. $(x + 3)(x + 4)$
12. $(x + 2)(x + 5)$
13. $(x + 6)(x + 1)$
14. $(x + 1)(x + 4)$
15. $(x + 2)(y + 4)$
16. $(x + 3)(y + 5)$
17. $(a + 5)(a - 3)$
18. $(a - 8)(a + 2)$
19. $(x - a)(y + b)$
20. $(x + a)(y - b)$
21. $(x + 6)(x - 6)$
22. $(x + 3)(x - 3)$
23. $(y - 2)(y + 2)$
24. $(y - 4)(y + 4)$
25. $(2x - 3)(x - 4)$
26. $(3x - 5)(x - 2)$
27. $(a + 2)(2a - 1)$
28. $(a - 6)(3a + 2)$
29. $(2x - 5)(3x - 2)$
30. $(3x + 6)(2x - 1)$
31. $(2x + 3)(a + 4)$
32. $(2x - 3)(a - 4)$
33. $(5x - 4)(5x + 4)$
34. $(6x + 5)(6x - 5)$
35. $(8x + 3)(x - 2)$
36. $(5x + 4)(3x - 2)$
37. $(1 - 2a)(3 - 4a)$
38. $(1 - 3a)(3 + 2a)$
39. $(7 - 6x)(8 - 5x)$
40. $(7 - 4x)(8 - 3x)$

Multiply the following.

41. $(x + 1)(x^2 + 3x - 4)$
42. $(x - 2)(x^2 + 3x - 4)$
43. $(a - 3)(a^2 - 3a + 2)$
44. $(a + 5)(a^2 + 2a + 3)$
45. $(x + 2)(x^2 - 2x + 4)$
46. $(x + 3)(x^2 - 3x + 9)$
47. $(2x + 1)(x^2 + 8x + 9)$
48. $(3x - 2)(x^2 - 7x + 8)$
49. $(5x^2 + 2x + 1)(x^2 - 3x + 5)$
50. $(2x^2 + x + 1)(x^2 - 4x + 3)$

51. The length of a rectangle is 5 more than twice the width. Write an expression for the area of the rectangle.
52. The length of a rectangle is 2 more than three times the width. Write an expression for the area of the rectangle.
53. The width and length of a rectangle are given by two consecutive integers. Write an expression for the area of the rectangle.
54. The width and length of a rectangle are given by two consecutive even integers. Write an expression for the area of the rectangle.
55. A store selling typewriter ribbons knows that the number of ribbons they can sell each week, x, is related to the price per ribbon, p, by the equation $x = 1200 - 100p$. Write an expression for the weekly revenue that involves only the variables R and p. (Remember: The equation for revenue is $R = xp$.)
56. A store selling small portable radios knows from past experience that the number of radios they can sell each week, x, is related to the price per radio, p, by the equation $x = 1300 - 100p$. Write an expression for the weekly revenue that involves only the variables R and p.

57. The relationship between the number of calculators a company sells per day, x, and the price of each calculator, p, is given by the equation $x = 1700 - 100p$. Write an expression for the daily revenue that involves only the variables R and p.

58. The relationship between the number of pencil sharpeners a company can sell each day, x, and the price of each sharpener, p, is given by the equation $x = 1800 - 100p$. Write an expression for the daily revenue that involves only the variables R and p.

Review Problems The problems that follow review material we covered in Section 3.4.

Graph the line with the given slope and y-intercept.

59. $m = \frac{2}{3}, b = -1$

60. $m = -\frac{1}{2}, b = 3$

Find the slope and y-intercept for each of the following equations by first writing them in the form $y = mx + b$.

61. $-2x + 4y = 8$

62. $5x - 3y = 15$

For each problem below, the slope and one point on a line are given. Find the equation.

63. $(-2, -6), m = 3$

64. $(4, 2), m = \frac{1}{2}$

Find the equation of the line that passes through each pair of points.

65. $(-3, -5), (3, 1)$

66. $(-1, -5), (2, 1)$

Section 4.6 Binomial Squares and Other Special Products

In this section we will combine the results of the last section with our definition of exponents to find some special products.

▼ **Example 1** Find the square of $(3x - 2)$.

Solution To square $3x - 2$, we multiply it by itself:

$$\begin{aligned}(3x - 2)^2 &= (3x - 2)(3x - 2) && \text{Definition of exponents}\\ &= 9x^2 - 6x - 6x + 4 && \text{FOIL method}\\ &= 9x^2 - 12x + 4 && \text{Combine similar terms}\end{aligned}$$ ▲

Notice that the first and last terms in the answer are the square of the first and last terms in the original problem and that the middle term is twice the product of the two terms in the original binomial.

▼ **Examples**

2. $(a + b)^2 = (a + b)(a + b)$
$= a^2 + 2ab + b^2$

3. $(a - b)^2 = (a - b)(a - b)$
$= a^2 - 2ab + b^2$ ▲

Binomial squares having the form of Examples 2 and 3 occur very frequently in algebra. It will be to your advantage to memorize the following rule for squaring a binomial.

> **Rule** The square of a binomial is the sum of the square of the first term, the square of the last term, and twice the product of the two original terms. In symbols, this rule is written as follows:
>
> | | | | | | |
> |---|---|---|---|---|---|
> | $(x + y)^2 =$ | x^2 | + | $2xy$ | + | y^2 |
> | | Square of first term | | Twice product of the two terms | | Square of last term |

▼ **Examples** Multiply using the preceding rule.

			First term squared		Twice their product		Last term squared		Answer
4.	$(x - 5)^2$	=	x^2	+	$2(x)(-5)$	+	25	=	$x^2 - 10x + 25$
5.	$(x + 2)^2$	=	x^2	+	$2(x)(2)$	+	4	=	$x^2 + 4x + 4$
6.	$(2x - 3)^2$	=	$4x^2$	+	$2(2x)(-3)$	+	9	=	$4x^2 - 12x + 9$
7.	$(5x - 4)^2$	=	$25x^2$	+	$2(5x)(-4)$	+	16	=	$25x^2 - 40x + 16$

▲

Another special product that occurs frequently is $(a + b)(a - b)$. The only difference in the two binomials is the sign between the two terms. The interesting thing about this type of product is that the middle term is always zero. Here are some examples:

▼ **Examples** Multiply using the FOIL method.

8. $(2x - 3)(2x + 3) = 4x^2 + 6x - 6x - 9$ FOIL method
$= 4x^2 - 9$

9. $(x - 5)(x + 5) = x^2 + 5x - 5x - 25$ FOIL method
$= x^2 - 25$

10. $(3x - 1)(3x + 1) = 9x^2 + 3x - 3x - 1$ FOIL method
$= 9x^2 - 1$ ▲

Notice that in each case the middle term is zero and therefore doesn't appear in the answer. The answers all turn out to be the difference of two squares. Here is a rule to help you memorize the result.

> **Rule** When multiplying two binomials that differ only in the sign between their terms, subtract the square of the last term from the square of the first term. Or
>
> $$(a - b)(a + b) = a^2 - b^2$$

Here are some problems that result in the difference of two squares.

▼ **Examples** Multiply using the preceding rule.

11. $(x + 3)(x - 3) = x^2 - 9$

12. $(a + 2)(a - 2) = a^2 - 4$

13. $(9a + 1)(9a - 1) = 81a^2 - 1$

14. $(2x - 5y)(2x + 5y) = 4x^2 - 25y^2$

15. $(3a - 7b)(3a + 7b) = 9a^2 - 49b^2$ ▲

Although all the problems in this section can be worked correctly using the methods in the last section, they can be done much faster if the two rules are *memorized.* Here is a summary of the two rules:

$$(a + b)^2 = (a + b)(a + b) = a^2 + 2ab + b^2$$
$$(a - b)^2 = (a - b)(a - b) = a^2 - 2ab + b^2$$
$$(a - b)(a + b) = a^2 - b^2$$

▼ **Example 16** If you deposit P dollars in an account with an interest rate r that is compounded annually, then the amount of money in that account at the end of two years is given by the formula

$$A = P(1 + r)^2$$

Expand the right side of this expression.

Solution To expand the right side we must first square $1 + r$ and then multiply the result by P.

$A = P(1 + r)^2$ Original formula
$= P(1 + 2r + r^2)$ Square $(1 + r)$
$= P + 2Pr + Pr^2$ Multiply through by P ▲

▼ **Example 17** Write an expression in symbols for the sum of the squares of three consecutive even integers. Then, simplify that expression.

Solution If we let $x =$ the first of the even integers, then $x + 2$ is the next consecutive even integer, and $x + 4$ is the one after that. An expression for the sum of their squares is

$x^2 + (x + 2)^2 + (x + 4)^2$	Sum of squares
$= x^2 + (x^2 + 4x + 4) + (x^2 + 8x + 16)$	Expand squares
$= 3x^2 + 12x + 20$	Add similar terms

▲

COMMON MISTAKES

A very common mistake when squaring binomials is to write

$$(a + b)^2 = a^2 + b^2 \qquad \text{Mistake}$$

It just isn't true. Exponents do *not* distribute over addition or subtraction. If we try it with 2 and 3 it becomes obvious:

$$(2 + 3)^2 \neq 2^2 + 3^2$$
$$25 \neq 13$$

Problem Set 4.6

Perform the indicated operations.

1. $(x - 2)^2$
2. $(x + 2)^2$
3. $(a + 3)^2$
4. $(a - 3)^2$
5. $(x - 5)^2$
6. $(x - 4)^2$
7. $(a - 7)^2$
8. $(a + 7)^2$
9. $(x + 10)^2$
10. $(x - 10)^2$
11. $(a + b)^2$
12. $(a - b)^2$
13. $(2x - 1)^2$
14. $(3x + 2)^2$
15. $(4a + 5)^2$
16. $(4a - 5)^2$
17. $(3x - 2)^2$
18. $(2x - 3)^2$
19. $(3a + 5b)^2$
20. $(5a - 3b)^2$
21. $(4x - 5y)^2$
22. $(5x + 4y)^2$
23. $(7m + 2n)^2$
24. $(2m - 7n)^2$
25. $(6x - 10y)^2$
26. $(10x + 6y)^2$
27. $(x^2 + 5)^2$
28. $(x^2 + 3)^2$
29. $(a^2 + 1)^2$
30. $(a^2 - 2)^2$
31. $(x^3 - 7)^2$
32. $(x^3 + 4)^2$
33. $(x - 3)(x + 3)$
34. $(x + 4)(x - 4)$
35. $(a + 5)(a - 5)$
36. $(a - 6)(a + 6)$

37. $(y - 1)(y + 1)$
38. $(y - 2)(y + 2)$
39. $(9 + x)(9 - x)$
40. $(10 - x)(10 + x)$
41. $(2x + 5)(2x - 5)$
42. $(3x + 5)(3x - 5)$
43. $(4x - 1)(4x + 1)$
44. $(6x + 1)(6x - 1)$
45. $(2a + 7)(2a - 7)$
46. $(3a + 10)(3a - 10)$
47. $(6 - 7x)(6 + 7x)$
48. $(7 - 6x)(7 + 6x)$
49. $(x^2 + 3)(x^2 - 3)$
50. $(x^2 + 2)(x^2 - 2)$
51. $(a^2 + 4)(a^2 - 4)$
52. $(a^2 + 9)(a^2 - 9)$

53. The formula for the difference of two squares can be used as a shortcut to multiplying certain whole numbers if they have the correct form. Use the difference of two squares formula to multiply 49(51) by first writing 49 as $(50 - 1)$, and 51 as $(50 + 1)$.

54. Use the difference of two squares formula to multiply 101(99) by first writing 101 as $(100 + 1)$ and 99 as $(100 - 1)$.

55. Evaluate the expression $(x + 3)^2$ and the expression $x^2 + 6x + 9$ for $x = 2$.

56. Evaluate the expression $x^2 - 25$ and the expression $(x - 5)(x + 5)$ for $x = 6$.

57. If \$100 is deposited in an account with interest rate r compounded annually, the amount of money in that account at the end of two years is $A = 100(1 + r)^2$. Expand the right side of this expression.

58. If \$100 is deposited in an account with an interest rate r that is compounded every six months, then the amount of money in that account at the end of one year is $A = 25(2 + r)^2$. Expand the right side of this expression.

59. Write an expression for the sum of the squares of two consecutive integers. Then, simplify that expression.

60. Write an expression for the sum of the squares of two consecutive odd integers. Then, simplify that expression.

61. Write an expression for the sum of the squares of three consecutive integers. Then, simplify that expression.

62. Write an expression for the sum of the squares of three consecutive odd integers. Then, simplify that expression.

Review Problems The problems below review material we covered in Sections 4.3 and 3.5. Reviewing the problems from Section 4.3 will help you understand the next section.

Simplify each expression (divide). [4.3]

63. $\dfrac{10x^3}{5x}$
64. $\dfrac{-15x^2}{5x}$

65. $\dfrac{15x^2y}{3xy}$
66. $\dfrac{21xy^2}{3xy}$

67. $\dfrac{35a^6b^8}{70a^2b^{10}}$
68. $\dfrac{75a^2b^6}{25a^4b^3}$

Solve each system by graphing. [3.5]

69. $x + y = 2$
$x - y = 4$

70. $x + y = 1$
$x - y = -3$

71. $y = 2x + 3$
$y = -2x - 1$

72. $y = 3x - 2$
$y = -2x + 3$

Section 4.7 Dividing a Polynomial by a Monomial

To divide a polynomial by a monomial, we will use the definition of division and apply the distributive property. Follow the steps in this example closely.

▼ **Example 1** Divide $10x^3 - 15x^2$ by $5x$.

Solution

$$\frac{10x^3 - 15x^2}{5x} = (10x^3 - 15x^2)\frac{1}{5x}$$ Division by $5x$ is the same as multiplication by $\frac{1}{5x}$

$$= 10x^3\left(\frac{1}{5x}\right) - 15x^2\left(\frac{1}{5x}\right)$$ Distribute $\frac{1}{5x}$ to both terms

$$= \frac{10x^3}{5x} - \frac{15x^2}{5x}$$ Multiplication by $\frac{1}{5x}$ is the same as division by $5x$

$$= 2x^2 - 3x$$ Division of monomials as done in Section 4.3 ▲

If we were to leave out the first steps, the problem would look like this:

$$\frac{10x^3 - 15x^2}{5x} = \frac{10x^3}{5x} - \frac{15x^2}{5x}$$
$$= 2x^2 - 3x$$

The problem is much shorter and clearer this way. You may leave out the first two steps from Example 1 when working problems in this section. They are part of Example 1 only to help show you why the following rule is true.

> **Rule** To divide a polynomial by a monomial, simply divide each term in the polynomial by the monomial.

Here are some further examples using our rule for division of a polynomial by a monomial.

▼ **Example 2** Divide: $\dfrac{3x^2 - 6}{3}$.

Solution We begin by writing the 3 in the denominator under each term in the numerator. Then, we simplify the result:

$$\frac{3x^2 - 6}{3} = \frac{3x^2}{3} - \frac{6}{3} \quad \text{Divide each term in the numerator by 3}$$
$$= x^2 - 2 \quad \text{Simplify}$$

▲

▼ **Example 3** Divide: $\dfrac{4x^2 - 2}{2}$.

Solution Dividing each term in the numerator by 2 we have

$$\frac{4x^2 - 2}{2} = \frac{4x^2}{2} - \frac{2}{2} \quad \text{Divide each term in the numerator by 2}$$
$$= 2x^2 - 1 \quad \text{Simplify}$$

▲

▼ **Example 4** Find the quotient of $27x^3 - 9x^2$ and $3x$.

Solution We are again asked to divide the first polynomial by the second one:

$$\frac{27x^3 - 9x^2}{3x} = \frac{27x^3}{3x} - \frac{9x^2}{3x} \quad \text{Divide each term by } 3x$$
$$= 9x^2 - 3x \quad \text{Simplify}$$

▲

▼ **Example 5** Divide: $(15x^2y - 21xy^2) \div -3xy$.

Solution This is the same type of problem we have shown in the first four examples; it is just worded a little differently. Note that when we divide each term in the first polynomial by $-3xy$, the negative sign must be taken into account:

$$\frac{15x^2y - 21xy^2}{-3xy} = \frac{15x^2y}{-3xy} - \frac{21xy^2}{-3xy} \quad \text{Divide each term by } -3xy$$
$$= -5x - (-7y) \quad \text{Simplify}$$
$$= -5x + 7y \quad \text{Simplify}$$

▲

▼ **Example 6** Divide: $\dfrac{10x^3 - 5x^2 + 20x}{10x^3}$.

Solution Proceeding as we have in the first five examples, we distribute the $10x^3$ under each term in the numerator:

$$\frac{10x^3 - 5x^2 + 20x}{10x^3} = \frac{10x^3}{10x^3} - \frac{5x^2}{10x^3} + \frac{20x}{10x^3} \quad \text{Divide each term by } 10x^3$$

$$= 1 - \frac{1}{2x} + \frac{2}{x^2} \quad \text{Simplify}$$

Note that our answer is not a polynomial. ▲

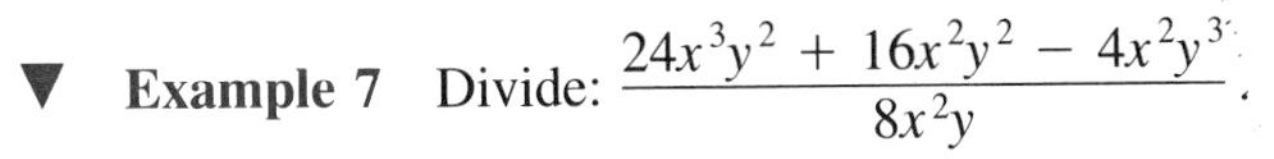

▼ **Example 7** Divide: $\dfrac{24x^3y^2 + 16x^2y^2 - 4x^2y^3}{8x^2y}$.

Solution Writing $8x^2y$ under each term in the numerator and then simplifying, we have

$$\frac{24x^3y^2 + 16x^2y^2 - 4x^2y^3}{8x^2y} = \frac{24x^3y^2}{8x^2y} + \frac{16x^2y^2}{8x^2y} - \frac{4x^2y^3}{8x^2y}$$

$$= 3xy + 2y - \frac{y^2}{2}$$ ▲

COMMON MISTAKE

From the examples in this section, it is clear that to divide a polynomial by a monomial, we must divide each term in the polynomial by the monomial. Often, students taking algebra for the first time will make the following mistake:

$$\frac{x + \not{2}}{\not{2}} = x + 1 \quad \text{Mistake}$$

The mistake here is in not dividing both terms in the numerator by 2. The correct way to divide $x + 2$ by 2 looks like this:

$$\frac{x + 2}{2} = \frac{x}{2} + \frac{2}{2} = \frac{x}{2} + 1 \quad \text{Correct}$$

Problem Set 4.7

Divide the following polynomials by $5x$.

1. $5x^2 - 10x$

2. $10x^3 - 15x$

3. $15x - 10x^3$

4. $50x^3 - 20x^2$

5. $25x^2y - 10xy$

6. $15xy^2 + 20x^2y$

7. $35x^5 - 30x^4 + 25x^3$

8. $40x^4 - 30x^3 + 20x^2$

9. $50x^5 - 25x^3 + 5x$

10. $75x^6 + 50x^3 - 25x$

Divide the following by $-2a$.

11. $8a^2 - 4a$

12. $a^3 - 6a^2$

13. $16a^5 + 24a^4$

14. $30a^6 + 20a^3$

15. $8ab + 10a^2$

16. $6a^2b - 10ab^2$

17. $12a^3b - 6a^2b^2 + 14ab^3$

18. $4ab^3 - 16a^2b^2 - 22a^3b$

19. $a^2 + 2ab + b^2$

20. $a^2b - 2ab^2 + b^3$

Perform the following divisions (find the following quotients).

21. $\dfrac{6x + 8y}{2}$

22. $\dfrac{9x - 3y}{3}$

23. $\dfrac{7y - 21}{-7}$

24. $\dfrac{14y - 12}{2}$

25. $\dfrac{10xy - 8x}{2x}$

26. $\dfrac{12xy^2 - 18x}{-6x}$

27. $\dfrac{x^2y - x^3y^2}{x}$

28. $\dfrac{x^2y - x^3y^2}{x^2}$

29. $\dfrac{x^2y - x^3y^2}{-x^2y}$

30. $\dfrac{ab + a^2b^2}{ab}$

31. $\dfrac{a^2b^2 - ab^2}{-ab^2}$

32. $\dfrac{a^2b^2c + ab^2c^2}{abc}$

33. $\dfrac{x^3 - 3x^2y + xy^2}{x}$

34. $\dfrac{x^2 - 3xy^2 + xy^3}{x}$

35. $\dfrac{10a^2 - 15a^2b + 25a^2b^2}{5a^2}$

36. $\dfrac{11a^2b^2 - 33ab}{-11ab}$

37. $\dfrac{26x^2y^2 - 13xy}{-13xy}$

38. $\dfrac{6x^2y^2 - 3xy}{6xy}$

39. $\dfrac{4x^2y^2 - 2xy}{4xy}$

40. $\dfrac{6x^2a + 12x^2b - 6x^2c}{36x^2}$

41. $\dfrac{5a^2x - 10ax^2 + 15a^2x^2}{20a^2x^2}$

42. $\dfrac{12ax - 9bx + 18cx}{6x^2}$

43. $\dfrac{16x^5 + 8x^2 + 12x}{12x^3}$

44. $\dfrac{27x^2 - 9x^3 - 18x^4}{-18x^3}$

45. Evaluate the expression $\dfrac{10x + 15}{5}$ and the expression $2x + 3$ when $x = 2$.

46. Evaluate the expression $\dfrac{6x^2 + 4x}{2x}$ and the expression $3x + 2$ when $x = 5$.

47. Show that the expression $\dfrac{3x + 8}{2}$ is not the same as the expression $3x + 4$ by replacing x with 10 in both expressions and simplifying the results.

48. Show that the expression $\frac{x + 10}{x}$ is not equal to 10 by replacing x with 5 and simplifying.

Review Problems The following problems review material we covered in Sections 3.6 and 3.7.

Solve each system of equations by the elimination method. [3.6]

49. $x + y = 6$
$x - y = 8$

50. $2x + y = 5$
$-x + y = -4$

51. $2x - 3y = -5$
$x + y = 5$

52. $2x - 4y = 10$
$3x - 2y = -1$

Solve each system by the substitution method. [3.7]

53. $x + y = 2$
$y = 2x - 1$

54. $2x - 3y = 4$
$x = 3y - 1$

55. $4x + 2y = 8$
$y = -2x + 4$

56. $4x + 2y = 8$
$y = -2x + 5$

Section 4.8 Dividing a Polynomial by a Polynomial

Since long division for polynomials is very similar to long division with whole numbers, we will begin by reviewing a division problem with whole numbers. You may realize when looking at Example 1 that you don't have a very good idea why you proceed as you do with long division. What you do know is that the process always works. We are going to approach the explanations in this section in much the same manner. That is, we won't always be sure why the steps we will use are important, only that they always produce the correct result.

▼ **Example 1** Divide: $27\overline{)3962}$.

Solution

$$\begin{array}{r l} 1 & \leftarrow \text{Estimate 27 into 39} \\ 27\overline{)3962} & \\ \underline{27} & \leftarrow \text{Multiply } 1 \times 27 = 27 \\ 12 & \leftarrow \text{Subtract } 39 - 27 = 12 \end{array}$$

$$\begin{array}{r l} 1 & \\ 27\overline{)3962} & \\ \underline{27\downarrow} & \\ 126 & \leftarrow \text{Bring down the 6} \end{array}$$

These are the four basic steps in long division. Estimate, multiply, subtract, and bring down the next term. To finish the problem, we simply perform the same four steps again:

$$\begin{array}{r l}
14 & \leftarrow 4 \text{ is the estimate} \\
27\overline{)3962} & \\
\underline{27} & \\
126 & \\
\underline{108\downarrow} & \leftarrow \text{Multiply to get } 108 \\
182 & \leftarrow \text{Subtract to get } 18, \\
 & \quad \text{then bring down the } 2
\end{array}$$

One more time.

$$\begin{array}{r l}
146 & \leftarrow 6 \text{ is the estimate} \\
27\overline{)3962} & \\
\underline{27} & \\
126 & \\
\underline{108} & \\
182 & \\
\underline{162} & \leftarrow \text{Multiply to get } 162 \\
20 & \leftarrow \text{Subtract to get } 20
\end{array}$$

Since there is nothing left to bring down we have our answer.

$$\frac{3962}{27} = 146 + \frac{20}{27} \quad \text{or} \quad 146\frac{20}{27}$$

▲

Here is how it works with polynomials:

▼ **Example 2** Divide: $\dfrac{x^2 - 5x + 8}{x - 3}$.

Solution

$$\begin{array}{r l}
x & \leftarrow \text{Estimate } x^2 \div x = x \\
x - 3\overline{)\; x^2 - 5x + 8} & \\
- \quad + & \\
\underline{\cancel{+}x^2 \cancel{-} 3x } & \leftarrow \text{Multiply } x(x - 3) = x^2 - 3x \\
-2x & \leftarrow \text{Subtract } (x^2 - 5x) - (x^2 - 3x) = -2x
\end{array}$$

$$\begin{array}{r l}
x & \\
x - 3\overline{)\; x^2 - 5x + 8} & \\
- \quad + & \\
\underline{\cancel{+}x^2 \cancel{-} 3x \downarrow} & \\
-2x + 8 & \leftarrow \text{Bring down the } 8
\end{array}$$

Notice that to subtract one polynomial from another, we add its opposite. That is why we change the signs on $x^2 - 3x$ and add what we get to $x^2 - 5x$. (To subtract the second polynomial, simply change the signs and add.)

We perform the same four steps again:

$$
\begin{array}{rrrrl}
 & & x & -\ 2 & \leftarrow -2 \text{ is the estimate } (-2x \div x = -2) \\
x - 3 \,\big) & x^2 & -\ 5x & +\ 8 & \\
 & - & + & \downarrow & \\
 & \cancel{+}x^2 & \cancel{-}\ 3x & & \\
\hline
 & & -\ 2x & +\ 8 & \\
 & & + & - & \\
 & & \cancel{-}\ 2x & \cancel{+}\ 6 & \leftarrow \text{Multiply } -2(x - 3) = -2x + 6 \\
\hline
 & & & 2 & \leftarrow \text{Subtract } (-2x + 8) - (-2x + 6) = 2
\end{array}
$$

Since there is nothing left to bring down, we have our answer:

$$\frac{x^2 - 5x + 8}{x - 3} = x - 2 + \frac{2}{x - 3}$$

To check our answer, we multiply $(x - 3)(x - 2)$ to get $x^2 - 5x + 6$. Then, adding on the remainder, 2, we have $x^2 - 5x + 8$. ▲

▼ **Example 3** Divide: $\dfrac{6x^2 - 11x - 14}{2x - 5}$.

Solution

$$
\begin{array}{rrrr}
 & & 3x & +\ 2 \\
2x - 5 \,\big) & 6x^2 & -\ 11x & -\ 14 \\
 & - & + & \downarrow \\
 & \cancel{+}6x^2 & \cancel{-}\ 15x & \\
\hline
 & & +\ 4x & -\ 14 \\
 & & - & + \\
 & & \cancel{+}\ 4x & \cancel{-}\ 10 \\
\hline
 & & & -\ 4
\end{array}
$$

$$\frac{6x^2 - 11x - 14}{2x - 5} = 3x + 2 + \frac{-4}{2x - 5}$$ ▲

One last step is sometimes necessary. The two polynomials in a division problem must both be in descending powers of the variable and cannot skip any powers from the highest power down to the constant term.

▼ **Example 4** Divide: $\dfrac{2x^3 - 3x + 2}{x - 5}$.

Solution The problem will be much less confusing if we write $2x^3 - 3x + 2$ as $2x^3 + 0x^2 - 3x + 2$. Adding $0x^2$ does not change our original problem.

$$
\begin{array}{r@{\;}l}
 & 2x^2 \\
x - 5 \big) & 2x^3 + 0x^2 - 3x + 2 \\
 & - \quad + \\
 & \cancel{+}2x^3 \cancel{-} 10x^2 \quad \downarrow \\
\hline
 & \qquad + 10x^2 - 3x
\end{array}
$$

← Estimate $2x^3 \div x = 2x^2$

← Multiply $2x^2(x - 5) = 2x^3 - 10x^2$

← Subtract $(2x^3 + 0x^2) - (2x^3 - 10x^2) = 10x^2$

Bring down the next term

Adding the term $0x^2$ gives us a column in which to write $10x^2$. (Remember, you can add and subtract only similar terms.)

Here is the completed problem:

$$
\begin{array}{r@{\;}l}
 & 2x^2 + 10x + 47 \\
x - 5 \big) & 2x^3 + 0x^2 - 3x + 2 \\
 & - \quad + \\
 & \cancel{+}2x^3 \cancel{-} 10x^2 \\
\hline
 & + 10x^2 - 3x \\
 & - \quad + \\
 & \cancel{+} 10x^2 \cancel{-} 50x \\
\hline
 & + 47x + 2 \\
 & - \quad + \\
 & + 47x \cancel{-} 235 \\
\hline
 & 237
\end{array}
$$

Our answer is $\dfrac{2x^3 - 3x + 2}{x - 5} = 2x^2 + 10x + 47 + \dfrac{237}{x - 5}$. ▲

As you can see, long division with polynomials is a mechanical process. Once you have done it correctly a couple of times, it becomes very easy to produce the correct answer.

Problem Set 4.8

Divide.

1. $\dfrac{x^2 - 5x + 6}{x - 3}$
2. $\dfrac{x^2 - 5x + 6}{x - 2}$
3. $\dfrac{a^2 + 9a + 20}{a + 5}$
4. $\dfrac{a^2 + 9a + 20}{a + 4}$

5. $\dfrac{x^2 - 6x + 9}{x - 3}$

6. $\dfrac{x^2 + 10x + 25}{x + 5}$

7. $\dfrac{2x^2 + 5x - 3}{2x - 1}$

8. $\dfrac{4x^2 + 4x - 3}{2x - 1}$

9. $\dfrac{2a^2 - 9a - 5}{2a + 1}$

10. $\dfrac{4a^2 - 8a - 5}{2a + 1}$

11. $\dfrac{x^2 + 5x + 8}{x + 3}$

12. $\dfrac{x^2 + 5x + 4}{x + 3}$

13. $\dfrac{a^2 + 3a + 2}{a + 5}$

14. $\dfrac{a^2 + 4a + 3}{a + 5}$

15. $\dfrac{x^2 + 2x + 1}{x - 2}$

16. $\dfrac{x^2 + 6x + 9}{x - 3}$

17. $\dfrac{x^2 + 5x - 6}{x + 1}$

18. $\dfrac{x^2 - x - 6}{x + 1}$

19. $\dfrac{a^2 + 3a + 1}{a + 2}$

20. $\dfrac{a^2 - a + 3}{a + 1}$

21. $\dfrac{2x^2 - 2x + 5}{2x + 4}$

22. $\dfrac{15x^2 + 19x - 4}{3x + 8}$

23. $\dfrac{6a^2 + 5a + 1}{2a + 3}$

24. $\dfrac{4a^2 + 4a + 3}{2a + 1}$

25. $\dfrac{6a^3 - 13a^2 - 4a + 15}{3a - 5}$

26. $\dfrac{2a^3 - a^2 + 3a + 2}{2a + 1}$

Fill in the missing terms in the numerator and then use long division to find the quotients (see Example 4).

27. $\dfrac{x^3 + 4x + 5}{x + 1}$

28. $\dfrac{x^3 + 4x^2 - 8}{x + 2}$

29. $\dfrac{x^3 - 1}{x - 1}$

30. $\dfrac{x^3 + 1}{x + 1}$

31. $\dfrac{x^3 - 8}{x - 2}$

32. $\dfrac{x^3 + 27}{x + 3}$

Review Problems The problems that follow review material we covered in Section 3.8.

Use systems of equations to solve the following word problems.

33. The sum of two numbers is 25. One of the numbers is four times the other. Find the numbers.

34. The sum of two numbers is 24. One of the numbers is 3 more than twice the other. Find the numbers.

35. Suppose you have a total of \$1200 invested in two accounts. One of the accounts pays 8% annual interest and the other pays 9% annual interest. If your total interest for the year is \$100, how much money did you invest in each of the accounts?
36. If you invest twice as much money in an account that pays 12% annual interest as you do in an account that pays 11% annual interest, how much do you have in each account if your total interest for a year is \$210?
37. If you have a total of \$160 in \$5 bills and \$10 bills, how many of each type of bill do you have if you have four more \$10 bills than \$5 bills?
38. Suppose you have 20 coins worth a total of \$2.80. If the coins are all nickels and quarters, how many of each type do you have?
39. How many gallons of 20% antifreeze solution and 60% antifreeze solution must be mixed to get 16 gallons of 35% antifreeze solution?
40. A chemist wants to obtain 80 liters of a solution that is 12% hydrochloric acid. How many liters of 10% hydrochloric acid solution and 20% hydrochloric acid solution should he mix to do so?

Chapter 4 Summary and Review

EXPONENTS: DEFINITION AND PROPERTIES [4.1, 4.2]

Integer exponents indicate repeated multiplications.

$a^r \cdot a^s = a^{r+s}$ To multiply with the same base you add exponents.

$\dfrac{a^r}{a^s} = a^{r-s}$ To divide with the same base you subtract exponents.

$(ab)^r = a^r \cdot b^r$ Exponents distribute over multiplication.

$\left(\dfrac{a}{b}\right)^r = \dfrac{a^r}{b^r}$ Exponents distribute over division.

$(a^r)^s = a^{r \cdot s}$ A power of a power is the product of the powers.

$a^{-r} = \dfrac{1}{a^r}$ Negative exponents imply reciprocals.

Examples

1.

(a) $2^3 = 2 \cdot 2 \cdot 2 = 8$

(b) $x^5 \cdot x^3 = x^{5+3} = x^8$

(c) $\dfrac{x^5}{x^3} = x^{5-3} = x^2$

(d) $(3x)^2 = 3^2 \cdot x^2 = 9x^2$

(e) $\left(\dfrac{2}{3}\right)^3 = \dfrac{2^3}{3^3} = \dfrac{8}{27}$

(f) $(x^5)^3 = x^{5 \cdot 3} = x^{15}$

(g) $3^{-2} = \dfrac{1}{3^2} = \dfrac{1}{9}$

MULTIPLICATION OF MONOMIALS [4.3]

To multiply two monomials, multiply coefficients and add exponents.

2. $(5x^2)(3x^4) = 15x^6$

DIVISION OF MONOMIALS [4.3]

To divide two monomials, divide coefficients and subtract exponents.

3. $\dfrac{12x^9}{4x^5} = 3x^4$

ADDING FRACTIONS WITH THE SAME DENOMINATORS [4.3]

We add fractions that have the same denominators by adding their numerators and then putting the result over that common denominator. The reason why we add fractions in this manner is based on the distributive property.

4. $\dfrac{2}{7} + \dfrac{3}{7} = \dfrac{2+3}{7}$

$= \dfrac{5}{7}$

SCIENTIFIC NOTATION [4.1, 4.2, 4.3]

A number is in scientific notation when it is written as the product of a number between 1 and 10 and an integer power of 10.

5. $768{,}000 = 7.68 \times 10^5$
$0.00039 = 3.9 \times 10^{-4}$

ADDITION OF POLYNOMIALS [4.4]

To add two polynomials, add coefficients of similar terms.

6. $(3x^2 - 2x + 1) + (2x^2 + 7x - 3)$
$= 5x^2 + 5x - 2$

SUBTRACTION OF POLYNOMIALS [4.4]

To subtract one polynomial from another, add the opposite of the second to the first.

7. $(3x + 5) - (4x - 3)$
$= 3x + 5 - 4x + 3$
$= -x + 8$

MULTIPLICATION OF POLYNOMIALS [4.5]

To multiply a polynomial by a monomial, we apply the distributive property. To multiply two binomials we use the FOIL method. In other situations we use the COLUMN method. Each method accomplishes the same result: To multiply any two polynomials, we multiply each term in the first polynomial by each term in the second polynomial.

8.

a. $2a^2(5a^2 + 3a - 2)$
$= 10a^4 + 6a^3 - 4a^2$

b. $(x + 2)(3x - 1)$
$= 3x^2 - x + 6x - 2$
$= 3x^2 + 5x - 2$

c.
$$\begin{array}{rrrr} & x^2 & -\ 5x & +\ 2 \\ & & 2x & -\ 3 \\ \hline & -\ 3x^2 & +\ 15x & -\ 6 \\ 2x^3 & -\ 10x^2 & +\ 4x & \\ \hline 2x^3 & -\ 13x^2 & +\ 19x & -\ 6 \end{array}$$

SPECIAL PRODUCTS [4.6]

$$\left.\begin{array}{l}(a + b)^2 = a^2 + 2ab + b^2 \\ (a - b)^2 = a^2 - 2ab + b^2\end{array}\right\}$$ Binomial squares

$(a + b)(a - b) = a^2 - b^2$ Difference of two squares

9. $(x + 3)^2 = x^2 + 6x + 9$
$(x - 3)^2 = x^2 - 6x + 9$
$(x + 3)(x - 3) = x^2 - 9$

DIVIDING A POLYNOMIAL BY A MONOMIAL [4.7]

To divide a polynomial by a monomial, divide each term in the polynomial by the monomial.

10. $\dfrac{12x^3 - 18x^2}{6x}$

$= 2x^2 - 3x$

LONG DIVISION WITH POLYNOMIALS [4.8]

Division with polynomials is similar to long division with whole numbers. The steps in the process are: estimate, multiply, subtract, and bring down the next term. The divisors in all the long division problems in this chapter were binomials.

11.
$$\begin{array}{r} x - 2 \\ x - 3 \overline{) \; x^2 - 5x + 8} \\ \underline{\not{+}x^2 \not{-} 3x} \\ -2x + 8 \\ \underline{\not{-}2x \not{+} 6} \\ 2 \end{array}$$

COMMON MISTAKES

1. If a term contains a variable that is raised to a power, then the exponent on the variable is associated only with that variable, unless there are parentheses. That is, the expression $3x^2$ means $3 \cdot x \cdot x$, not $3x \cdot 3x$. It is a mistake to write $3x^2$ as $9x^2$. The only way to end up with $9x^2$ is to start with $(3x)^2$.

2. It is a mistake to add nonsimilar terms. For example, $2x$ and $3x^2$ are nonsimilar terms, and therefore cannot be combined. That is, $2x + 3x^2 \neq 5x^3$. If you were to substitute 10 for x in the preceding expression, you would see that the two sides are not equal.

3. It is a mistake to distribute exponents over sums and differences. That is, $(a + b)^2 \neq a^2 + b^2$. Convince yourself of this by letting $a = 2$ and $b = 3$ and then simplifying both sides.

4. Another common mistake can occur when dividing a polynomial by a monomial. Here is an example:

$$\frac{x + \not{2}}{\not{2}} = x + 1 \qquad \text{Mistake}$$

The mistake here is in not dividing both terms in the numerator by 2. The correct way to divide $x + 2$ by 2 looks like this:

$$\frac{x + 2}{2} = \frac{x}{2} + \frac{2}{2} \qquad \text{Correct}$$

$$= \frac{x}{2} + 1$$

Chapter 4 Test

Simplify each of the following expressions. [4.1]

1. $(-3)^4$
2. $(\frac{3}{4})^2$
3. $(3x^3)^2(2x^4)^3$

Simplify each expression. Write all answers with positive exponents only. [4.2]

4. 3^{-2}
5. $(3a^4b^2)^0$
6. $\dfrac{a^{-3}}{a^{-5}}$
7. $\dfrac{(x^{-2})^3(x^{-3})^{-5}}{(x^{-4})^{-2}}$
8. Write 0.0278 in scientific notation. [4.2]
9. Write 2.43×10^5 in expanded form. [4.2]

Simplify. Write all answers with positive exponents only. [4.3]

10. $\dfrac{35x^2y^4z}{70x^6y^2z}$
11. $\dfrac{(6a^2b)(9a^3b^2)}{18a^4b^3}$
12. $\dfrac{24x^7}{3x^2} + \dfrac{14x^9}{7x^4}$
13. $\dfrac{(2.4 \times 10^5)(4.5 \times 10^{-2})}{1.2 \times 10^{-6}}$

Combine by adding or subtracting as indicated. [4.3]

14. $\frac{9}{7} + \frac{5}{7}$
15. $8x^5 - 4x^5 + 3x^5$

Add and subtract as indicated. [4.4]

16. $8x^2 - 4x + 6x + 2$
17. $(5x^2 - 3x + 4) - (2x^2 - 7x - 2)$
18. Subtract $3x - 4$ from $6x - 8$. [4.4]
19. Find the value of $2y^2 - 3y - 4$ when y is -2. [4.4]

Multiply. [4.5]

20. $2a^2(3a^2 - 5a + 4)$
21. $(x + 2)(x + 3)$
22. $(4x - 5)(2x + 3)$
23. $(x - 3)(x^2 + 3x + 9)$

Multiply. [4.6]

24. $(x + 5)^2$
25. $(3a - 2b)^2$
26. $(3x - 4y)(3x + 4y)$
27. $(a^2 - 3)(a^2 + 3)$
28. Divide $10x^3 + 15x^2 - 5x$ by $5x$. [4.7]

Divide. [4.8]

29. $\dfrac{8x^2 - 6x - 5}{2x - 3}$
30. $\dfrac{3x^3 - 2x + 1}{x - 3}$

Factoring

In this chapter we will be doing the reverse of some of the things we did in the last chapter. We are going to do what is called *factoring*.

The two concepts from our previous work that are necessary for success in this chapter are (1) the distributive property and (2) multiplication of binomials. The more familiar you are with the idea of the distributive property and the process of multiplying binomials, the more easily you will understand factoring.

Section 5.1 Factoring Integers

The following diagram shows the relationship between multiplication and factoring:

Multiplication

$$\text{Factors} \rightarrow 3 \cdot 4 = 12 \leftarrow \text{Product}$$

Factoring

When we read the problem from left to right, we say the product of 3 and 4 is 12. Or we multiply 3 and 4 to get 12. When we read the problem in the other direction, from right to left, we say we have *factored* 12 into 3 times 4, or 3 and 4 are *factors* of 12.

The number 12 can be factored still further:

$$\begin{aligned} 12 &= 4 \cdot 3 \\ &= 2 \cdot 2 \cdot 3 \\ &= 2^2 \cdot 3 \end{aligned}$$

The numbers 2 and 3 are called *prime factors* of 12, because neither of them can be factored any further.

DEFINITION If a and b represent integers, then a is said to be a *factor* (or divisor) of b if a divides b evenly—that is, if a divides b with no remainder.

DEFINITION A *prime* number is any positive integer larger than 1 whose only positive factors (divisors) are itself and 1.

Here is a list of the first few prime numbers:

Prime numbers = {2, 3, 5, 7, 11, 13, 17, 19, 23, 29, 31, 37, 41, . . .}

DEFINITION An integer greater than 1 that is not prime is said to be *composite*.

The number 15 is not prime since it has factors of 3 and 5. That is,

$$15 = 3 \cdot 5$$

Since 15 is not prime, it is a composite number.

When a number is not prime, we can factor it into the product of prime numbers. To factor a number into the product of primes, we simply factor it until it cannot be factored further.

▼ **Example 1** Factor the number 60 into the product of prime numbers.

Solution We begin by writing 60 as the product of any two positive integers whose product is 60, like 6 and 10:

$$60 = 6 \cdot 10$$

We then factor these numbers:

$$\begin{aligned} 60 &= 6 \cdot 10 \\ &= (2 \cdot 3) \cdot (2 \cdot 5) \\ &= 2 \cdot 2 \cdot 3 \cdot 5 \\ &= 2^2 \cdot 3 \cdot 5 \end{aligned}$$ ▲

Note that it is customary to write the prime factors of a number in order from smallest to largest.

▼ **Example 2** Factor the number 630 into the product of primes.

Solution Let's begin by writing 630 as the product of 63 and 10:

$$\begin{aligned} 630 &= 63 \cdot 10 \\ &= (7 \cdot 9) \cdot (2 \cdot 5) \\ &= 7 \cdot 3 \cdot 3 \cdot 2 \cdot 5 \\ &= 2 \cdot 3^2 \cdot 5 \cdot 7 \end{aligned}$$

It makes no difference which two numbers we start with, as long as their product is 630. We will always get the same result because a number has only one set of prime factors.

$$\begin{aligned} 630 &= 18 \cdot 35 \\ &= 3 \cdot 6 \cdot 5 \cdot 7 \\ &= 3 \cdot 2 \cdot 3 \cdot 5 \cdot 7 \\ &= 2 \cdot 3^2 \cdot 5 \cdot 7 \end{aligned}$$

▲

Note There are some "tricks" to finding the divisors of a number. For instance, if a number ends in 0 or 5, then it is divisible by 5. If a number ends in an even number (0, 2, 4, 6, or 8) then it is divisible by 2. A number is divisible by 3 if the sum of its digits is divisible by 3. For example, 921 is divisible by 3 because the sum of its digits is $9 + 2 + 1 = 12$ which is divisible by 3.

▼ **Example 3** Factor 525 into the product of primes.

Solution Since 525 ends in 5, it is divisible by 5:

$$\begin{aligned} 525 &= 5 \cdot 105 \\ &= 5 \cdot 5 \cdot 21 \\ &= 5 \cdot 5 \cdot 3 \cdot 7 \\ &= 3 \cdot 5^2 \cdot 7 \end{aligned}$$

▲

When we have factored a number into the product of its prime factors, we not only know what prime numbers divide the original number, we also know all of the other numbers that divide it as well. For instance, if we were to factor 210 into its prime factors, we would have $210 = 2 \cdot 3 \cdot 5 \cdot 7$, which means that 2, 3, 5, and 7 divide 210, as well as any combination of products of 2, 3, 5, and 7. That is, since 3 and 7 divide 210, then so does their product 21. Since 3, 5, and 7 each divide 210, then so does their product 105:

21 divides 210

$$210 = 2 \cdot 3 \cdot 5 \cdot 7$$

105 divides 210

At this point, you may be wondering what this kind of factoring can be used for. Although there are many ways in which factoring is used in arithmetic and algebra, one simple application is in reducing fractions to lowest terms.

Recall that we reduce fractions to lowest terms by dividing the numerator and denominator by the same number. We can use the prime factorization of numbers to help us reduce fractions with large numerators and denominators. Here is an example:

▼ **Example 4** Reduce $\frac{210}{231}$ to lowest terms.

Solution First we factor 210 and 231 into the product of prime factors. Then we reduce to lowest terms by dividing the numerator and denominator by any factors they have in common.

$$\frac{210}{231} = \frac{2 \cdot 3 \cdot 5 \cdot 7}{3 \cdot 7 \cdot 11}$$ Factor the numerator and denominator completely

$$= \frac{2 \cdot \cancel{3} \cdot 5 \cdot \cancel{7}}{\cancel{3} \cdot \cancel{7} \cdot 11}$$ Divide the numerator and denominator by $3 \cdot 7$

$$= \frac{2 \cdot 5}{11}$$

$$= \frac{10}{11}$$ ▲

The small lines we have drawn through the factors that are common to the numerator and denominator are used to indicate that we have divided the numerator and denominator by those factors.

Problem Set 5.1

Label each of the following numbers as *prime* or *composite*. If a number is composite, then list at least two of its factors besides itself and 1.

1. 48 **2.** 72 **3.** 37 **4.** 23
5. 33 **6.** 39 **7.** 221 **8.** 169
9. 29 **10.** 598 **11.** 156 **12.** 171

13. 53 **14.** 420 **15.** 3150 **16.** 1023
17. 1024 **18.** 43 **19.** 21 **20.** 543

Factor the following into the product of primes. When the number has been factored completely, write its prime factors from smallest to largest.

21. 35 **22.** 70 **23.** 128 **24.** 256
25. 144 **26.** 288 **27.** 38 **28.** 63
29. 105 **30.** 210 **31.** 180 **32.** 900
33. 385 **34.** 1925 **35.** 121 **36.** 546
37. 420 **38.** 598 **39.** 620 **40.** 2310

Reduce each fraction to lowest terms by first factoring the numerator and denominator into the product of prime factors, and then dividing out any factors they have in common.

41. $\frac{105}{165}$ **42.** $\frac{165}{385}$ **43.** $\frac{525}{735}$ **44.** $\frac{550}{735}$

45. $\frac{385}{455}$ **46.** $\frac{385}{735}$ **47.** $\frac{322}{345}$ **48.** $\frac{266}{285}$

49. $\frac{205}{369}$ **50.** $\frac{111}{185}$ **51.** $\frac{215}{344}$ **52.** $\frac{279}{310}$

53. Factor 6^3 into the product of prime factors by first factoring 6 and then raising each of its factors to the third power.
54. Factor 12^2 into the product of prime factors by first factoring 12 and then raising each of its factors to the second power.
55. Factor $9^4 \cdot 16^2$ into the product of prime factors by first factoring 9 and 16 completely.
56. Factor $10^2 \cdot 12^3$ into the product of prime factors by first factoring 10 and 12 completely.
57. Simplify the expression $3 \cdot 8 + 3 \cdot 7 + 3 \cdot 5$ and then factor the result into the product of primes. (Notice one of the factors of the answer is 3.)
58. Simplify the expression $5 \cdot 4 + 5 \cdot 9 + 5 \cdot 3$ and then factor the result into the product of primes.

Review Problems The problems below review material we covered in Section 4.5. Reviewing these problems will help you understand the next section.

Multiply.

59. $3(x - 5)$ **60.** $4(x + 1)$
61. $5x^2(x - 3)$ **62.** $5x^2(x + 3)$
63. $6xy(x^2 - 3xy + 2y^2)$ **64.** $3a^2b(1 - 2ab + 3ab^2)$
65. $(x + 2)(y + 3)$ **66.** $(x + 5)(y + 4)$
67. $(x + y)(a + b)$ **68.** $(x - y)(a + b)$
69. $(2x + 3)(a + 4)$ **70.** $(3x - 2)(a + 5)$

Section 5.2 The Greatest Common Factor and Factoring by Grouping

In this section we will apply the distributive property to polynomials to factor from them what is called the greatest common factor.

DEFINITION The *greatest common factor* for a polynomial is the largest monomial that divides (is a factor of) each term of the polynomial.

We use the term *largest monomial* to mean the monomial with the greatest coefficient and highest power of the variable.

▼ **Example 1** Find the greatest common factor for the polynomial

$$3x^5 + 12x^2$$

Solution The terms of the polynomial are $3x^5$ and $12x^2$. The largest number that divides the coefficients is 3, and the highest power of x that is a factor of x^5 and x^2 is x^2. Therefore, the greatest common factor for $3x^5 + 12x^2$ is $3x^2$. That is, $3x^2$ is the largest monomial that divides each term of $3x^5 + 12x^2$. ▲

▼ **Example 2** Find the greatest common factor for

$$8a^3b^2 + 16a^2b^3 + 20a^3b^3$$

Solution The largest number that divides each of the coefficients is 4. The highest power of the variable that is a factor of a^3b^2, a^2b^3, and a^3b^3 is a^2b^2. The greatest common factor for $8a^3b^2 + 16a^2b^3 + 20a^3b^3$ is $4a^2b^2$. It is the largest monomial that is a factor of each term. ▲

Once we have recognized the greatest common factor of a polynomial, we can apply the distributive property and factor it out of each term. We rewrite the polynomial as the product of its greatest common factor with the polynomial that remains after the greatest common factor has been factored from each term.

▼ **Example 3** Factor the greatest common factor from $3x - 15$.

Solution The greatest common factor for the terms $3x$ and 15 is 3. We can rewrite both $3x$ and 15 so that the greatest common factor, 3, is showing in each term. It is important to realize that $3x$ means $3 \cdot x$. The 3 and the x are not "stuck" together.

$$3x - 15 = 3 \cdot x - 3 \cdot 5$$

Now, applying the distributive property we have

$$3 \cdot x - 3 \cdot 5 = 3(x - 5)$$

To check a factoring problem like this, we can multiply 3 and $x - 5$ to get $3x - 15$, which is what we started with. Factoring is simply a procedure by which we change sums and differences into products. In this case, we changed the difference, $3x - 15$, into the product $3(x - 5)$. Note, however, that we have not changed the meaning or value of the expression. The expression we end up with is equal to the expression we started with. ▲

▼ **Example 4** Factor the greatest common factor from

$$5x^3 - 15x^2$$

Solution The greatest common factor is $5x^2$. We rewrite the polynomial as

$$5x^3 - 15x^2 = 5x^2 \cdot x - 5x^2 \cdot 3$$

Then, we apply the distributive property to get

$$5x^2 \cdot x - 5x^2 \cdot 3 = 5x^2(x - 3)$$

To check our work, we simply multiply $5x^2$ and $(x - 3)$ to get $5x^3 - 15x^2$, which is our original polynomial. ▲

▼ **Example 5** Factor the greatest common factor from

$$16x^5 - 20x^4 + 8x^3$$

Solution The greatest common factor is $4x^3$. We rewrite the polynomial so we can see the greatest common factor, $4x^3$, in each term; then we apply the distributive property to factor it out:

$$\begin{aligned} 16x^5 - 20x^4 + 8x^3 &= 4x^3 \cdot 4x^2 - 4x^3 \cdot 5x + 4x^3 \cdot 2 \\ &= 4x^3(4x^2 - 5x + 2) \end{aligned}$$ ▲

▼ **Example 6** Factor the greatest common factor from

$$6x^3y - 18x^2y^2 + 12xy^3$$

Solution The greatest common factor is $6xy$. We rewrite the polynomial in terms of $6xy$ and then apply the distributive property as follows:

$$\begin{aligned} 6x^3y - 18x^2y^2 + 12xy^3 &= 6xy \cdot x^2 - 6xy \cdot 3xy + 6xy \cdot 2y^2 \\ &= 6xy(x^2 - 3xy + 2y^2) \end{aligned}$$ ▲

▼ **Example 7** Factor the greatest common factor from

$$3a^2b - 6a^3b^2 + 9a^3b^3$$

Solution The greatest common factor is $3a^2b$:

$$\begin{aligned} 3a^2b - 6a^3b^2 + 9a^3b^3 &= 3a^2b(1) - 3a^2b(2ab) + 3a^2b(3ab^2) \\ &= 3a^2b(1 - 2ab + 3ab^2) \end{aligned}$$ ▲

COMMON MISTAKES

Notice in Example 7 that we write the first term, $3a^2b$, as $3a^2b(1)$ before we apply the distributive property. It is a very common mistake when first factoring this type of problem to forget the 1 and write

$$3a^2b - 6a^3b^2 + 9a^3b^3 = 3a^2b(-2ab + 3ab^2) \qquad \text{Mistake}$$

The mistake is obvious when we multiply the right side and notice we get something different from what we started with.

We never need to make a mistake with factoring, since we can always multiply our results and compare them with our original polynomial. They must be identical.

Factoring by Grouping

Many polynomials have no greatest common factor other than the number 1. Some of these polynomials can be factored by grouping the terms together in pairs. For example, the polynomial $xy + 3x + 2y + 6$ can be factored if we factor an x from the first two terms and a 2 from the last two terms:

$$xy + 3x + 2y + 6 = x(y + 3) + 2(y + 3)$$

The expression on the right can be thought of as having two terms: $x(y + 3)$ and $2(y + 3)$. Each of these expressions contains the common factor $y + 3$, which can be factored out using the distributive property:

$$x(y + 3) + 2(y + 3) = (y + 3)(x + 2)$$

This last expression is in factored form. The process we used to obtain it is called factoring by grouping. Here are some additional examples:

▼ **Example 8** Factor $ax + bx + ay + by$.

Solution We begin by factoring x from the first two terms and y from the last two terms:

$$\begin{aligned} ax + bx + ay + by &= x(a + b) + y(a + b) \\ &= (a + b)(x + y) \end{aligned}$$

To convince yourself that we have factored correctly, multiply the two factors $(a + b)$ and $(x + y)$. ▲

▼ **Example 9** Factor by grouping: $3ax - 2a + 15x - 10$.

Solution First, we factor a from the first two terms and 5 from the last two terms. Then, we factor $3x - 2$ from the remaining two expressions:

$$\begin{aligned} 3ax - 2a + 15x - 10 &= a(3x - 2) + 5(3x - 2) \\ &= (3x - 2)(a + 5) \end{aligned}$$

Again, multiplying $(3x - 2)$ and $(a + 5)$ will convince you that these are the correct factors. ▲

Problem Set 5.2

Factor the following by taking out the greatest common factor.

1. $15x + 25$
2. $14x + 21$
3. $6a + 9$
4. $8a + 10$
5. $4x - 8y$
6. $9x - 12y$
7. $3x^2 - 6x - 9$
8. $2x^2 + 6x + 4$
9. $3a^2 - 3a - 60$
10. $2a^2 - 18a + 28$
11. $24y^2 - 52y + 24$
12. $18y^2 + 48y + 32$
13. $9x^2 - 8x^3$
14. $7x^3 - 4x^2$
15. $13a^2 - 26a^3$
16. $5a^2 - 10a^3$
17. $21x^2y - 28xy^2$
18. $30xy^2 - 25x^2y$
19. $22a^2b^2 - 11ab^2$
20. $15x^3 - 25x^2 + 30x$
21. $7x^3 + 21x^2 - 28x$
22. $16x^4 - 20x^2 - 16x$
23. $121y^4 - 11x^4$
24. $25a^4 - 5b^4$
25. $100x^4 - 50x^3 + 25x^2$
26. $36x^5 + 72x^3 - 81x^2$
27. $8a^2 + 16b^2 + 32c^2$
28. $9a^2 - 18b^2 - 27c^2$
29. $4a^2b - 16ab^2 + 32a^2b^2$
30. $5ab^2 + 10a^2b^2 + 15a^2b$
31. $121a^3b^2 - 22a^2b^3 + 33a^3b^3$
32. $20a^4b^3 - 18a^3b^4 + 22a^4b^4$
33. $12x^2y^3 - 72x^5y^3 - 36x^4y^4$
34. $49xy - 21x^2y^2 + 35x^3y^3$

Factor by grouping.

35. $xy + 5x + 3y + 15$
36. $xy + 2x + 4y + 8$
37. $xy + 6x + 2y + 12$
38. $xy + 2y + 6x + 12$
39. $ab + 7a + 3b + 21$
40. $ab + 3b + 7a + 21$
41. $ax - bx + ay - by$
42. $ax - ay + bx - by$
43. $2ax + 6x + 5a + 15$
44. $6ax + 42x + 2a + 14$
45. $3xb - 4b + 6x - 8$
46. $3xb - 4b + 15x - 20$
47. $x^2 + ax + 2x + 2a$
48. $x^2 + ax + 3x + 3a$
49. $x^2 + ax + bx + ab$
50. $x^2 - ax + bx - ab$

51. The greatest common factor of the binomial $3x + 6$ is 3. The greatest common factor of the binomial $2x + 4$ is 2. What is the greatest common factor of their product, $(3x + 6)(2x + 4)$, when it has been multiplied out?

52. The greatest common factors of the binomials $4x + 2$ and $5x + 10$ are 2 and 5, respectively. What is the greatest common factor of their product, $(4x + 2)(5x + 10)$, when it has been multiplied out?

53. The following factorization is incorrect. Find the mistake and correct the right-hand side:

$$12x^2 + 6x + 3 = 3(4x^2 + 2x)$$

54. Find the mistake in the following factorization and then rewrite the right-hand side correctly:

$$10x^2 + 2x + 6 = 2(5x^2 + 3)$$

55. An arrow is shot straight up in the air with a velocity of 48 feet per second. The equation that gives the height of the arrow t seconds later is

$$h = 48t - 16t^2$$

Rewrite the equation by factoring the greatest common factor from the right side of the equation. Then, find h when t is 3.

56. A bullet is fired straight up in the air with a velocity of 128 feet per second. The equation that gives the height of the bullet at any time t is

$$h = 128t - 16t^2$$

Write the equation again with the right side in factored form. Then, find h when t is 8.

57. If you invest $1,000 in an account with an annual interest rate of r compounded annually, the amount of money you have in the account after one year is

$$A = 1{,}000 + 1{,}000r$$

Write this formula again with the right side in factored form. Then, find the amount of money in this account at the end of one year if the interest rate is 12%.

58. If you invest P dollars in an account with an annual interest rate of 8% compounded annually, then the amount of money in that account after one year is given by the formula

$$A = P + .08P$$

Rewrite this formula with the right side in factored form and then find the amount of money in the account at the end of one year if $500 was the initial investment.

Review Problems The problems below review material we covered in Section 4.5. Reviewing these problems will help you with the next section.

Multiply using the FOIL method.

59. $(x + 3)(x + 4)$

60. $(x - 5)(x + 2)$

61. $(x + 7)(x - 2)$

62. $(x + 7)(x + 2)$

63. $(x - 7)(x + 2)$

64. $(x - 7)(x - 2)$

65. $(x - 3)(x + 2)$

66. $(x + 3)(x - 2)$

67. $(x + 3)(x^2 - 3x + 9)$

68. $(x - 2)(x^2 - 2x + 4)$

69. $(2x + 1)(x^2 + 4x - 3)$

70. $(3x + 2)(x^2 - 2x - 4)$

Section 5.3 Factoring Trinomials

In this section we will factor trinomials in which the coefficient of the squared term is 1. The more familiar we are with multiplication of binomials, the easier factoring trinomials will be.

Recall multiplication of binomials from Chapter 4:

$$(x + 3)(x + 4) = x^2 + 7x + 12$$
$$(x - 5)(x + 2) = x^2 - 3x - 10$$

The first term in the answer is the product of the first terms in each binomial. The last term in the answer is the product of the last terms in each binomial. The middle term in the answer comes from adding the product of the outside terms with the product of the inside terms.

Let's let a and b represent real numbers and look at the product of $(x + a)$ and $(x + b)$:

$$\begin{aligned}(x + a)(x + b) &= x^2 + ax + bx + ab \\ &= x^2 + (a + b)x + ab\end{aligned}$$

The coefficient of the middle term is the sum of a and b. The last term is the product of a and b. Writing this as a factoring problem, we have

$$x^2 + \underset{\text{Sum}}{(a + b)}x + \underset{\text{Product}}{ab} = (x + a)(x + b)$$

To factor a trinomial in which the coefficient of x^2 is 1, we need only find the numbers a and b whose sum is the coefficient of the middle term and whose product is the constant term (last term).

▼ **Example 1** Factor: $x^2 + 8x + 12$.

Solution The coefficient of x^2 is 1. We need two numbers whose sum is 8 and whose product is 12. The numbers are 6 and 2:

$$x^2 + 8x + 12 = (x + 6)(x + 2)$$

We can easily check our work by multiplying $(x + 6)$ and $(x + 2)$.

Check:
$$\begin{aligned}(x + 6)(x + 2) &= x^2 + 6x + 2x + 12 \\ &= x^2 + 8x + 12\end{aligned}$$ ▲

▼ **Example 2** Factor: $x^2 - 2x - 15$.

Solution The coefficient of x^2 is again 1. We need to find a pair of numbers whose sum is -2 and whose product is -15. Here are all the possibilities for products that are -15:

Products	Sums
$-1(15) = -15$	$-1 + 15 = 14$
$1(-15) = -15$	$1 + (-15) = -14$
$-5(3) = -15$	$-5 + 3 = -2$
$5(-3) = -15$	$5 + (-3) = 2$

The third line gives us what we want. The factors of $x^2 - 2x - 15$ are $(x - 5)$ and $(x + 3)$:

$$x^2 - 2x - 15 = (x - 5)(x + 3)$$ ▲

▼ **Example 3** Factor: $2x^2 + 10x - 28$.

Solution The coefficient of x^2 is 2. We begin by factoring out the greatest common factor, which is 2:

$$2x^2 + 10x - 28 = 2(x^2 + 5x - 14)$$

Now we factor the remaining trinomial by finding a pair of numbers whose sum is 5 and whose product is -14. Here are the possibilities:

Products	Sums
$-1(14) = -14$	$-1 + 14 = 13$
$1(-14) = -14$	$1 + (-14) = -13$
$-7(2) = -14$	$-7 + 2 = -5$
$7(-2) = -14$	$7 + (-2) = 5$

From the last line, we see that the factors of $x^2 + 5x - 14$ are $(x + 7)$ and $(x - 2)$. Here is the complete problem:

$$\begin{aligned} 2x^2 + 10x - 28 &= 2(x^2 + 5x - 14) \\ &= 2(x + 7)(x - 2) \end{aligned}$$ ▲

Note In Example 3, we began by factoring out the greatest common factor. The first step in factoring any trinomial is to look for the greatest common factor. If the trinomial in question has a greatest common factor other than 1, we factor it out first and then try to factor the trinomial that remains.

▼ **Example 4** Factor: $3x^3 - 3x^2 - 18x$.

Solution We begin by factoring out the greatest common factor, which is $3x$. Then, we factor the remaining trinomial. Without showing the table of products and sums as we did in Examples 2 and 3, here is the complete problem:

$$3x^3 - 3x^2 - 18x = 3x(x^2 - x - 6)$$
$$= 3x(x - 3)(x + 2)$$

▲

▼ **Example 5** Factor: $x^2 + 8xy + 12y^2$.

Solution This time we need two expressions whose product is $12y^2$ and whose sum is $8y$. The two expressions are $6y$ and $2y$ (see Example 1 in this section):

$$x^2 + 8xy + 12y^2 = (x + 6y)(x + 2y)$$

You should convince yourself that these factors are correct by finding their product. ▲

▼ **Example 6** A ball is tossed into the air with an upward velocity of 16 feet per second from the top of a building 32 feet high. The equation that gives the height of the ball above the ground at any time t is

$$h = 32 + 16t - 16t^2$$

Factor the right side of this equation and then find h when t is 2.

Solution We begin by factoring out the greatest common factor, 16. Then, we factor the trinomial that remains:

$$h = 32 + 16t - 16t^2$$
$$h = 16(2 + t - t^2)$$
$$h = 16(2 - t)(1 + t)$$

Letting $t = 2$ in the equation, we have

$$h = 16(2 - 2)(1 + 2)$$
$$= 16(0)(3)$$
$$= 0$$

▲

When t is 2, h is 0.

Problem Set 5.3

Factor the following trinomials.

1. $x^2 + 7x + 12$
2. $x^2 + 7x + 10$
3. $x^2 + 3x + 2$
4. $x^2 + 7x + 6$
5. $a^2 + 10a + 21$
6. $a^2 - 7a + 12$
7. $x^2 - 7x + 10$
8. $x^2 - 3x + 2$
9. $y^2 - 10y + 21$
10. $y^2 - 7y + 6$
11. $x^2 - x - 12$
12. $x^2 - 4x - 5$

13. $y^2 + y - 12$
14. $y^2 + 3y - 18$
15. $x^2 + 5x - 14$
16. $x^2 - 5x - 24$
17. $r^2 - 8r - 9$
18. $r^2 - r - 2$
19. $x^2 - x - 30$
20. $x^2 + 8x + 12$
21. $a^2 + 15a + 56$
22. $a^2 - 9a + 20$
23. $y^2 - y - 42$
24. $y^2 + y - 42$
25. $x^2 + 13x + 42$
26. $x^2 - 13x + 42$

Factor the following problems completely. First, factor out the greatest common factor, then factor the remaining trinomial.

27. $2x^2 + 6x + 4$
28. $3x^2 - 6x - 9$
29. $3a^2 - 3a - 60$
30. $2a^2 - 18a + 28$
31. $100x^2 - 500x + 600$
32. $100x^2 - 900x + 2000$
33. $100p^2 - 1300p + 4000$
34. $100p^2 - 1200p + 3200$
35. $x^4 - x^3 - 12x^2$
36. $x^4 - 11x^3 + 24x^2$
37. $2r^3 + 4r^2 - 30r$
38. $5r^3 + 45r^2 + 100r$
39. $2y^4 - 6y^3 - 8y^2$
40. $3r^3 - 3r^2 - 6r$
41. $x^5 + 4x^4 + 4x^3$
42. $x^5 + 13x^4 + 42x^3$
43. $3y^4 - 12y^3 - 15y^2$
44. $5y^4 - 10y^3 + 5y^2$
45. $4x^4 - 52x^3 + 144x^2$
46. $3x^3 - 3x^2 - 18x$

Factor the following trinomials.

47. $x^2 + 5xy + 6y^2$
48. $x^2 - 5xy + 6y^2$
49. $x^2 - 9xy + 20y^2$
50. $x^2 + 9xy + 20y^2$
51. $a^2 + 2ab - 8b^2$
52. $a^2 - 2ab - 8b^2$
53. $a^2 - 10ab + 25b^2$
54. $a^2 + 6ab + 9b^2$
55. $a^2 + 10ab + 25b^2$
56. $a^2 - 6ab + 9b^2$
57. $x^2 + 2xa - 48a^2$
58. $x^2 - 3xa - 10a^2$
59. $x^2 - 5xb - 36b^2$
60. $x^2 - 13xb + 36b^2$

61. If one of the factors of $x^2 + 24x + 128$ is $x + 8$, what is the other factor?
62. If one factor of $x^2 + 260x + 2500$ is $x + 10$, what is the other factor?
63. What polynomial, when factored, gives $(4x + 3)(x - 1)$?
64. What polynomial factors to $(4x - 3)(x + 1)$?
65. If an arrow is shot into the air with a velocity of 48 feet per second from the top of a building 64 feet high, the equation that gives the height of the arrow at any time t is

$$h = 64 + 48t - 16t^2$$

Factor the right side of this equation and then find h when t is 4.

66. A bullet is fired into the air with an initial upward velocity of 64 feet per second from the top of a building 80 feet high. The equation that gives the height of the bullet at any time t is

$$h = 80 + 64t - 16t^2$$

Factor the right side of this equation and then find h when t is 5.

67. A company can manufacture x hundred items for a total cost of $C = 800 + 700x - 100x^2$. Factor the right side of this equation completely and then find C when x is 3 and when x is 8.

68. A company can manufacture x hundred items for a total cost of $C = 700 + 600x - 100x^2$. Factor the right side of this equation completely and then find C when x is 3 and when x is 7.

Review Problems The problems below review material we covered in Sections 4.4 and 4.5. Reviewing the problems from Section 4.5 will help you with the next section.

Multiply, using the FOIL method. [4.5]

69. $(6a + 1)(a + 2)$

70. $(6a - 1)(a - 2)$

71. $(3a + 2)(2a + 1)$

72. $(3a - 2)(2a - 1)$

73. $(6a + 2)(a + 1)$

74. $(3a + 1)(2a + 2)$

Subtract. [4.4]

75. $(5x^2 + 5x - 4) - (3x^2 - 2x + 7)$

76. $(7x^4 - 4x^2 - 5) - (2x^4 - 4x^2 + 5)$

77. Subtract $4x - 5$ from $7x + 3$.

78. Subtract $3x + 2$ from $-6x + 1$.

79. Subtract $2x^2 - 4x$ from $5x^2 - 5$.

80. Subtract $6x^2 + 3$ from $2x^2 - 4x$.

Section 5.4 More Trinomials to Factor

We will now consider trinomials whose greatest common factor is 1 and whose leading coefficient (the coefficient of the squared term) is a number other than one.

Suppose we want to factor the trinomial $2x^2 - 5x - 3$. We know the factors will be a pair of binomials. The product of their first terms is $2x^2$ and the product of their last terms is -3. Let us list all the possible factors along with the trinomial that would result if we were to multiply them together. Remember, the middle term comes from the product of the inside terms plus the product of the outside terms:

Binomial Factors	First Term	Middle Term	Last Term
$(2x - 3)(x + 1)$	$2x^2$	$-x$	-3
$(2x + 3)(x - 1)$	$2x^2$	$+x$	-3
$(2x - 1)(x + 3)$	$2x^2$	$+5x$	-3
$(2x + 1)(x - 3)$	$2x^2$	$-5x$	-3

We can see from the last line that the factors of $2x^2 - 5x - 3$ are $(2x + 1)(x - 3)$. There is no straightforward way, as there was in the last section, in which to find the factors other than by trial and error or by simply listing all the possibilities. We look for possible factors that, when multiplied, will give the correct first and last terms, and then we see if we can adjust them to give the correct middle term.

▼ **Example 1** Factor: $6a^2 + 7a + 2$.

Solution We list all the possible pairs of factors that, when multiplied together, give a trinomial whose first term is $6a^2$ and whose last term is $+2$:

Binomial Factors	First Term	Middle Term	Last Term
$(6a + 1)(a + 2)$	$6a^2$	$+13a$	$+2$
$(6a - 1)(a - 2)$	$6a^2$	$-13a$	$+2$
$(3a + 2)(2a + 1)$	$6a^2$	$+7a$	$+2$
$(3a - 2)(2a - 1)$	$6a^2$	$-7a$	$+2$

The factors of $6a^2 + 7a + 2$ are $(3a + 2)$ and $(2a + 1)$.

Check: $$(3a + 2)(2a + 1) = 6a^2 + 7a + 2$$ ▲

Notice in the preceding list that we did not include the factors $(6a + 2)$ and $(a + 1)$. We do not need to try these, since the first factor has a 2 common to each term and so could be factored again, giving $2(3a + 1)(a + 1)$. Since our original trinomial, $6a^2 + 7a + 2$, did *not* have a greatest common factor of 2, neither of its factors will.

▼ **Example 2** Factor: $4x^2 - x - 3$.

Solution We list all the possible factors that, when multiplied, give a trinomial whose first term is $4x^2$ and whose last term is -3:

Binomial Factors	First Term	Middle Term	Last Term
$(4x + 1)(x - 3)$	$4x^2$	$-11x$	-3
$(4x - 1)(x + 3)$	$4x^2$	$+11x$	-3
$(4x + 3)(x - 1)$	$4x^2$	$-x$	-3
$(4x - 3)(x + 1)$	$4x^2$	$+x$	-3
$(2x + 1)(2x - 3)$	$4x^2$	$-4x$	-3
$(2x - 1)(2x + 3)$	$4x^2$	$+4x$	-3

The third line shows that the factors are $(4x + 3)$ and $(x - 1)$.

Check: $$(4x + 3)(x - 1) = 4x^2 - x - 3.$$ ▲

You will find that the more practice you have at factoring this type of trinomial, the faster you will get the correct factors. You will pick up some shortcuts along the way or maybe come across a system of eliminating some factors as possibilities. Whatever works best for you is the method you should use. Factoring is a very important tool and you must become good at it.

▼ **Example 3** Factor: $12y^3 + 10y^2 - 12y$.

Solution We begin by factoring out the greatest common factor, $2y$:

$$12y^3 + 10y^2 - 12y = 2y(6y^2 + 5y - 6)$$

We now list all possible factors of a trinomial with the first term $6y^2$ and last term -6, along with the associated middle terms:

Possible Factors	Middle Term when Multiplied
$(3y + 2)(2y - 3)$	$-5y$
$(3y - 2)(2y + 3)$	$+5y$
$(6y + 1)(y - 6)$	$-35y$
$(6y - 1)(y + 6)$	$+35y$

The second line gives the correct factors. The complete problem is

$$\begin{aligned} 12y^3 + 10y^2 - 12y &= 2y(6y^2 + 5y - 6) \\ &= 2y(3y - 2)(2y + 3) \end{aligned}$$ ▲

▼ **Example 4** Factor: $30x^2y - 5xy^2 - 10y^3$.

Solution The greatest common factor is $5y$:

$$\begin{aligned} 30x^2y - 5xy^2 - 10y^3 &= 5y(6x^2 - xy - 2y^2) \\ &= 5y(2x + y)(3x - 2y) \end{aligned}$$ ▲

Problem Set 5.4

Factor the following trinomials.

1. $2x^2 + 7x + 3$

2. $2x^2 + 5x + 3$

3. $2a^2 - a - 3$

4. $2a^2 + a - 3$

5. $3x^2 + 2x - 5$

6. $3x^2 - 2x - 5$

7. $3y^2 - 14y - 5$
8. $3y^2 + 14y - 5$
9. $6x^2 + 13x + 6$
10. $6x^2 - 13x + 6$
11. $4x^2 - 12xy + 9y^2$
12. $4x^2 + 12xy + 9y^2$
13. $4y^2 - 11y - 3$
14. $4y^2 + y - 3$
15. $20x^2 - 41x + 20$
16. $20x^2 + 9x - 20$
17. $20a^2 + 48ab - 5b^2$
18. $20a^2 + 29ab + 5b^2$
19. $20x^2 - 21x - 5$
20. $20x^2 - 48x - 5$
21. $12m^2 + 16m - 3$
22. $12m^2 + 20m + 3$
23. $20x^2 + 37x + 15$
24. $20x^2 + 13x - 15$
25. $12a^2 - 25ab + 12b^2$
26. $12a^2 + 7ab - 12b^2$
27. $3x^2 - xy - 14y^2$
28. $3x^2 + 19xy - 14y^2$
29. $14x^2 + 29x - 15$
30. $14x^2 + 11x - 15$
31. $6x^2 - 43x + 55$
32. $6x^2 - 7x - 55$
33. $15t^2 - 67t + 38$
34. $15t^2 - 79t - 34$

Factor each of the following completely. Look first for the greatest common factor.

35. $4x^2 + 2x - 6$
36. $6x^2 - 51x + 63$
37. $24a^2 - 50a + 24$
38. $18a^2 + 48a + 32$
39. $10x^3 - 23x^2 + 12x$
40. $10x^4 + 7x^3 - 12x^2$
41. $6x^4 - 11x^3 - 10x^2$
42. $6x^3 + 19x^2 + 10x$
43. $10a^3 - 6a^2 - 4a$
44. $6a^3 + 15a^2 + 9a$
45. $15x^3 - 102x^2 - 21x$
46. $2x^4 - 24x^3 + 64x^2$
47. $35y^3 - 60y^2 - 20y$
48. $14y^4 - 32y^3 + 8y^2$
49. $15a^4 - 2a^3 - a^2$
50. $10a^5 - 17a^4 + 3a^3$
51. $24x^2y - 6xy - 45y$
52. $8x^2y^2 + 26xy^2 + 15y^2$
53. $12x^2y - 34xy^2 + 14y^3$
54. $12x^2y - 46xy^2 + 14y^3$
55. Evaluate the expression $2x^2 + 7x + 3$ and the expression $(2x + 1)(x + 3)$ for $x = 2$.
56. Evaluate the expression $2a^2 - a - 3$ and the expression $(2a - 3)(a + 1)$ for $a = 5$.
57. What polynomial factors to $(2x + 3)(2x - 3)$?
58. What polynomial factors to $(5x + 4)(5x - 4)$?
59. What polynomial factors to $(x + 3)(x - 3)(x^2 + 9)$?
60. What polynomial factors to $(x + 2)(x - 2)(x^2 + 4)$?

Review Problems The following problems review material we covered in Sections 4.6 and 4.7. Reviewing the problems from Section 4.6 will help you understand the next section.

Multiply. [4.6]

61. $(x + 3)(x - 3)$
62. $(x + 5)(x - 5)$
63. $(6a + 1)(6a - 1)$
64. $(4a + 5)(4a - 5)$
65. $(x + 4)^2$
66. $(x - 5)^2$
67. $(2x + 3)^2$
68. $(2x - 3)^2$

Divide. [4.7]

69. $\dfrac{12x^6 - 18x^4 + 24x^3}{6x^2}$

70. $\dfrac{15x^5 + 10x^4 - 25x^3}{5x^3}$

71. $\dfrac{a^3b^2 - a^4b^5}{a^2b^2}$

72. $\dfrac{4a^4b^3 - 8a^3b^4}{2a^2b^2}$

Section 5.5 The Difference of Two Squares

In Chapter 4 we listed the following three special products:

$$\begin{aligned}(a + b)^2 &= (a + b)(a + b) = a^2 + 2ab + b^2\\ (a - b)^2 &= (a - b)(a - b) = a^2 - 2ab + b^2\\ &\ (a + b)(a - b) = a^2 - b^2\end{aligned}$$

Since factoring is the reverse of multiplication, we can also consider the three special products as three special factorizations:

$$\begin{aligned}a^2 + 2ab + b^2 &= (a + b)^2\\ a^2 - 2ab + b^2 &= (a - b)^2\\ a^2 - b^2 &= (a + b)(a - b)\end{aligned}$$

▼ **Example 1** Factor: $16x^2 - 25$.

Solution We can see that the first term is a perfect square and the last term is also. This fact becomes even more obvious if we rewrite the problem as

$$16x^2 - 25 = (4x)^2 - (5)^2$$

The first term is the square of the quantity $4x$ and the last term is the square of 5. The completed problem looks like this:

$$\begin{aligned}16x^2 - 25 &= (4x)^2 - (5)^2\\ &= (4x + 5)(4x - 5)\end{aligned}$$

To check our results, we multiply:

$$\begin{aligned}(4x + 5)(4x - 5) &= 16x^2 + 20x - 20x - 25\\ &= 16x^2 - 25\end{aligned}$$

▲

▼ **Example 2** Factor: $36a^2 - 1$.

Solution We rewrite the two terms to show they are perfect squares and then factor. Remember, 1 is its own square, $1^2 = 1$.

$$\begin{aligned}36a^2 - 1 &= (6a)^2 - (1)^2\\ &= (6a + 1)(6a - 1)\end{aligned}$$

To check our results, we multiply:

$$\begin{aligned}(6a + 1)(6a - 1) &= 36a^2 + 6a - 6a - 1\\ &= 36a^2 - 1\end{aligned}$$

▲

▼ **Example 3** Factor: $x^4 - y^4$.

Solution We can see that x^4 is the perfect square $(x^2)^2$, and y^4 is $(y^2)^2$.

$$\begin{aligned}x^4 - y^4 &= (x^2)^2 - (y^2)^2\\ &= (x^2 - y^2)(x^2 + y^2)\end{aligned}$$

The factor $(x^2 - y^2)$ is itself the difference of two squares and therefore can be factored again. The factor $(x^2 + y^2)$ is the *sum* of two squares and cannot be factored again. The complete problem looks like this:

$$\begin{aligned}x^4 - y^4 &= (x^2)^2 - (y^2)^2\\ &= (x^2 - y^2)(x^2 + y^2)\\ &= (x + y)(x - y)(x^2 + y^2)\end{aligned}$$

▲

Note If you think the sum of two squares, $x^2 + y^2$, factors, you should try it. Write down the factors you think it has and then multiply them using the FOIL method. You won't get $x^2 + y^2$.

▼ **Example 4** Factor: $25x^2 - 60x + 36$.

Solution Although this trinomial can be factored by the method we used in Section 5.4, we notice that the first and last terms are the perfect squares $(5x)^2$ and $(6)^2$. Before going through the method for factoring trinomials by listing all possible factors, we can check to see if $25x^2 - 60x + 36$ factors to $(5x - 6)^2$. We need only multiply to check:

$$\begin{aligned}(5x - 6)^2 &= (5x - 6)(5x - 6)\\ &= 25x^2 - 30x - 30x + 36\\ &= 25x^2 - 60x + 36\end{aligned}$$

The trinomial $25x^2 - 60x + 36$ factors to $(5x - 6)(5x - 6) = (5x - 6)^2$.

▲

▼ **Example 5** Factor: $m^2 + 14m + 49$.

Solution Since the first and last terms are perfect squares, we can try the factors $(m + 7)(m + 7)$:

$$\begin{aligned}(m + 7)^2 &= (m + 7)(m + 7)\\ &= m^2 + 7m + 7m + 49\\ &= m^2 + 14m + 49\end{aligned}$$

The factors of $m^2 + 14m + 49$ are $(m + 7)(m + 7) = (m + 7)^2$.

▲

Note As we have indicated before, perfect square trinomials like the ones in Examples 4 and 5 can be factored by the methods developed in previous sections. Recognizing that they factor to binomial squares simply saves time in factoring.

▼ **Example 6** Factor: $5x^2 + 30x + 45$.

Solution We begin by factoring out the greatest common factor, which is 5. Then, we notice that the trinomial that remains is a perfect square trinomial:

$$\begin{aligned} 5x^2 + 30x + 45 &= 5(x^2 + 6x + 9) \\ &= 5(x + 3)^2 \end{aligned}$$

▲

Problem Set 5.5

Factor the following.

1. $x^2 - 9$
2. $x^2 - 25$
3. $a^2 - 36$
4. $a^2 - 64$
5. $x^2 - 49$
6. $x^2 - 121$
7. $4a^2 - 16$
8. $4a^2 + 16$
9. $9x^2 + 25$
10. $16x^2 - 36$
11. $25x^2 - 169$
12. $x^2 - y^2$
13. $9a^2 - 16b^2$
14. $49a^2 - 25b^2$
15. $9 - m^2$
16. $16 - m^2$
17. $25 - 4x^2$
18. $36 - 49y^2$
19. $2x^2 - 18$
20. $3x^2 - 27$
21. $32a^2 - 128$
22. $3a^3 - 48a$
23. $8x^2y - 18y$
24. $50a^2b - 72b$
25. $a^4 - b^4$
26. $a^4 - 16$
27. $16m^4 - 81$
28. $81 - m^4$
29. $3x^3y - 75xy^3$
30. $2xy^3 - 8x^3y$

Factor the following.

31. $x^2 - 2x + 1$
32. $x^2 - 6x + 9$
33. $x^2 + 2x + 1$
34. $x^2 + 6x + 9$
35. $a^2 - 10a + 25$
36. $a^2 + 10a + 25$
37. $y^2 + 4y + 4$
38. $y^2 - 8y + 16$
39. $x^2 - 4x + 4$
40. $x^2 + 8x + 16$
41. $m^2 - 12m + 36$
42. $m^2 + 12m + 36$
43. $4a^2 + 12a + 9$
44. $9a^2 - 12a + 4$
45. $49x^2 - 14x + 1$
46. $64x^2 - 16x + 1$
47. $9y^2 - 30y + 25$
48. $25y^2 + 30y + 9$
49. $x^2 + 10xy + 25y^2$
50. $25x^2 + 10xy + y^2$
51. $9a^2 + 6ab + b^2$
52. $9a^2 - 6ab + b^2$

Factor the following by first factoring out the greatest common factor.

53. $3a^2 + 18a + 27$

54. $4a^2 - 16a + 16$

55. $2x^2 + 20xy + 50y^2$

56. $3x^2 + 30xy + 75y^2$

57. $5x^3 + 30x^2y + 45xy^2$

58. $12x^2y - 36xy^2 + 27y^3$

59. Find a value for b so that the polynomial $x^2 + bx + 49$ factors to $(x + 7)^2$.

60. Find a value of b so that the polynomial $x^2 + bx + 81$ factors to $(x + 9)^2$.

61. Find the value of c for which the polynomial $x^2 + 10x + c$ factors to $(x + 5)^2$.

62. Find the value of a for which the polynomial $ax^2 + 12x + 9$ factors to $(2x + 3)^2$.

63. If \$1,000 is invested in an account with interest rate r compounded annually, the amount of money in that account at the end of two years is given by the formula

$$A = 1{,}000(1 + 2r + r^2)$$

Rewrite the right side of this formula in factored form and then find the amount of money in this account if the interest rate is 12%.

64. If \$600 is invested in an account with an annual interest rate r that is compounded annually, the amount of money in that account at the end of two years is given by the formula

$$A = 600 + 1200r + 600r^2$$

Factor the right side of this formula and then find the amount of money in this account if the interest rate is 10%.

Review Problems The following problems review material we covered in Section 4.8.

Use long division to divide.

65. $\dfrac{x^2 - 5x + 8}{x - 3}$

66. $\dfrac{x^2 - 5x + 8}{x - 2}$

67. $\dfrac{x^2 + 5x + 6}{x + 2}$

68. $\dfrac{x^2 + 7x + 12}{x + 4}$

69. $\dfrac{6x^2 + 5x + 3}{2x + 3}$

70. $\dfrac{4x^2 + 8x + 3}{2x + 1}$

71. $\dfrac{x^3 - 8}{x - 2}$

72. $\dfrac{x^3 + 27}{x + 3}$

Section 5.6 Factoring: A General Review

In this section we will review the different methods of factoring that we have presented in the previous sections of this chapter. This section is important because it will give you an opportunity to factor a variety of polynomials. Prior to this section, the polynomials you have worked with have been grouped together according to the method used to factor them. That is, in Section 5.5, all the polynomials you factored were either the difference of two squares or perfect square trinomials. What usually happens in a situation like this is that you will become proficient at factoring the kind of polynomial you are working with at the time, but may have trouble when given a variety of polynomials to factor.

We begin this section by giving a checklist that can be used to factor polynomials of any type. When you have finished this section and the problem set that follows, you want to be proficient enough at factoring so that the checklist is second nature to you.

Checklist for Factoring a Polynomial

Step 1: If the polynomial has a greatest common factor other than 1, then factor out the greatest common factor.

Step 2: If the polynomial has two terms (it is a binomial) then see if it is the difference of two squares. Remember, if it is the sum of two squares, it will not factor.

Step 3: If the polynomial has three terms (a trinomial), then it is either a perfect square trinomial that will factor into the square of a binomial, or it is not a perfect square trinomial, in which case you use the trial and error method developed in Section 5.4.

Step 4: If the polynomial has more than three terms, try to factor it by grouping.

Step 5: As a final check, look and see if any of the factors you have written can be factored further. If you have overlooked a common factor, you can catch it here.

Here are some examples illustrating how we use the checklist. There are no new factoring problems in this section. The problems here are all similar to the problems you have seen before. What is different is that they are all mixed up.

▼ **Example 1** Factor: $2x^5 - 8x^3$.

Solution First, we check to see if the greatest common factor is other than 1. Since the greatest common factor is $2x^3$, we begin by factoring it. Once we have done so, we notice that the binomial that remains is the difference of two squares, which we factor according to the formula

$$a^2 - b^2 = (a + b)(a - b).$$

$$2x^5 - 8x^3 = 2x^3(x^2 - 4)$$ Factor out the greatest common factor $2x^3$

$$= 2x^3(x + 2)(x - 2)$$ Factor the difference of two squares ▲

Note that the greatest common factor, $2x^3$, that we factored from each term in the first step of Example 1 remains as part of the answer to the problem. That is because it is one of the factors of the original binomial. Remember, the expression we end up with when factoring must be equal to the expression we start with. We can't just drop a factor and expect the resulting expression to be equal to the original expression.

▼ **Example 2** Factor: $3x^4 - 18x^3 + 27x^2$.

Solution Step 1 is to factor out the greatest common factor, $3x^2$. After we have done so, we notice that the trinomial that remains is a perfect square trinomial, which will factor as the square of a binomial:

$$3x^4 - 18x^3 + 27x^2 = 3x^2(x^2 - 6x + 9)$$ Factor out $3x^2$

$$= 3x^2(x - 3)^2$$ $x^2 - 6x + 9$ is the square of $x - 3$ ▲

▼ **Example 3** Factor: $y^3 + 25y$.

Solution We begin by factoring out the y that is common to both terms. The binomial that remains after we have done so is the sum of two squares that does not factor. So, after the first step, we are finished:

$$y^3 + 25y = y(y^2 + 25)$$ Factor out the greatest common factor y, then notice that $y^2 + 25$ cannot be factored further ▲

▼ **Example 4** Factor: $6a^2 - 11a + 4$.

Solution Here we have a trinomial that does not have a greatest common factor other than 1. Since it is not a perfect square trinomial, we factor it by trial and error. That is, we look for binomial factors the product of whose first terms is $6a^2$ and the product of whose last terms is 4. Then, we look for the combination of these types of binomials whose product gives us a middle term of $-11a$. Without showing all the different possibilities, here is the answer:

$$6a^2 - 11a + 4 = (3a - 4)(2a - 1)$$ ▲

▼ **Example 5** Factor: $6x^3 - 12x^2 - 48x$.

Solution This trinomial has a greatest common factor of $6x$. The trinomial that remains after the $6x$ has been factored from each term must be factored by trial and error:

$$\begin{aligned} 6x^3 - 12x^2 - 48x &= 6x(x^2 - 2x - 8) \\ &= 6x(x - 4)(x + 2) \end{aligned}$$ ▲

▼ **Example 6** Factor: $2ab^5 + 8ab^4 + 2ab^3$.

Solution The greatest common factor is $2ab^3$. We begin by factoring it from each term. After that, we find that the trinomial that remains cannot be factored further:

$$2ab^5 + 8ab^4 + 2ab^3 = 2ab^3(b^2 + 4b + 1)$$ ▲

▼ **Example 7** Factor: $xy + 8x + 3y + 24$.

Solution Since our polynomial has four terms, we try factoring by grouping:

$$\begin{aligned} xy + 8x + 3y + 24 &= x(y + 8) + 3(y + 8) \\ &= (y + 8)(x + 3) \end{aligned}$$ ▲

Problem Set 5.6

Factor each of the following polynomials completely. That is, once you are finished factoring, none of the factors you obtain should be factorable. Also, note that the even-numbered problems are not necessarily similar to the odd-numbered problems that precede them in this problem set.

1. $x^2 - 81$
2. $x^2 - 18x + 81$
3. $x^2 + 2x - 15$
4. $15x^2 + 11x - 6$
5. $x^2 + 6x + 9$
6. $12x^2 - 11x + 2$
7. $y^2 - 10y + 25$
8. $21y^2 - 25y - 4$
9. $2a^3b + 6a^2b + 2ab$
10. $6a^2 - ab - 15b^2$
11. $x^2 + x + 1$
12. $2x^2 - 4x + 2$
13. $12a^2 - 75$
14. $18a^2 - 50$
15. $9x^2 - 12xy + 4y^2$
16. $x^3 - x^2$
17. $4x^3 + 16xy^2$
18. $16x^2 + 49y^2$
19. $2y^3 + 20y^2 + 50y$
20. $3y^2 - 9y - 30$
21. $a^6 + 4a^4b^2$
22. $5a^2 - 45b^2$
23. $xy + 3x + 4y + 12$
24. $xy + 7x + 6y + 42$
25. $x^4 - 16$
26. $x^4 - 81$
27. $xy - 5x + 2y - 10$
28. $xy - 7x + 3y - 21$

29. $5a^2 + 10ab + 5b^2$
30. $3a^3b^2 + 15a^2b^2 + 3ab^2$
31. $x^2 + 49$
32. $16 - x^4$
33. $3x^2 + 15xy + 18y^2$
34. $3x^2 + 27xy + 54y^2$
35. $2x^2 + 15x - 38$
36. $2x^2 + 7x - 85$
37. $100x^2 - 300x + 200$
38. $100x^2 - 400x + 300$
39. $x^2 - 64$
40. $9x^2 - 4$
41. $x^2 + 3x + ax + 3a$
42. $x^2 + 4x + bx + 4b$
43. $49a^7 - 9a^5$
44. $a^4 - 1$
45. $49x^2 + 9y^2$
46. $12x^4 - 62x^3 + 70x^2$
47. $25a^3 + 20a^2 + 3a$
48. $36a^4 - 100a^2$
49. $xa - xb + ay - by$
50. $xy - bx + ay - ab$
51. $48a^4b - 3a^2b$
52. $18a^4b^2 - 12a^3b^3 + 8a^2b^4$
53. $20x^4 - 45x^2$
54. $16x^3 + 16x^2 + 3x$
55. $3x^2 + 35xy - 82y^2$
56. $3x^2 + 37xy - 86y^2$
57. $16x^5 - 44x^4 + 30x^3$
58. $16x^2 + 16x - 1$
59. $2x^2 + 2ax + 3x + 3a$
60. $2x^2 + 2ax + 5x + 5a$
61. $y^4 - 1$
62. $25y^7 - 16y^5$
63. $12x^4y^2 + 36x^3y^3 + 27x^2y^4$
64. $16x^3y^2 - 4xy^2$

Review Problems The problems that follow review material we covered in Sections 2.3 and 4.1. Reviewing the problems from Section 2.3 will help you understand the next section.

Solve each equation. [2.3]

65. $3x - 6 = 9$
66. $5x - 1 = 14$
67. $2x + 3 = 0$
68. $4x - 5 = 0$
69. $4x + 3 = 0$
70. $3x - 1 = 0$

Simplify, using the properties of exponents. [4.1]

71. $x^8 \cdot x^7$
72. $(x^5)^2$
73. $(3x^3)^2(2x^4)^3$
74. $(5x^2y)^2(4xy^3)^2$
75. Write the number 57,600 in scientific notation.
76. Write the number 4.3×10^5 in expanded form.

Section 5.7 Quadratic Equations

In this section we will use the methods of factoring developed in previous sections, along with a special property of 0, to solve quadratic equations.

DEFINITION Any equation that can be put in the form $ax^2 + bx + c = 0$, where a, b, and c are real numbers ($a \neq 0$), is called a *quadratic equation*; $ax^2 + bx + c = 0$ is called *standard form* for a quadratic equation:

$$\begin{array}{ccccc} \text{an } x^2 \text{ term} & & \text{an } x \text{ term} & & \text{and a constant term} \\ a\,(\text{variable})^2 & + & b\,(\text{variable}) & + & (\text{absence of the variable}) = 0 \end{array}$$

The number zero has a special property. If we multiply two numbers and the product is 0, then one or both of the original two numbers must be 0. In symbols, this property looks like this:

Zero-Factor Property

Let a and b represent real numbers. If $a \cdot b = 0$, then $a = 0$ or $b = 0$.

Suppose we want to solve the quadratic equation $x^2 + 5x + 6 = 0$. We can factor the left side into $(x + 2)(x + 3)$. Then, we have

$$\begin{aligned} x^2 + 5x + 6 &= 0 \\ (x + 2)(x + 3) &= 0 \end{aligned}$$

Now $(x + 2)$ and $(x + 3)$ both represent real numbers. Their product is 0; therefore, either $(x + 3)$ is 0 or $(x + 2)$ is 0. Either way, we have a solution to our equation. We use the property of zero stated above to finish the problem:

$$\begin{aligned} x^2 + 5x + 6 &= 0 \\ (x + 2)(x + 3) &= 0 \end{aligned}$$

$$\begin{array}{rcr} x + 2 = 0 & \text{or} & x + 3 = 0 \\ x = -2 & \text{or} & x = -3 \end{array}$$

Our solution set is $\{-2, -3\}$. Our equation has two solutions. To check our solutions, we have to check each one separately to see that they both produce a true statement when used in place of the variable:

$$\begin{array}{lr} \text{When} & x = -3 \\ \text{the equation} & x^2 + 5x + 6 = 0 \\ \text{becomes} & (-3)^2 + 5(-3) + 6 \stackrel{?}{=} 0 \\ & 9 + (-15) + 6 = 0 \\ & 0 = 0 \end{array}$$

$$\begin{array}{lr} \text{When} & x = -2 \\ \text{the equation} & x^2 + 5x + 6 = 0 \\ \text{becomes} & (-2)^2 + 5(-2) + 6 \stackrel{?}{=} 0 \\ & 4 + (-10) + 6 = 0 \\ & 0 = 0 \end{array}$$

We have solved a quadratic equation by replacing it with two linear equations in one variable.

Steps to Solve a Quadratic Equation by Factoring

Step 1: Put the equation in standard form, that is, 0 on one side and decreasing powers of the variable on the other.
Step 2: Factor completely.
Step 3: Use the zero-factor property to set each variable factor from Step 2 to 0.
Step 4: Solve each equation produced in Step 3.
Step 5: Check each solution, if necessary.

▼ **Example 1** Solve the following equation: $2x^2 - 5x = 12$.

Solution

Step 1: We begin by adding -12 to both sides, so the equation is in standard form:

$$2x^2 - 5x = 12$$
$$2x^2 - 5x - 12 = 0$$

Step 2: We factor the left side completely:

$$(2x + 3)(x - 4) = 0$$

Step 3: We set each factor to 0:

$$2x + 3 = 0 \quad \text{or} \quad x - 4 = 0$$

Step 4: Solve each of the equations from Step 3:

$$\begin{aligned} 2x + 3 &= 0 \\ 2x &= -3 \\ x &= -\frac{3}{2} \end{aligned} \qquad \begin{aligned} x - 4 &= 0 \\ x &= 4 \end{aligned}$$

Step 5: Substitute each solution into $2x^2 - 5x = 12$ to check:

Check: $-\frac{3}{2}$

$$\begin{aligned} 2\left(-\frac{3}{2}\right)^2 - 5\left(-\frac{3}{2}\right) &\stackrel{?}{=} 12 \\ 2\left(\frac{9}{4}\right) + 5\left(\frac{3}{2}\right) &= 12 \\ \frac{9}{2} + \frac{15}{2} &= 12 \\ \frac{24}{2} &= 12 \\ 12 &= 12 \end{aligned}$$

Check: 4

$$\begin{aligned} 2(4)^2 - 5(4) &\stackrel{?}{=} 12 \\ 2(16) - 20 &= 12 \\ 32 - 20 &= 12 \\ 12 &= 12 \end{aligned}$$

▲

▼ **Example 2** Solve for a: $16a^2 - 25 = 0$.

Solution The equation is already in standard form:

$$
\begin{aligned}
16a^2 - 25 &= 0 \\
(4a - 5)(4a + 5) &= 0 && \text{Factor left side} \\
4a - 5 = 0 \quad \text{or} \quad 4a + 5 &= 0 && \text{Set each factor to 0} \\
4a = 5 \qquad\qquad 4a &= -5 && \text{Solve the resulting equations} \\
a = \frac{5}{4} \qquad\qquad a &= -\frac{5}{4}
\end{aligned}
$$

▲

▼ **Example 3** Solve: $4x^2 = 8x$.

Solution We begin by adding $-8x$ to each side of the equation to put it in standard form. Then, we factor the left side of the equation by factoring out the greatest common factor:

$$
\begin{aligned}
4x^2 &= 8x \\
4x^2 - 8x &= 0 && \text{Add } -8x \text{ to each side} \\
4x(x - 2) &= 0 && \text{Factor the left side} \\
4x = 0 \quad \text{or} \quad x - 2 &= 0 && \text{Set each factor to 0} \\
x = 0 \quad \text{or} \qquad x &= 2 && \text{Solve the resulting equations}
\end{aligned}
$$

The solutions are 0 and 2. ▲

▼ **Example 4** Solve: $x(2x + 3) = 44$.

Solution We must multiply out the left side first and then put the equation in standard form:

$$
\begin{aligned}
x(2x + 3) &= 44 \\
2x^2 + 3x &= 44 && \text{Multiply on the left side} \\
2x^2 + 3x - 44 &= 0 && \text{Add } -44 \text{ to each side} \\
(2x + 11)(x - 4) &= 0 && \text{Factor the left side} \\
2x + 11 = 0 \quad \text{or} \quad x - 4 &= 0 && \text{Set each factor to 0} \\
2x = -11 \quad \text{or} \qquad x &= 4 && \text{Solve the resulting equations} \\
x = -\frac{11}{2}
\end{aligned}
$$

The two solutions are $-\frac{11}{2}$ and 4. ▲

▼ **Example 5** Solve for x: $5^2 = x^2 + (x + 1)^2$.

Solution Before we can put this equation in standard form we must square the binomial. Remember, to square a binomial, we use the formula $(a + b)^2 = a^2 + 2ab + b^2$.

$$
\begin{aligned}
5^2 &= x^2 + (x+1)^2 && \\
25 &= x^2 + x^2 + 2x + 1 && \text{Expand } 5^2 \text{ and } (x+1)^2 \\
25 &= 2x^2 + 2x + 1 && \text{Simplify the right side} \\
0 &= 2x^2 + 2x - 24 && \text{Add } -25 \text{ to each side} \\
0 &= 2(x^2 + x - 12) && \text{Begin factoring} \\
0 &= 2(x+4)(x-3) && \text{Factor completely} \\
x + 4 = 0 \quad &\text{or} \quad x - 3 = 0 && \text{Set each variable factor to 0} \\
x = -4 \quad &\text{or} \quad x = 3 &&
\end{aligned}
$$

Note, in the second to the last line, that we do not set 2 equal to 0. That is because 2 can never be 0. It is always 2. We only use the Zero-Factor property to set variable factors to 0 because they are the only factors that can possibly be 0.

Also notice that it makes no difference which side of the equation is 0 when we write the equation in standard form. ▲

Although the equation in the next example is not a quadratic equation, it can be solved by the method shown in the first five examples.

▼ **Example 6** Solve: $24x^3 = -10x^2 + 6x$ for x.

Solution First we write the equation in standard form:

$$
\begin{aligned}
24x^3 + 10x^2 - 6x &= 0 && \text{Standard form} \\
2x(12x^2 + 5x - 3) &= 0 && \text{Factor out } 2x \\
2x(3x - 1)(4x + 3) &= 0 && \text{Factor remaining trinomial}
\end{aligned}
$$

$$2x = 0 \quad \text{or} \quad 3x - 1 = 0 \quad \text{or} \quad 4x + 3 = 0 \qquad \text{Set factors to 0}$$

$$x = 0 \quad \text{or} \quad x = \frac{1}{3} \quad \text{or} \quad x = -\frac{3}{4} \qquad \text{Solutions}$$ ▲

Problem Set 5.7

The following equations are already in factored form. Use the special property with zero to set the factors to zero and solve.

1. $(x + 2)(x - 1) = 0$
2. $(x + 3)(x + 2) = 0$
3. $(a - 4)(a - 5) = 0$
4. $(a + 6)(a - 1) = 0$
5. $x(x + 1)(x - 3) = 0$
6. $x(2x + 1)(x - 5) = 0$
7. $(3x + 2)(2x + 3) = 0$
8. $(4x - 5)(x - 6) = 0$
9. $m(3m + 4)(3m - 4) = 0$
10. $m(2m - 5)(3m - 1) = 0$
11. $2y(3y + 1)(5y + 3) = 0$
12. $3y(2y - 3)(3y - 4) = 0$

Solve the following equations.

13. $x^2 + 3x + 2 = 0$
14. $x^2 - x - 6 = 0$
15. $x^2 - 9x + 20 = 0$
16. $x^2 + 2x - 3 = 0$

17. $a^2 - 2a - 24 = 0$

18. $a^2 - 11a + 30 = 0$

19. $100x^2 - 500x + 600 = 0$

20. $100x^2 - 300x + 200 = 0$

21. $x^2 = -6x - 9$

22. $x^2 = 10x - 25$

23. $a^2 - 16 = 0$

24. $a^2 - 36 = 0$

25. $2x^2 + 5x - 12 = 0$

26. $3x^2 + 14x - 5 = 0$

27. $9x^2 + 12x + 4 = 0$

28. $12x^2 - 24x + 9 = 0$

29. $a^2 + 25 = 10a$

30. $a^2 + 16 = 8a$

31. $2x^2 = 3x + 20$

32. $6x^2 = x + 2$

33. $3m^2 = 20 - 7m$

34. $2m^2 = -18 + 15m$

35. $4x^2 - 49 = 0$

36. $16x^2 - 25 = 0$

37. $x^2 + 6x = 0$

38. $x^2 - 8x = 0$

39. $x^2 - 3x = 0$

40. $x^2 + 5x = 0$

41. $2x^2 = 8x$

42. $2x^2 = 10x$

43. $3x^2 = 15x$

44. $5x^2 = 15x$

45. $1400 = 400 + 700x - 100x^2$

46. $2700 = 700 + 900x - 100x^2$

47. $6x^2 = -5x + 4$

48. $9x^2 = 12x - 4$

49. $x(2x - 3) = 20$

50. $x(3x - 5) = 12$

51. $t(t + 2) = 80$

52. $t(t + 2) = 99$

53. $4000 = (1300 - 100p)p$

54. $3200 = (1200 - 100p)p$

55. $x(14 - x) = 48$

56. $x(12 - x) = 32$

57. $(x + 5)^2 = 2x + 9$

58. $(x + 7)^2 = 2x + 13$

59. $(y - 6)^2 = y - 4$

60. $(y + 4)^2 = y + 6$

61. $10^2 = (x + 2)^2 + x^2$

62. $15^2 = (x + 3)^2 + x^2$

63. $2x^3 + 11x^2 + 12x = 0$

64. $3x^3 + 17x^2 + 10x = 0$

65. $4y^3 - 2y^2 - 30y = 0$

66. $9y^3 + 6y^2 - 24y = 0$

67. $8x^3 + 16x^2 = 10x$

68. $24x^3 - 22x^2 = -4x$

69. $20a^3 = -18a^2 + 18a$

70. $12a^3 = -2a^2 + 10a$

Review Problems The following problems review material we covered in Section 3.8 and 4.2.

The following word problems are taken from the book *Academic Algebra,* written by William J. Milne and published by the American Book Company in 1901. Solve each problem. [3.8]

71. A bicycle and a suit cost \$90. How much did each cost, if the bicycle cost five times as much as the suit?

72. A man bought a cow and a calf for \$36, paying eight times as much for the cow as for the calf. What was the cost of each?

73. A house and a lot cost \$3000. If the house cost four times as much as the lot, what was the cost of each?

74. A plumber and two helpers together earned \$7.50 per day. How much did each earn per day, if the plumber earned four times as much as each helper?

Use the properties of exponents to simplify each expression. [4.2]

75. 2^{-3}

76. 5^{-2}

77. $\dfrac{x^5}{x^{-3}}$

78. $\dfrac{x^{-2}}{x^{-5}}$

79. $\dfrac{(x^2)^3}{(x^{-3})^4}$

80. $\dfrac{(x^2)^{-4}(x^{-2})^3}{(x^{-3})^{-5}}$

81. Write the number 0.0056 in scientific notation.
82. Write the number 2.34×10^{-4} in expanded form.

Section 5.8 Word Problems

We will use the same steps in solving word problems in this section that we have used in the past. The equation that describes the situation will turn out to be a quadratic equation.

Number Problems

▼ **Example 1** The product of two consecutive odd integers is 63. Find the integers.

Solution

Step 1: Let x = the first odd integer, then $x + 2$ = the second odd integer.

Step 2: An equation that describes the situation is

$$x(x + 2) = 63 \qquad \text{(Their product is 63)}$$

Step 3: We solve the equation.

$$\begin{aligned} x(x + 2) &= 63 \\ x^2 + 2x &= 63 \\ x^2 + 2x - 63 &= 0 \\ (x - 7)(x + 9) &= 0 \\ x - 7 = 0 \quad \text{or} \quad x + 9 &= 0 \\ x = 7 \quad \text{or} \quad x &= -9 \end{aligned}$$

If the first odd integer is 7, the next odd integer is $7 + 2 = 9$. If the first odd integer is -9, the next consecutive odd integer is $-9 + 2 = -7$. We have two pairs of consecutive odd integers that are solutions. They are 7, 9 and -9, -7.

Step 4: We check to see that their products are 63:

$$\begin{aligned} 7(9) &= 63 \\ -7(-9) &= 63 \end{aligned}$$

▲

Suppose we know that the sum of two numbers is 50. We want to find a way to represent each number using only one variable. If we let x represent one of the two numbers, how can we represent the other? Let's suppose for a moment that x turns out to be 30. Then, the other number will be 20 because their sum is 50. That is, if two numbers add up to 50 and one of them is 30, then the other must be $50 - 30 = 20$. Generalizing this to any number x, we see that, if two numbers have a sum of 50, and one of the numbers is x, then the other must be $50 - x$. The table that follows shows some additional examples:

If two numbers have a sum of	and one of them is	then the other must be
50	x	$50 - x$
100	x	$100 - x$
10	y	$10 - y$
12	n	$12 - n$

Now, let's look at an example that uses this idea.

▼ **Example 2** The sum of two numbers is 13. Their product is 40. Find the numbers.

Solution If we let x represent one of the numbers, then $13 - x$ must be the other number because their sum is 13. Since their product is 40, we can write

$$
\begin{aligned}
x(13 - x) &= 40 && \text{The product of the two numbers is 40} \\
13x - x^2 &= 40 && \text{Multiply left side} \\
x^2 - 13x &= -40 && \text{Multiply both sides by } -1 \text{ and reverse order of terms of left side} \\
x^2 - 13x + 40 &= 0 && \text{Add 40 to each side} \\
(x - 8)(x - 5) &= 0 && \text{Factor left side}
\end{aligned}
$$

$$
\begin{aligned}
x - 8 = 0 \quad &\text{or} \quad x - 5 = 0 \\
x = 8 \quad & \quad\quad\;\; x = 5
\end{aligned}
$$

The two solutions are 8 and 5. If x is 8, then the other number is $13 - x = 13 - 8 = 5$. Similarly, if x is 5, the other number is $13 - x = 13 - 5 = 8$. Therefore, the two numbers we are looking for are 8 and 5. Their sum is 13 and their product is 40. ▲

Geometry Problems

Many word problems dealing with area can best be described algebraically using quadratic equations.

The area of a rectangle is length times width:

Area = length · width
$A = L \cdot W$

W
Width
L
Length

The area of a triangle is one-half the product of the base and height:

$$\text{Area} = \frac{1}{2}(\text{base} \cdot \text{height})$$

$$A = \frac{1}{2}(b \cdot h)$$

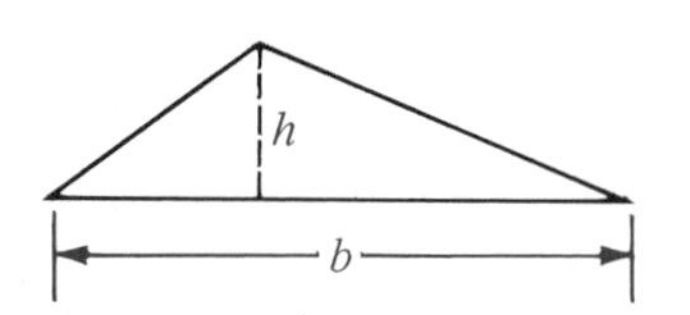

▼ **Example 3** The length of a rectangle is 3 inches more than twice the width. The area is 44 square inches (sq. in.). Find the dimensions (find the length and width).

Solution Let x = the width of the rectangle. From this, $2x + 3$ = the length of the rectangle because the length is three more than twice the width.

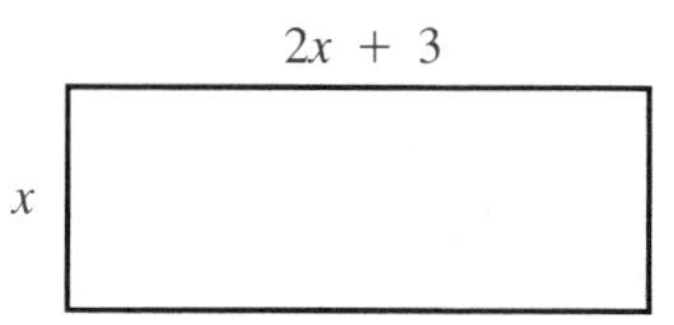

Since the area is 44 sq. in., an equation that describes the situation is

$$x(2x + 3) = 44 \qquad (\text{Length} \cdot \text{width} = \text{area})$$

We now solve the equation:

$$x(2x + 3) = 44$$
$$2x^2 + 3x = 44$$
$$2x^2 + 3x - 44 = 0$$
$$(2x + 11)(x - 4) = 0$$
$$2x + 11 = 0 \quad \text{or} \quad x - 4 = 0$$
$$x = -\frac{11}{2} \quad \text{or} \quad x = 4$$

The solution $x = -\frac{11}{2}$ cannot be used, since length and width are always given in positive units. The width is 4. The length is three more than twice the width or $2(4) + 3 = 11$:

Width = 4 inches
Length = 11 inches

The solutions check in the original problem, since $4(11) = 44$. ▲

Another application of quadratic equations involves the Pythagorean theorem, an important theorem from geometry. This theorem gives the relationship between the sides of any right triangle. A right triangle is a triangle in which one of the angles is 90°. We will state the Pythagorean theorem here, but we will not prove it.

Pythagorean Theorem

In any right triangle, the square of the longest side (called the hypotenuse) is equal to the sum of the squares of the other two sides (called legs).

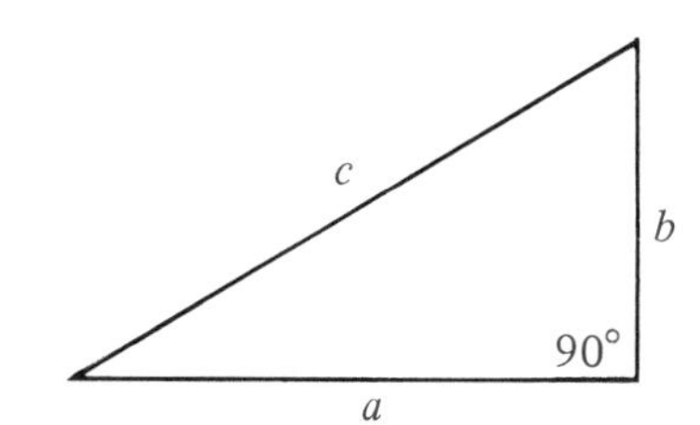

$$c^2 = a^2 + b^2$$

▼ **Example 4** The hypotenuse of a right triangle is 5 inches, while the lengths of the two legs (the other two sides) are given by two consecutive integers. Find the lengths of the two legs.

Solution If we let x = the length of the shortest side, then the other side must be $x + 1$. A diagram of the triangle looks like this:

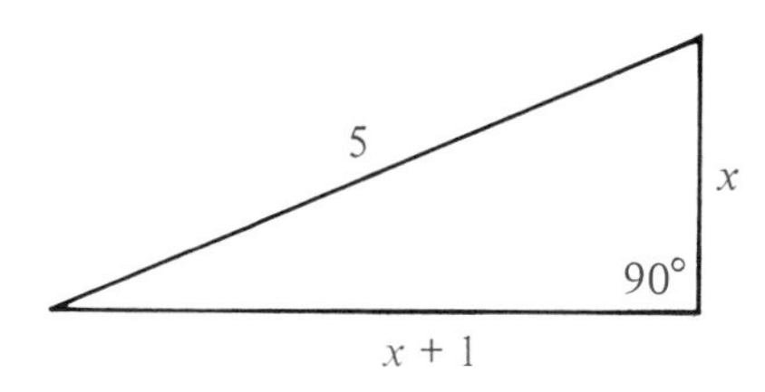

The Pythagorean theorem tells us that the square of the longest side, 5^2, is equal to the sum of the squares of the two shorter sides, $x^2 + (x + 1)^2$. Here is the equation:

$5^2 = x^2 + (x + 1)^2$	Pythagorean theorem
$25 = x^2 + x^2 + 2x + 1$	Expand 5^2 and $(x + 1)^2$
$25 = 2x^2 + 2x + 1$	Simplify the right side
$0 = 2x^2 + 2x - 24$	Add -25 to each side
$0 = 2(x^2 + x - 12)$	Begin factoring
$0 = 2(x + 4)(x - 3)$	Factor completely
$x + 4 = 0$ or $x - 3 = 0$	Set variable factors to 0
$x = -4$ or $x = 3$	

Since a triangle cannot have a side with a negative number for its length, we cannot not use -4. Therefore, the shortest side must be 3 inches. The next side is $x + 1 = 3 + 1 = 4$ inches. Since the hypotenuse is 5, we can check our solutions in the Pythagorean theorem:

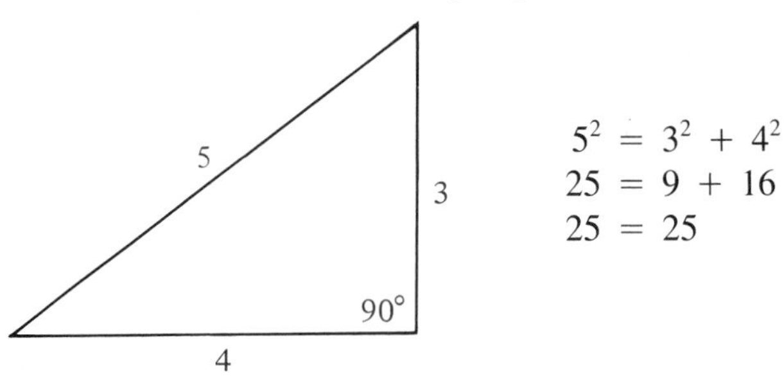

▲

Business Problems

▼ **Example 5** A company can manufacture x hundred items for a total cost of $C = 300 + 500x - 100x^2$. How many items were manufactured if the total cost is \$900?

Solution We are looking for x when C is 900. We begin by substituting 900 for C in the cost equation. Then we solve for x:

$$\begin{aligned} \text{When } C &= 900 \\ \text{the equation } \quad C &= 300 + 500x - 100x^2 \\ \text{becomes } \quad 900 &= 300 + 500x - 100x^2 \end{aligned}$$

We can write this equation in standard form by adding -300, $-500x$, and $100x^2$ to each side. The result of doing so looks like this:

$$\begin{aligned} 100x^2 - 500x + 600 &= 0 \\ 100(x^2 - 5x + 6) &= 0 && \text{Begin factoring} \\ 100(x - 2)(x - 3) &= 0 && \text{Factor completely} \\ x - 2 = 0 \quad \text{or} \quad x - 3 &= 0 && \text{Set variable factors to 0} \\ x = 2 \quad \text{or} \quad x &= 3 \end{aligned}$$

Our solutions are 2 and 3, which means that the company can manufacture 2 hundred items or 3 hundred items for a total cost of \$900. ▲

▼ **Example 6** A manufacturer of small portable radios knows that the number of radios she can sell each week is related to the price of the radios by the equation $x = 1300 - 100p$ (x is the number of radios and p is the price per radio). What price should she charge for the radios in order to have a weekly revenue of \$4,000?

Solution First, we must find the revenue equation. The equation for total revenue is $R = xp$ where x is the number of units sold and p is the price per unit. Since we want R in terms of p, we substitute $1300 - 100p$ for x in the equation $R = xp$:

$$\begin{aligned} &\text{If} & R &= xp \\ &\text{and} & x &= 1300 - 100p \\ &\text{then} & R &= (1300 - 100p)p \end{aligned}$$

We want to find p when R is 4,000. Substituting 4,000 for R in the equation gives us

$$4{,}000 = (1300 - 100p)p$$

If we multiply out the right side we have

$$4{,}000 = 1300p - 100p^2$$

To write this equation in standard form we add $100p^2$ and $-1300p$ to each side.

$$\begin{aligned} 100p^2 - 1300p + 4{,}000 &= 0 && \text{Add } 100p^2 \text{ and } -1300p \text{ to each side} \\ 100(p^2 - 13p + 40) &= 0 && \text{Begin factoring} \\ 100(p - 5)(p - 8) &= 0 && \text{Factor completely} \\ p - 5 = 0 \quad \text{or} \quad p - 8 &= 0 && \text{Set variable factors to 0} \\ p = 5 \quad \text{or} \quad p &= 8 \end{aligned}$$

If she sells the radios for \$5 each or for \$8 each she will have a weekly revenue of \$4,000. ▲

Problem Set 5.8

Solve the following word problems. Be sure to show the equation used.

Number Problems

1. The product of two consecutive even integers is 80. Find the two integers.
2. The product of two consecutive integers is 72. Find the two integers.
3. The product of two consecutive odd integers is 99. Find the two integers.
4. The product of two consecutive integers is 132. Find the two integers.
5. The product of two consecutive even integers is ten less than five times their sum. Find the two integers.
6. The product of two consecutive odd integers is one less than four times their sum. Find the two integers.
7. The sum of two numbers is 14. Their product is 48. Find the numbers.
8. The sum of two numbers is 12. Their product is 32. Find the numbers.
9. One number is two more than five times another. Their product is 24. Find the numbers.
10. One number is one more than twice another. Their product is 55. Find the numbers.

11. One number is four times another. Their product is four times their sum. Find the numbers.
12. One number is two more than twice another. Their product is two more than twice their sum. Find the numbers.

Geometry Problems

13. The length of a rectangle is one more than the width. The area is 12 sq. in. Find the dimensions.
14. The length of a rectangle is three more than twice the width. The area is 44 sq. in. Find the dimensions.
15. The height of a triangle is twice the base. The area is 9 sq. in. Find the base.
16. The height of a triangle is two more than twice the base. The area is 20 sq. ft. Find the base.
17. The hypotenuse of a right triangle is 10 inches. The length of the two legs are given by two consecutive even integers. Find the lengths of the two legs.
18. The hypotenuse of a right triangle is 15 inches. One of the legs is 3 inches more than the other. Find the lengths of the two legs.
19. The shorter leg of a right triangle is 5 meters. The hypotenuse is 1 meter longer than the longer leg. Find the length of the longer leg.
20. The shorter leg of a right triangle is 12 yards. If the hypotenuse is 20 yards, how long is the other leg?

Business Problems

21. A company can manufacture x hundred items for a total cost of $C = 400 + 700x - 100x^2$. Find x if the total cost is \$1400.
22. If the total cost, C, of manufacturing x hundred items is given by the equation $C = 700 + 900x - 100x^2$, find x when C is \$2700.
23. The total cost, C, of manufacturing x hundred video tapes is given by the equation $C = 600 + 1000x - 100x^2$. Find x if the total cost is \$2200.
24. The total cost, C, of manufacturing x hundred pen and pencil sets is given by the equation $C = 500 + 800x - 100x^2$. Find x when C is \$1700.
25. A company that manufactures typewriter ribbons knows that the number of ribbons they can sell each week, x, is related to the price per ribbon, p, by the equation $x = 1200 - 100p$. At what price should they sell the ribbons if they want the weekly revenue to be \$3200? (Remember: The equation for revenue is $R = xp$.)
26. A company manufactures diskettes for home computers. They know from past experience that the number of diskettes they can sell each day, x, is related to the price per diskette, p, by the equation $x = 800 - 100p$. At what price should they sell their diskettes if they want the daily revenue to be \$1200?
27. The relationship between the number of calculators a company sells per week, x, and the price of each calculator, p, is given by the equation $x = 1700 - 100p$. At what price should the calculators be sold if the weekly revenue is to be \$7,000?
28. The relationship between the number of pencil sharpeners a company can sell each week, x, and the price of each sharpener, p, is given by the equation $x = 1800 - 100p$. At what price should the sharpeners be sold if the weekly revenue is to be \$7200?

Review Problems The problems that follow review material we covered in Section 4.3.

Multiply.

29. $(6a^2b)(7a^3b^2)$

30. $(2a^3b^2)(6ab)$

Divide.

31. $\dfrac{12x^3y^5}{6xy^3}$

32. $\dfrac{25x^4y^8}{5x^3y}$

Simplify.

33. $\dfrac{(5x^4y^4)(10x^3y^3)}{2x^2y^7}$

34. $\dfrac{(4x^3y^2)(9x^4y^{10})}{3x^5y^2}$

35. $(2 \times 10^5)(3 \times 10^{-8})$

36. $\dfrac{8 \times 10^{-8}}{2 \times 10^{-3}}$

37. $\dfrac{45a^6}{9a^3} - \dfrac{15a^8}{5a^5}$

38. $\dfrac{27a^{10}}{9a^6} + \dfrac{12a^{12}}{6a^8}$

Chapter 5 Summary and Review

FACTORING [5.1]

Factoring is the reverse of multiplication.

Multiplication

Factors → $3 \cdot 5 = 15$ ← Product

Factoring

GREATEST COMMON FACTOR [5.2]

The largest monomial that divides each term of a polynomial is called the greatest common factor for that polynomial. We begin all factorizations by factoring out the greatest common factor.

FACTORING TRINOMIALS [5.3, 5.4]

One method of factoring a trinomial is to list all pairs of binomials the product of whose first terms gives the first term of the trinomial and the product of whose last terms gives the last term of the trinomial. We then choose the pair that gives the correct middle term for the original trinomial.

Examples

1. The number 150 can be factored into the product of prime numbers:

$$\begin{aligned} 150 &= 15 \cdot 10 \\ &= (3 \cdot 5)(2 \cdot 5) \\ &= 2 \cdot 3 \cdot 5^2 \end{aligned}$$

2. $$\begin{aligned} &8x^4 - 10x^3 + 6x^2 \\ &= 2x^2 \cdot 4x^2 - 2x^2 \cdot 5x + 2x^2 \cdot 3 \\ &= 2x^2(4x^2 - 5x + 3) \end{aligned}$$

3. $$\begin{aligned} x^2 + 5x + 6 &= (x + 2)(x + 3) \\ x^2 - 5x + 6 &= (x - 2)(x - 3) \\ 6x^2 - x - 2 &= (2x + 1)(3x - 2) \\ 6x^2 + 7x + 2 &= (2x + 1)(3x + 2) \end{aligned}$$

SPECIAL FACTORIZATIONS [5.5]

$$a^2 + 2ab + b^2 = (a + b)^2$$
$$a^2 - 2ab + b^2 = (a - b)^2$$
$$a^2 - b^2 = (a + b)(a - b)$$

4. $x^2 + 10x + 25 = (x + 5)^2$
$x^2 - 10x + 25 = (x - 5)^2$
$x^2 - 25 = (x + 5)(x - 5)$

CHECKLIST FOR FACTORING A POLYNOMIAL [5.6]

Step 1: If the polynomial has a greatest common factor other than 1, then factor out the greatest common factor.

Step 2: If the polynomial has two terms (it is a binomial) then see if it is the difference of two squares. Remember, if it is the sum of two squares, it will not factor.

Step 3: If the polynomial has three terms (a trinomial), then it is either a perfect square trinomial that will factor into the square of a binomial, or it is not a perfect square trinomial, in which case you use the trial and error method developed in Section 5.4.

Step 4: If the polynomial has more than three terms, then try to factor it by grouping.

Step 5: As a final check, look and see if any of the factors you have written can be factored further. If you have overlooked a common factor, you can catch it here.

5.

a. $2x^5 - 8x^3 = 2x^3(x^2 - 4)$
$= 2x^3(x + 2)(x - 2)$

b. $3x^4 - 18x^3 + 27x^2$
$= 3x^2(x^2 - 6x + 9)$
$= 3x^2(x - 3)^2$

c. $6x^3 - 12x^2 - 48x$
$= 6x(x^2 - 2x - 8)$
$= 6x(x - 4)(x + 2)$

d. $x^2 + ax + bx + ab$
$= x(x + a) + b(x + a)$
$= (x + a)(x + b)$

TO SOLVE A QUADRATIC EQUATION [5.7]

Step 1: Write the equation in standard form: $ax^2 + bx + c = 0$.

Step 2: Factor completely.

Step 3: Set each variable factor equal to zero.

Step 4: Solve the equations found in Step 3.

Step 5: Check solutions, if necessary.

6. Solve $x^2 - 6x = -8$.

$$x^2 - 6x + 8 = 0$$
$$(x - 4)(x - 2) = 0$$
$$x - 4 = 0 \quad \text{or} \quad x - 2 = 0$$
$$x = 4 \quad \text{or} \quad x = 2$$

Both solutions check.

THE PYTHAGOREAN THEOREM [5.8]

In any right triangle, the square of the longest side (called the hypotenuse) is equal to the sum of the squares of the other two sides (called legs).

7. The hypotenuse of a right triangle is 5 inches, while the lengths of the two legs (the other two sides) are given by two consecutive integers. Find the lengths of the two legs.

If we let $x =$ the length of the shortest side, then the other side must be $x + 1$. The Pythagorean theorem tells us that the square of the longest side, 5^2, is equal to

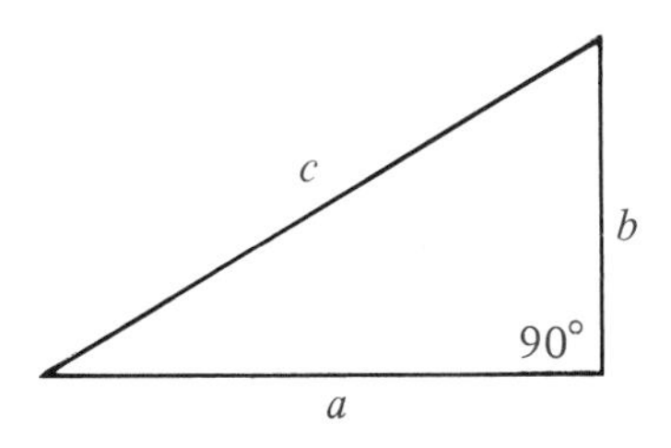

$$c^2 = a^2 + b^2$$

the sum of the squares of the two shorter sides, $x^2 + (x + 1)^2$.

$$5^2 = x^2 + (x + 1)^2$$
$$25 = x^2 + x^2 + 2x + 1$$
$$25 = 2x^2 + 2x + 1$$
$$0 = 2x^2 + 2x - 24$$
$$0 = 2(x^2 + x - 12)$$
$$0 = 2(x + 4)(x - 3)$$
$$x + 4 = 0 \quad \text{or} \quad x - 3 = 0$$
$$x = -4 \quad \text{or} \quad x = 3$$

Since a triangle cannot have a side with a negative number for its length, we cannot use -4. Therefore, the shortest side must be 3 inches. The next side is $x + 1 = 3 + 1 = 4$ inches.

COMMON MISTAKES

Try to apply the zero-factor property to other numbers. For example, consider the equation $(x - 3)(x + 4) = 18$. A fairly common mistake is to attempt to solve it with the following steps:

$$(x - 3)(x + 4) = 18$$
$$x - 3 = 18 \quad \text{or} \quad x + 4 = 18$$
$$x = 21 \quad \text{or} \quad x = 14$$

These are obviously not solutions, as a quick check will verify:

Check $x = 21$

$$(21 - 3)(21 + 4) \stackrel{?}{=} 18$$
$$18 \cdot 25 = 18$$
$$450 = 18$$

false statements ⟷

Check $x = 14$

$$(14 - 3)(14 + 4) \stackrel{?}{=} 18$$
$$11 \cdot 18 = 18$$
$$198 = 18$$

The mistake is in setting each factor equal to 18. It is not necessarily true that since the product of two numbers is 18 that either one of them is itself 18.

The correct solution looks like this:

$$(x - 3)(x + 4) = 18$$
$$x^2 + x - 12 = 18$$
$$x^2 + x - 30 = 0$$
$$(x + 6)(x - 5) = 0$$
$$x + 6 = 0 \quad \text{or} \quad x - 5 = 0$$
$$x = -6 \quad \text{or} \quad x = 5$$

To avoid this mistake, remember that before you factor a quadratic equation, you must write it in standard form. It is only in standard form when 0 is on one side and decreasing powers of the variable are on the other.

Chapter 5 Test

Write each term as the product of prime numbers. [5.1]

1. 35
2. 77
3. 128
4. 324

Factor the following completely. [5.2 through 5.6]

5. $x^2 - 5x + 6$
6. $x^2 - x - 6$
7. $a^2 - 16$
8. $x^2 + 25$
9. $x^4 - 81$
10. $27x^2 - 75y^2$
11. $xy + 4x + 7y + 28$
12. $x^2 - bx + 5x - 5b$
13. $4a^2 + 22a + 10$
14. $3m^2 - 3m - 18$
15. $6y^2 + 7y - 5$
16. $12x^3 - 14x^2 - 10x$

Solve the following quadratic equations. [5.7]

17. $x^2 + 7x + 12 = 0$
18. $x^2 - 4x + 4 = 0$
19. $x^2 - 36 = 0$
20. $x^2 = x + 20$
21. $x^2 - 11x = -30$
22. $y^3 = 16y$
23. $2a^2 = a + 15$
24. $30x^3 - 20x^2 = 10x$

Solve the following word problems. Be sure to show the equation used. [5.8]

25. Two numbers have a sum of 20. Their product is 64. Find the numbers.
26. The product of two consecutive odd integers is seven more than their sum. Find the integers.
27. The length of a rectangle is five more than three times the width. The area is 42 sq. ft. Find the dimensions.
28. One leg of a right triangle is 2 more than twice the other. The hypotenuse is 13 meters. Find the lengths of the two legs.
29. A company can manufacture x hundred items for a total cost of $C = 200 + 500x - 100x^2$. How many items can be manufactured if the total cost is to be \$800?
30. A manufacturer knows that the number of items he can sell each week, x, is related to the price of each item, p, by the equation $x = 900 - 100p$. What price should he charge for each item in order to have a weekly revenue of \$1800? (Remember: $R = xp$.)

Rational Expressions

Rational expressions are to polynomials what fractions are to integers. As you will see, the problems you will work with fractions are very similar to the problems you will work with rational expressions.

Most important to your success in this chapter is your ability to factor polynomials. There are many problems in this chapter that are impossible to solve without factoring.

Section 6.1 Reducing Rational Expressions to Lowest Terms

In Chapter 1, we defined the set of rational numbers to be the set of all numbers that could be put in the form $\frac{a}{b}$, where a and b are integers and $b \neq 0$:

$$\text{Rational numbers} = \left\{\frac{a}{b} \,\middle|\, a \text{ and } b \text{ are integers, } b \neq 0\right\}$$

A *rational expression* is any expression that can be put in the form $\frac{P}{Q}$, where P and Q are polynomials and $Q \neq 0$:

$$\text{Rational expressions} = \left\{\frac{P}{Q} \,\middle|\, P \text{ and } Q \text{ are polynomials, } Q \neq 0\right\}$$

Each of the following is an example of a rational expression:

$$\frac{2x + 3}{x} \qquad \frac{x^2 - 6x + 9}{x^2 - 4} \qquad \frac{5}{x^2 + 6} \qquad \frac{2x^2 + 3x + 4}{2}$$

For the rational expression

$$\frac{x^2 - 6x + 9}{x^2 - 4}$$

the polynomial on top, $x^2 - 6x + 9$, is called the numerator, and the polynomial on the bottom, $x^2 - 4$, is called the denominator. The same is true of the other rational expressions.

We must be careful that we do not use a value of the variable that will give us a denominator of zero. Remember, division by zero is not defined.

▼ **Examples** State the restrictions on the variable in the following rational expressions:

1. $\dfrac{x + 2}{x - 3}$

Solution The variable x can be any real number except $x = 3$, since, when $x = 3$, the denominator is $3 - 3 = 0$. We state this restriction by writing $x \neq 3$.

2. $\dfrac{5}{x^2 - x - 6}$

Solution If we factor the denominator, we have $x^2 - x - 6 = (x - 3)(x + 2)$. If either of the factors is zero, the whole denominator is zero. Our restrictions are $x \neq 3$ and $x \neq -2$, since either one makes $x^2 - x + 6 = 0$. ▲

We will not always list each restriction on a rational expression, but keep in mind that no rational expression can have a denominator of zero.

The fundamental property of rational expressions is listed next. We will use these two properties many times in this chapter.

Fundamental Property of Rational Expressions

If $\dfrac{P}{Q}$ is a rational expression ($Q \neq 0$) and K is any nonzero quantity, then

$$\frac{P}{Q} = \frac{P \cdot K}{Q \cdot K} \quad \text{which is equivalent to} \quad \frac{P \cdot K}{Q \cdot K} = \frac{P}{Q}$$

In words: Multiplying (or dividing) the numerator and denominator of a rational expression by the same nonzero quantity always yields an equivalent rational expression.

Note that we have stated our fundamental property for division as well as multiplication. The reason for this is that division is equivalent to multiplication by the reciprocal.

We can use the fundamental property to reduce rational expressions to lowest terms. Since this process is almost identical to the process of reducing fractions to lowest terms, let's recall how the fraction $\frac{6}{15}$ is reduced to lowest terms:

$$\frac{6}{15} = \frac{2 \cdot 3}{5 \cdot 3} \qquad \text{Factor numerator and denominator}$$

$$= \frac{2 \cdot \cancel{3}}{5 \cdot \cancel{3}} \qquad \text{Divide out the common factor, 3}$$

$$= \frac{2}{5} \qquad \text{Reduce to lowest terms}$$

The same procedure applies to reducing rational expressions to lowest terms. The process is summarized in the following rule.

Rule To reduce a rational expression to lowest terms, first factor each expression completely and then divide both the numerator and denominator by any factors they have in common.

▼ **Example 3** Reduce $\frac{x^2 - 9}{x^2 + 5x + 6}$ to lowest terms.

Solution We begin by factoring:

$$\frac{x^2 - 9}{x^2 + 5x + 6} = \frac{(x - 3)(x + 3)}{(x + 2)(x + 3)}$$

Notice that both polynomials contain the factor $(x + 3)$. If we divide the numerator by $(x + 3)$, we are left with $(x - 3)$. If we divide the denominator by $(x + 3)$, we are left with $(x + 2)$. The complete problem looks like this:

$$\frac{x^2 - 9}{x^2 + 5x + 6} = \frac{(x - 3)\cancel{(x + 3)}}{(x + 2)\cancel{(x + 3)}} \qquad \text{Factor the numerator and denominator completely}$$

$$= \frac{x - 3}{x + 2} \qquad \text{Divide out the common factor, } x + 3$$ ▲

It is convenient to draw a line through the factors as we divide them out. It is especially helpful when the problems become longer.

▼ **Example 4** Reduce to lowest terms: $\dfrac{10a + 20}{5a^2 - 20}$.

Solution We begin by factoring out the greatest common factor from the numerator and denominator:

$$\frac{10a + 20}{5a^2 - 20} = \frac{10(a + 2)}{5(a^2 - 4)}$$ Factor the greatest common factor from the numerator and denominator

$$= \frac{10\cancel{(a + 2)}}{5\cancel{(a + 2)}(a - 2)}$$ Factor the denominator as the difference of two squares

$$= \frac{2}{a - 2}$$ Divide out the common factors 5 and $(a + 2)$ ▲

▼ **Example 5** Reduce $\dfrac{2x^3 + 2x^2 - 24x}{x^3 + 2x^2 - 8x}$ to lowest terms.

Solution We begin by factoring the numerator and denominator completely. Then, we divide out all factors common to the numerator and denominator. Here is what it looks like:

$$\frac{2x^3 + 2x^2 - 24x}{x^3 + 2x^2 - 8x} = \frac{2x(x^2 + x - 12)}{x(x^2 + 2x - 8)}$$ Factor out the greatest common factor first

$$= \frac{2\cancel{x}(x - 3)\cancel{(x + 4)}}{\cancel{x}(x - 2)\cancel{(x + 4)}}$$ Factor the remaining trinomials

$$= \frac{2(x - 3)}{x - 2}$$ Divide out the factors common to the numerator and denominator ▲

▼ **Example 6** Reduce $\dfrac{x - 5}{x^2 - 25}$ to lowest terms.

Solution

$$\frac{x - 5}{x^2 - 25} = \frac{\cancel{(x - 5)}}{\cancel{(x - 5)}(x + 5)}$$ Factor numerator and denominator completely

$$= \frac{1}{x + 5}$$ Divide out the common factor, $(x - 5)$ ▲

You may have noticed that many light bulbs are rated in lumens as well as in watts. For example, a 100-watt light bulb may be rated at 1500 lumens. A lumen is a measure of the amount of light given off by the bulb.

To find the intensity of light, E, that falls on a surface d feet away from a bulb that gives off L lumens, you use the formula

$$E = \frac{7L}{88d^2}$$

To have enough light to read comfortably, it is commonly thought that E should be approximately 10 lumens per square foot on the surface on which you are reading.

▼ **Example 7** A lamp hangs 6 feet above a desk. If a 100-watt bulb rated at 1500 lumens is placed in the lamp, calculate E for the light that falls on the surface of the desk.

Solution We substitute 1500 for L and 6 for d in the formula

$$E = \frac{7L}{88d^2}$$

to obtain

$$E = \frac{7 \cdot 1500}{88 \cdot 6^2}$$

Factoring the numerator and denominator and dividing out any factors they have in common, we have

$$\begin{aligned} E &= \frac{7(\cancel{2} \cdot \cancel{2} \cdot \cancel{3} \cdot 5 \cdot 5 \cdot 5)}{(\cancel{2} \cdot \cancel{2} \cdot 2 \cdot 11)(2 \cdot \cancel{3})(2 \cdot 3)} \\ &= \frac{7 \cdot 5 \cdot 5 \cdot 5}{2 \cdot 11 \cdot 2 \cdot 2 \cdot 3} \\ &= \frac{875}{264} \quad \text{lumens/square foot} \end{aligned}$$

If we divide 875 by 264, we get approximately 3.3 lumens/square foot. If 10 lumens per square foot is considered ideal for a reading surface, then 3.3 lumens per square foot may be too dim. The bulb should be replaced by a bulb with a higher lumen rating, or the height of the lamp should be lowered. ▲

COMMON MISTAKES

You should look over Example 6 closely and see that you understand each step. There are two common mistakes with this type of problem. The first arises in dividing the numerator $(x - 5)$ by $(x - 5)$. Many times first-year algebra students will give the result as 0. It is not 0 but 1. That is, $(x - 5)$ goes into $(x - 5)$ once.

The second mistake is usually made by trying to do what some people mistake for canceling:

$$\frac{x-5}{x^2-25} = \frac{\cancel{x}-\cancel{5}}{\cancel{x^2}-\cancel{25}} \quad \text{Mistake}$$
$$\qquad\qquad x \quad 5$$
$$= \frac{1}{x-5}$$

But, if we replaced x by 10, we would have

$$\frac{10-5}{100-25} = \frac{1}{10-5}$$
$$\frac{5}{75} = \frac{1}{5}$$
$$\frac{1}{15} = \frac{1}{5}$$

which is not true.

To avoid this mistake, remember that we can divide out (sometimes called canceling) *factors* common to the numerator and denominator of a rational expression. Neither x nor 5 is a factor of either the numerator or denominator in the problem and they therefore cannot be divided out.

Problem Set 6.1

Reduce the following rational expressions to lowest terms, if possible. Also, specify any restrictions on the variable in problems 1 through 10.

1. $\frac{5}{5x-10}$

2. $\frac{-4}{2x-8}$

3. $\frac{-8}{8x+16}$

4. $\frac{3}{6x-12}$

5. $\frac{a-3}{a^2-9}$

6. $\frac{a+4}{a^2-16}$

7. $\frac{x+5}{x^2-25}$

8. $\frac{x-2}{x^2-4}$

9. $\frac{2a}{10a^2}$

10. $\frac{22a^3}{11a}$

11. $\frac{2x^2-8}{4}$

12. $\frac{5x-10}{x-2}$

13. $\frac{35m^2}{7m^4}$

14. $\frac{27m^5n^2}{9m^2n^3}$

15. $\dfrac{2x - 10}{3x - 6}$

16. $\dfrac{4x - 8}{x - 2}$

17. $\dfrac{10a + 20}{5a + 10}$

18. $\dfrac{11a + 33}{6a + 18}$

19. $\dfrac{5x^2 - 5}{4x + 4}$

20. $\dfrac{7x^2 - 28}{2x + 4}$

21. $\dfrac{x - 3}{x^2 - 6x + 9}$

22. $\dfrac{x^2 - 10x + 25}{x - 5}$

23. $\dfrac{y + 2}{y^2 - y - 6}$

24. $\dfrac{y + 1}{y^2 - 2y - 3}$

25. $\dfrac{x + 5}{x^2 + 8x + 15}$

26. $\dfrac{x + 3}{x^2 + 8x + 15}$

27. $\dfrac{a - 3}{a^2 - 8a + 15}$

28. $\dfrac{a + 3}{a^2 - 2a - 15}$

29. $\dfrac{3x - 2}{9x^2 - 4}$

30. $\dfrac{2x - 3}{4x^2 - 9}$

31. $\dfrac{x^2 + 8x + 15}{x^2 + 5x + 6}$

32. $\dfrac{x^2 - 8x + 15}{x^2 - x - 6}$

33. $\dfrac{2m^3 - 2m^2 - 12m}{m^2 - 5m + 6}$

34. $\dfrac{2m^3 + 4m^2 - 6m}{m^2 - m - 12}$

35. $\dfrac{3x^2 - 3x}{2(x - 1)}$

36. $\dfrac{16y^2 - 4}{4y - 2}$

37. $\dfrac{x^3 + 3x^2 - 4x}{x^3 - 16x}$

38. $\dfrac{4x^3 - 10x^2 + 6x}{2x^3 + x^2 - 3x}$

39. $\dfrac{3a^2 - 8a + 4}{9a^2 - 4}$

40. $\dfrac{3a^2 - 8a + 5}{4a^2 - 5a + 1}$

41. $\dfrac{4x^2 - 12x + 9}{4x^2 - 9}$

42. $\dfrac{5x^2 + 18x - 8}{5x^2 + 13x - 6}$

43. $\dfrac{x + 3}{x^4 - 81}$

44. $\dfrac{x^2 + 9}{x^4 - 81}$

To reduce each of the following rational expressions to lowest terms, you will have to use factoring by grouping. Be sure to factor each numerator and denominator completely before dividing out any common factors. (Remember, factoring by grouping takes two steps.)

45. $\dfrac{xy + 3x + 2y + 6}{xy + 3x + 5y + 15}$

46. $\dfrac{xy + 7x + 4y + 28}{xy + 3x + 4y + 12}$

47. $\dfrac{x^2 - 3x + ax - 3a}{x^2 - 3x + bx - 3b}$

48. $\dfrac{x^2 - 6x + ax - 6a}{x^2 - 7x + ax - 7a}$

49. $\frac{xy + bx + ay + ab}{xy + bx + 3y + 3b}$

50. $\frac{x^2 + 5x + ax + 5a}{x^2 + 5x + bx + 5b}$

51. Replace x with 5 and y with 4 in the expression

$$\frac{x^2 - y^2}{x - y}$$

and simplify the result. Is the result equal to $5 - 4$ or $5 + 4$?

52. Replace x with 2 in the expression

$$\frac{x^3 - 1}{x - 1}$$

and simplify the result. Your answer should be equal to what you would get if you replaced x with 2 in $x^2 + x + 1$.

For the next two problems, use the formula $E = \frac{7L}{88d^2}$ where E is a measure of the intensity of light that falls on a surface d feet below a light bulb that gives off L lumens of light.

53. A light bulb with a rating of 630 lumens is in a lamp 4 feet above the surface of a desk. Find the intensity of light E that falls on the desk.

54. The lamp in Problem 53 is raised another 2 feet above the desk. Find E for light that falls on the surface of the desk with the lamp in its new position.

55. A man deposits money in an account with an interest rate r compounded annually. At the end of two years he finds that the account has a total of \$1254.40 in it. The amount of money P that he originally deposited in the account is given by the formula

$$P = \frac{1254.40}{1 + 2r + r^2}$$

Factor the denominator of this expression and then use it to find the amount of money he originally deposited, if the interest rate was 12%.

56. Two years after a woman deposits money in an account, the bank sends her a statement indicating that she has a total of \$726 in the account. The amount of money she originally deposited is given by the formula

$$P = \frac{726}{1 + 2r + r^2}$$

where r is the annual rate of interest. If r is 10%, was her original investment \$500 or \$600?

Review Problems The problems below review material we covered in Sections 1.7 and 5.1. Reviewing these problems will help you understand the next section.

Multiply or divide as indicated. [1.7]

57. $\frac{3}{8} \cdot \frac{5}{2}$

58. $\frac{4}{5} \cdot \frac{2}{3}$

59. $\frac{3}{10} \cdot \frac{5}{9}$

60. $\frac{2}{7} \cdot \frac{21}{30}$

61. $\frac{2}{3} \div \frac{1}{3}$

62. $\frac{3}{4} \div \frac{1}{4}$

63. $\frac{4}{5} \div \frac{9}{10}$

64. $\frac{1}{6} \div \frac{3}{12}$

Factor into the product of prime factors. [5.1]

65. 75

66. 210

67. 225

68. 630

Section 6.2 Multiplication and Division of Rational Expressions

Recall that to multiply two fractions, we simply multiply numerators and multiply denominators and then reduce to lowest terms, if possible:

$$\frac{3}{4} \cdot \frac{10}{21} = \frac{30}{84} \begin{matrix} \leftarrow \text{Multiply numerators} \\ \leftarrow \text{Multiply denominators} \end{matrix}$$

$$= \frac{5}{14} \leftarrow \text{Reduce to lowest terms}$$

The same problem can also be done by factoring numerators and denominators first and then dividing out the factors they have in common:

$$\frac{3}{4} \cdot \frac{10}{21} = \frac{3}{2 \cdot 2} \cdot \frac{2 \cdot 5}{3 \cdot 7} \qquad \text{Factor}$$

$$= \frac{\cancel{3} \cdot \cancel{2} \cdot 5}{\cancel{2} \cdot 2 \cdot \cancel{3} \cdot 7} \qquad \begin{matrix} \text{Multiply numerators} \\ \text{Multiply denominators} \end{matrix}$$

$$= \frac{5}{14} \qquad \text{Divide out common factors}$$

We can apply the second process to the product of two rational expressions, as the following example illustrates.

▼ **Example 1** Multiply: $\frac{x - 2}{x + 3} \cdot \frac{x^2 - 9}{2x - 4}$.

Solution We begin by factoring numerators and denominators as much as possible. Then, we multiply the numerators and denominators. The last step consists of dividing out all factors common to the numerator and denominator:

$$\frac{x - 2}{x + 3} \cdot \frac{x^2 - 9}{2x - 4} = \frac{x - 2}{x + 3} \cdot \frac{(x - 3)(x + 3)}{2(x - 2)} \qquad \text{Factor completely}$$

$$= \frac{\cancel{(x - 2)}(x - 3)\cancel{(x + 3)}}{\cancel{(x + 3)}(2)\cancel{(x - 2)}} \qquad \text{Multiply numerators and denominators}$$

$$= \frac{x - 3}{2} \qquad \text{Divide out common factors}$$

▲

In Chapter 1 we defined division as the equivalent of multiplication by the reciprocal. This is how it looks with fractions:

$$\frac{4}{5} \div \frac{8}{9} = \frac{4}{5} \cdot \frac{9}{8} \qquad \text{Division as multiplication by the reciprocal}$$

$$= \frac{\cancel{2} \cdot \cancel{2} \cdot 3 \cdot 3}{5 \cdot \cancel{2} \cdot \cancel{2} \cdot 2} \qquad \text{Factor and divide out common factors}$$

$$= \frac{9}{10}$$

The same idea holds for division with rational expressions. The rational expression that follows the division symbol is called the *divisor*; to divide, we multiply by the reciprocal of the divisor.

▼ **Example 2** Divide: $\dfrac{3x - 9}{x^2 - x - 20} \div \dfrac{x^2 + 2x - 15}{x^2 - 25}$.

Solution We begin by taking the reciprocal of the divisor and writing the problem again in terms of multiplication. We then factor, multiply, and, finally, divide out all factors common to the numerator and denominator of the resulting expression. The complete problem looks like this:

$$\frac{3x - 9}{x^2 - x - 20} \div \frac{x^2 + 2x - 15}{x^2 - 25}$$

$$= \frac{3x - 9}{x^2 - x - 20} \cdot \frac{x^2 - 25}{x^2 + 2x - 15} \qquad \text{Multiply by the reciprocal of the divisor}$$

$$= \frac{3(x - 3)}{(x + 4)(x - 5)} \cdot \frac{(x - 5)(x + 5)}{(x + 5)(x - 3)} \qquad \text{Factor}$$

$$= \frac{3\cancel{(x - 3)}\cancel{(x - 5)}\cancel{(x + 5)}}{(x + 4)\cancel{(x - 5)}\cancel{(x + 5)}\cancel{(x - 3)}} \qquad \text{Divide out common factors}$$

$$= \frac{3}{x + 4}$$

▲

As you can see, factoring is the single most important tool we use in working with rational expressions. Most of the work we have done or will do with rational expressions is most easily accomplished if the rational expressions are in factored form. Here are some more examples of multiplication and division with rational expressions.

▼ **Examples**

3. Multiply: $\dfrac{3a + 6}{a^2} \cdot \dfrac{a}{2a + 4}$.

Solution

$$\frac{3a+6}{a^2}\cdot\frac{a}{2a+4}$$

$$=\frac{3(a+2)}{a^2}\cdot\frac{a}{2(a+2)}$$ Factor completely

$$=\frac{3(a+2)a}{a^2(2)(a+2)}$$ Multiply

$$=\frac{3}{2a}$$ Divide numerator and denominator by common factors $a(a+2)$

4. Divide: $\dfrac{x^2+7x+12}{x^2-16} \div \dfrac{x^2+6x+9}{2x-8}$.

Solution

$$\frac{x^2+7x+12}{x^2-16}\div\frac{x^2+6x+9}{2x-8}$$

$$=\frac{x^2+7x+12}{x^2-16}\cdot\frac{2x-8}{x^2+6x+9}$$ Division is multiplication by the reciprocal

$$=\frac{(x+3)(x+4)(2)(x-4)}{(x-4)(x+4)(x+3)(x+3)}$$ Factor and multiply

$$=\frac{2}{x+3}$$ Divide out common factors ▲

In Example 4, we factored and multiplied the two expressions in a single step. This saves writing the problem one extra time.

▼ **Example 5** Multiply: $(x^2-49)\left(\dfrac{x+4}{x+7}\right)$.

Solution We can think of the polynomial x^2-49 as having a denominator of 1. Thinking of x^2-49 in this way allows us to proceed as we did in previous examples:

$$(x^2-49)\left(\frac{x+4}{x+7}\right)=\frac{x^2-49}{1}\cdot\frac{x+4}{x+7}$$ Write x^2-49 with denominator 1

$$=\frac{(x+7)(x-7)(x+4)}{x+7}$$ Factor and multiply

$$=(x-7)(x+4)$$ Divide out common factors

We can leave the answer in this form or multiply to get $x^2 - 3x - 28$. In this section, let's agree to leave our answers in factored form. ▲

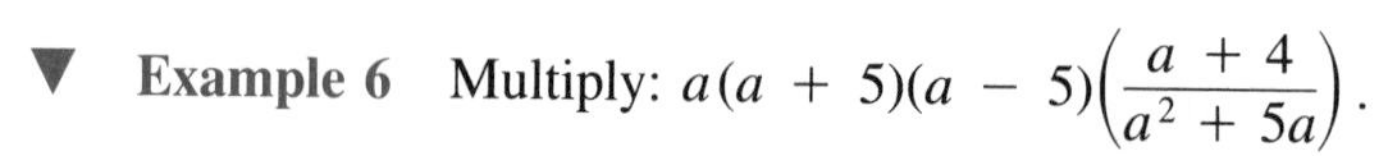

▼ **Example 6** Multiply: $a(a + 5)(a - 5)\left(\dfrac{a + 4}{a^2 + 5a}\right)$.

Solution We can think of the expression $a(a + 5)(a - 5)$ as having a denominator of 1:

$$a(a + 5)(a - 5)\left(\frac{a + 4}{a^2 + 5a}\right)$$

$$= \frac{a(a + 5)(a - 5)}{1} \cdot \frac{a + 4}{a^2 + 5a}$$

$$= \frac{\cancel{a(a + 5)}(a - 5)(a + 4)}{\cancel{a(a + 5)}} \qquad \text{Factor and multiply}$$

$$= (a - 5)(a + 4) \qquad \text{Divide out common factors}$$

▲

Problem Set 6.2

Multiply or divide as indicated. Be sure to reduce all answers to lowest terms. (That is, the numerator and denominator of the answer should not have any factors in common.)

1. $\dfrac{x + y}{3} \cdot \dfrac{6}{x + y}$

2. $\dfrac{x - 1}{x + 1} \cdot \dfrac{5}{x - 1}$

3. $\dfrac{2x + 10}{x^2} \cdot \dfrac{x^3}{4x + 20}$

4. $\dfrac{3x^4}{3x - 6} \cdot \dfrac{x - 2}{x^2}$

5. $\dfrac{9}{2a - 8} \div \dfrac{3}{a - 4}$

6. $\dfrac{8}{a^2 - 25} \div \dfrac{16}{a + 5}$

7. $\dfrac{x + 1}{x^2 - 9} \div \dfrac{2x + 2}{x + 3}$

8. $\dfrac{11}{x - 2} \div \dfrac{22}{2x^2 - 8}$

9. $\dfrac{a^2 + 5a}{7a} \cdot \dfrac{4a^2}{a^2 + 4a}$

10. $\dfrac{4a^2 + 4a}{a^2 - 25} \cdot \dfrac{a^2 - 5a}{8a}$

11. $\dfrac{y^2 - 5y + 6}{2y + 4} \div \dfrac{2y - 6}{y + 2}$

12. $\dfrac{y^2 - 7y}{3y^2 - 48} \div \dfrac{y^2 - 9}{y^2 - 7y + 12}$

13. $\dfrac{2x - 8}{x^2 - 4} \cdot \dfrac{x^2 + 6x + 8}{x - 4}$

14. $\dfrac{x^2 + 5x + 1}{7x - 7} \cdot \dfrac{x - 1}{x^2 + 5x + 1}$

15. $\dfrac{x - 1}{x^2 - x - 6} \cdot \dfrac{x^2 + 5x + 6}{x^2 - 1}$

16. $\dfrac{x^2 - 3x - 10}{x^2 - 4x + 3} \cdot \dfrac{x^2 - 5x + 6}{x^2 - 3x - 10}$

17. $\frac{a^2 + 10a + 25}{a + 5} \div \frac{a^2 - 25}{a - 5}$

18. $\frac{a^2 + a - 2}{a^2 + 5a + 6} \div \frac{a - 1}{a}$

19. $\frac{y - 5}{y^2 + 3y + 2} \div \frac{y^2 - 5y + 6}{y^2 - 2y - 3}$

20. $\frac{y - 5}{y^2 + 7y + 12} \div \frac{y^2 - 7y + 10}{y^2 + 9y + 18}$

21. $\frac{x^2 + 10x + 21}{x^2 + 2x - 35} \cdot \frac{x^2 - 25}{x^2 - 2x - 15}$

22. $\frac{x^2 - 13x + 42}{x^2 + 10x + 21} \cdot \frac{x^2 + x - 6}{x^2 - 4}$

23. $\frac{x^2 + 5x + 6}{x^2 + 6x + 8} \cdot \frac{x^2 + 9x + 20}{x^2 + 8x + 15}$

24. $\frac{x^2 - 1}{x^2 + 3x + 2} \cdot \frac{x^2 - 4}{x^2 - 3x + 2}$

25. $\frac{2a^2 + 7a + 3}{a^2 - 16} \div \frac{4a^2 + 8a + 3}{2a^2 - 5a - 12}$

26. $\frac{3a^2 + 7a - 20}{a^2 + 3a - 4} \div \frac{3a^2 - 2a - 5}{a^2 - 2a + 1}$

27. $\frac{4y^2 - 12y + 9}{y^2 - 36} \div \frac{2y^2 - 5y + 3}{y^2 + 5y - 6}$

28. $\frac{5y^2 - 6y + 1}{y^2 - 1} \div \frac{16y^2 - 9}{4y^2 + 7y + 3}$

29. $\frac{x^2 - 1}{x^2 + 7x + 10} \cdot \frac{x^2 + 5x + 6}{x + 1} \cdot \frac{x + 5}{x^2 + 2x - 3}$

30. $\frac{4x^2 - 1}{x - 5} \cdot \frac{x^2 - 2x - 15}{2x^2 - x - 1} \cdot \frac{x - 1}{x^2 - 9}$

Multiply the following expressions using the method shown in Examples 5 and 6 in this section.

31. $(5x - 5)\left(\frac{3}{x - 1}\right)$

32. $(4x - 8)\left(\frac{x}{x - 2}\right)$

33. $(x^2 - 9)\left(\frac{2}{x + 3}\right)$

34. $(x^2 - 9)\left(\frac{-3}{x - 3}\right)$

35. $a(a + 5)(a - 5)\left(\frac{2}{a^2 - 25}\right)$

36. $a(a^2 - 4)\left(\frac{a}{a + 2}\right)$

37. $(x^2 - x - 6)\left(\frac{x + 1}{x - 3}\right)$

38. $(x^2 - 2x - 8)\left(\frac{x + 3}{x - 4}\right)$

39. $(x^2 - 4x - 5)\left(\frac{-2x}{x + 1}\right)$

40. $(x^2 - 6x + 8)\left(\frac{4x}{x - 2}\right)$

Each of the following problems involves some factoring by grouping. Remember, before you can divide out factors common to the numerators and denominators of a product, you must factor completely.

41. $\frac{x^2 - 9}{x^2 - 3x} \cdot \frac{2x + 10}{xy + 5x + 3y + 15}$

42. $\frac{x^2 - 16}{x^2 - 4x} \cdot \frac{3x + 18}{xy + 6x + 4y + 24}$

43. $\frac{2x^2 + 4x}{x^2 - y^2} \cdot \frac{x^2 + 3x + xy + 3y}{x^2 + 5x + 6}$

44. $\frac{x^2 - 25}{3x^2 + 3xy} \cdot \frac{x^2 + 4x + xy + 4y}{x^2 + 9x + 20}$

Review Problems The following problems review material we covered in Section 4.3. Reviewing these problems will help you in the next section.

Add the following fractions.

45. $\frac{4}{2} + \frac{5}{2}$

46. $\frac{2}{3} + \frac{8}{3}$

47. $\frac{6}{8} + \frac{2}{8}$

48. $\frac{3}{7} + \frac{4}{7}$

49. $\frac{4}{15} + \frac{6}{15}$

50. $\frac{5}{14} + \frac{2}{14}$

Simplify each term, then add.

51. $\dfrac{10x^4}{2x^2} + \dfrac{12x^6}{3x^4}$

52. $\dfrac{32x^8}{8x^3} + \dfrac{27x^7}{3x^2}$

53. $\dfrac{12a^2b^5}{3ab^3} + \dfrac{14a^4b^7}{7a^3b^5}$

54. $\dfrac{16a^3b^2}{4ab} + \dfrac{25a^6b^5}{5a^4b^4}$

Section 6.3 Addition and Subtraction of Rational Expressions

In Section 4.3 we combined fractions having the same denominator by combining their numerators and putting the result over the common denominator. We use the same process to add two rational expressions with the same denominator.

▼ **Examples**

1. Add: $\dfrac{5}{x} + \dfrac{3}{x}$.

Solution Adding numerators, we have

$$\frac{5}{x} + \frac{3}{x} = \frac{8}{x}$$

2. Add: $\dfrac{x}{x^2 - 9} + \dfrac{3}{x^2 - 9}$.

Solution Since both expressions have the same denominator, we add numerators and reduce to lowest terms:

$$\frac{x}{x^2 - 9} + \frac{3}{x^2 - 9} = \frac{x + 3}{x^2 - 9}$$

$$= \frac{\cancel{x + 3}}{(\cancel{x + 3})(x - 3)}$$

$$= \frac{1}{x - 3}$$

Reduce to lowest terms by factoring the denominator and then dividing out the common factor $x + 3$ ▲

It is the distributive property that allows us to add rational expressions by simply adding numerators. Here is Example 1 again, this time showing the use of the distributive property:

$$\frac{5}{x} + \frac{3}{x} = 5 \cdot \frac{1}{x} + 3 \cdot \frac{1}{x}$$ Dividing by x is the same as multiplying by its reciprocal, $\frac{1}{x}$

$$= (5 + 3) \cdot \frac{1}{x}$$ Distributive property

$$= 8 \cdot \frac{1}{x}$$

$$= \frac{8}{x}$$

It is not necessary to show the distributive property when adding rational expressions. The only reason for showing it here is so you can see that rational expressions cannot be combined by addition unless they have the same denominator. Because of this property, we must begin all addition problems involving rational expressions by first making sure all the expressions have the same denominator.

DEFINITION The *least common denominator* (LCD) for a set of denominators is the simplest quantity that is exactly divisible by all the denominators.

The LCD for the fractions $\frac{1}{4}$ and $\frac{2}{3}$ is 12, because 12 is the smallest positive number that is divisible by both 4 and 3. Similarly, the LCD for the rational expressions $\frac{1}{x}$ and $\frac{2}{3}$ is the expression $3x$, because $3x$ is the simplest expression that both 3 and x divide evenly. Not all LCDs are so easy to find. Sometimes we have to factor each denominator and then build the LCD from the factors.

▼ **Example 3** Add: $\frac{1}{10} + \frac{3}{14}$.

Solution

Step 1: Find the LCD for 10 and 14. To do so, we factor each denominator and build the LCD from the factors:

$$\left.\begin{aligned} 10 &= 2 \cdot 5 \\ 14 &= 2 \cdot 7 \end{aligned}\right\} \quad \text{LCD} = 2 \cdot 5 \cdot 7 = 70$$

We know the LCD is divisible by 10 because it contains the factors 2 and 5. It is also divisible by 14 because it contains the factors 2 and 7.

Step 2: Change to equivalent fractions that each have denominator 70. To accomplish this task, we multiply the numerator and denominator of each fraction by the factor of the LCD that is not also a factor of its denominator:

Original Fractions		*Denominators in Factored Form*		*Multiply by Factor Needed to Obtain LCD*		*These Have the Same Value as the Original Fractions*
$\frac{1}{10}$	=	$\frac{1}{2 \cdot 5}$	=	$\frac{1}{2 \cdot 5} \cdot \frac{\mathbf{7}}{\mathbf{7}}$	=	$\frac{7}{70}$
$\frac{3}{14}$	=	$\frac{3}{2 \cdot 7}$	=	$\frac{3}{2 \cdot 7} \cdot \frac{\mathbf{5}}{\mathbf{5}}$	=	$\frac{15}{70}$

The fraction $\frac{7}{70}$ has the same value as the fraction $\frac{1}{10}$. Similarly, the fractions $\frac{15}{70}$ and $\frac{3}{14}$ are equivalent; they have the same value.

Step 3: Add numerators and put the result over the LCD:

$$\frac{7}{70} + \frac{15}{70} = \frac{7 + 15}{70} = \frac{22}{70}$$

Step 4: Reduce to lowest terms:

$$\frac{22}{70} = \frac{11}{35} \quad \text{Divide numerator and denominator by 2}$$

▲

The main idea in adding fractions is to write each fraction again with the LCD for a denominator. Once we have done that, we simply add numerators. The same process can be used to add rational expressions, as the next example illustrates.

▼ **Example 4** Subtract: $\frac{3}{x} - \frac{1}{2}$.

Solution

Step 1: The LCD for x and 2 is $2x$. It is the smallest expression divisible by x and by 2.

Step 2: To change to equivalent expressions with the denominator $2x$, we multiply the first fraction by $\frac{2}{2}$ and the second by $\frac{x}{x}$:

$$\frac{3}{x} \cdot \frac{\mathbf{2}}{\mathbf{2}} = \frac{6}{2x}$$

$$\frac{1}{2} \cdot \frac{\boldsymbol{x}}{\boldsymbol{x}} = \frac{x}{2x}$$

Step 3: Subtracting numerators of the rational expressions in Step 2, we have

$$\frac{6}{2x} - \frac{x}{2x} = \frac{6 - x}{2x}$$

Step 4: Since $6 - x$ and $2x$ do not have any factors in common, we cannot reduce any further.

This is how the complete problem looks:

$$\frac{3}{x} - \frac{1}{2} = \frac{3}{x} \cdot \frac{\mathbf{2}}{\mathbf{2}} - \frac{1}{2} \cdot \frac{\boldsymbol{x}}{\boldsymbol{x}}$$

$$= \frac{6}{2x} - \frac{x}{2x}$$

$$= \frac{6 - x}{2x}$$

▲

▼ **Example 5** Add: $\dfrac{5}{2x - 6} + \dfrac{x}{x - 3}$

Solution If we factor $2x - 6$, we have $2x - 6 = 2(x - 3)$. We need only multiply the second rational expression in our problem by $\frac{2}{2}$ to have two expressions with the same denominator:

$$\frac{5}{2x - 6} + \frac{x}{x - 3} = \frac{5}{2(x - 3)} + \frac{x}{x - 3}$$

$$= \frac{5}{2(x - 3)} + \frac{\mathbf{2}}{\mathbf{2}}\left(\frac{x}{x - 3}\right)$$

$$= \frac{5}{2(x - 3)} + \frac{2x}{2(x - 3)}$$

$$= \frac{2x + 5}{2(x - 3)}$$

▲

▼ **Example 6** Add: $\dfrac{1}{x + 4} + \dfrac{8}{x^2 - 16}$.

Solution After writing each denominator in factored form, we find that the least common denominator is $(x + 4)(x - 4)$. To change the first fraction to an equivalent fraction with the common denominator, we multiply its numerator and denominator by $x - 4$:

$$\frac{1}{x + 4} + \frac{8}{x^2 - 16}$$

$$= \frac{1}{x + 4} + \frac{8}{(x + 4)(x - 4)} \quad \text{Factor each denominator}$$

$$= \frac{1}{x + 4} \cdot \frac{\boldsymbol{x - 4}}{\boldsymbol{x - 4}} + \frac{8}{(x + 4)(x - 4)} \quad \text{Change to equivalent fractions}$$

$$= \frac{x - 4}{(x + 4)(x - 4)} + \frac{8}{(x + 4)(x - 4)} \quad \text{Simplify}$$

$$= \frac{x + 4}{(x + 4)(x - 4)}$$ Add numerators

$$= \frac{1}{x - 4}$$ Divide out common factor $x + 4$

Note that in the last step we reduced the fraction to lowest terms by dividing out the common factor of $x + 4$. The last thing you should do in any addition or subtraction problem involving fractions is to see if your answer can be reduced further. ▲

▼ **Example 7** Add: $\dfrac{2}{x^2 + 5x + 6} + \dfrac{x}{x^2 - 9}$.

Solution

Step 1: We factor each denominator and build the LCD from the factors:

$$\left.\begin{aligned} x^2 + 5x + 6 &= (x + 2)(x + 3) \\ x^2 - 9 &= (x + 3)(x - 3) \end{aligned}\right\} \quad \text{LCD} = (x + 2)(x + 3)(x - 3)$$

Step 2: Change to equivalent rational expressions:

$$\frac{2}{x^2 + 5x + 6} = \frac{2}{(x + 2)(x + 3)} \cdot \frac{\mathbf{(x - 3)}}{\mathbf{(x - 3)}} = \frac{2x - 6}{(x + 2)(x + 3)(x - 3)}$$

$$\frac{x}{x^2 - 9} = \frac{x}{(x + 3)(x - 3)} \cdot \frac{\mathbf{(x + 2)}}{\mathbf{(x + 2)}} = \frac{x^2 + 2x}{(x + 2)(x + 3)(x - 3)}$$

Step 3: Add numerators of the rational expressions produced in Step 2:

$$\frac{2x - 6}{(x + 2)(x + 3)(x - 3)} + \frac{x^2 + 2x}{(x + 2)(x + 3)(x - 3)}$$

$$= \frac{x^2 + 4x - 6}{(x + 2)(x + 3)(x - 3)}$$

The numerator and denominator do not have any factors in common. ▲

▼ **Example 8** Subtract: $\dfrac{x}{x^2 - 25} - \dfrac{2}{3x + 15}$.

Solution We begin by factoring each denominator:

$$\frac{x}{x^2 - 25} - \frac{2}{3x + 15} = \frac{x}{(x + 5)(x - 5)} + \frac{-2}{3(x + 5)}$$

The LCD is $3(x + 5)(x - 5)$. Completing the problem, we have

$$= \frac{\mathbf{3}}{\mathbf{3}} \cdot \frac{x}{(x+5)(x-5)} + \frac{-2}{3(x+5)} \cdot \frac{\mathbf{(x-5)}}{\mathbf{(x-5)}}$$

$$= \frac{3x}{3(x+5)(x-5)} + \frac{-2x+10}{3(x+5)(x-5)}$$

$$= \frac{x+10}{3(x+5)(x-5)}$$ ▲

Notice in the first step that we replaced subtraction by addition of the opposite. There seems to be less chance for error when this is done on longer problems.

▼ **Example 9** Write an expression for the sum of a number and its reciprocal and then simplify that expression.

Solution If we let $x =$ the number, then its reciprocal is $\frac{1}{x}$. To find the sum of the number and its reciprocal we add them:

$$x + \frac{1}{x}$$

The first term, x, can be thought of as having a denominator of 1. Since the denominators are 1 and x, the least common denominator is x. Here is the complete problem:

$$x + \frac{1}{x} = \frac{x}{1} + \frac{1}{x} \qquad \text{Write } x \text{ as } \frac{x}{1}$$

$$= \frac{x}{1} \cdot \frac{\mathbf{x}}{\mathbf{x}} + \frac{1}{x} \qquad \text{The LCD is } x$$

$$= \frac{x^2}{x} + \frac{1}{x}$$

$$= \frac{x^2+1}{x} \qquad \text{Add numerators}$$ ▲

Problem Set 6.3

Find the following sums and differences.

1. $\frac{3}{x} + \frac{4}{x}$

2. $\frac{5}{x} + \frac{3}{x}$

3. $\frac{9}{a} - \frac{5}{a}$

4. $\frac{8}{a} - \frac{7}{a}$

5. $\frac{1}{x+1} + \frac{x}{x+1}$

6. $\frac{x}{x-3} - \frac{3}{x-3}$

7. $\frac{y^2}{y - 1} - \frac{1}{y - 1}$

8. $\frac{y^2}{y + 3} - \frac{9}{y + 3}$

9. $\frac{x^2}{x + 2} + \frac{4x + 4}{x + 2}$

10. $\frac{x^2 - 6x}{x - 3} + \frac{9}{x - 3}$

11. $\frac{x^2}{x - 2} - \frac{4x - 4}{x - 2}$

12. $\frac{x^2}{x - 5} - \frac{10x - 25}{x - 5}$

13. $\frac{x + 2}{x + 6} - \frac{x - 4}{x + 6}$

14. $\frac{x + 5}{x + 2} - \frac{x + 3}{x + 2}$

15. $\frac{y}{2} - \frac{2}{y}$

16. $\frac{3}{y} + \frac{y}{3}$

17. $\frac{1}{2} + \frac{a}{3}$

18. $\frac{2}{3} + \frac{2a}{5}$

19. $\frac{x}{x + 1} + \frac{3}{4}$

20. $\frac{x}{x - 3} + \frac{1}{3}$

21. $\frac{3}{x - 2} + \frac{2}{5x - 10}$

22. $\frac{5}{2x - 6} + \frac{3}{x - 3}$

23. $\frac{4}{3x + 12} - \frac{x}{x + 4}$

24. $\frac{2}{5x - 25} - \frac{x}{x - 5}$

25. $\frac{6}{x(x - 2)} + \frac{3}{x}$

26. $\frac{10}{x(x + 5)} - \frac{2}{x}$

27. $\frac{4}{a} - \frac{12}{a^2 + 3a}$

28. $\frac{5}{a} + \frac{20}{a^2 - 4a}$

29. $\frac{2}{x + 5} + \frac{1}{x^2 - 25}$

30. $\frac{4}{x^2 - 1} + \frac{3}{x + 1}$

31. $\frac{1}{x - 3} - \frac{6}{x^2 - 9}$

32. $\frac{1}{x - 1} - \frac{2}{x^2 - 1}$

33. $\frac{5a}{a^2 - a - 6} - \frac{2}{a - 3}$

34. $\frac{3a}{a^2 - 5a - 6} - \frac{4}{a + 1}$

35. $\frac{2}{2x - 6} - \frac{5}{x^2 - 9}$

36. $\frac{x}{3x - 3} - \frac{1}{x^2 - 2x + 1}$

37. $\frac{5}{a} + \frac{2}{a + 1}$

38. $\frac{a}{a - 3} + \frac{7}{a}$

39. $\frac{8}{x^2 - 16} - \frac{7}{x^2 - x - 12}$

40. $\frac{6}{x^2 - 9} - \frac{5}{x^2 - x - 6}$

41. $\frac{4y}{y^2 + 6y + 5} - \frac{3y}{y^2 + 5y + 4}$

42. $\frac{3y}{y^2 + 7y + 10} - \frac{2y}{y^2 + 6y + 8}$

43. Write an expression for the sum of a number and twice its reciprocal. Then, simplify that expression. (If the reciprocal of a number is $\frac{1}{x}$, then twice that is $\frac{2}{x}$, not $\frac{1}{2x}$.)

44. Write an expression for the sum of a number and three times its reciprocal. Then, simplify that expression.

45. One number is twice another. Write an expression for the sum of their reciprocals. Then, simplify that expression. (Hint: the numbers are x and $2x$. Their reciprocals are respectively $\frac{1}{x}$ and $\frac{1}{2x}$.)

46. One number is three times another. Write an expression for the sum of their reciprocals. Then, simplify that expression.

Review Problems The following problems review material we covered in Sections 2.4 and 5.7. Reviewing these problems will help you understand the next section.

Solve each equation. [2.4]

47. $2x + 3(x - 3) = 6$

48. $4x - 2(x - 5) = 6$

49. $x - 3(x + 3) = x - 3$

50. $x - 4(x + 4) = x - 4$

51. $7 - 2(3x + 1) = 4x + 3$

52. $8 - 5(2x - 1) = 2x + 4$

Solve each quadratic equation. [5.7]

53. $x^2 + 5x + 6 = 0$

54. $x^2 - 5x + 6 = 0$

55. $x^2 - x = 6$

56. $x^2 + x = 6$

57. $x^2 - 5x = 0$

58. $x^2 - 6x = 0$

Section 6.4 Equations Involving Rational Expressions

The first step in solving an equation that contains one or more rational expressions is to find the LCD for all denominators in the equation. Once the LCD has been found, we multiply both sides of the equation by it. The resulting equation should be equivalent to the original one (unless we inadvertently multiplied by zero) and free from any denominators except the number 1.

▼ **Example 1** Solve $\dfrac{x}{3} + \dfrac{5}{2} = \dfrac{1}{2}$ for x.

Solution The LCD for 3 and 2 is 6. If we multiply both sides by 6, we have

$$6\left(\frac{x}{3} + \frac{5}{2}\right) = 6\left(\frac{1}{2}\right) \quad \text{Multiply both sides by 6}$$

$$6\left(\frac{x}{3}\right) + 6\left(\frac{5}{2}\right) = 6\left(\frac{1}{2}\right) \quad \text{Distributive property}$$

$$2x + 15 = 3$$

$$2x = -12$$

$$x = -6$$

We can check our solution by replacing x with -6 in the original equation:

$$-\frac{6}{3} + \frac{5}{2} \stackrel{?}{=} \frac{1}{2}$$

$$\frac{1}{2} = \frac{1}{2}$$

▲

Multiplying both sides of an equation containing fractions by the LCD clears the equation of all denominators, because the LCD has the property that all the denominators will divide it evenly.

▼ **Example 2** Solve for x: $\dfrac{3}{x - 1} = \dfrac{3}{5}$.

Solution The LCD for $(x - 1)$ and 5 is $5(x - 1)$. Multiplying both sides by $5(x - 1)$ we have

$$\begin{aligned} 5(x - 1) \cdot \frac{3}{x - 1} &= 5(x - 1) \cdot \frac{3}{5} \\ 5 \cdot 3 &= (x - 1) \cdot 3 \\ 15 &= 3x - 3 \\ 18 &= 3x \\ 6 &= x \end{aligned}$$

If we substitute $x = 6$ into the original equation, we have

$$\frac{3}{6 - 1} \stackrel{?}{=} \frac{3}{5}$$

$$\frac{3}{5} = \frac{3}{5}$$

The solution set is $\{6\}$. ▲

▼ **Example 3** Solve: $1 - \dfrac{5}{x} = \dfrac{-6}{x^2}$.

Solution The LCD is x^2. Multiplying both sides by x^2, we have

$$x^2\left(1 - \frac{5}{x}\right) = x^2\left(\frac{-6}{x^2}\right) \quad \text{Multiply both sides by } x^2$$

$$x^2(1) - x^2\left(\frac{5}{x}\right) = x^2\left(\frac{-6}{x^2}\right) \quad \text{Apply distributive property to the left side}$$

$$x^2 - 5x = -6 \quad \text{Simplify each side}$$

We have a quadratic equation, which we write in standard form, factor, and solve as we did in Section 5.7.

$$
\begin{aligned}
x^2 - 5x + 6 &= 0 && \text{Standard form} \\
(x - 2)(x - 3) &= 0 && \text{Factor} \\
x - 2 = 0 \quad \text{or} \quad x - 3 &= 0 && \text{Set factors equal to 0} \\
x = 2 \quad \text{or} \quad x &= 3
\end{aligned}
$$

The two possible solutions are 2 and 3. Checking each in the original equation, we find they both give true statements. They are both solutions to the original equation.

Check $x = 2$	Check $x = 3$
$1 - \frac{5}{2} \stackrel{?}{=} \frac{-6}{4}$	$1 - \frac{5}{3} \stackrel{?}{=} \frac{-6}{9}$
$\frac{2}{2} - \frac{5}{2} = -\frac{3}{2}$	$\frac{3}{3} - \frac{5}{3} = -\frac{2}{3}$
$-\frac{3}{2} = -\frac{3}{2}$	$-\frac{2}{3} = -\frac{2}{3}$

▲

▼ **Example 4** Solve: $\dfrac{x}{x^2 - 9} - \dfrac{3}{x - 3} = \dfrac{1}{x + 3}$.

Solution The factors of $x^2 - 9$ are $(x + 3)(x - 3)$. The LCD, then, is $(x + 3)(x - 3)$:

$$
(x + 3)(x - 3) \cdot \frac{x}{(x + 3)(x - 3)} + (x + 3)(x - 3) \cdot \frac{-3}{x - 3} = (x + 3)(x - 3) \cdot \frac{1}{x + 3}
$$

$$
\begin{aligned}
x + (x + 3)(-3) &= (x - 3)1 \\
x + (-3x) + (-9) &= x - 3 \\
-2x - 9 &= x - 3 \\
-3x &= 6 \\
x &= -2
\end{aligned}
$$

The solution is $x = -2$. It checks when replaced for x in the original equation. ▲

▼ **Example 5** Solve: $\dfrac{x}{x - 3} + \dfrac{3}{2} = \dfrac{3}{x - 3}$.

Solution We begin by multiplying each term on both sides of the equation by $2(x - 3)$:

$$
2(x - 3) \cdot \frac{x}{x - 3} + 2(x - 3) \cdot \frac{3}{2} = 2(x - 3) \cdot \frac{3}{x - 3}
$$

$$
\begin{aligned}
2x + (x - 3) \cdot 3 &= 2 \cdot 3 \\
2x + 3x - 9 &= 6 \\
5x - 9 &= 6 \\
5x &= 15 \\
x &= 3
\end{aligned}
$$

Our only possible solution is $x = 3$. If we substitute $x = 3$ into our original equation, we get

$$\frac{3}{3-3} + \frac{3}{2} \stackrel{?}{=} \frac{3}{3-3}$$
$$\frac{3}{0} + \frac{3}{2} = \frac{3}{0}$$

Two of the terms are undefined, so the equation is meaningless. What has happened is that we multiplied both sides of the original equation by zero. The equation produced by doing this is not equivalent to our original equation. We must always check our solution when we multiply both sides of an equation by an expression containing the variable in order to make sure we have not multiplied both sides by zero.

Our original equation has no solutions. That is, there is no real number x such that

$$\frac{x}{x-3} + \frac{3}{2} = \frac{3}{x-3}$$

The solution set is $\emptyset$. ▲

▼ **Example 6** Solve $\dfrac{a+4}{a^2+5a} = \dfrac{-2}{a^2-25}$ for a.

Solution Factoring each denominator, we have

$$a^2 + 5a = a(a+5)$$
$$a^2 - 25 = (a+5)(a-5)$$

The LCD is $a(a+5)(a-5)$. Multiplying both sides of the equation by the LCD gives us

$$\cancel{a(a+5)}(a-5) \cdot \frac{a+4}{\cancel{a(a+5)}} = \frac{-2}{\cancel{(a+5)(a-5)}} \cdot a\cancel{(a+5)(a-5)}$$
$$(a-5)(a+4) = -2a$$
$$a^2 - a - 20 = -2a$$

The result is a quadratic equation, which we write in standard form, factor, and solve:

$$a^2 + a - 20 = 0 \qquad \text{Add } 2a \text{ to both sides}$$
$$(a+5)(a-4) = 0 \qquad \text{Factor}$$
$$a + 5 = 0 \quad \text{or} \quad a - 4 = 0 \qquad \text{Set each factor to 0}$$
$$a = -5 \quad \text{or} \quad a = 4$$

The two possible solutions are -5 and 4. There is no problem with the

4. It checks when substituted for a in the original equation. However, -5 is not a solution. Substituting -5 into the original equation gives

$$\frac{-5+4}{(-5)^2+5(-5)} \stackrel{?}{=} \frac{-2}{(-5)^2-25}$$

$$\frac{-1}{0} = \frac{-2}{0}$$

This indicates -5 is not a solution. The solution is 4. ▲

Problem Set 6.4

Solve the following equations. Be sure to check each answer in the original equation if you multiply both sides by an expression that contains the variable.

1. $\frac{x}{3} + \frac{1}{2} = -\frac{1}{2}$

2. $\frac{x}{2} + \frac{4}{3} = -\frac{2}{3}$

3. $\frac{4}{a} = \frac{1}{5}$

4. $\frac{2}{3} = \frac{6}{a}$

5. $\frac{3}{x} + 1 = \frac{2}{x}$

6. $\frac{4}{x} + 3 = \frac{1}{x}$

7. $\frac{3}{a} - \frac{2}{a} = \frac{1}{5}$

8. $\frac{7}{a} + \frac{1}{a} = 2$

9. $\frac{3}{x} + 2 = \frac{1}{2}$

10. $\frac{5}{x} + 3 = \frac{4}{3}$

11. $\frac{1}{y} - \frac{1}{2} = -\frac{1}{4}$

12. $\frac{3}{y} - \frac{4}{5} = -\frac{1}{5}$

13. $1 - \frac{8}{x} = \frac{-15}{x^2}$

14. $1 - \frac{3}{x} = \frac{-2}{x^2}$

15. $\frac{x}{2} - \frac{4}{x} = -\frac{7}{2}$

16. $\frac{x}{2} - \frac{5}{x} = -\frac{3}{2}$

17. $\frac{x-3}{2} + \frac{2x}{3} = \frac{5}{6}$

18. $\frac{x-2}{3} + \frac{5x}{2} = 5$

19. $\frac{x+1}{3} + \frac{x-3}{4} = \frac{1}{6}$

20. $\frac{x+2}{3} + \frac{x-1}{5} = -\frac{3}{5}$

21. $\frac{6}{x+2} = \frac{3}{5}$

22. $\frac{4}{x+3} = \frac{1}{2}$

23. $\frac{3}{y-2} = \frac{2}{y-3}$

24. $\frac{5}{y+1} = \frac{4}{y+2}$

25. $\frac{x}{x-2} + \frac{2}{3} = \frac{2}{x-2}$

26. $\frac{x}{x-5} + \frac{1}{5} = \frac{5}{x-5}$

27. $\frac{x}{x-2}+\frac{3}{2}=\frac{9}{2(x-2)}$

28. $\frac{x}{x+1}+\frac{4}{5}=\frac{-14}{5(x+1)}$

29. $\frac{5}{x+2}+\frac{1}{x+3}=\frac{-1}{x^2+5x+6}$

30. $\frac{3}{x-1}+\frac{2}{x+3}=\frac{-3}{x^2+2x-3}$

31. $\frac{8}{x^2-4}+\frac{3}{x+2}=\frac{1}{x-2}$

32. $\frac{10}{x^2-25}-\frac{1}{x-5}=\frac{3}{x+5}$

33. $\frac{a}{2}+\frac{3}{a-3}=\frac{a}{a-3}$

34. $\frac{a}{2}+\frac{4}{a-4}=\frac{a}{a-4}$

35. $\frac{6}{y^2-4}=\frac{4}{y^2+2y}$

36. $\frac{2}{y^2-9}=\frac{5}{y^2-3y}$

37. $\frac{2}{a^2-9}=\frac{3}{a^2+a-12}$

38. $\frac{2}{a^2-1}=\frac{6}{a^2-2a-3}$

39. $\frac{3x}{x-5}-\frac{2x}{x+1}=\frac{-42}{x^2-4x-5}$

40. $\frac{4x}{x-4}-\frac{3x}{x-2}=\frac{-3}{x^2-6x+8}$

41. $\frac{2x}{x+2}=\frac{x}{x+3}-\frac{3}{x^2+5x+6}$

42. $\frac{3x}{x-4}=\frac{2x}{x-3}+\frac{6}{x^2-7x+12}$

Review Problems The problems that follow review material we covered in Sections 2.6 and 5.8. Reviewing these problems will help you with the next section.

Solve each word problem. [2.6]

43. If twice the difference of a number and three were decreased by five, the result would be three. Find the number.

44. If three times the sum of a number and two were increased by six, the result would be 27. Find the number.

45. The length of a rectangle is five more than twice the width. The perimeter is 34 inches. Find the length and width.

46. The length of a rectangle is two more than three times the width. The perimeter is 44 feet. Find the length and width.

Solve each problem. Be sure to show the equation that describes the situation. [5.8]

47. The product of two consecutive even integers is 48. Find the two integers.

48. The product of two consecutive odd integers is 35. Find the two integers.

49. The hypotenuse (the longest side) of a right triangle is 10 inches, while the lengths of the two legs (the other two sides) are given by two consecutive even integers. Find the lengths of the two legs.

50. One leg of a right triangle is two more than twice the other. If the hypotenuse is 13 feet, find the lengths of the two legs.

Section 6.5 Word Problems

In this section we will solve word problems whose equations involve rational expressions. Like the other word problems we have encountered, the more you work with them the easier they become.

▼ **Example 1** One number is twice another. The sum of their reciprocals is $\frac{9}{2}$. Find the two numbers.

Solution Let x = the smaller number. The larger number then must be $2x$. Their reciprocals are $\frac{1}{x}$ and $\frac{1}{2x}$, respectively. An equation that describes the situation is

$$\frac{1}{x} + \frac{1}{2x} = \frac{9}{2}$$

We can multiply both sides by the LCD $2x$ and then solve the resulting equation:

$$2x\left(\frac{1}{x}\right) + 2x\left(\frac{1}{2x}\right) = 2x\left(\frac{9}{2}\right)$$

$$2 + 1 = 9x$$

$$3 = 9x$$

$$x = \frac{3}{9}$$

$$x = \frac{1}{3}$$

The smaller number is $\frac{1}{3}$. The other number is twice as large, or $\frac{2}{3}$. If we add their reciprocals, we have

$$\frac{3}{1} + \frac{3}{2} = \frac{6}{2} + \frac{3}{2}$$

$$= \frac{9}{2}$$

The solutions check with the original problem. ▲

▼ **Example 2** A boat travels 30 miles up a river in the same amount of time it takes to travel 50 miles down the same river. If the current is 5 miles per hour, what is the speed of the boat in still water?

Solution The easiest way to work a problem like this is with a table. The top row of the table is labeled with d for distance, r for rate, and t for time. The left column of the table is labeled with the two trips upstream and downstream. Here is what the table looks like:

	d	r	t
upstream			
downstream			

The next step is to read the problem over again and fill in as much of the table as we can with the information in the problem. The distance the boat travels upstream is 30 and the distance downstream is 50. Since we are asked for the speed of the boat in still water, we will let that be x. If the speed of the boat in still water is x, then its speed upstream (against the current) must be $x - 5$, and its speed downstream (with the current) must be $x + 5$. Putting these four quantities into the appropriate positions in the table we have

	d	r	t
upstream	30	$x - 5$	
downstream	50	$x + 5$	

The last two positions in the table are filled in by using the equation $d = r \cdot t$. Since we want to fill in positions that are in the time category, we use the equivalent form of the equation, $t = \frac{d}{r}$, that is, time is equal to distance divided by rate:

	d	r	t
upstream	30	$x - 5$	$\frac{30}{x - 5}$
downstream	50	$x + 5$	$\frac{50}{x + 5}$

The last step is to read the problem again and look for a relationship between two of the squares in the table. Reading the problem again we find that the time for the trip upstream is equal to the time for the trip downstream. Setting these two quantities equal to each other, we have our equation:

$$\text{time (downstream)} = \text{time (upstream)}$$

$$\frac{50}{x + 5} = \frac{30}{x - 5}$$

The LCD is $(x + 5)(x - 5)$. We multiply both sides of the equation by the LCD to clear it of all denominators. Here is the solution:

$$\begin{aligned} (x + 5)(x - 5)\frac{50}{x + 5} &= (x + 5)(x - 5) \cdot \frac{30}{x - 5} \\ (x - 5)50 &= (x + 5)30 \\ 50x - 250 &= 30x + 150 \\ 20x &= 400 \\ x &= 20 \end{aligned}$$

The speed of the boat in still water is 20 mph. ▲

▼ **Example 3** An inlet pipe can fill a water tank in 10 hours, while an outlet pipe can empty the same tank in 15 hours. By mistake, both pipes are left open. How long will it take to fill the water tank with both pipes open?

Solution Let x = amount of time to fill the tank with both pipes open.

One method of solving this type of problem is to think in terms of how much of the job is done by a pipe in 1 hour:

1. If the inlet pipe fills the tank in 10 hours, then in 1 hour the inlet pipe fills $\frac{1}{10}$ of the tank.
2. If the outlet pipe empties the tank in 15 hours, then in 1 hour the outlet pipe empties $\frac{1}{15}$ of the tank.
3. If it takes x hours to fill the tank with both pipes open, then in 1 hour the tank is $\frac{1}{x}$ full.

Here is how we set up the equation. *In one hour,*

$$\underbrace{\frac{1}{10}}_{\text{Amount of water let in by inlet pipe}} - \underbrace{\frac{1}{15}}_{\text{Amount of water let out by outlet pipe}} = \underbrace{\frac{1}{x}}_{\text{Total amount of water in tank}}$$

The LCD for our equation is $30x$. We multiply both sides by the LCD and solve:

$$30x\left(\frac{1}{10}\right) - 30x\left(\frac{1}{15}\right) = 30x\left(\frac{1}{x}\right)$$

$$3x - 2x = 30$$

$$x = 30$$

It takes 30 hours with both pipes open to fill the tank. ▲

Note In solving a problem like the one in Example 3, we have to assume that the thing doing the work (whether it is a pipe, a person, or a machine) is working at a constant rate. That is, as much work gets done in the first hour as is done in the last hour and any other hour in between.

Problem Set 6.5

Number Problems

1. One number is three times as large as another. The sum of their reciprocals is $\frac{16}{3}$. Find the two numbers.

2. If $\frac{3}{5}$ is added to twice the reciprocal of a number, the result is 1. Find the number.

3. The sum of a number and its reciprocal is $\frac{13}{6}$. Find the number.

4. The sum of a number with 10 times its reciprocal is 7. Find the number.
5. If a certain number is added to both the numerator and denominator of the fraction $\frac{7}{9}$, the result is $\frac{5}{7}$. Find the number.
6. The numerator of a certain fraction is two more than the denominator. If $\frac{1}{3}$ is added to the fraction, the result is 2. Find the fraction.
7. The sum of the reciprocals of two consecutive even integers is $\frac{5}{12}$. Find the integers.
8. The sum of the reciprocals of two consecutive integers is $\frac{7}{12}$. Find the two integers.

Motion Problems

9. A boat travels 26 miles up a river in the same amount of time it takes to travel 38 miles down the same river. If the current is 3 mph, what is the speed of the boat in still water?

	d	r	t
upstream			
downstream			

10. A boat can travel 9 miles up a river in the same amount of time it takes to travel 11 miles down the same river. If the current is 2 mph, what is the speed of the boat in still water?

	d	r	t
upstream			
downstream			

11. An airplane flying against the wind travels 140 miles in the same amount of time it would take the same plane to travel 160 miles with the wind. If the wind speed is a constant 20 mph, how fast would the plane travel in still air?
12. An airplane flying against the wind travels 500 miles in the same amount of time that it would take to travel 600 miles with the wind. If the speed of the wind is 50 mph, what is the speed of the plane in still air?
13. One plane can travel 20 mph faster than another. One of them goes 285 miles in the same time it takes the other to go 255 miles. What are their speeds?
14. One car travels 300 miles in the same amount of time it takes a second car traveling 5 mph slower than the first to go 275 miles. What are the speeds of the cars?

Work Problems

15. An inlet pipe can fill a pool in 12 hours, while an outlet pipe can empty it in 15 hours. If both pipes are left open, how long will it take to fill the pool?
16. A water tank can be filled in 20 hours by an inlet pipe and emptied in 25 hours by an outlet pipe. How long will it take to fill the tank if both pipes are left open?
17. A bathtub can be filled by the cold water faucet in 10 minutes and by the hot water faucet in 12 minutes. How long does it take to fill the tub if both faucets are open?
18. A water faucet can fill a sink in 6 minutes, while the drain can empty it in 4 minutes. If the sink is full, how long will it take to empty if both the faucet and the drain are open?

Review Problems The following problems review material we covered in Sections 3.6 and 3.7.

Solve each system of equations by the elimination method. [3.6]

19. $2x + y = 3$
$3x - y = 7$

20. $3x - y = -6$
$4x + y = -8$

21. $4x - 5y = 1$
$x - 2y = -2$

22. $6x - 4y = 2$
$2x + y = 10$

Solve by the substitution method. [3.7]

23. $5x + 2y = 7$
$y = 3x - 2$

24. $-7x - 5y = -1$
$y = x + 5$

25. $2x - 3y = 4$
$x = 2y + 1$

26. $4x - 5y = 2$
$x = 2y - 1$

Section 6.6 Complex Fractions

A complex fraction is a fraction or rational expression that contains other fractions in its numerator or denominator. Each of the following is a complex fraction:

$$\frac{\frac{1}{2}}{\frac{2}{3}}, \quad \frac{x + \frac{1}{y}}{y + \frac{1}{x}}, \quad \frac{\frac{a + 1}{a^2 - 9}}{\frac{2}{a + 3}}$$

We will begin this section by simplifying the first of these complex fractions. Before we do, though, let's agree on some vocabulary. So that we won't have to use phrases such as the numerator of the denominator, let's call the numerator of a complex fraction the *top* and the denominator of a complex fraction the *bottom*.

▼ **Example 1** Simplify: $\frac{\frac{1}{2}}{\frac{2}{3}}$.

Solution There are two methods we can use to solve this problem.

METHOD 1 We can multiply the top and bottom of this complex fraction by the LCD for both fractions. In this case, the LCD is 6:

$$\frac{\frac{1}{2}}{\frac{2}{3}} = \frac{6 \cdot \frac{1}{2}}{6 \cdot \frac{2}{3}} = \frac{3}{4}$$

METHOD 2 We can treat this as a division problem. Instead of dividing by $\frac{2}{3}$, we can multiply by its reciprocal $\frac{3}{2}$:

$$\frac{\frac{1}{2}}{\frac{2}{3}} = \frac{1}{2} \cdot \frac{3}{2} = \frac{3}{4}$$

Using either method, we obtain the same result. ▲

▼ **Example 2** Simplify:

$$\frac{\frac{2x^3}{y^2}}{\frac{4x}{y^5}}$$

Solution

METHOD 1 The LCD for each rational expression is y^5. Multiplying the top and bottom of the complex fraction by y^5, we have

$$\frac{\frac{2x^3}{y^2}}{\frac{4x}{y^5}} = \frac{y^5 \cdot \frac{2x^3}{y^2}}{y^5 \cdot \frac{4x}{y^5}} = \frac{2x^3y^3}{4x} = \frac{x^2y^3}{2}$$

METHOD 2 Instead of dividing by $\frac{4x}{y^5}$, we can multiply by its reciprocal, $\frac{y^5}{4x}$:

$$\frac{\frac{2x^3}{y^2}}{\frac{4x}{y^5}} = \frac{2x^3}{y^2} \cdot \frac{y^5}{4x} = \frac{x^2y^3}{2}$$

Again the result is the same, whether we use Method 1 or Method 2. ▲

▼ **Example 3** Simplify:

$$\frac{x + \frac{1}{y}}{y + \frac{1}{x}}$$

Solution To apply Method 2 as we did in the first two examples, we would have to simplify the top and bottom separately to obtain a single rational expression for both before we could multiply by the reciprocal.

It is much easier, in this case, to multiply the top and bottom by the LCD xy:

$$\frac{x + \frac{1}{y}}{y + \frac{1}{x}} = \frac{xy\left(x + \frac{1}{y}\right)}{xy\left(y + \frac{1}{x}\right)} \qquad \text{Multiply top and bottom by } xy$$

$$= \frac{xy \cdot x + xy \cdot \frac{1}{y}}{xy \cdot y + xy \cdot \frac{1}{x}} \qquad \text{Distributive property}$$

$$= \frac{x^2y + x}{xy^2 + y} \qquad \text{Simplify}$$

We can factor an x from $x^2y + x$ and a y from $xy^2 + y$ and then reduce to lowest terms:

$$= \frac{x(xy + 1)}{y(xy + 1)}$$

$$= \frac{x}{y}$$

▲

▼ **Example 4** Simplify:

$$\frac{1 - \frac{4}{x^2}}{1 - \frac{1}{x} - \frac{6}{x^2}}$$

Solution The easiest way to simplify this complex fraction is to multiply the top and bottom by the LCD, x^2:

$$\frac{1 - \frac{4}{x^2}}{1 - \frac{1}{x} - \frac{6}{x^2}} = \frac{x^2\left(1 - \frac{4}{x^2}\right)}{x^2\left(1 - \frac{1}{x} - \frac{6}{x^2}\right)} \qquad \text{Multiply top and bottom by } x^2$$

$$= \frac{x^2 \cdot 1 - x^2 \cdot \frac{4}{x^2}}{x^2 \cdot 1 - x^2 \cdot \frac{1}{x} - x^2 \cdot \frac{6}{x^2}} \qquad \text{Distributive property}$$

$$= \frac{x^2 - 4}{x^2 - x - 6} \qquad \text{Simplify}$$

$$= \frac{(x - 2)\cancel{(x + 2)}}{(x - 3)\cancel{(x + 2)}} \qquad \text{Factor}$$

$$= \frac{x - 2}{x - 3} \qquad \text{Reduce}$$

▲

Problem Set 6.6

Simplify each complex fraction.

1. $\dfrac{\frac{3}{4}}{\frac{1}{8}}$
2. $\dfrac{\frac{1}{3}}{\frac{5}{6}}$
3. $\dfrac{\frac{2}{3}}{4}$
4. $\dfrac{5}{\frac{1}{2}}$
5. $\dfrac{\frac{x^2}{y}}{\frac{x}{y^3}}$
6. $\dfrac{\frac{x^5}{y^3}}{\frac{x^2}{y^8}}$
7. $\dfrac{\frac{4x^3}{y^6}}{\frac{8x^2}{y^7}}$
8. $\dfrac{\frac{6x^4}{y}}{\frac{2x}{y^5}}$
9. $\dfrac{y + \frac{1}{x}}{x + \frac{1}{y}}$
10. $\dfrac{y - \frac{1}{x}}{x - \frac{1}{y}}$
11. $\dfrac{1 + \frac{1}{a}}{1 - \frac{1}{a}}$
12. $\dfrac{\frac{1}{a} - 1}{\frac{1}{a} + 1}$
13. $\dfrac{\frac{x + 1}{x^2 - 9}}{\frac{2}{x + 3}}$
14. $\dfrac{\frac{3}{x - 5}}{\frac{x + 1}{x^2 - 25}}$
15. $\dfrac{\frac{1}{a + 2}}{\frac{1}{a^2 - a - 6}}$
16. $\dfrac{\frac{1}{a^2 + 5a + 6}}{\frac{1}{a + 3}}$
17. $\dfrac{1 - \frac{9}{y^2}}{1 - \frac{1}{y} - \frac{6}{y^2}}$
18. $\dfrac{1 - \frac{4}{y^2}}{1 - \frac{2}{y} - \frac{8}{y^2}}$
19. $\dfrac{\frac{1}{y} + \frac{1}{x}}{\frac{1}{xy}}$
20. $\dfrac{\frac{1}{xy}}{\frac{1}{y} - \frac{1}{x}}$
21. $\dfrac{1 - \frac{1}{a^2}}{1 - \frac{1}{a}}$
22. $\dfrac{1 + \frac{1}{a}}{1 - \frac{1}{a^2}}$
23. $\dfrac{\frac{1}{10x} - \frac{y}{10x^2}}{\frac{1}{10} - \frac{y}{10x}}$
24. $\dfrac{\frac{1}{2x} + \frac{y}{2x^2}}{\frac{1}{4} + \frac{y}{4x}}$
25. $\dfrac{\frac{1}{a + 1} + 2}{\frac{1}{a + 1} + 3}$
26. $\dfrac{\frac{2}{a + 1} + 3}{\frac{3}{a + 1} + 4}$

Review Problems The following problems review material we covered in Section 2.7.

Solve each inequality.

27. $2x + 3 < 5$
28. $3x - 2 > 7$
29. $-3x \leq 21$
30. $-5x \geq -10$
31. $-2x + 8 > -4$
32. $-4x - 1 < 11$
33. $4 - 2(x + 1) \geq -2$
34. $6 - 2(x + 3) \leq -8$

Section 6.7 Ratio and Proportion

We will begin this section with the definition of the ratio of two numbers.

DEFINITION If a and b are any two numbers, $b \neq 0$, then the ratio of a and b is

$$\frac{a}{b}$$

(sometimes written $a:b$).

As you can see, ratios are another name for fractions or rational numbers. They are a way of comparing quantities. Since we can also think of $\frac{a}{b}$ as the quotient of a and b, ratios are also quotients. The following table gives some ratios in words and as fractions.

Ratio	As a Fraction	In Lowest Terms
25 to 75	$\frac{25}{75}$	$\frac{1}{3}$
8 to 2	$\frac{8}{2}$	$\frac{4}{1}$ With ratios it is common to leave the 1 in the denominator
20 to 16	$\frac{20}{16}$	$\frac{5}{4}$
$\frac{2}{5}$ to $\frac{3}{5}$	$\frac{\frac{2}{5}}{\frac{3}{5}}$	$\frac{2}{3}$ Found by multiplying $\frac{2}{5}$ by $\frac{5}{3}$

The table shows that ratios can involve all kinds of numbers—whole numbers, fractions, and complex fractions.

▼ **Example 1** A solution of hydrochloric acid (HCl) and water contains 49 milliliters of water and 21 milliliters of HCl. Find the ratio of HCl to water and of HCl to the total volume of the solution.

Solution The ratio of HCl to water is 21 to 49 or

$$\frac{21}{49} = \frac{3}{7}$$

The amount of total solution is $49 + 21 = 70$ milliliters. Therefore, the ratio of HCl to total solution is 21 to 70, or

$$\frac{21}{70} = \frac{3}{10}$$ ▲

When two ratios are equal, we say they form a *proportion*.

DEFINITION A proportion is a statement that two ratios are equal. If $\frac{a}{b}$ and $\frac{c}{d}$ are two equal ratios, then the statement

$$\frac{a}{b} = \frac{c}{d}$$

is called a proportion.

Each of the four numbers in a proportion is called a *term* of the proportion. We number the terms as follows:

$$\begin{array}{rcl} \text{First term} \rightarrow a & = & c \leftarrow \text{Third term} \\ \text{Second term} \rightarrow b & & d \leftarrow \text{Fourth term} \end{array}$$

The first and fourth terms are called the *extremes*, and the second and third terms are called the *means*:

Means $\frac{a}{b} = \frac{c}{d}$ Extremes

For example, in the proportion

$$\frac{3}{8} = \frac{12}{32}$$

the extremes are 3 and 32, and the means are 8 and 12.

Means-Extremes Property

If a, b, c, and d are real numbers with $b \neq 0$ and $d \neq 0$, then

$$\text{if} \quad \frac{a}{b} = \frac{c}{d}$$

$$\text{then} \quad ad = bc$$

In words: In any proportion, the product of the extremes is equal to the product of the means.

This property of proportions comes from the multiplication property of equality. We can use it to solve for a missing term in a proportion.

▼ **Example 2** Solve the proportion $\frac{3}{x} = \frac{6}{7}$ for x.

Solution We could solve for x by using the method developed in Section 6.4; that is, multiplying both sides by the LCD $7x$. Instead, let's use our new means-extremes property:

$\frac{3}{x} = \frac{6}{7}$	Extremes are 3 and 7; means are x and 6
$21 = 6x$	Product of extremes = product of means
$\frac{21}{6} = x$	Divide both sides by 6
$x = \frac{7}{2}$	Reduce to lowest terms ▲

▼ **Example 3** Solve for x: $\frac{x+1}{2} = \frac{3}{x}$.

Solution Again, we want to point out that we could solve for x by using the method described in Section 6.4. Using the means-extremes property is simply an alternative to the method developed in Section 6.4.

$\frac{x+1}{2} = \frac{3}{x}$	Extremes are $x + 1$ and x; means are 2 and 3
$x^2 + x = 6$	Product of extremes = product of means
$x^2 + x - 6 = 0$	Standard form for a quadratic equation
$(x + 3)(x - 2) = 0$	Factor
$x + 3 = 0$ or $x - 2 = 0$	Set factors equal to 0
$x = -3$ or $x = 2$	

This time we have two solutions, -3 and 2. ▲

▼ **Example 4** A manufacturer knows that during a production run, 8 out of every 100 parts produced by a certain machine will be defective. If the machine produces 1450 parts, how many can be expected to be defective?

Solution The ratio of defective parts to total parts produced is $\frac{8}{100}$. If we let x represent the number of defective parts out of the total of 1450 parts, then we can write this ratio again as $\frac{x}{1450}$. This gives us a proportion to solve:

Defective parts in numerator → x; Total parts in denominator → 1450

$\frac{x}{1450} = \frac{8}{100}$	Extremes are x and 100; means are 1450 and 8
$100x = 11{,}600$	Product of extremes = Product of means
$x = 116$	

The manufacturer can expect 116 defective parts out of the total of 1450 parts if the machine usually produces 8 defective parts for every 100 parts it produces. ▲

Problem Set 6.7

Write each ratio as a fraction in lowest terms.

1. 8 to 6 **2.** 6 to 8 **3.** 200 to 250 **4.** 250 to 200

5. 32 to 4 **6.** 4 to 32 **7.** $\frac{3}{7}$ to $\frac{5}{7}$ **8.** $\frac{3}{5}$ to $\frac{7}{5}$

9. $\frac{2}{3}$ to $\frac{3}{4}$ **10.** $\frac{3}{4}$ to $\frac{9}{12}$ **11.** $15x^2$ to $3x^3$ **12.** $8x^4$ to $4x^7$

The following table gives the number of kilocalories of energy contained in an average serving of a number of foods.

Food	Kcal of Energy
Whole wheat bread	55
White bread	60
Ice cream	205
Cherry pie	420

13. Find the ratio of the number of kilocalories of energy in white bread to the number of kilocalories of energy in whole wheat bread.

14. Give the ratio of the number of kilocalories of energy in ice cream to the number of kilocalories of energy in cherry pie.

15. If a person has two servings of whole wheat bread for breakfast and that night has a midnight snack of cherry pie and ice cream, what is the ratio of breakfast kilocalories consumed to the number of kilocalories in the midnight snack?

16. How many servings of white bread equal one serving of cherry pie in terms of kilocalories of energy? (Find the ratio of the number of kilocalories of energy in cherry pie to the number of kilocalories of energy in white bread.)

17. If there are 60 minutes in 1 hour, what is the ratio of 20 minutes to 2 hours?

18. If there are 3 feet in 1 yard, what is the ratio of 2 feet to 3 yards?

Solve each of the following proportions.

19. $\frac{x}{2} = \frac{6}{12}$ **20.** $\frac{x}{4} = \frac{6}{8}$ **21.** $\frac{2}{5} = \frac{4}{x}$ **22.** $\frac{3}{8} = \frac{9}{x}$

23. $\frac{10}{20} = \frac{20}{x}$ **24.** $\frac{15}{60} = \frac{60}{x}$ **25.** $\frac{a}{3} = \frac{5}{12}$ **26.** $\frac{a}{5} = \frac{7}{20}$

27. $\frac{2}{x} = \frac{6}{7}$ **28.** $\frac{4}{x} = \frac{6}{7}$ **29.** $\frac{x+1}{3} = \frac{4}{x}$ **30.** $\frac{x+1}{6} = \frac{7}{x}$

31. $\frac{x}{2} = \frac{8}{x}$

32. $\frac{x}{9} = \frac{4}{x}$

33. $\frac{4}{a + 2} = \frac{a}{2}$

34. $\frac{3}{a + 2} = \frac{a}{5}$

35. $\frac{1}{x} = \frac{x - 5}{6}$

36. $\frac{1}{x} = \frac{x - 6}{7}$

37. A baseball player gets 6 hits in the first 18 games of the season. If he continues hitting at the same rate, how many hits will he get in the first 45 games?

38. A basketball player makes 8 of 12 free throws in the first game of the season. If she shoots with the same accuracy in the second game, how many of the 15 free throws she attempts will she make?

39. A solution contains 12 milliliters of alcohol and 16 milliliters of water. If another solution is to have the same concentration of alcohol in water but is to contain 28 milliliters of water, how much alcohol must it contain?

40. A solution contains 15 milliliters of HCl and 42 milliliters of water. If another solution is to have the same concentration of HCl in water but is to contain 140 milliliters of water, how much HCl must it contain?

41. If 100 grams of ice cream contain 13 grams of fat, how much fat is in 350 grams of ice cream?

42. A 6-ounce serving of grapefruit juice contains 159 grams of water. How many grams of water are in 20 ounces of grapefruit juice?

43. A map is drawn so that every 3.5 inches on the map corresponds to an actual distance of 100 miles. If the actual distance between two cities is 420 miles, how far apart are they on the map?

44. The scale on a map indicates that 1 inch on the map corresponds to an actual distance of 105 miles. Two cities are 4.5 inches apart on the map. What is the actual distance between the two cities?

45. A man drives his car 245 miles in 5 hours. At this rate, how far will he travel in 7 hours?

46. An airplane flies 1380 miles in 3 hours. How far will it fly in 5 hours?

Review Problems The following problems review material we covered in Sections 6.1, 6.2, and 6.3.

Reduce to lowest terms. [6.1]

47. $\frac{x^2 - x - 6}{x^2 - 9}$

48. $\frac{xy + 5x + 3y + 15}{x^2 + ax + 3x + 3a}$

Multiply and divide. [6.2]

49. $\frac{x^2 - 25}{x + 4} \cdot \frac{2x + 8}{x^2 - 9x + 20}$

50. $\frac{3x + 6}{x^2 + 4x + 3} \div \frac{x^2 + x - 2}{x^2 + 2x - 3}$

Add and subtract. [6.3]

51. $\frac{x}{x^2 - 16} + \frac{4}{x^2 - 16}$

52. $\frac{2}{x^2 - 1} - \frac{5}{x^2 + 3x - 4}$

Chapter 6 Summary and Review

Examples

1. We can reduce $\frac{6}{8}$ to lowest terms by dividing the numerator and denominator by their greatest common factor 2:

$$\frac{6}{8} = \frac{\cancel{2} \cdot 3}{\cancel{2} \cdot 4} = \frac{3}{4}$$

RATIONAL NUMBERS [6.1]

Any number that can be put in the form $\frac{a}{b}$, where a and b are integers ($b \neq 0$), is called a rational number.

Multiplying or dividing the numerator and denominator of a rational number by the same nonzero number never changes the value of the rational number.

2. We reduce rational expressions to lowest terms by factoring the numerator and denominator and then dividing out any factors they have in common:

$$\frac{x-3}{x^2-9} = \frac{\cancel{x-3}}{\cancel{(x-3)}(x+3)} = \frac{1}{x+3}$$

RATIONAL EXPRESSIONS [6.1]

Any expression of the form $\frac{P}{Q}$, where P and Q are polynomials ($Q \neq 0$), is a rational expression.

Multiplying or dividing the numerator and denominator of a rational expression by the same nonzero quantity always produces a rational expression equivalent to the original one.

3. $\frac{x-1}{x^2+2x-3} \cdot \frac{x^2-9}{x-2}$

$$= \frac{\cancel{x-1}}{\cancel{(x+3)}\cancel{(x-1)}} \cdot \frac{(x-3)\cancel{(x+3)}}{x-2}$$

$$= \frac{x-3}{x-2}$$

MULTIPLICATION [6.2]

To multiply two rational numbers or two rational expressions, multiply numerators, multiply denominators, and divide out any factors common to the numerator and denominator:

For rational numbers: $\frac{a}{b}$ and $\frac{c}{d}$, $\frac{a}{b} \cdot \frac{c}{d} = \frac{ac}{bd}$

For rational expressions: $\frac{P}{Q}$ and $\frac{R}{S}$, $\frac{P}{Q} \cdot \frac{R}{S} = \frac{PR}{QS}$

4. $\frac{2x}{x^2-25} \div \frac{4}{x-5}$

$$= \frac{2x}{\cancel{(x-5)}(x+5)} \cdot \frac{\cancel{x-5}}{4}$$

$$= \frac{x}{2(x+5)}$$

DIVISION [6.2]

To divide by a rational number or rational expression, simply multiply by its reciprocal:

For rational numbers: $\frac{a}{b}$ and $\frac{c}{d}$, $\frac{a}{b} \div \frac{c}{d} = \frac{a}{b} \cdot \frac{d}{c}$

For rational expressions: $\frac{P}{Q}$ and $\frac{R}{S}$, $\frac{P}{Q} \div \frac{R}{S} = \frac{P}{Q} \cdot \frac{S}{R}$

5. $\frac{3}{x-1} + \frac{x}{2}$

$$= \frac{3}{x-1} \cdot \mathbf{\frac{2}{2}} + \frac{x}{2} \cdot \mathbf{\frac{x-1}{x-1}}$$

$$= \frac{6}{2(x-1)} + \frac{x^2-x}{2(x-1)}$$

$$= \frac{x^2-x+6}{2(x-1)}$$

ADDITION [6.3]

To add two rational numbers or rational expressions, find a common denominator, change each expression to an equivalent expression having the common denominator, then add numerators and reduce if possible:

For rational numbers: $\frac{a}{c}$ and $\frac{b}{c}$, $\frac{a}{c} + \frac{b}{c} = \frac{a+b}{c}$

For rational expressions: $\frac{P}{S}$ and $\frac{Q}{S}$, $\frac{P}{S} + \frac{Q}{S} = \frac{P+Q}{S}$

SUBTRACTION [6.3]

To subtract a rational number or rational expression, simply add its opposite:

For rational numbers: $\frac{a}{c}$ and $\frac{b}{c}$, $\frac{a}{c} - \frac{b}{c} = \frac{a}{c} + \left(\frac{-b}{c}\right)$

For rational expressions: $\frac{P}{S}$ and $\frac{Q}{S}$, $\frac{P}{S} - \frac{Q}{S} = \frac{P}{S} + \left(\frac{-Q}{S}\right)$

6. $\frac{x}{x^2-4} - \frac{2}{x^2-4}$

$$= \frac{x-2}{x^2-4}$$

$$= \frac{\cancel{x-2}}{\cancel{(x-2)}(x+2)}$$

$$= \frac{1}{x+2}$$

EQUATIONS [6.4]

To solve equations involving rational expressions, first find the least common denominator (LCD) for all denominators. Then multiply both sides by the LCD and solve as usual. Be sure to check all solutions in the original equation to be sure there are no undefined terms.

7. Solve $\frac{1}{2} + \frac{3}{x} = 5$.

$$2x\left(\frac{1}{2}\right) + 2x\left(\frac{3}{x}\right) = 2x(5)$$

$$x + 6 = 10x$$

$$x = \frac{2}{3}$$

COMPLEX FRACTIONS [6.6]

A rational expression that contains a fraction in its numerator or denominator is called a complex fraction. The most common method of simplifying a complex fraction is to multiply the top and bottom by the LCD for all denominators.

8. $\frac{1 - \frac{4}{x}}{x - \frac{16}{x}} = \frac{x\left(1 - \frac{4}{x}\right)}{x\left(x - \frac{16}{x}\right)}$

$$= \frac{x-4}{x^2-16}$$

$$= \frac{\cancel{x-4}}{\cancel{(x-4)}(x+4)}$$

$$= \frac{1}{x+4}$$

RATIO AND PROPORTION [6.7]

The ratio of a to b is

$$\frac{a}{b}$$

Two equal ratios form a proportion. In the proportion

$$\frac{a}{b} = \frac{c}{d}$$

a and d are the *extremes*, and b and c are the *means*. In any proportion, the product of the extremes is equal to the product of the means.

9. Solve for x: $\frac{3}{x} = \frac{5}{20}$

$$3 \cdot 20 = 5 \cdot x$$

$$60 = 5x$$

$$x = 12$$

COMMON MISTAKES

1. Trying to reduce by dividing top and bottom of a rational expression by a quantity that is not a factor of both the top and bottom:

$$\frac{\overset{2}{\cancel{x^2} - \cancel{4x} + \cancel{4}}}{\underset{3\quad 2}{\cancel{x^2} - \cancel{6x} + \cancel{8}}} \qquad \text{Mistake}$$

This makes no sense at all. The numerator and denominator must be factored completely before any factors common to the numerator and denominator can be recognized:

$$\frac{x^2 - 4x + 4}{x^2 - 6x + 8} = \frac{\cancel{(x - 2)}(x - 2)}{\cancel{(x - 2)}(x - 4)}$$

$$= \frac{x - 2}{x - 4}$$

2. Forgetting to check solutions to equations involving rational expressions. If you multiply both sides of an equation by an expression that contains the variable, you must check your solutions to be sure you have not multiplied by zero.

Chapter 6 Test

Reduce to lowest terms. [6.1]

1. $\frac{x^2 - 16}{x^2 - 8x + 16}$

2. $\frac{10a + 20}{5a^2 + 20a + 20}$

3. $\frac{xy + 7x + 5y + 35}{x^2 + ax + 5x + 5a}$

Multiply and divide as indicated. [6.2]

4. $\frac{3x - 12}{4} \cdot \frac{8}{2x - 8}$

5. $\frac{x^2 - 49}{x + 1} \div \frac{x + 7}{x^2 - 1}$

6. $\frac{x^2 - 3x - 10}{x^2 - 8x + 15} \div \frac{3x^2 + 2x - 8}{x^2 + x - 12}$

7. $(x^2 - 9)\left(\frac{x + 2}{x + 3}\right)$

Add and subtract as indicated. [6.3]

8. $\frac{3}{x - 2} - \frac{6}{x - 2}$

9. $\frac{x}{x^2 - 9} + \frac{4}{4x - 12}$

10. $\dfrac{2x}{x^2 - 1} + \dfrac{x}{x^2 - 3x + 2}$

Solve the following equations. [6.4]

11. $\dfrac{7}{5} = \dfrac{x + 2}{3}$

12. $\dfrac{10}{x + 4} = \dfrac{6}{x} - \dfrac{4}{x}$

13. $\dfrac{3}{x - 2} - \dfrac{4}{x + 1} = \dfrac{5}{x^2 - x - 2}$

Solve the following problems. [6.5]

14. The sum of a number and its reciprocal is $\frac{25}{12}$. Find the number.

15. A boat travels 26 miles up a river in the same amount of time it takes to travel 34 miles down the same river. If the current is 2 mph, what is the speed of the boat in still water.

16. An inlet pipe can fill a pool in 15 hours, while an outlet pipe can empty it in 12 hours. If the pool is full and both pipes are open, how long will it take to empty?

Solve the following problems involving ratio and proportion. [6.7]

17. A solution of alcohol and water contains 27 milliliters of alcohol and 54 milliliters of water. What is the ratio of alcohol to water and the ratio of alcohol to total volume?

18. A manufacturer knows that during a production run, 8 out of every 100 parts produced by a certain machine will be defective. If the machine produces 1650 parts, how many can be expected to be defective?

Simplify each complex fraction. [6.6]

19. $\dfrac{1 + \dfrac{1}{x}}{1 - \dfrac{1}{x}}$

20. $\dfrac{1 - \dfrac{16}{x^2}}{1 - \dfrac{2}{x} - \dfrac{8}{x^2}}$

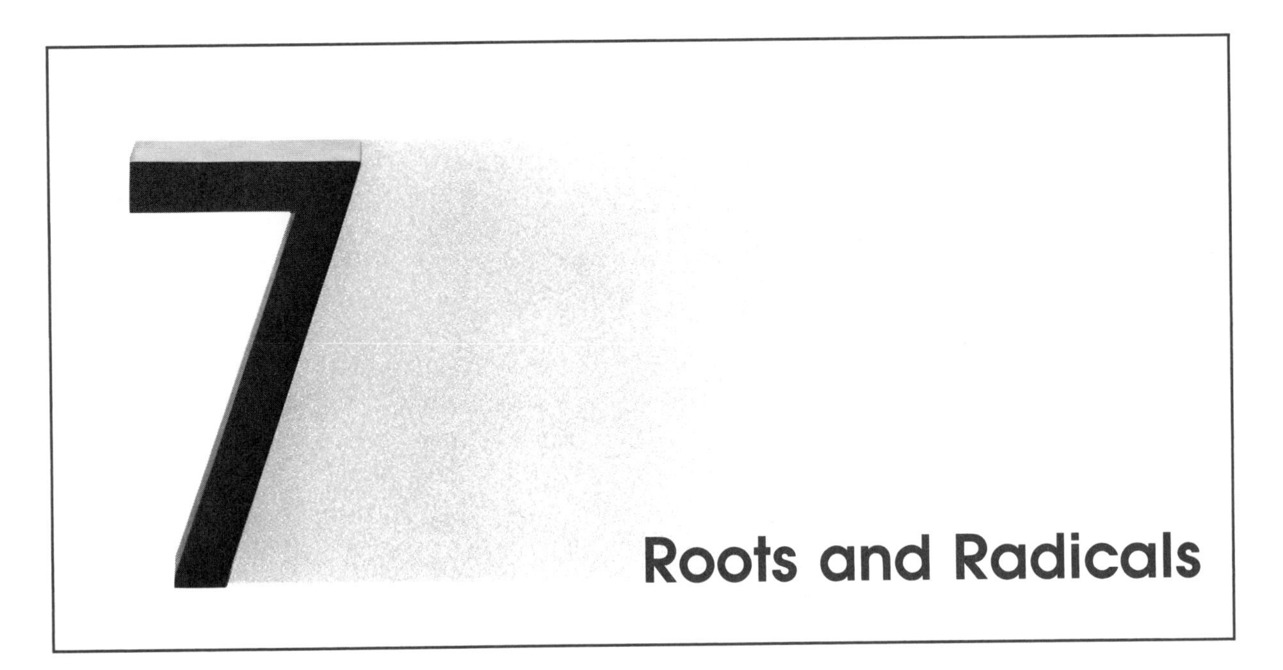

7 Roots and Radicals

Chapter 7 is concerned with operations on and simplification of radicals. Most of the radical expressions we will work with will involve square roots. Occasionally, we will encounter higher roots. Finding the square root of a number is the reverse of raising a number to the second power.

Since much of what we will do in this chapter is the reverse of what we did in Chapter 4, the more familiar we are with the properties of exponents, the better equipped we will be to handle this chapter.

The main ideas necessary for success in Chapter 7 are (1) properties of exponents and (2) operations on polynomials.

Section 7.1 Definitions and Common Roots

In Chapter 4 we developed notation (exponents) that would take us from a number to its square. If we wanted the square of 5, we wrote $5^2 = 25$. In this section, we will use another type of notation that will take us in the reverse direction—from the square of a number back to the number itself.

In general, we are interested in going from a number, say, 49, back to the number we squared to get 49. Since the square of 7 is 49, we say 7 is a square root of 49. The notation we use looks like this:

$$\sqrt{49}$$

Notation: In the expression $\sqrt{49}$, 49 is called the *radicand,* $\sqrt{\ }$ is the *radical sign,* and the complete expression $\sqrt{49}$ is called the *radical.*

DEFINITION If x represents any positive real number, then the expression $\sqrt{x}$ is the *positive square root* of x. It is the *positive* number we square to get x.

The expression $-\sqrt{x}$ is the *negative square root of* x. It is the *negative* number we square to get x.

Square Roots of Positive Numbers

Every positive number has two square roots, one positive and the other negative. Some books refer to the positive square root of a number as the principal root.

▼ **Example 1** The positive square root of 25 is 5 and can be written $\sqrt{25} = 5$. The negative square root of 25 is -5 and can be written $-\sqrt{25} = -5$. ▲

If we want to consider the negative square root of a number, we must put a negative sign in front of the radical. It is a common mistake to think of $\sqrt{25}$ as meaning either 5 or -5. The expression $\sqrt{25}$ means the *positive* square root of 25, which is 5. If we want the negative square root, we write $-\sqrt{25}$ to begin with.

▼ **Examples** Find the following roots.

2. $\sqrt{49} = 7$ 7 is the positive number we square to get 49

3. $-\sqrt{49} = -7$ -7 is the negative number we square to get 49

4. $\sqrt{121} = 11$ 11 is the positive number we square to get 121 ▲

▼ **Example 5** The positive square root of 17 is written $\sqrt{17}$. The negative square root of 17 is written $-\sqrt{17}$. ▲

We have no other exact representation for the two roots in Example 5. Since 17 itself is not a perfect square (the square of an integer), its two square roots, $\sqrt{17}$ and $-\sqrt{17}$, are irrational numbers. They have a place on the real number line but cannot be written as the ratio of two integers. The square roots of any number that is not itself a perfect square are irrational numbers.

▼ **Example 6**

Number	Positive Square Root	Negative Square Root	Roots Are
9	3	-3	Rational numbers
36	6	-6	Rational numbers
7	$\sqrt{7}$	$-\sqrt{7}$	Irrational numbers
22	$\sqrt{22}$	$-\sqrt{22}$	Irrational numbers
100	10	-10	Rational numbers

▲

Square Root of Zero

The number 0 is the only real number with one square root. It is also its own square root: $\sqrt{0} = 0$

Square Roots of Negative Numbers

Negative numbers have square roots, but their square roots are not real numbers. They do not have a place on the real number line. We will consider square roots of negative numbers later in the book.

▼ **Example 7** The expression $\sqrt{-4}$ does not represent a real number, since there is no real number we can square and end up with -4. The same is true of square roots of any negative number. ▲

Other Roots

There are many other roots of numbers besides square roots, although square roots seem to be the most commonly used. The cube root of a number is the number we cube (raise to the third power) to get the original number. The cube root of 8 is 2, since $2^3 = 8$. The cube root of 27 is 3, since $3^3 = 27$. The notation for cube roots looks like this:

The 3 is called the *index* → $\sqrt[3]{8} = 2$

We can go as high as we want with roots. The fourth root of 16 is 2 because $2^4 = 16$. We can write this in symbols as $\sqrt[4]{16} = 2$.

Here is a list of the most common roots. They are the roots that will come up most often in the remainder of the book and should be memorized.

Square Root		Cube Roots	Fourth Roots
$\sqrt{1} = 1$	$\sqrt{49} = 7$	$\sqrt[3]{1} = 1$	$\sqrt[4]{1} = 1$
$\sqrt{4} = 2$	$\sqrt{64} = 8$	$\sqrt[3]{8} = 2$	$\sqrt[4]{16} = 2$
$\sqrt{9} = 3$	$\sqrt{81} = 9$	$\sqrt[3]{27} = 3$	$\sqrt[4]{81} = 3$
$\sqrt{16} = 4$	$\sqrt{100} = 10$	$\sqrt[3]{64} = 4$	$\sqrt[4]{625} = 5$
$\sqrt{25} = 5$	$\sqrt{121} = 11$	$\sqrt[3]{125} = 5$	
$\sqrt{36} = 6$	$\sqrt{144} = 12$		

With even roots—square roots, fourth roots, sixth roots, and so on—we cannot have negative numbers *under* the radical sign. With odd roots, negative numbers under the radical sign do not cause problems.

▼ **Examples** Find the following roots, if possible.

8. $\sqrt[3]{-8} = -2$ Because $(-2)^3 = -8$

9. $\sqrt[3]{-27} = -3$ Because $(-3)^3 = -27$

10. $\sqrt{-4}$ Not a real number since there is no real number whose square is -4

11. $-\sqrt{4} = -2$ Because -2 is the negative number we square to get 4

12. $\sqrt[4]{-16}$ Not a real number since there is no real number that can be raised to the fourth power to obtain -16

13. $-\sqrt[4]{16} = -2$ Because -2 is the negative number we raise to the fourth power to get 16 ▲

Variables Under the Radical Sign

In this chapter, unless we say otherwise, we will assume that all variables that appear under a radical sign represent positive numbers. That way we can simplify expressions involving radicals that contain variables. Here are some examples.

▼ **Example 14** Simplify: $\sqrt{49x^2}$.

Solution We are looking for the expression we square to get $49x^2$. Since the square of 7 is 49 and the square of x is x^2, we can square $7x$ and get $49x^2$:

$$\sqrt{49x^2} = 7x \qquad \text{Because } (7x)^2 = 49x^2$$ ▲

▼ **Example 15** Simplify: $\sqrt{16a^2b^2}$.

Solution We want an expression whose square is $16a^2b^2$. That expression is $4ab$:

$$\sqrt{16a^2b^2} = 4ab \qquad \text{Because } (4ab)^2 = 16a^2b^2$$ ▲

▼ **Example 16** Simplify: $\sqrt[3]{125a^3}$.

Solution We are looking for the expression we cube to get $125a^3$. That expression is $5a$:

$$\sqrt[3]{125a^3} = 5a \qquad \text{Because } (5a)^3 = 125a^3$$ ▲

▼ **Example 17** Simplify: $\sqrt{x^6}$.

Solution The number we square to obtain x^6 is x^3.

$$\sqrt{x^6} = x^3 \qquad \text{Because } (x^3)^2 = x^6$$ ▲

▼ **Example 18** Simplify: $\sqrt[4]{16a^8b^4}$.

Solution The number we raise to the fourth power to obtain $16a^8b^4$ is $2a^2b$.

$$\sqrt[4]{16a^8b^4} = 2a^2b \qquad \text{Because } (2a^2b)^4 = 16a^8b^4$$ ▲

There are many application problems involving radicals that require decimal approximations. When we need a decimal approximation to a square root, we can use either the table of square roots at the back of the book or a calculator.

▼ **Example 19** If you invest P dollars in an account and after two years the account has A dollars in it, then the annual rate of return, r, on the money you originally invested is given by the formula

$$r = \frac{\sqrt{A} - \sqrt{P}}{\sqrt{P}}$$

Suppose you pay \$65 for a coin collection and find that the same coins sell for \$84 two years later. Find the annual rate of return on your investment.

Solution Substituting $A = 84$ and $P = 65$ in the formula we have

$$r = \frac{\sqrt{84} - \sqrt{65}}{\sqrt{65}}$$

From either the table of square roots or a calculator, we find that $\sqrt{84} = 9.165$ and $\sqrt{65} = 8.062$. Using these numbers in our formula gives us

$$\begin{aligned} r &= \frac{9.165 - 8.062}{8.062} \\ &= \frac{1.103}{8.062} \\ &= .137 \text{ or } 13.7\% \end{aligned}$$

To earn as much as this in savings account that compounds interest once a year, you would have to find an account that paid 13.7% in annual interest. ▲

Problem Set 7.1

Find the following roots. If the root does not exist as a real number, write "not a real number."

1. $\sqrt{9}$
2. $\sqrt{16}$
3. $-\sqrt{9}$
4. $-\sqrt{16}$
5. $\sqrt{-25}$
6. $\sqrt{-36}$
7. $-\sqrt{144}$
8. $\sqrt{256}$
9. $\sqrt{625}$
10. $-\sqrt{625}$
11. $\sqrt{-49}$
12. $\sqrt{-169}$
13. $-\sqrt{64}$
14. $-\sqrt{25}$
15. $-\sqrt{100}$
16. $\sqrt{121}$
17. $\sqrt{1225}$
18. $-\sqrt{1681}$
19. $\sqrt[4]{1}$
20. $-\sqrt[4]{81}$
21. $\sqrt[3]{-8}$
22. $\sqrt[3]{125}$
23. $-\sqrt[3]{125}$
24. $-\sqrt[3]{-8}$
25. $\sqrt[3]{-1}$
26. $-\sqrt[3]{-1}$
27. $\sqrt[3]{-27}$
28. $-\sqrt[3]{27}$
29. $-\sqrt[4]{16}$
30. $\sqrt[4]{-16}$

Assume all variables are positive and find the following roots.

31. $\sqrt{x^2}$
32. $\sqrt{a^2}$
33. $\sqrt{9x^2}$
34. $\sqrt{25x^2}$
35. $\sqrt{x^2y^2}$
36. $\sqrt{a^2b^2}$
37. $\sqrt{(a+b)^2}$
38. $\sqrt{(x+y)^2}$
39. $\sqrt{49x^2y^2}$
40. $\sqrt{81x^2y^2}$
41. $\sqrt[3]{x^3}$
42. $\sqrt[3]{a^3}$
43. $\sqrt[3]{8x^3}$
44. $\sqrt[3]{27x^3}$
45. $\sqrt{x^4}$
46. $\sqrt{x^6}$
47. $\sqrt{36a^6}$
48. $\sqrt{64a^4}$
49. $\sqrt{25a^8b^4}$
50. $\sqrt{16a^4b^8}$
51. $\sqrt[3]{x^6}$
52. $\sqrt[3]{x^9}$
53. $\sqrt[3]{27a^{12}}$
54. $\sqrt[3]{8a^6}$
55. $\sqrt[4]{x^8}$
56. $\sqrt[4]{x^{12}}$

Simplify each expression.

57. $\sqrt{9}+\sqrt{16}$
58. $\sqrt{64}+\sqrt{36}$
59. $\sqrt{9+16}$
60. $\sqrt{64+36}$
61. $\sqrt{144}+\sqrt{25}$
62. $\sqrt{25}-\sqrt{16}$
63. $\sqrt{144+25}$
64. $\sqrt{25-16}$

65. We know that the trinomial x^2+6x+9 is the square of the binomial $x+3$. That is, $x^2+6x+9=(x+3)^2$. Use this fact to find $\sqrt{x^2+6x+9}$.
66. Use the fact that $x^2+10x+25=(x+5)^2$ to find $\sqrt{x^2+10x+25}$.
67. Replace x with 4 in the expression $\sqrt{x^2+9}$ and in the expression $x+3$. Simplify each result.
68. Replace x with 8 in the expression $\sqrt{x^2+36}$ and in the expression $x+6$. Simplify both results.

In the next four problems you will need the following formula. The formula gives the annual rate of return, r, on an investment of P dollars that grows to A dollars over a two-year period of time:

$$r = \frac{\sqrt{A} - \sqrt{P}}{\sqrt{P}}$$

69. If you invest \$50 in silver coins and two years later you find the coins are worth \$65, what is the annual rate of return on your investment?

70. If you pay \$35 for some rare stamps and two years later find that the same stamps cost \$45, what is the annual rate of return on your investment?

71. Suppose you invest \$500 in the stock market and two years later your stocks are worth \$600. Find the annual rate of return on your stocks.

72. Suppose you invest \$700 in the stock market and two years later your stocks are worth \$800. Find the annual rate of return on your stocks.

Review Problems The following problems review material we covered in Section 6.1.

Reduce each expression to lowest terms.

73. $\frac{x^2 - 16}{x + 4}$

74. $\frac{x - 5}{x^2 - 25}$

75. $\frac{10a + 20}{5a^2 - 20}$

76. $\frac{8a - 16}{4a^2 - 16}$

77. $\frac{2x^2 - 5x - 3}{x^2 - 3x}$

78. $\frac{x^2 - 5x}{3x^2 - 13x - 10}$

79. $\frac{xy + 3x + 2y + 6}{xy + 3x + ay + 3a}$

80. $\frac{xy + 5x + 4y + 20}{x^2 + bx + 4x + 4b}$

Section 7.2 Properties of Radicals

In this section we will consider the first part of what is called simplified form for radical expressions. A radical expression is any expression containing a radical, whether it is a square root, cube root, or higher root. Simplified form for a radical expression is the form that is easiest to work with. The first step in putting a radical expression in simplified form is to take as much out from under the radical sign as possible. To do this, we must first develop two properties of radicals in general.

Consider the following two problems:

$$\sqrt{9 \cdot 16} = \sqrt{144} = 12$$
$$\sqrt{9} \cdot \sqrt{16} = 3 \cdot 4 = 12$$

Since the answers to both are equal, the original problems must also be equal. That is, $\sqrt{9 \cdot 16} = \sqrt{9} \cdot \sqrt{16}$. We can generalize this property as follows.

Property 1 for Radicals

If x and y represent nonnegative real numbers, then it is always true that

$$\sqrt{xy} = \sqrt{x}\sqrt{y}$$

In words: The square root of a product is the product of the square roots.

We can use this property to simplify radical expressions.

▼ **Example 1** Simplify: $\sqrt{20}$.

Solution To simplify $\sqrt{20}$, we want to take as much out from under the radical sign as possible. We begin by looking for the largest perfect square that is a factor of 20. The largest perfect square that divides 20 is 4, so we write 20 as $4 \cdot 5$:

$$\sqrt{20} = \sqrt{4 \cdot 5}$$

Next we apply the first property of radicals and write

$$\sqrt{4 \cdot 5} = \sqrt{4}\sqrt{5}$$

And, since $\sqrt{4} = 2$, we have

$$\sqrt{4}\sqrt{5} = 2\sqrt{5}$$

The expression $2\sqrt{5}$ is the simplified form of $\sqrt{20}$, since we have taken as much out from under the radical sign as possible. ▲

Note Working a problem like the one in Example 1 depends on recognizing the largest perfect square that divides (is a factor of) the number under the radical sign. The set of perfect squares is the set

$$\{1, 4, 9, 16, 25, 36, \ldots\}$$

To simplify an expression like $\sqrt{20}$, we must first find the largest number in this set that is a factor of the number under the radical sign.

▼ **Example 2** Simplify: $\sqrt{75}$.

Solution Since 25 is the largest perfect square that divides 75, we have

$$\begin{aligned} \sqrt{75} &= \sqrt{25 \cdot 3} && \text{Factor 75 into } 25 \cdot 3 \\ &= \sqrt{25}\,\sqrt{3} && \text{Property 1 for radicals} \\ &= 5\sqrt{3} && \sqrt{25} = 5 \end{aligned}$$

The expression $5\sqrt{3}$ is the simplified form for $\sqrt{75}$, since we have taken as much out from under the radical sign as possible. ▲

The next two examples involve square roots of expressions that contain variables. Remember, we are assuming that all variables that appear under a radical sign represent positive numbers.

▼ **Example 3** Simplify: $\sqrt{25x^3}$.

Solution The largest perfect square that is a factor of $25x^3$ is $25x^2$. We write $25x^3$ as $25x^2 \cdot x$ and apply Property 1:

$$\begin{aligned} \sqrt{25x^3} &= \sqrt{25x^2 \cdot x} && \text{Factor } 25x^3 \text{ into } 25x^2 \cdot x \\ &= \sqrt{25x^2}\,\sqrt{x} && \text{Property 1 for radicals} \\ &= 5x\sqrt{x} && \sqrt{25x^2} = 5x \end{aligned}$$

▲

▼ **Example 4** Simplify: $\sqrt{18y^4}$.

Solution The largest perfect square that is a factor of $18y^4$ is $9y^4$. We write $18y^4$ as $9y^4 \cdot 2$ and apply Property 1:

$$\begin{aligned} \sqrt{18y^4} &= \sqrt{9y^4 \cdot 2} && \text{Factor } 18y^4 \text{ into } 9y^4 \cdot 2 \\ &= \sqrt{9y^4}\,\sqrt{2} && \text{Property 1 for radicals} \\ &= 3y^2\sqrt{2} && \sqrt{9y^4} = 3y^2 \end{aligned}$$

▲

▼ **Example 5** Simplify: $3\sqrt{32}$.

Solution We want to get as much out from under $\sqrt{32}$ as possible. Since 16 is the largest perfect square that divides 32, we have

$$\begin{aligned} 3\sqrt{32} &= 3\sqrt{16 \cdot 2} && \text{Factor 32 into } 16 \cdot 2 \\ &= 3\sqrt{16}\,\sqrt{2} && \text{Property 1 for radicals} \\ &= 3 \cdot 4\sqrt{2} && \sqrt{16} = 4 \\ &= 12\sqrt{2} && 3 \cdot 4 = 12 \end{aligned}$$

▲

Although we have stated Property 1 for radicals in terms of square roots only, it holds for higher roots as well. If we were to state property 1 again for cube roots, it would look like this:

$$\sqrt[3]{xy} = \sqrt[3]{x}\,\sqrt[3]{y}$$

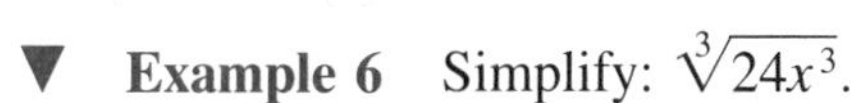

▼ **Example 6** Simplify: $\sqrt[3]{24x^3}$.

Solution Since we are simplifying a cube root, we look for the largest perfect cube that is a factor of $24x^3$. Since 8 is a perfect cube, the largest perfect cube that is a factor of $24x^3$ is $8x^3$:

$$\begin{aligned} \sqrt[3]{24x^3} &= \sqrt[3]{8x^3 \cdot 3} && \text{Factor } 24x^3 \text{ into } 8x^3 \cdot 3 \\ &= \sqrt[3]{8x^3}\,\sqrt[3]{3} && \text{Property 1 for radicals} \\ &= 2x\sqrt[3]{3} && \sqrt[3]{8x^3} = 2x \end{aligned}$$ ▲

The second property of radicals has to do with division.The property becomes apparent when we consider the following two problems:

$$\sqrt{\frac{64}{16}} = \sqrt{4} = 2$$

$$\frac{\sqrt{64}}{\sqrt{16}} = \frac{8}{4} = 2$$

Since the answers in each case are equal, the original problems must be also:

$$\sqrt{\frac{64}{16}} = \frac{\sqrt{64}}{\sqrt{16}}$$

Here is the property in general.

Property 2 for Radicals

If x and y both represent nonnegative real numbers and $y \neq 0$, then it is always true that

$$\sqrt{\frac{x}{y}} = \frac{\sqrt{x}}{\sqrt{y}}$$

In words: The square root of a quotient is the quotient of the square roots.

Although we have stated Property 2 for square roots only, it holds for higher roots as well.

We can use Property 2 for radicals in much the same way as we used Property 1 to simplify radical expressions.

▼ **Example 7** Simplify: $\sqrt{\frac{49}{81}}$.

Solution We begin by applying Property 2 for radicals to separate the fraction into two separate radicals. Then, we simplify each radical separately:

$$\sqrt{\frac{49}{81}} = \frac{\sqrt{49}}{\sqrt{81}} \qquad \text{Property 2 for radicals}$$

$$= \frac{7}{9} \qquad \sqrt{49} = 7 \text{ and } \sqrt{81} = 9$$

▲

▼ **Example 8** Simplify: $\sqrt[4]{\frac{81}{16}}$.

Solution Remember, although Property 2 has been stated in terms of square roots, it holds for higher roots as well. Proceeding as we did in Example 7, we have

$$\sqrt[4]{\frac{81}{16}} = \frac{\sqrt[4]{81}}{\sqrt[4]{16}} \qquad \text{Property 2}$$

$$= \frac{3}{2} \qquad \sqrt[4]{81} = 3 \text{ and } \sqrt[4]{16} = 2$$

▲

▼ **Example 9** Simplify: $\sqrt{\frac{50}{49}}$.

Solution Applying Property 2 for radicals and then simplifying each resulting radical separately, we have

$$\sqrt{\frac{50}{49}} = \frac{\sqrt{50}}{\sqrt{49}} \qquad \text{Property 2 for radicals}$$

$$= \frac{\sqrt{25 \cdot 2}}{7} \qquad \text{Factor: } 50 = 25 \cdot 2, \quad \sqrt{49} = 7$$

$$= \frac{\sqrt{25}\sqrt{2}}{7} \qquad \text{Property 1 for radicals}$$

$$= \frac{5\sqrt{2}}{7} \qquad \sqrt{25} = 5$$

▲

▼ **Example 10** Simplify: $\sqrt{\frac{12x^2}{25}}$.

Solution Proceeding as we have in the previous two examples, we use Property 2 for radicals to separate the numerator and denominator into two separate radicals. Then, we simplify each radical separately:

$$\sqrt{\frac{12x^2}{25}} = \frac{\sqrt{12x^2}}{\sqrt{25}} \qquad \text{Property 2 for radicals}$$

$$= \frac{\sqrt{4x^2 \cdot 3}}{5} \qquad \text{Factor: } 12x^2 = 4x^2 \cdot 3, \quad \sqrt{25} = 5$$

$$= \frac{\sqrt{4x^2}\sqrt{3}}{5} \qquad \text{Property 1 for radicals}$$

$$= \frac{2x\sqrt{3}}{5} \qquad \sqrt{4x^2} = 2x$$

▲

Note Although this last expression may not look simpler than the expression we started with, it is in simplified form for radicals. Radical expressions that are in simplified form are not necessarily simpler looking expressions. But generally, they are easier to work with when they are in simplified form.

▼ **Example 11** Simplify: $\sqrt{\dfrac{50x^3y^2}{49}}$.

Solution We begin by taking the square roots of $50x^3y^2$ and 49 separately and then writing $\sqrt{49}$ as 7:

$$\sqrt{\frac{50x^3y^2}{49}} = \frac{\sqrt{50x^3y^2}}{\sqrt{49}} \qquad \text{Property 2}$$

$$= \frac{\sqrt{50x^3y^2}}{7} \qquad \sqrt{49} = 7$$

To simplify the numerator of this last expression, we determine that the largest perfect square that is a factor of $50x^3y^2$ is $25x^2y^2$. Continuing, we have

$$= \frac{\sqrt{25x^2y^2 \cdot 2x}}{7} \qquad \text{Factor } 50x^3y^2 \text{ into } 25x^2y^2 \cdot 2x$$

$$= \frac{\sqrt{25x^2y^2}\sqrt{2x}}{7} \qquad \text{Property 1}$$

$$= \frac{5xy\sqrt{2x}}{7} \qquad \sqrt{25x^2y^2} = 5xy$$

▲

COMMON MISTAKES

The two properties we have developed in this section are the two main properties of radicals. The properties hold for multiplication and division. There are no similar properties for addition and subtraction. That is, in general,

$$\sqrt{x + y} \neq \sqrt{x} + \sqrt{y}$$

The square root of a sum is not, in general, the sum of square roots. It just doesn't work. If we try it with numbers 16 and 9, we can see what is wrong:

$$\sqrt{16 + 9} \stackrel{?}{=} \sqrt{16} + \sqrt{9}$$

$$\sqrt{25} \stackrel{?}{=} 4 + 3$$

$$5 \neq 7$$

Although it is obvious that the property doesn't hold for addition, it is a very common mistake for beginning algebra students to try using it.

Problem Set 7.2

Use Property 1 for radicals to simplify the following radical expressions as much as possible. Assume that all variables represent positive numbers.

1. $\sqrt{8}$ **2.** $\sqrt{18}$ **3.** $\sqrt{12}$ **4.** $\sqrt{27}$

5. $\sqrt[3]{24}$ **6.** $\sqrt[3]{54}$ **7.** $\sqrt{50x^2}$ **8.** $\sqrt{32x^2}$

9. $\sqrt{45a^2b^2}$ **10.** $\sqrt{128a^2b^2}$ **11.** $\sqrt[3]{54x^3}$ **12.** $\sqrt[3]{128x^3}$

13. $\sqrt{32x^4}$ **14.** $\sqrt{48x^4}$ **15.** $\sqrt{80}$ **16.** $\sqrt{125}$

17. $\sqrt{28x^3}$ **18.** $\sqrt{54x^3}$ **19.** $\sqrt[3]{8x^4}$ **20.** $\sqrt[3]{8x^5}$

21. $\sqrt[3]{27a^5}$ **22.** $\sqrt[3]{27a^4}$ **23.** $5\sqrt{45a^3}$ **24.** $3\sqrt{300a^3}$

25. $3\sqrt{50xy^2}$ **26.** $4\sqrt{18xy^2}$ **27.** $7\sqrt{12x^2y}$ **28.** $6\sqrt{20x^2y}$

Use Property 2 for radicals to simplify each of the following. Assume all variables represent positive numbers.

29. $\sqrt{\frac{16}{25}}$ **30.** $\sqrt{\frac{81}{64}}$ **31.** $\sqrt{\frac{4}{9}}$ **32.** $\sqrt{\frac{49}{16}}$

33. $\sqrt[3]{\frac{8}{27}}$ **34.** $\sqrt[3]{\frac{64}{27}}$ **35.** $\sqrt[4]{\frac{16}{81}}$ **36.** $\sqrt[4]{\frac{81}{16}}$

37. $\sqrt{\frac{100x^2}{25}}$ **38.** $\sqrt{\frac{100x^2}{4}}$ **39.** $\sqrt{\frac{81a^2b^2}{9}}$ **40.** $\sqrt{\frac{64a^2b^2}{16}}$

41. $\sqrt[3]{\frac{27x^3}{8y^3}}$ **42.** $\sqrt[3]{\frac{125x^3}{64y^3}}$

Use combinations of Properties 1 and 2 for radicals to simplify the following problems as much as possible. Assume all variables represent positive numbers.

43. $\sqrt{\frac{50}{9}}$ **44.** $\sqrt{\frac{32}{49}}$ **45.** $\sqrt{\frac{75}{25}}$ **46.** $\sqrt{\frac{300}{4}}$

47. $\sqrt{\frac{128}{49}}$ **48.** $\sqrt{\frac{32}{64}}$ **49.** $\sqrt{\frac{288x}{25}}$ **50.** $\sqrt{\frac{28y}{81}}$

51. $\sqrt{\frac{54a^2}{25}}$ **52.** $\sqrt{\frac{243a^2}{49}}$ **53.** $\frac{3\sqrt{50}}{2}$ **54.** $\frac{5\sqrt{48}}{3}$

55. $\frac{7\sqrt{28y^2}}{3}$ **56.** $\frac{9\sqrt{243x^2}}{2}$ **57.** $\frac{5\sqrt{72a^2b^2}}{\sqrt{36}}$ **58.** $\frac{2\sqrt{27a^2b^2}}{\sqrt{9}}$

59. $\frac{6\sqrt{8x^2y}}{\sqrt{4}}$ **60.** $\frac{5\sqrt{32xy^2}}{\sqrt{25}}$ **61.** $\frac{8\sqrt{12x^2y^3}}{\sqrt{100}}$ **62.** $\frac{6\sqrt{18x^3y^2}}{\sqrt{81}}$

63. Use the table of squares and square roots at the back of the book to find decimal approximations for $\sqrt{12}$ and $2\sqrt{3}$. Round your answers to three places past the decimal point.

64. Use the table at the back of the book to find decimal approximations for $\sqrt{50}$ and $5\sqrt{2}$. Round your answers to three places past the decimal point.

65. The formula that gives the number of seconds, t, it takes for an object to reach the ground when dropped from the top of a building h feet high is

$$t = \sqrt{\frac{h}{16}}$$

If a rock is dropped from the top of a building 25 feet high, how long will it take for the rock to hit the ground?

66. How long will it take for an object dropped from a building 100 feet high to reach the ground?

Review Problems The following problems review material we covered in Section 6.2.

Multiply and divide as indicated.

67. $\dfrac{8x}{x^2 - 5x} \cdot \dfrac{x^2 - 25}{4x^2 + 4x}$

68. $\dfrac{x^2 + 4x}{4x^2} \cdot \dfrac{7x}{x^2 + 5x}$

69. $\dfrac{x^2 + 3x - 4}{3x^2 + 7x - 20} \div \dfrac{x^2 - 2x + 1}{3x^2 - 2x - 5}$

70. $\dfrac{x^2 - 16}{2x^2 + 7x + 3} \div \dfrac{2x^2 - 5x - 12}{4x^2 + 8x + 3}$

71. $(x^2 - 36)\left(\dfrac{x + 3}{x - 6}\right)$

72. $(x^2 - 49)\left(\dfrac{x + 5}{x + 7}\right)$

Section 7.3 Simplified Form for Radicals

A radical expression is in simplified form if it has three special characteristics. Radical expressions that are in simplified form are generally easier to work with.

DEFINITION A radical expression is in *simplified form* if:

1. There are no perfect squares that are factors of the quantity under the square root sign, no perfect cubes of the quantity under the cube root sign, and so on. We want as little as possible under the radical sign.
2. There are no fractions under the radical sign.
3. There are no radicals in the denominator.

A radical expression that has these three characteristics is said to be in simplified form. As we will see, simplified form is not always the least complicated expression. In many cases, the simplified expression looks more complicated than the original expression. The important thing about simplified form for radicals is that simplified expressions are easier to work with.

The tools we will use to put radical expressions into simplified form are the properties of radicals. We list the properties again for clarity:

Properties of Radicals

If a and b represent any two positive real numbers, then it is always true that

1. $\sqrt{a}\sqrt{b} = \sqrt{a \cdot b}$
2. $\dfrac{\sqrt{a}}{\sqrt{b}} = \sqrt{\dfrac{a}{b}}$
3. $\sqrt{a}\sqrt{a} = (\sqrt{a})^2 = a$ — This property comes directly from the definition of radicals

The following examples illustrate how we put a radical expression into simplified form using the three properties of radicals. Although the properties are stated for square roots only, they hold for all roots. [Property 3 written for cube roots would be $\sqrt[3]{a}\,\sqrt[3]{a}\,\sqrt[3]{a} = (\sqrt[3]{a})^3 = a$.]

▼ **Example 1** Put $\sqrt{\frac{1}{2}}$ into simplified form.

Solution The expression $\sqrt{\frac{1}{2}}$ is not in simplified form because there is a fraction under the radical sign. We can change this by applying Property 2 for radicals:

$$\sqrt{\frac{1}{2}} = \frac{\sqrt{1}}{\sqrt{2}} \qquad \text{Property 2 or radicals}$$

$$= \frac{1}{\sqrt{2}} \qquad \sqrt{1} = 1$$

The expression $\dfrac{1}{\sqrt{2}}$ is not in simplified form because there is a radical sign in the denominator. If we multiply the numerator and denominator of $\dfrac{1}{\sqrt{2}}$ by $\sqrt{2}$, the denominator becomes $\sqrt{2} \cdot \sqrt{2} = 2$:

$$\frac{1}{\sqrt{2}} = \frac{1}{\sqrt{2}} \cdot \frac{\sqrt{\mathbf{2}}}{\sqrt{\mathbf{2}}} \qquad \text{Multiply numerator and denominator by } \sqrt{\mathbf{2}}$$

$$= \frac{\sqrt{2}}{2} \qquad 1 \cdot \sqrt{2} = \sqrt{2},\quad \sqrt{2} \cdot \sqrt{2} = \sqrt{4} = 2$$

▲

If we check the expression $\frac{\sqrt{2}}{2}$ against our definition of simplified form for radicals, we find that all three rules hold. There are no perfect squares that are factors of 2. There are no fractions under the radical sign. No radicals appear in the denominator. The expression $\frac{\sqrt{2}}{2}$, therefore, must be in simplified form.

▼ **Example 2** Write $\sqrt{\frac{2}{3}}$ in simplified form.

Solution We proceed as we did in Example 1:

$$\sqrt{\frac{2}{3}} = \frac{\sqrt{2}}{\sqrt{3}} \qquad \text{Use Property 2 to separate radicals}$$

$$= \frac{\sqrt{2}}{\sqrt{3}} \cdot \frac{\sqrt{\mathbf{3}}}{\sqrt{\mathbf{3}}} \qquad \text{Multiply by } \frac{\sqrt{\mathbf{3}}}{\sqrt{\mathbf{3}}} \text{ to remove the radical from the denominator}$$

$$= \frac{\sqrt{6}}{3} \qquad \begin{aligned}\sqrt{2}\cdot\sqrt{3} &= \sqrt{6}\\ \sqrt{3}\cdot\sqrt{3} &= \sqrt{9} = 3\end{aligned}$$

▲

▼ **Example 3** Put the expression $\frac{6\sqrt{20}}{2\sqrt{5}}$ into simplified form.

Solution Although there are many ways to begin this problem, we notice that 20 is divisible by 5. Using Property 2 for radicals as the first step, we can quickly put the expression into simplified form:

$$\frac{6\sqrt{20}}{2\sqrt{5}} = \frac{6}{2}\sqrt{\frac{20}{5}} \qquad \text{Property 2 for radicals}$$

$$= 3\sqrt{4} \qquad \frac{20}{5} = 4$$

$$= 3 \cdot 2 \qquad \sqrt{4} = 2$$

$$= 6$$

▲

▼ **Example 4** Simplify: $\sqrt{\frac{4x^3y^2}{3}}$.

Solution We begin by separating the numerator and denominator and then taking the perfect squares out of the numerator:

$$\sqrt{\frac{4x^3y^2}{3}} = \frac{\sqrt{4x^3y^2}}{\sqrt{3}} \qquad \text{Property for 2 for radicals}$$

$$= \frac{\sqrt{4x^2y^2}\sqrt{x}}{\sqrt{3}} \qquad \text{Property 1 for radicals}$$

$$= \frac{2xy\sqrt{x}}{\sqrt{3}} \qquad \sqrt{4x^2y^2} = 2xy$$

The only thing keeping our expression from being in simplified form is the $\sqrt{3}$ in the denominator. We can take care of this by multiplying the numerator and denominator by $\sqrt{3}$:

$$\frac{2xy\sqrt{x}}{\sqrt{3}} = \frac{2xy\sqrt{x}}{\sqrt{3}} \cdot \frac{\sqrt{\mathbf{3}}}{\sqrt{\mathbf{3}}} \qquad \text{Multiply numerator and denominator by } \sqrt{\mathbf{3}}$$

$$= \frac{2xy\sqrt{3x}}{3} \qquad \sqrt{3} \cdot \sqrt{3} = \sqrt{9} = 3$$

▲

Although the final expression may look more complicated than the original expression, it is in simplified form. The last step is called *rationalizing the denominator*. We have taken the radical out of the denominator and replaced it with a rational number.

Our last two examples involve rationalizing denominators that contain cube roots.

▼ **Example 5** Simplify: $\sqrt[3]{\frac{2}{3}}$.

Solution We can apply Property 2 first to separate the cube roots:

$$\sqrt[3]{\frac{2}{3}} = \frac{\sqrt[3]{2}}{\sqrt[3]{3}}$$

To write this expression in simplified form, we must remove the radical from the denominator. Since the radical is a cube root, we will need to multiply it by an expression that will give us a perfect cube under that cube root. We can accomplish this by multiplying the numerator and denominator by $\sqrt[3]{9}$. Here is what it looks like:

$$\frac{\sqrt[3]{2}}{\sqrt[3]{3}} = \frac{\sqrt[3]{2}}{\sqrt[3]{3}} \cdot \frac{\sqrt[3]{\mathbf{9}}}{\sqrt[3]{\mathbf{9}}}$$

$$= \frac{\sqrt[3]{18}}{\sqrt[3]{27}} \qquad \begin{aligned} \sqrt[3]{2} \cdot \sqrt[3]{9} &= \sqrt[3]{18} \\ \sqrt[3]{3} \cdot \sqrt[3]{9} &= \sqrt[3]{27} \end{aligned}$$

$$= \frac{\sqrt[3]{18}}{3} \qquad \sqrt[3]{27} = 3$$

To see why multiplying numerator and denominator by $\sqrt[3]{9}$ works in this example, you must first convince yourself that multiplying numerator and denominator by $\sqrt[3]{3}$ would not have worked. ▲

▼ **Example 6** Simplify: $\sqrt[3]{\frac{1}{4}}$.

Solution We begin by separating the numerator and denominator:

$$\sqrt[3]{\frac{1}{4}} = \frac{\sqrt[3]{1}}{\sqrt[3]{4}} \qquad \text{Property 2 for radicals}$$
$$= \frac{1}{\sqrt[3]{4}} \qquad \sqrt[3]{1} = 1$$

To rationalize the denominator, we need to have a perfect cube under the cube root sign. If we multiply numerator and denominator by $\sqrt[3]{2}$, we will have $\sqrt[3]{4} \cdot \sqrt[3]{2} = \sqrt[3]{8}$ in the denominator:

$$\frac{1}{\sqrt[3]{4}} = \frac{1}{\sqrt[3]{4}} \cdot \frac{\sqrt[3]{2}}{\sqrt[3]{2}} \qquad \text{Multiply numerator and denominator by } \sqrt[3]{2}$$
$$= \frac{\sqrt[3]{2}}{\sqrt[3]{8}} \qquad \sqrt[3]{4} \cdot \sqrt[3]{2} = \sqrt[3]{8}$$
$$= \frac{\sqrt[3]{2}}{2} \qquad \sqrt[3]{8} = 2$$

The final expression has no radical sign in the denominator and therefore is in simplified form. ▲

Problem Set 7.3

Put each of the following radical expressions into simplified form. Assume all variables represent positive numbers.

1. $\sqrt{\frac{1}{2}}$
2. $\sqrt{\frac{1}{5}}$
3. $\sqrt{\frac{1}{3}}$
4. $\sqrt{\frac{1}{6}}$
5. $\sqrt{\frac{2}{5}}$
6. $\sqrt{\frac{3}{7}}$
7. $\sqrt{\frac{3}{2}}$
8. $\sqrt{\frac{5}{3}}$
9. $\sqrt{\frac{20}{3}}$
10. $\sqrt{\frac{32}{5}}$
11. $\sqrt{\frac{45}{6}}$
12. $\sqrt{\frac{48}{7}}$
13. $\sqrt{\frac{20}{5}}$
14. $\sqrt{\frac{12}{3}}$
15. $\frac{\sqrt{21}}{\sqrt{3}}$
16. $\frac{\sqrt{21}}{\sqrt{7}}$
17. $\frac{\sqrt{35}}{\sqrt{7}}$
18. $\frac{\sqrt{35}}{\sqrt{5}}$
19. $\frac{10\sqrt{15}}{5\sqrt{3}}$
20. $\frac{4\sqrt{12}}{8\sqrt{3}}$
21. $\frac{6\sqrt{21}}{3\sqrt{7}}$
22. $\frac{8\sqrt{50}}{16\sqrt{2}}$
23. $\frac{6\sqrt{35}}{12\sqrt{5}}$
24. $\frac{8\sqrt{35}}{16\sqrt{7}}$
25. $\sqrt{\frac{4x^2y^2}{2}}$
26. $\sqrt{\frac{9x^2y^2}{3}}$
27. $\sqrt{\frac{5x^2y}{3}}$
28. $\sqrt{\frac{7x^2y}{5}}$
29. $\sqrt{\frac{16a^4}{5}}$
30. $\sqrt{\frac{25a^4}{7}}$
31. $\sqrt{\frac{72a^5}{5}}$
32. $\sqrt{\frac{12a^5}{5}}$

33. $\sqrt{\dfrac{20x^2y^3}{3}}$ **34.** $\sqrt{\dfrac{27x^2y^3}{2}}$ **35.** $\dfrac{2\sqrt{20x^2y^3}}{3}$ **36.** $\dfrac{5\sqrt{27x^3y^2}}{2}$

37. $\dfrac{6\sqrt{54a^2b^3}}{5}$ **38.** $\dfrac{7\sqrt{75a^3b^2}}{6}$ **39.** $\dfrac{3\sqrt{72x^4}}{\sqrt{2x}}$ **40.** $\dfrac{2\sqrt{45x^4}}{\sqrt{5x}}$

41. $\sqrt[3]{\dfrac{1}{2}}$ **42.** $\sqrt[3]{\dfrac{1}{4}}$ **43.** $\sqrt[3]{\dfrac{1}{9}}$ **44.** $\sqrt[3]{\dfrac{1}{3}}$

45. $\sqrt[3]{\dfrac{3}{2}}$ **46.** $\sqrt[3]{\dfrac{7}{9}}$ **47.** $\sqrt[3]{\dfrac{3}{4}}$ **48.** $\sqrt[3]{\dfrac{5}{3}}$

49. The table of square roots at the back of the book gives the four-digit decimal approximation of $\sqrt{2}$ as 1.414. Use this number to find approximations for the expressions $\frac{1}{\sqrt{2}}$ and $\frac{\sqrt{2}}{2}$. Round your answer to three places past the decimal point.

50. Find the decimal approximation for $\sqrt{3}$ in the table of squares and square roots at the back of the book and use it to find decimal approximations for $\frac{1}{\sqrt{3}}$ and $\frac{\sqrt{3}}{3}$. Round your answers to three places past the decimal point.

51. The higher you are above the surface of the earth, the farther you can see. The formula that gives the approximate distance in miles, d, that you can see from a height of h feet above the surface of the earth is

$$d = \sqrt{\frac{3h}{2}}$$

How far can you see from a window that is 24 feet above the ground? (This is assuming that your view is unobstructed.)

52. How far can you see from the window of an airplane flying at 6,000 feet?

Review Problems The following problems review material we covered in Sections 2.1 and 6.3. Reviewing the problems from Section 2.1 will help you understand the next section.

Use the distributive property to combine the following. [2.1]

53. $3x + 7x$ **54.** $3x - 7x$

55. $15x + 8x$ **56.** $15x - 8x$

57. $7a - 3a + 6a$ **58.** $25a + 3a - a$

Add and subtract as indicated. [6.3]

59. $\dfrac{x^2}{x+5} + \dfrac{10x+25}{x+5}$ **60.** $\dfrac{x^2}{x-3} - \dfrac{9}{x-3}$

61. $\dfrac{a}{3} + \dfrac{2}{5}$ **62.** $\dfrac{4}{a} + \dfrac{2}{3}$

63. $\dfrac{6}{a^2-9} - \dfrac{5}{a^2-a-6}$ **64.** $\dfrac{4a}{a^2+6a+5} - \dfrac{3a}{a^2+5a+4}$

Section 7.4 Addition and Subtraction of Radical Expressions

To add two or more radical expressions we apply the distributive property. Adding radical expressions is very similar to adding similar terms of polynomials.

▼ **Example 1** Combine terms in the expression $3\sqrt{5} - 7\sqrt{5}$.

Solution The two terms $3\sqrt{5}$ and $7\sqrt{5}$ each have $\sqrt{5}$ in common. Since $3\sqrt{5}$ means 3 times $\sqrt{5}$, or $3 \cdot \sqrt{5}$, we apply the distributive property:

$$\begin{aligned} 3\sqrt{5} - 7\sqrt{5} &= (3 - 7)\sqrt{5} && \text{Distributive property} \\ &= -4\sqrt{5} && 3 - 7 = -4 \end{aligned}$$

▲

Since we use the distributive property to add radical expressions, each expression must contain exactly the same radical.

▼ **Example 2** Combine terms in the expression $7\sqrt{2} - 3\sqrt{2} + 6\sqrt{2}$.

Solution

$$\begin{aligned} 7\sqrt{2} - 3\sqrt{2} + 6\sqrt{2} &= (7 - 3 + 6)\sqrt{2} && \text{Distributive property} \\ &= 10\sqrt{2} && \text{Addition} \end{aligned}$$

▲

In Examples 1 and 2, each term was a radical expression in simplified form. If one or more terms is not in simplified form, we must put it into simplified form and then combine terms, if possible. It is not always possible to combine terms containing radicals. Occasionally two or more of the terms will not have a radical in common. If there is a possibility of combining terms, it will always become apparent when each term is in simplified form.

> **Rule** To combine two or more radical expressions, put each expression in simplified form and then apply the distributive property, if possible.

▼ **Example 3** Combine terms in the expression $3\sqrt{50} + 2\sqrt{32}$.

Solution We begin by putting each term into simplified form:

$$\begin{aligned} 3\sqrt{50} + 2\sqrt{32} &= 3\sqrt{25}\sqrt{2} + 2\sqrt{16}\sqrt{2} && \text{Property 1 for radicals} \\ &= 3 \cdot 5\sqrt{2} + 2 \cdot 4\sqrt{2} && \sqrt{25} = 5 \text{ and } \sqrt{16} = 4 \\ &= 15\sqrt{2} + 8\sqrt{2} && \text{Multiplication} \end{aligned}$$

Applying the distributive property to the last line, we have

$$\begin{aligned} 15\sqrt{2} + 8\sqrt{2} &= (15 + 8)\sqrt{2} && \text{Distributive property} \\ &= 23\sqrt{2} && 15 + 8 = 23 \end{aligned}$$

▲

▼ **Example 4** Combine terms in the expression $5\sqrt{75} + \sqrt{27} - \sqrt{3}$.

Solution

$$\begin{aligned} &5\sqrt{75} + \sqrt{27} - \sqrt{3} \\ &= 5\sqrt{25}\,\sqrt{3} + \sqrt{9}\,\sqrt{3} - \sqrt{3} && \text{Property 1 for radicals} \\ &= 5 \cdot 5\sqrt{3} + 3\sqrt{3} - \sqrt{3} && \sqrt{25} = 5 \text{ and } \sqrt{9} = 3 \\ &= 25\sqrt{3} + 3\sqrt{3} - \sqrt{3} && 5 \cdot 5 = 25 \\ &= (25 + 3 - 1)\sqrt{3} && \text{Distributive property} \\ &= 27\sqrt{3} && \text{Addition} \end{aligned}$$

▲

The most time-consuming part of combining most radical expressions is simplifying each term in the expression. Once this has been done, applying the distributive property is simple and fast.

▼ **Example 5** Simplify: $a\sqrt{12} + 5\sqrt{3a^2}$.

Solution We must assume that a represents a positive number. Then, we simplify each term in the expression by putting it in simplified form for radicals:

$$\begin{aligned} a\sqrt{12} + 5\sqrt{3a^2} &= a\sqrt{4}\,\sqrt{3} + 5\sqrt{a^2}\,\sqrt{3} && \text{Property 1 for radicals} \\ &= a \cdot 2\sqrt{3} + 5 \cdot a\sqrt{3} && \sqrt{4} = 2 \text{ and } \sqrt{a^2} = a \\ &= 2a\sqrt{3} + 5a\sqrt{3} && \text{Commutative property} \\ &= (2a + 5a)\sqrt{3} && \text{Distributive property} \\ &= 7a\sqrt{3} && \text{Addition} \end{aligned}$$

▲

▼ **Example 6** Combine terms in the expression

$\sqrt{20x^3} - 3x\sqrt{45x} + 10\sqrt{25x^2}$. (Assume x is a positive real number.)

Solution

$$\begin{aligned} &\sqrt{20x^3} - 3x\sqrt{45x} + 10\sqrt{25x^2} \\ &= \sqrt{4x^2}\,\sqrt{5x} - 3x\sqrt{9}\,\sqrt{5x} + 10\sqrt{25x^2} \\ &= 2x\sqrt{5x} - 3x \cdot 3\sqrt{5x} + 10 \cdot 5x \\ &= 2x\sqrt{5x} - 9x\sqrt{5x} + 50x \end{aligned}$$

Each term is now in simplified form. The best we can do next is to combine the first two terms. The last term does not have the common radical $\sqrt{5x}$.

$$2x\sqrt{5x} - 9x\sqrt{5x} + 50x = (2x - 9x)\sqrt{5x} + 50x$$
$$= -7x\sqrt{5x} + 50x$$

We have, in any case, succeeded in reducing the number of terms in our original problem. ▲

Problem Set 7.4

In each of the following problems, simplify each term, if necessary, and then use the distributive property to combine terms, if possible.

1. $3\sqrt{2} + 4\sqrt{2}$
2. $7\sqrt{3} + 2\sqrt{3}$
3. $9\sqrt{5} - 7\sqrt{5}$
4. $6\sqrt{7} - 10\sqrt{7}$
5. $\sqrt{3} + 6\sqrt{3}$
6. $\sqrt{2} + 10\sqrt{2}$
7. $6\sqrt{5} - \sqrt{5}$
8. $9\sqrt{11} - \sqrt{11}$
9. $14\sqrt{13} - \sqrt{13}$
10. $-2\sqrt{6} - 9\sqrt{6}$
11. $-3\sqrt{10} + 9\sqrt{10}$
12. $11\sqrt{11} + \sqrt{11}$
13. $5\sqrt{5} + \sqrt{5}$
14. $\sqrt{6} - 10\sqrt{6}$
15. $\sqrt{8} + 2\sqrt{2}$
16. $\sqrt{20} + 3\sqrt{5}$
17. $3\sqrt{3} - \sqrt{27}$
18. $4\sqrt{5} - \sqrt{80}$
19. $5\sqrt{12} - 10\sqrt{48}$
20. $3\sqrt{300} - 5\sqrt{27}$
21. $-\sqrt{75} - \sqrt{3}$
22. $5\sqrt{20} + 8\sqrt{80}$
23. $4\sqrt{75} - 8\sqrt{12}$
24. $5\sqrt{24} + 4\sqrt{150}$
25. $8\sqrt{8} + 5\sqrt{75}$
26. $9\sqrt{54} - 8\sqrt{24}$
27. $\sqrt{27} - 2\sqrt{12} + \sqrt{3}$
28. $\sqrt{20} + 3\sqrt{45} - \sqrt{5}$
29. $\sqrt{72} - \sqrt{8} + \sqrt{50}$
30. $\sqrt{24} - \sqrt{54} - \sqrt{150}$
31. $5\sqrt{7} + 2\sqrt{28} - 4\sqrt{63}$
32. $3\sqrt{3} - 5\sqrt{27} + 8\sqrt{75}$
33. $6\sqrt{48} - 2\sqrt{12} + 5\sqrt{27}$
34. $5\sqrt{50} + 8\sqrt{12} - \sqrt{32}$
35. $6\sqrt{48} - \sqrt{72} - 3\sqrt{300}$
36. $7\sqrt{44} - 8\sqrt{99} + \sqrt{176}$

All variables in the following problems represent positive real numbers. Simplify each term and combine, if possible.

37. $\sqrt{x^3} + x\sqrt{x}$
38. $2\sqrt{x} - 2\sqrt{4x}$
39. $5\sqrt{3a^2} - a\sqrt{3}$
40. $6a\sqrt{a} + 7\sqrt{a^3}$
41. $5\sqrt{8x^3} + x\sqrt{50x}$
42. $2\sqrt{27x^2} - x\sqrt{48}$
43. $3\sqrt{75x^3y} - 2x\sqrt{3xy}$
44. $9\sqrt{24x^3y^2} - 5x\sqrt{54xy^2}$

45. $\sqrt{20ab^2} - b\sqrt{45a}$
46. $4\sqrt{a^3b^2} - 5a\sqrt{ab^2}$
47. $9\sqrt{18x^3} - 2x\sqrt{48x}$
48. $8\sqrt{72x^2} - x\sqrt{8}$
49. $7\sqrt{50x^2y} + 8x\sqrt{8y} - 7\sqrt{32x^2y}$
50. $6\sqrt{44x^3y^3} - 8x\sqrt{99xy^3} - 6y\sqrt{176x^3y}$
51. Use the table of squares and square roots at the back of the book to find a decimal approximation of the expression $\sqrt{5} + \sqrt{3}$. Is the answer equal to the decimal approximation of $\sqrt{8}$ listed in the table?
52. Find a decimal approximation of the expression $\sqrt{5} - \sqrt{3}$ using the table at the back of the book. Is the answer equal to the decimal approximation of $\sqrt{2}$ listed in the table?
53. The statement below is false. Correct the right side to make the statement true.

$$4\sqrt{3} + 5\sqrt{3} = 9\sqrt{6}$$

54. The statement below is false. Correct the right side to make the statement true.

$$7\sqrt{5} - 3\sqrt{5} = 4\sqrt{25}$$

Review Problems The following problems review material we covered in Sections 4.6 and 6.4. Reviewing the problems from Section 4.6 will help you in the next section.

Multiply. [4.6]

55. $(3x + y)^2$
56. $(2x - 3y)^2$
57. $(3x - 4y)(3x + 4y)$
58. $(7x + 2y)(7x - 2y)$

Solve each equation. [6.4]

59. $\frac{x}{3} - \frac{1}{2} = \frac{5}{2}$
60. $\frac{3}{x} + \frac{1}{5} = \frac{4}{5}$
61. $1 - \frac{5}{x} = \frac{-6}{x^2}$
62. $1 - \frac{1}{x} = \frac{6}{x^2}$
63. $\frac{a}{a - 4} - \frac{a}{2} = \frac{4}{a - 4}$
64. $\frac{a}{a - 3} - \frac{a}{2} = \frac{3}{a - 3}$

Section 7.5 Multiplication and Division of Radicals

In this section we will look at multiplication and division of expressions that contain radicals. As you will see, multiplication of expressions that contain radicals is very similar to multiplication of polynomials. The division problems in this section are just an extension of the work we did previously when we rationalized denominators.

▼ **Example 1** Multiply: $(3\sqrt{5})(2\sqrt{7})$.

Solution We can rearrange the order and grouping of the numbers in this product by applying the commutative and associative properties. Following that, we apply Property 1 for radicals and multiply:

$$
\begin{aligned}
(3\sqrt{5})(2\sqrt{7}) &= (3 \cdot 2)(\sqrt{5}\,\sqrt{7}) && \text{Commutative and associative properties} \\
&= (3 \cdot 2)(\sqrt{5 \cdot 7}) && \text{Property 1 for radicals} \\
&= 6\sqrt{35} && \text{Multiplication}
\end{aligned}
$$

In actual practice, it is not necessary to show either of the first two steps, although you may want to show them on the first few problems you work, just to be sure you understand them. ▲

▼ **Example 2** Multiply: $\sqrt{5}(\sqrt{2} + \sqrt{5})$.

Solution

$$
\begin{aligned}
\sqrt{5}(\sqrt{2} + \sqrt{5}) &= \sqrt{5} \cdot \sqrt{2} + \sqrt{5} \cdot \sqrt{5} && \text{Distributive property} \\
&= \sqrt{10} + 5 && \text{Multiplication}
\end{aligned}
$$

▲

▼ **Example 3** Multiply: $3\sqrt{2}(2\sqrt{5} + 5\sqrt{3})$.

Solution

$$
\begin{aligned}
&3\sqrt{2}(2\sqrt{5} + 5\sqrt{3}) \\
&\quad = 3\sqrt{2} \cdot 2\sqrt{5} + 3\sqrt{2} \cdot 5\sqrt{3} && \text{Distributive property} \\
&\quad = 3 \cdot 2 \cdot \sqrt{2}\,\sqrt{5} + 3 \cdot 5\sqrt{2}\,\sqrt{3} && \text{Commutative property} \\
&\quad = 6\sqrt{10} + 15\sqrt{6}
\end{aligned}
$$

▲

Each item in the last line is in simplified form, so the problem is complete.

▼ **Example 4** Multiply: $(\sqrt{5} + 2)(\sqrt{5} + 7)$.

Solution We multiply using the FOIL method we used to multiply binomials:

$$
\begin{aligned}
(\sqrt{5} + 2)(\sqrt{5} + 7) &= \underset{F}{\sqrt{5} \cdot \sqrt{5}} + \underset{O}{7\sqrt{5}} + \underset{I}{2\sqrt{5}} + \underset{L}{14} \\
&= 5 + 9\sqrt{5} + 14 \\
&= 19 + 9\sqrt{5}
\end{aligned}
$$

We must be careful not to try to simplify further by adding 19 and 9. We can add only radical expressions that have a common radical part; 19 and $9\sqrt{5}$ are not similar. ▲

▼ **Example 5** Multiply: $(\sqrt{x} + 3)(\sqrt{x} - 7)$.

Solution Remember, we are assuming that any variables that appear under a radical represent positive numbers.

$$\begin{aligned}(\sqrt{x} + 3)(\sqrt{x} - 7) &= \underset{F}{\sqrt{x}\sqrt{x}} - \underset{O}{7\sqrt{x}} + \underset{I}{3\sqrt{x}} - \underset{L}{21} \\ &= x - 4\sqrt{x} - 21\end{aligned}$$

▲

▼ **Example 6** Expand and simplify: $(\sqrt{3} - 2)^2$.

Solution Multiplying $\sqrt{3} - 2$ times itself, we have

$$\begin{aligned}(\sqrt{3} - 2)^2 &= (\sqrt{3} - 2)(\sqrt{3} - 2) \\ &= \sqrt{3}\sqrt{3} - 2\sqrt{3} - 2\sqrt{3} + 4 \\ &= 3 - 4\sqrt{3} + 4 \\ &= 7 - 4\sqrt{3}\end{aligned}$$

▲

▼ **Example 7** Multiply: $(\sqrt{5} + \sqrt{2})(\sqrt{5} - \sqrt{2})$.

Solution We can apply the formula $(x + y)(x - y) = x^2 - y^2$ to obtain

$$\begin{aligned}(\sqrt{5} + \sqrt{2})(\sqrt{5} - \sqrt{2}) &= (\sqrt{5})^2 - (\sqrt{2})^2 \\ &= 5 - 2 \\ &= 3\end{aligned}$$

We could also have multiplied the two expressions using the FOIL method. If we were to do so, the work would look like this:

$$\begin{aligned}&(\sqrt{5} + \sqrt{2})(\sqrt{5} - \sqrt{2}) \\ &\quad = \underset{F}{\sqrt{5}\sqrt{5}} - \underset{O}{\sqrt{2}\sqrt{5}} + \underset{I}{\sqrt{2}\sqrt{5}} - \underset{L}{\sqrt{2}\sqrt{2}} \\ &\quad = 5 - \sqrt{10} + \sqrt{10} - 2 \\ &\quad = 5 - 2 \\ &\quad = 3\end{aligned}$$

In either case, the product is 3. Also, the expressions $\sqrt{5} + \sqrt{2}$ and $\sqrt{5} - \sqrt{2}$ are called *conjugates* of each other. ▲

▼ **Example 8** Multiply: $(\sqrt{a} + \sqrt{b})(\sqrt{a} - \sqrt{b})$.

Solution We can apply the formula $(x + y)(x - y) = x^2 - y^2$ to obtain

$$\begin{aligned}(\sqrt{a} + \sqrt{b})(\sqrt{a} - \sqrt{b}) &= (\sqrt{a})^2 - (\sqrt{b})^2 \\ &= a - b\end{aligned}$$

▲

▼ **Example 9** Rationalize the denominator in the expression

$$\frac{\sqrt{3}}{\sqrt{3} - \sqrt{2}}$$

Solution To remove the two radicals in the denominator, we must multiply both the numerator and denominator by $\sqrt{3} + \sqrt{2}$. That way, when we multiply $\sqrt{3} - \sqrt{2}$ and $\sqrt{3} + \sqrt{2}$ we will obtain the difference of two squares in the denominator:

$$\begin{aligned}
\frac{\sqrt{3}}{\sqrt{3} - \sqrt{2}} &= \frac{\sqrt{3}}{(\sqrt{3} - \sqrt{2})} \cdot \frac{(\sqrt{3} + \sqrt{2})}{(\sqrt{3} + \sqrt{2})} \\
&= \frac{\sqrt{3}\sqrt{3} + \sqrt{3}\sqrt{2}}{(\sqrt{3})^2 - (\sqrt{2})^2} \\
&= \frac{3 + \sqrt{6}}{3 - 2} \\
&= \frac{3 + \sqrt{6}}{1} \\
&= 3 + \sqrt{6}
\end{aligned}$$

▲

▼ **Example 10** Rationalize the denominator in the expression $\frac{2}{5 - \sqrt{3}}$.

Solution We use the same procedure as in Example 9. Multiply the numerator and denominator by the conjugate of the denominator, which is $5 + \sqrt{3}$:

$$\begin{aligned}
\left(\frac{2}{5 - \sqrt{3}}\right)\left(\frac{5 + \sqrt{3}}{5 + \sqrt{3}}\right) &= \frac{10 + 2\sqrt{3}}{5^2 - (\sqrt{3})^2} \\
&= \frac{10 + 2\sqrt{3}}{25 - 3} \\
&= \frac{10 + 2\sqrt{3}}{22}
\end{aligned}$$

The numerator and denominator of this last expression have a factor of 2 in common. We can reduce to lowest terms by dividing out the common factor 2. Continuing, we have

$$\begin{aligned}
&= \frac{\cancel{2}(5 + \sqrt{3})}{\cancel{2} \cdot 11} \\
&= \frac{5 + \sqrt{3}}{11}
\end{aligned}$$

The final expression is in simplified form. ▲

▼ **Example 11** Rationalize the denominator in the expression

$$\frac{\sqrt{2} + \sqrt{3}}{\sqrt{2} - \sqrt{3}}$$

Solution We remove the two radicals in the denominator by multiplying both the numerator and denominator by the conjugate of $\sqrt{2} - \sqrt{3}$, which is $\sqrt{2} + \sqrt{3}$:

$$\begin{aligned}\frac{\sqrt{2} + \sqrt{3}}{\sqrt{2} - \sqrt{3}} &= \left(\frac{\sqrt{2} + \sqrt{3}}{\sqrt{2} - \sqrt{3}}\right)\frac{(\sqrt{2} + \sqrt{3})}{(\sqrt{2} + \sqrt{3})} \\ &= \frac{\sqrt{2}\sqrt{2} + \sqrt{2}\sqrt{3} + \sqrt{3}\sqrt{2} + \sqrt{3}\sqrt{3}}{(\sqrt{2})^2 - (\sqrt{3})^2} \\ &= \frac{2 + \sqrt{6} + \sqrt{6} + 3}{2 - 3} \\ &= \frac{5 + 2\sqrt{6}}{-1} \\ &= -(5 + 2\sqrt{6}) \quad \text{or} \quad -5 - 2\sqrt{6}\end{aligned}$$

▲

Problem Set 7.5

Perform the following multiplications. All answers should be in simplified form for radical expressions.

1. $\sqrt{3}\sqrt{2}$
2. $\sqrt{5}\sqrt{6}$
3. $\sqrt{6}\sqrt{2}$
4. $\sqrt{6}\sqrt{3}$
5. $(2\sqrt{3})(5\sqrt{7})$
6. $(3\sqrt{2})(4\sqrt{5})$
7. $(4\sqrt{3})(2\sqrt{6})$
8. $(7\sqrt{6})(3\sqrt{2})$
9. $\sqrt{2}(\sqrt{3} - 1)$
10. $\sqrt{3}(\sqrt{5} + 2)$
11. $\sqrt{2}(\sqrt{3} + \sqrt{2})$
12. $\sqrt{5}(\sqrt{7} - \sqrt{5})$
13. $\sqrt{3}(2\sqrt{2} + \sqrt{3})$
14. $\sqrt{11}(3\sqrt{2} - \sqrt{11})$
15. $\sqrt{3}(2\sqrt{3} - \sqrt{5})$
16. $\sqrt{7}(\sqrt{14} - \sqrt{7})$
17. $2\sqrt{3}(\sqrt{2} + \sqrt{5})$
18. $3\sqrt{2}(\sqrt{3} + \sqrt{2})$
19. $(\sqrt{2} + 1)^2$
20. $(\sqrt{5} - 4)^2$
21. $(\sqrt{x} + 3)^2$
22. $(\sqrt{x} - 4)^2$
23. $(5 - \sqrt{2})^2$
24. $(2 + \sqrt{5})^2$
25. $(\sqrt{a} - 5)^2$
26. $(\sqrt{a} + 7)^2$
27. $(3 + \sqrt{7})^2$
28. $(3 - \sqrt{2})^2$
29. $(\sqrt{5} + 3)(\sqrt{5} + 2)$
30. $(\sqrt{7} + 4)(\sqrt{7} - 5)$
31. $(\sqrt{2} - 5)(\sqrt{2} + 6)$
32. $(\sqrt{3} + 8)(\sqrt{3} - 2)$
33. $(\sqrt{3} + 1)(\sqrt{2} + 3)$
34. $(\sqrt{5} - 2)(\sqrt{3} + 4)$

35. $(\sqrt{x} + 6)(\sqrt{x} - 6)$

36. $(\sqrt{x} + 7)(\sqrt{x} - 7)$

37. $(\sqrt{a} + 2)(\sqrt{a} - 4)$

38. $(\sqrt{a} + 6)(\sqrt{a} - 3)$

39. $(\sqrt{5} - 2)(\sqrt{5} + 2)$

40. $(\sqrt{6} - 3)(\sqrt{6} + 3)$

41. $(2\sqrt{7} + 3)(3\sqrt{7} - 4)$

42. $(3\sqrt{5} + 1)(4\sqrt{5} + 3)$

43. $(2\sqrt{x} + 4)(3\sqrt{x} + 2)$

44. $(3\sqrt{x} + 5)(4\sqrt{x} + 2)$

45. $(7\sqrt{a} + 2\sqrt{b})(7\sqrt{a} - 2\sqrt{b})$

46. $(3\sqrt{a} - 2\sqrt{b})(3\sqrt{a} + 2\sqrt{b})$

Rationalize the denominator. All answers should be expressed in simplified form.

47. $\dfrac{\sqrt{3}}{\sqrt{5} - \sqrt{2}}$

48. $\dfrac{\sqrt{2}}{\sqrt{6} + \sqrt{3}}$

49. $\dfrac{\sqrt{5}}{\sqrt{5} + \sqrt{2}}$

50. $\dfrac{\sqrt{7}}{\sqrt{7} - \sqrt{2}}$

51. $\dfrac{8}{3 - \sqrt{5}}$

52. $\dfrac{10}{5 + \sqrt{5}}$

53. $\dfrac{\sqrt{3} + \sqrt{2}}{\sqrt{3} - \sqrt{2}}$

54. $\dfrac{\sqrt{5} - \sqrt{2}}{\sqrt{5} + \sqrt{2}}$

55. $\dfrac{\sqrt{7} - \sqrt{3}}{\sqrt{7} + \sqrt{3}}$

56. $\dfrac{\sqrt{11} - \sqrt{6}}{\sqrt{11} + \sqrt{6}}$

57. $\dfrac{\sqrt{x} + 2}{\sqrt{x} - 2}$

58. $\dfrac{\sqrt{x} - 3}{\sqrt{x} + 3}$

59. $\dfrac{\sqrt{5} - \sqrt{2}}{\sqrt{5} + \sqrt{3}}$

60. $\dfrac{\sqrt{7} - \sqrt{3}}{\sqrt{5} + \sqrt{2}}$

61. The statement below is incorrect. Correct the right side to make the statement true.

$$2(3\sqrt{5}) = 6\sqrt{15}$$

62. The statement below is incorrect. Correct the right side to make the statement true.

$$5(2\sqrt{6}) = 10\sqrt{30}$$

63. The statement below is incorrect. What is missing from the right side?

$$(\sqrt{3} + 7)^2 = 3 + 49$$

64. The statement below is incorrect. What is missing from the right side?

$$(\sqrt{5} + \sqrt{2})^2 = 5 + 2$$

Review Problems The following problems review material we covered in Sections 5.7 and 6.7. Reviewing these problems will help you understand the next section.

Solve each equation. [5.7]

65. $x^2 + 5x - 6 = 0$

66. $x^2 + 5x + 6 = 0$

67. $x^2 - 3x = 0$

68. $x^2 + 5x = 0$

Solve each proportion. [6.7]

69. $\frac{x}{3} = \frac{27}{x}$

70. $\frac{x}{2} = \frac{8}{x}$

71. $\frac{x}{5} = \frac{3}{x + 2}$

72. $\frac{x}{2} = \frac{4}{x + 2}$

73. Suppose you drive your car 375 miles in 15 hours. At this rate, how far will you drive in 20 hours?

74. If 2.5 inches on a map correspond to an actual distance of 100 miles, then how many miles apart are two cities that are 3.5 inches apart on the map?

Section 7.6 Equations Involving Radicals

In order to solve equations that contain one or more radical expressions we need an additional property. From our work with exponents, we know that if two quantities are equal, then so are the squares of those quantities. That is, for real numbers a and b

$$\text{if} \quad a = b$$
$$\text{then} \quad a^2 = b^2$$

The only problem with squaring both sides of an equation is that occasionally we will change a false statement into a true statement. Let's take the false statement $3 = -3$ as an example. Squaring both sides, we have

$$3 = -3 \qquad \text{A false statement}$$
$$(3)^2 = (-3)^2 \qquad \text{Square both sides}$$
$$9 = 9 \qquad \text{A true statement}$$

We can avoid this problem by always checking our solutions if, at any time during the process of solving an equation, we have squared both sides of the equation. Here is how the property is stated.

Squaring Property of Equality

We can square both sides of an equation any time it is convenient to do so, as long as we check all solutions in the original equation.

▼ **Example 1** Solve for x: $\sqrt{x + 1} = 7$.

Solution In order to solve this equation by our usual methods, we must first eliminate the radical sign. We can accomplish this by squaring both sides of the equation:

$$\begin{aligned} \sqrt{x+1} &= 7 \\ (\sqrt{x+1})^2 &= 7^2 && \text{Square both sides} \\ x + 1 &= 49 \\ x &= 48 \end{aligned}$$

To check our solution, we substitute $x = 48$ into the original equation:

$$\begin{aligned} \sqrt{48+1} &\stackrel{?}{=} 7 \\ \sqrt{49} &= 7 \\ 7 &= 7 && \text{A true statement} \end{aligned}$$

The solution checks. ▲

▼ **Example 2** Solve for x: $\sqrt{2x - 3} = -9$.

Solution We square both sides and proceed as in Example 1:

$$\begin{aligned} \sqrt{2x-3} &= -9 \\ (\sqrt{2x-3})^2 &= (-9)^2 && \text{Square both sides} \\ 2x - 3 &= 81 \\ 2x &= 84 \\ x &= 42 \end{aligned}$$

Checking our solution in the original equation, we have

$$\begin{aligned} \sqrt{2(42) - 3)} &\stackrel{?}{=} -9 \\ \sqrt{84 - 3} &= -9 \\ \sqrt{81} &= -9 \\ 9 &= -9 && \text{A false statement} \end{aligned}$$

Our solution does not check because we end up with a false statement. ▲

Squaring both sides of the equation has produced what is called an extraneous solution. This happens occasionally when we use the squaring property of equality. We can always eliminate extraneous solutions by checking each solution with the original equation.

Note In Example 2, when we checked $x = 42$ in the original equation, we found that it is not a solution to the equation. Actually, it was apparent from the beginning that the equation had no solution. That is, no matter what x is, the equation

$$\sqrt{2x - 3} = -9$$

can never be true because the left side is a positive number (or zero) for any value of x, while the right side is always negative.

▼ **Example 3** Solve for a: $\sqrt{3a - 2} + 3 = 5$.

Solution Before we can square both sides to eliminate the radical, we must isolate the radical on the left side of the equation. To do so, we add -3 to both sides:

$$\begin{aligned} \sqrt{3a - 2} + 3 &= 5 && \text{Add } -3 \text{ to both sides} \\ \sqrt{3a - 2} &= 2 && \text{Square both sides} \\ (\sqrt{3a - 2})^2 &= 2^2 \\ 3a - 2 &= 4 \\ 3a &= 6 \\ a &= 2 \end{aligned}$$

Checking $a = 2$ in the original equation, we have

$$\begin{aligned} \sqrt{3 \cdot 2 - 2} + 3 &\stackrel{?}{=} 5 \\ \sqrt{4} + 3 &= 5 \\ 2 + 3 &= 5 \\ 5 &= 5 && \text{A true statement} \end{aligned}$$

▲

▼ **Example 4** Solve for x: $\sqrt{x + 15} = x + 3$.

Solution We begin by squaring both sides:

$$\begin{aligned} (\sqrt{x + 15})^2 &= (x + 3)^2 && \text{Square both sides} \\ x + 15 &= x^2 + 6x + 9 \end{aligned}$$

We have a quadratic equation. We put it into standard form by adding $-x$ and -15 to both sides. Then we factor and solve as usual:

$$\begin{aligned} 0 &= x^2 + 5x - 6 && \text{Standard form} \\ 0 &= (x + 6)(x - 1) && \text{Factor} \\ x + 6 = 0 \quad &\text{or} \quad x - 1 = 0 && \text{Set factors equal to 0} \\ x = -6 \quad &\text{or} \quad x = 1 \end{aligned}$$

We check each solution in the original equation:

Check -6	Check 1
$\sqrt{-6 + 15} \stackrel{?}{=} -6 + 3$	$\sqrt{1 + 15} \stackrel{?}{=} 1 + 3$
$\sqrt{9} = -3$	$\sqrt{16} = 4$
$3 = -3$	$4 = 4$
A false statement	A true statement

Since $x = -6$ does not check in the original equation, it cannot be a solution. The only solution is $x = 1$. ▲

Problem Set 7.6

Solve each equation by applying the squaring property of equality. Be sure to check all solutions in the original equation.

1. $\sqrt{x + 1} = 2$
2. $\sqrt{x - 3} = 4$
3. $\sqrt{x + 5} = 7$
4. $\sqrt{x + 8} = 5$
5. $\sqrt{x - 9} = -6$
6. $\sqrt{x + 10} = -3$
7. $\sqrt{x - 5} = -4$
8. $\sqrt{x + 7} = -5$
9. $\sqrt{x - 8} = 0$
10. $\sqrt{x - 9} = 0$
11. $\sqrt{2x + 1} = 3$
12. $\sqrt{2x - 5} = 7$
13. $\sqrt{2x - 3} = -5$
14. $\sqrt{3x - 8} = -4$
15. $\sqrt{3x + 6} = 2$
16. $\sqrt{5x - 1} = 5$
17. $2\sqrt{x} = 10$
18. $3\sqrt{x} = 9$
19. $3\sqrt{a} = 6$
20. $2\sqrt{a} = 12$
21. $\sqrt{3x + 4} - 3 = 2$
22. $\sqrt{2x - 1} + 2 = 5$
23. $\sqrt{5y - 4} - 2 = 4$
24. $\sqrt{3y + 1} + 7 = 2$
25. $\sqrt{2x + 1} + 5 = 2$
26. $\sqrt{6x - 8} - 1 = 3$
27. $\sqrt{x + 3} = x - 3$
28. $\sqrt{x - 3} = x - 3$
29. $\sqrt{a + 2} = a + 2$
30. $\sqrt{a + 10} = a - 2$
31. $\sqrt{2x + 9} = x + 5$
32. $\sqrt{x + 6} = x + 4$
33. $\sqrt{y - 4} = y - 6$
34. $\sqrt{2y + 13} = y + 7$
35. $\sqrt{5x + 1} = x + 1$
36. $\sqrt{6x + 1} = x - 1$
37. The sum of a number and two is equal to the positive square root of eight times the number. Find the number.
38. The sum of twice a number and one is equal to three times the positive square root of the number. Find the number.
39. The difference of a number and three is equal to twice the positive square root of the number. Find the number.
40. The difference of a number and two is equal to the positive square root of the number. Find the number.
41. The number of seconds, T, it takes the pendulum of a grandfather clock to swing through one complete cycle is given by the formula

$$T = \frac{11}{7}\sqrt{\frac{L}{2}}$$

where L is the length, in feet, of the pendulum. Find how long the pendulum must be for one complete cycle to take 2 seconds by substituting 2 for T in the formula and then solving for L.

42. How long must the pendulum on a grandfather clock be if one complete cycle is to take 1 second?

Review Problems The problems that follow review material we covered in Sections 6.5 and 6.6.

Simplify each complex fraction. [6.6]

43. $\dfrac{\frac{2}{5}}{\frac{4}{15}}$

44. $\dfrac{\frac{x^4}{y^2}}{\frac{x^3}{y^5}}$

45. $\dfrac{1 - \frac{1}{a}}{1 + \frac{1}{a}}$

46. $\dfrac{1 - \frac{16}{x^2}}{1 - \frac{2}{x} - \frac{24}{x^2}}$

Solve each word problem. [6.5]

47. The sum of a number and its reciprocal is $\frac{10}{3}$. Find the number.

48. One number is twice another. The sum of their reciprocals is $\frac{1}{2}$. Find the two numbers.

49. Suppose a person can jog four miles an hour faster than she can walk. If she jogs 12 miles in the same amount of time it takes her to walk 6 miles, how fast can she walk?

50. A boat can travel 6 miles up a river in the same amount of time it takes to travel 14 miles down the same river. If the speed of the current is 2 miles per hour, what is the speed of the boat in still water?

51. A hot water faucet can fill a sink in 4 minutes, while the cold water faucet can fill the sink in 3 minutes. If both faucets are open, how long will it take to fill the sink?

52. A hose can fill a swimming pool in 6 hours, while the drain can empty the pool in 18 hours. If the hose is on and the drain is open, how long will it take to fill the pool?

Chapter 7 Summary and Review

ROOTS [7.1]

Every positive real number x has two square roots, one positive and one negative. The positive square root is written $\sqrt{x}$. The negative square root of x is written $-\sqrt{x}$. In both cases the square root of x is a number we square to get x.

The cube root of x is written $\sqrt[3]{x}$ and is the number we cube to get x.

Examples

1. The two square roots of 9 are 3 and -3:

$$\sqrt{9} = 3 \quad \text{and} \quad -\sqrt{9} = -3$$

NOTATION [7.1]

In the expression $\sqrt[3]{8}$, 8 is called the *radicand*, 3 is the *index*, $\sqrt{\ }$ is called the *radical sign*, and the whole expression $\sqrt[3]{8}$ is called the *radical*.

2. Index → $\sqrt[3]{24}$ ← Radicand; ↙ Radical sign; Radical (the whole expression)

PROPERTIES OF RADICALS [7.2]

If a and b represent positive real numbers, then

1. $\sqrt{a}\sqrt{b} = \sqrt{ab}$ — The product of the square root is the square root of the product
2. $\dfrac{\sqrt{a}}{\sqrt{b}} = \sqrt{\dfrac{a}{b}}$ — The quotient of the square roots is the square root of the quotient
3. $\sqrt{a} \cdot \sqrt{a} = (\sqrt{a})^2 = a$ — This property shows that squaring and square roots are inverse operations

3.

(a) $\sqrt{3} \cdot \sqrt{2} = \sqrt{3 \cdot 2} = \sqrt{6}$

(b) $\dfrac{\sqrt{12}}{\sqrt{3}} = \sqrt{\dfrac{12}{3}} = \sqrt{4} = 2$

(c) $\sqrt{5} \cdot \sqrt{5} = (\sqrt{5})^2 = 5$

SIMPLIFIED FORM FOR RADICALS [7.3]

A radical expression is in simplified form if:

1. There are no perfect squares as factors of the quantity under the square root sign.
2. There are no fractions under the radical sign.
3. There are no radicals in the denominator.

4. Simplify $\sqrt{20}$ and $\sqrt{\frac{2}{3}}$.

$\sqrt{20} = \sqrt{4 \cdot 5} = \sqrt{4}\sqrt{5} = 2\sqrt{5}$

$$\sqrt{\frac{2}{3}} = \frac{\sqrt{2}}{\sqrt{3}} = \frac{\sqrt{2}\,\mathbf{\sqrt{3}}}{\sqrt{3}\,\mathbf{\sqrt{3}}} = \frac{\sqrt{6}}{3}$$

ADDITION AND SUBTRACTION OF RADICAL EXPRESSIONS [7.4]

We add and subtract radical expressions by using the distributive property to combine terms that have the same radical parts. If the radicals are not in simplified form, we begin by writing them in simplified form and then combining similar terms, if possible.

5.

(a) $5\sqrt{7} + 3\sqrt{7} = 8\sqrt{7}$

(b) $2\sqrt{18} - 3\sqrt{50}$
$= 2 \cdot 3\sqrt{2} - 3 \cdot 5\sqrt{2}$
$= 6\sqrt{2} - 15\sqrt{2}$
$= -9\sqrt{2}$

MULTIPLICATION OF RADICAL EXPRESSIONS [7.5]

We multiply radical expressions by applying the distributive property or the FOIL method.

6.

(a) $\sqrt{3}(\sqrt{5} - \sqrt{3}) = \sqrt{15} - 3$

(b) $(\sqrt{7} + 3)(\sqrt{7} - 5)$
$= 7 - 5\sqrt{7} + 3\sqrt{7} - 15$
$= -8 - 2\sqrt{7}$

DIVISION OF RADICAL EXPRESSIONS [7.5]

To divide by an expression like $\sqrt{5} - \sqrt{3}$, we multiply the numerator and denominator by its conjugate, $\sqrt{5} + \sqrt{3}$. This process is also called rationalizing the denominator.

7. $$\frac{7}{\sqrt{5} - \sqrt{3}} = \frac{7}{\sqrt{5} - \sqrt{3}} \cdot \frac{\sqrt{5} + \sqrt{3}}{\sqrt{5} + \sqrt{3}} = \frac{7\sqrt{5} + 7\sqrt{3}}{2}$$

SQUARING PROPERTY OF EQUALITY [7.6]

We are free to square both sides of an equation whenever it is convenient, as long as we check all solutions in the original equation. We must check solutions because squaring both sides of an equation occasionally produces extraneous solutions.

8. Solve $\sqrt{x - 3} = 2$

$$(\sqrt{x - 3})^2 = 2^2$$
$$x - 3 = 4$$
$$x = 7$$

The solution checks in the original equation.

COMMON MISTAKES

1. A very common mistake with radicals is to think of $\sqrt{25}$ as representing both the positive and negative square roots of 25. The notation $\sqrt{25}$ stands for the *positive* square root of 25. If we want the negative square root of 25, we write $-\sqrt{25}$.

2. The most common mistake when working with radicals is to try to apply a property similar to Property 2 for radicals involving addition instead of multiplication. Here is an example:

$$\sqrt{16 + 9} = \sqrt{16} + \sqrt{9} \qquad \text{Mistake}$$

Although this example looks like it may be true, it isn't. If we carry it out further, the mistake becomes obvious:

$$\sqrt{16 + 9} \stackrel{?}{=} \sqrt{16} + \sqrt{9}$$
$$\sqrt{25} = 4 + 3$$
$$5 = 7 \qquad \text{False}$$

3. It is a mistake to try and simplify expressions like $2 + 3\sqrt{7}$. The 2 and 3 cannot be combined because the terms they appear in are not similar. Therefore, $2 + 3\sqrt{7} \neq 5\sqrt{7}$. The expression $2 + 3\sqrt{7}$ cannot be simplified further.

Chapter 7 Test

Find the following roots. [7.1]

1. $\sqrt{16}$

2. $-\sqrt{36}$

3. The square roots of 49

4. $\sqrt[3]{27}$

5. $\sqrt[3]{-8}$

6. $-\sqrt[4]{81}$

Put the following expressions into simplified form. [7.2, 7.3]

7. $\sqrt{75}$

8. $\sqrt{32}$

9. $\sqrt{\frac{2}{3}}$

10. $\frac{1}{\sqrt[3]{4}}$

11. $3\sqrt{50x^2}$

12. $\sqrt{\frac{12x^2y^3}{5}}$

Combine. [7.4]

13. $5\sqrt{12} - 2\sqrt{27}$

14. $2x\sqrt{18} + 5\sqrt{2x^2}$

Multiply. [7.5]

15. $\sqrt{3}(\sqrt{5} - 2)$

16. $(\sqrt{5} + 7)(\sqrt{5} - 8)$

17. $(\sqrt{x} + 6)(\sqrt{x} - 6)$

18. $(\sqrt{5} - \sqrt{3})^2$

Divide. (Rationalize the denominator.) [7.5]

19. $\frac{\sqrt{7} - \sqrt{3}}{\sqrt{7} + \sqrt{3}}$

20. $\frac{\sqrt{x}}{\sqrt{x} + 5}$

Solve the following equations. [7.6]

21. $\sqrt{x + 5} = 2$

22. $\sqrt{2x + 1} + 2 = 7$

23. $\sqrt{3x + 1} + 6 = 2$

24. $\sqrt{2x - 3} = x - 3$

25. The difference of a number and four is equal to three times the positive square root of the number. Find the number.

In this chapter we will develop a method of solving quadratic equations that can be used on all quadratic equations, regardless of whether or not they are factorable. We will then apply this method to the general quadratic equation $ax^2 + bx + c = 0$ and arrive at what is known as the quadratic formula. We will end the chapter by considering the graphs of second-degree equations.

To be successful in this chapter you should have a working knowledge of (1) square roots, (2) binomial squares, and (3) graphing by the use of tables.

Section 8.1 More Quadratic Equations

Consider the equation $x^2 = 9$. Inspection shows that there are two solutions, $x = 3$ and $x = -3$. We arrive at these solutions by taking the square root of both sides of the original equation. Since we are taking the square roots ourselves, we must be sure to include both the positive and the negative square roots. When all the work is shown, the problem looks like this:

$$\begin{aligned} x^2 &= 9 \\ \sqrt{x^2} &= \pm\sqrt{9} && \text{Take the square root of both sides} \\ x &= \pm 3 && \pm\sqrt{9} = \pm 3 \end{aligned}$$

The notation $x = \pm 3$ is shorthand for the expression $x = 3$ or $x = -3$.

We can take the square root of both sides of an equation any time we think it is helpful. We must make sure, however, that we include both the positive and the negative square roots.

▼ **Example 1** Solve for x: $x^2 = 7$.

Solution

$$\begin{aligned} x^2 &= 7 \\ \sqrt{x^2} &= \pm\sqrt{7} && \text{Take the square root of both sides} \\ x &= \pm\sqrt{7} \end{aligned}$$

The two solutions are $\sqrt{7}$ and $-\sqrt{7}$. ▲

▼ **Example 2** Solve for y: $3y^2 = 60$.

Solution We begin by dividing both sides by 3 (which is the same as multiplying both sides by $\frac{1}{3}$). Then, we take the square root of each side:

$$\begin{aligned} 3y^2 &= 60 \\ y^2 &= 20 && \text{Divide each side by 3} \\ \sqrt{y^2} &= \pm\sqrt{20} && \text{Take the square root of each side} \\ y &= \pm 2\sqrt{5} && \sqrt{20} = \sqrt{4 \cdot 5} = \sqrt{4}\sqrt{5} = 2\sqrt{5} \end{aligned}$$

Our two solutions are $2\sqrt{5}$ and $-2\sqrt{5}$. Each of them will yield a true statement when used in place of the variable in the original equation $3y^2 = 60$. ▲

▼ **Example 3** Solve for a: $(a + 3)^2 = 16$.

Solution We proceed as we did in the last example. The square root of $(a + 3)^2$ is $(a + 3)$:

$$\begin{aligned} (a + 3)^2 &= 16 \\ a + 3 &= \pm 4 \end{aligned}$$

At this point we add -3 to both sides to get

$$a = -3 \pm 4$$

which we can write as

$$\begin{aligned} a &= -3 + 4 && \text{or} && a = -3 - 4 \\ a &= 1 && \text{or} && a = -7 \end{aligned}$$

Our solutions are 1 and -7. ▲

▼ **Example 4** Solve for x: $(3x - 2)^2 = 25$.

Solution

$$(3x - 2)^2 = 25$$
$$3x - 2 = \pm 5$$

Adding 2 to both sides, we have

$$3x = 2 \pm 5$$

Dividing both sides by 3 gives us

$$x = \frac{2 \pm 5}{3}$$

We separate this last equation into two separate statements:

$$x = \frac{2 + 5}{3} \quad \text{or} \quad x = \frac{2 - 5}{3}$$

$$x = \frac{7}{3} \quad \text{or} \quad x = \frac{-3}{3} = -1$$

▲

▼ **Example 5** Solve for y: $(4y - 5)^2 = 6$.

Solution

$$(4y - 5)^2 = 6$$
$$4y - 5 = \pm\sqrt{6}$$
$$4y = 5 \pm \sqrt{6} \qquad \text{Add 5 to both sides}$$
$$y = \frac{5 \pm \sqrt{6}}{4} \qquad \text{Divide both sides by 4}$$

Since $\sqrt{6}$ is irrational, we cannot simplify the expression further. The solution set is $\left\{\frac{5 + \sqrt{6}}{4}, \frac{5 - \sqrt{6}}{4}\right\}$.

▲

▼ **Example 6** Solve for x: $(2x + 6)^2 = 8$.

Solution

$$(2x + 6)^2 = 8$$
$$2x + 6 = \pm\sqrt{8}$$
$$2x + 6 = \pm 2\sqrt{2} \qquad \sqrt{8} = \sqrt{4 \cdot 2} = 2\sqrt{2}$$
$$2x = -6 \pm 2\sqrt{2} \qquad \text{Add } -6 \text{ to both sides}$$
$$x = \frac{-6 \pm 2\sqrt{2}}{2} \qquad \text{Divide each side by 2}$$

We can reduce this last expression to lowest terms by factoring a 2 from each term in the numerator and then dividing that 2 by the 2 in the denominator. This is equivalent to dividing each term in the numerator by the 2 in the denominator. Here is what it looks like:

$$x = \frac{\cancel{2}(-3 \pm \sqrt{2})}{\cancel{2}} \qquad \text{Factor a 2 from each term in the numerator}$$

$$x = -3 \pm \sqrt{2} \qquad \text{Divide numerator and denominator by 2}$$

The two solutions are $-3 + \sqrt{2}$ and $-3 - \sqrt{2}$.

We can check our two solutions in the original equation. Let's check our first solution, $-3 + \sqrt{2}$.

When $x = -3 + \sqrt{2}$

the equation $(2x + 6)^2 = 8$

becomes

$$[2(-3 + \sqrt{2}) + 6]^2 \stackrel{?}{=} 8$$

$$(-6 + 2\sqrt{2} + 6)^2 = 8$$

$$(2\sqrt{2})^2 = 8$$

$$4 \cdot 2 = 8$$

$$8 = 8 \qquad \text{A true statement}$$

The second solution, $-3 - \sqrt{2}$, checks also. ▲

▼ **Example 7** If an object is dropped from a height of h feet, the amount of time in seconds it will take for the object to reach the ground (neglecting the resistance of air) is given by the formula

$$h = 16t^2$$

Solve this formula for t.

Solution To solve for t, we begin by taking the square root of each side of the formula:

$$h = 16t^2 \qquad \text{Original formula}$$

$$\pm\sqrt{h} = 4t \qquad \text{Take square root of each side}$$

$$\pm\frac{\sqrt{h}}{4} = t \qquad \text{Divide each side by 4}$$

Since t represents the time it takes for the object to fall h feet, t will never be negative. Therefore, the formula that gives t in terms of h is

$$t = \frac{\sqrt{h}}{4}$$

Whenever we are solving an applied problem like this one and we obtain a result that includes the $\pm$ sign, we must ask ourselves if the result can actually be negative. If it cannot be, we delete the negative result and use only the positive result. ▲

Problem Set 8.1

Solve each of the following equations using the methods learned in this section.

1. $x^2 = 9$
2. $x^2 = 16$
3. $a^2 = 25$
4. $a^2 = 36$
5. $y^2 = 8$
6. $y^2 = 75$
7. $2x^2 = 100$
8. $2x^2 = 54$
9. $3a^2 = 54$
10. $2a^2 = 64$
11. $(x + 2)^2 = 4$
12. $(x - 3)^2 = 16$
13. $(x + 1)^2 = 25$
14. $(x + 3)^2 = 64$
15. $(a - 5)^2 = 75$
16. $(a - 4)^2 = 32$
17. $(y + 1)^2 = 50$
18. $(y - 5)^2 = 27$
19. $(2x + 1)^2 = 25$
20. $(3x - 2)^2 = 16$
21. $(4a - 5)^2 = 36$
22. $(2a + 6)^2 = 64$
23. $(3y - 1)^2 = 12$
24. $(5y - 4)^2 = 12$
25. $(6x + 2)^2 = 27$
26. $(8x - 1)^2 = 20$
27. $(3x - 9)^2 = 27$
28. $(2x + 8)^2 = 32$
29. $(3x + 6)^2 = 45$
30. $(5x - 10)^2 = 75$
31. $(2y - 4)^2 = 8$
32. $(4y - 6)^2 = 48$

33. Check the solution $x = -1 + 5\sqrt{2}$ in the equation $(x + 1)^2 = 50$.

34. Check the solution $x = -8 + 2\sqrt{6}$ in the equation $(x + 8)^2 = 24$.

35. The square of the sum of a number and 3 is 16. Find the number. (There are two solutions.)

36. The square of the sum of twice a number and 3 is 25. Find the number. (There are two solutions.)

37. If you invest \$100 in an account with interest rate r compounded annually, the amount of money, A, in the account after two years is given by the formula

$$A = 100(1 + r)^2$$

Solve this formula for r.

38. If you invest P dollars in an account with interest rate r compounded annually, the amount of money, A, in the account after two years is given by the formula

$$A = P(1 + r)^2$$

Solve this formula for r.

Review Problems The following problems review material we covered in Sections 4.6, 5.5, and 7.1. Reviewing the problems from Sections 4.6 and 5.5 will help you understand the material in the next section.

Multiply. [4.6]

39. $(x + 3)^2$
40. $(x - 3)^2$
41. $(x - 5)^2$
42. $(x + 5)^2$

Factor. [5.5]

43. $x^2 - 12x + 36$
44. $x^2 + 12x + 36$
45. $x^2 + 4x + 4$
46. $x^2 - 4x + 4$

Find the following roots. [7.1]

47. $\sqrt[3]{8}$
48. $\sqrt[3]{27}$
49. $\sqrt[4]{16}$
50. $\sqrt[4]{81}$

Section 8.2 Completing the Square

In this section we will develop a method of solving quadratic equations that works whether or not the equation can be factored. Since we will be working with the individual terms of trinomials, we need some new definitions so that we can keep our vocabulary straight.

DEFINITION In the trinomial $2x^2 + 3x + 4$, the first term, $2x^2$, is called the *quadratic term*; the middle term, $3x$, is called the *linear term*; and the last term, 4, is called the *constant term*.

Now consider the following list of perfect square trinomials and their corresponding binomial squares.

$$\begin{aligned} x^2 + 6x + 9 &= (x + 3)^2 \\ x^2 - 8x + 16 &= (x - 4)^2 \\ x^2 - 10x + 25 &= (x - 5)^2 \\ x^2 + 12x + 36 &= (x + 6)^2 \end{aligned}$$

In each case the coefficient of x^2 is 1. A more important observation, however, stems from the relationship between the linear terms (middle terms) and the constant terms (last terms). Notice that the constant in each case is the square of half the coefficient of x in the linear term. That is:

1. For the first trinomial, $x^2 + 6x + 9$, the last term, 9, is the square of half the coefficient of the middle term: $9 = (\frac{6}{2})^2$.
2. For the second trinomial, $x^2 - 8x + 16$, $16 = [\frac{1}{2}(-8)]^2$.
3. For the trinomial $x^2 - 10x + 25$, it also holds: $25 = [\frac{1}{2}(-10)]^2$.

Check and see that it also works for the last trinomial.

In summary, then, for every perfect square trinomial in which the coefficient of x^2 is 1, the last term is always the square of half the coefficient of the linear term. We can use this fact to build our own perfect square trinomials.

▼ **Examples** Write the correct last term to each of the following expressions so each becomes a perfect square trinomial.

1. $x^2 - 2x$

Solution The coefficient of the linear term is -2. If we take the square of half of -2, we get 1. Adding the 1 as the last term, we have the perfect square trinomial:

$$x^2 - 2x + 1 = (x - 1)^2$$

2. $x^2 + 18x$

Solution Half of 18 is 9, the square of which is 81. If we add 81 at the end, we have

$$x^2 + 18x + 81 = (x + 9)^2$$

3. $x^2 + 3x$

Solution Half of 3 is $\frac{3}{2}$, the square of which is $\frac{9}{4}$:

$$x^2 + 3x + \frac{9}{4} = \left(x + \frac{3}{2}\right)^2$$

▲

We can use this procedure, along with the method developed in Section 8.1, to solve some quadratic equations.

▼ **Example 4** Solve $x^2 - 6x + 5 = 0$ by completing the square.

Solution We begin by adding -5 to both sides of the equation. We want just $x^2 - 6x$ on the left side so that we can add on our own last term to get a perfect square trinomial:

$$x^2 - 6x + 5 = 0$$
$$x^2 - 6x = -5 \qquad \text{Add } -5 \text{ to both sides}$$

Now we can add 9 to both sides and the left side will be a perfect square:

$$x^2 - 6x + \mathbf{9} = -5 + \mathbf{9}$$
$$(x - 3)^2 = 4$$

The last line is in the form of the equations we solved in Section 8.1. Taking the square root of both sides, we have

$$x - 3 = \pm 2$$
$$x = 3 \pm 2 \qquad \text{Add 3 to both sides}$$
$$x = 3 + 2 \quad \text{or} \quad x = 3 - 2$$
$$x = 5 \quad \text{or} \quad x = 1$$

The two solutions are 5 and 1. ▲

The preceding method of solution is called *completing the square*.

▼ **Example 5** Solve by completing the square: $2x^2 + 16x - 18 = 0$.

Solution We begin by moving the constant term to the other side:

$$2x^2 + 16x - 18 = 0$$
$$2x^2 + 16x = 18 \quad \text{Add 18 to both sides}$$

In order to complete the square, we must be sure the coefficient of x^2 is 1. To accomplish this, we divide both sides by 2:

$$\frac{2x^2}{2} + \frac{16x}{2} = \frac{18}{2}$$
$$x^2 + 8x = 9$$

We now complete the square by adding the square of half the coefficient of the linear term to both sides:

$$x^2 + 8x + \mathbf{16} = 9 + \mathbf{16} \quad \text{Add } \mathbf{16} \text{ to both sides}$$
$$(x + 4)^2 = 25$$
$$x + 4 = \pm 5 \quad \text{Take the square root of both sides}$$
$$x = -4 \pm 5 \quad \text{Add } -4 \text{ to both sides}$$
$$x = -4 + 5 \quad \text{or} \quad x = -4 - 5$$
$$x = 1 \quad \text{or} \quad x = -9$$

The solution set arrived at by completing the square is $\{1, -9\}$. ▲

We will now summarize the preceding examples by listing the steps involved in solving quadratic equations by completing the square.

To Solve a Quadratic Equation by Completing the Square

Step 1: Put the equation in the form $ax^2 + bx = c$.
Step 2: Make sure the coefficient of the squared term is 1. If it is not 1, simply divide both sides by whatever it is.
Step 3: Add the square of half the coefficient of the linear term to both sides of the equation.
Step 4: Write the left-hand side of the equation as a binomial square and solve, using the methods developed in Section 8.1.

Here is one final example.

▼ **Example 6** Solve for y: $3y^2 - 9y + 3 = 0$.

Solution

$$3y^2 - 9y + 3 = 0$$
$$3y^2 - 9y = -3 \quad \text{Add } -3 \text{ to both sides}$$

$$y^2 - 3y = -1 \qquad \text{Divide by 3}$$

$$y^2 - 3y + \frac{9}{4} = -1 + \frac{9}{4} \qquad \text{Complete the square}$$

$$\left(y - \frac{3}{2}\right)^2 = \frac{5}{4} \qquad -1 + \tfrac{9}{4} = -\tfrac{4}{4} + \tfrac{9}{4} = \tfrac{5}{4}$$

$$y - \frac{3}{2} = \pm\frac{\sqrt{5}}{2} \qquad \text{Square root of both sides}$$

$$y = \frac{3}{2} \pm \frac{\sqrt{5}}{2} \qquad \text{Add } \tfrac{3}{2} \text{ to both sides}$$

$$y = \frac{3}{2} + \frac{\sqrt{5}}{2} \quad \text{or} \quad y = \frac{3}{2} - \frac{\sqrt{5}}{2}$$

$$y = \frac{3 + \sqrt{5}}{2} \quad \text{or} \quad y = \frac{3 - \sqrt{5}}{2}$$

The solutions are $\frac{3 + \sqrt{5}}{2}$ and $\frac{3 - \sqrt{5}}{2}$ which can be written in a shorter form as

$$\frac{3 \pm \sqrt{5}}{2}$$

▲

Note We can use a calculator or the table at the back of the book to get decimal approximations to these solutions. If we use the approximation $\sqrt{5} \cong 2.236$, then

$$\frac{3 + \sqrt{5}}{2} \cong \frac{3 + 2.236}{2} = \frac{5.236}{2} = 2.618$$

$$\frac{3 - \sqrt{5}}{2} \cong \frac{3 - 2.236}{2} = \frac{0.764}{2} = 0.382$$

Problem Set 8.2

Give the correct last term for each of the following expressions to ensure that the resulting trinomial is a perfect square trinomial.

1. $x^2 + 6x$
2. $x^2 - 10x$
3. $x^2 + 2x$
4. $x^2 + 14x$
5. $y^2 - 8y$
6. $y^2 + 12y$
7. $y^2 - 2y$
8. $y^2 - 6y$
9. $x^2 + 16x$
10. $x^2 - 4x$
11. $a^2 - 3a$
12. $a^2 + 5a$
13. $x^2 - 7x$
14. $x^2 - 9x$
15. $y^2 + y$
16. $y^2 - y$
17. $x^2 - \frac{3}{2}x$
18. $x^2 + \frac{2}{3}x$

Solve each of the following equations by completing the square. Follow the steps given at the end of this section.

19. $x^2 + 4x = 12$
20. $x^2 - 2x = 8$
21. $x^2 - 6x = 16$
22. $x^2 + 12x = -27$
23. $a^2 + 2a = 3$
24. $a^2 - 8a = -7$
25. $x^2 - 10x = 0$
26. $x^2 + 4x = 0$
27. $y^2 + 2y - 15 = 0$
28. $y^2 - 10y - 11 = 0$
29. $x^2 + 4x - 3 = 0$
30. $x^2 + 6x + 5 = 0$
31. $x^2 - 4x = 4$
32. $x^2 + 4x = -1$
33. $a^2 = 7a + 8$
34. $a^2 = 3a + 1$
35. $4x^2 + 8x - 4 = 0$
36. $3x^2 + 12x + 6 = 0$
37. $2x^2 + 2x - 4 = 0$
38. $4x^2 + 4x - 3 = 0$
39. $4x^2 + 8x + 1 = 0$
40. $3x^2 + 6x + 2 = 0$
41. $2x^2 - 2x = 1$
42. $3x^2 - 3x = 1$
43. $4a^2 - 4a + 1 = 0$
44. $2a^2 + 4a + 1 = 0$
45. $3y^2 - 9y = 2$
46. $5y^2 - 10y = 4$

47. The two solutions to Problem 29 are $-2 + \sqrt{7}$ and $-2 - \sqrt{7}$. The table of square roots at the back of the book gives the decimal approximation of $\sqrt{7}$ as 2.646. Use this approximation for $\sqrt{7}$ to find decimal approximations of $-2 + \sqrt{7}$ and $-2 - \sqrt{7}$.

48. The solutions to Problem 41 are $\frac{1 + \sqrt{3}}{2}$ and $\frac{1 - \sqrt{3}}{2}$. Use the decimal approximation of $\sqrt{3}$ given in the table at the back of the book to find decimal approximations of these two numbers.

49. One of the solutions to the equation in Problem 35 is $-1 + \sqrt{2}$. Check this solution in the original equation.

50. Check the solution $x = -1 - \sqrt{2}$ from Problem 35 in the original equation, $4x^2 + 8x - 4 = 0$.

51. Find the sum of the two solutions to the equation in Problem 29. (Add $-2 + \sqrt{7}$ and $-2 - \sqrt{7}$.)

52. Find the product of the two solutions to the equation in Problem 29.

Review Problems The following problems review material we covered in Sections 2.1 and 7.2. Reviewing the problems from section 2.1 will help you with the next section.

Find the value of each expression if $a = 2$, $b = 4$, and $c = -3$. [2.1]

53. $2a$
54. b^2
55. $4ac$
56. $b^2 - 4ac$
57. $\sqrt{b^2 - 4ac}$
58. $-b + \sqrt{b^2 - 4ac}$

Put each expression in simplified form for radicals. [7.2]

59. $\sqrt{12}$
60. $\sqrt{50x^2}$

61. $\sqrt{20x^2y^3}$

62. $3\sqrt{48x^4}$

63. $\sqrt{\frac{81}{25}}$

64. $\frac{6\sqrt{8x^2y}}{\sqrt{9}}$

Section 8.3 The Quadratic Formula

In this section we will derive the quadratic formula. It is one formula that you will use in almost all types of mathematics. We will first state the formula as a theorem and then prove it. The proof is based on the method of completing the square developed in the last section.

The Quadratic Theorem

For any quadratic equation in the form $ax^2 + bx + c = 0$, when a, b, and c are real numbers and $a \neq 0$, the two solutions are

$$x = \frac{-b + \sqrt{b^2 - 4ac}}{2a} \quad \text{and} \quad x = \frac{-b - \sqrt{b^2 - 4ac}}{2a}$$

PROOF We will prove the theorem by completing the square on

$$ax^2 + bx + c = 0$$

Adding $-c$ to both sides, we have

$$ax^2 + bx = -c$$

To make the coefficient of x^2 one, we divide both sides by a:

$$\frac{ax^2}{a} + \frac{bx}{a} = -\frac{c}{a}$$

$$x^2 + \frac{b}{a}x = -\frac{c}{a}$$

Now, to complete the square, we add the square of half of $\frac{b}{a}$ to both sides:

$$x^2 + \frac{b}{a}x + \left(\frac{b}{2a}\right)^2 = -\frac{c}{a} + \left(\frac{b}{2a}\right)^2 \qquad \frac{1}{2} \text{ of } \frac{b}{a} \text{ is } \frac{b}{2a}$$

Let's simplify the right side separately:

$$-\frac{c}{a} + \left(\frac{b}{2a}\right)^2 = -\frac{c}{a} + \frac{b^2}{4a^2}$$

The common denominator is $4a^2$. We multiply the numerator and denominator of $-\frac{c}{a}$ by $4a$ to give it the common denominator. Then we combine numerators:

$$\frac{\mathbf{4a}}{\mathbf{4a}}\left(-\frac{c}{a}\right) + \frac{b^2}{4a^2} = -\frac{4ac}{4a^2} + \frac{b^2}{4a^2}$$
$$= \frac{-4ac + b^2}{4a^2}$$
$$= \frac{b^2 - 4ac}{4a^2}$$

Now back to the equation. We use our simplified expression for the right side:

$$x^2 + \frac{b}{a}x + \left(\frac{b}{2a}\right)^2 = \frac{b^2 - 4ac}{4a^2}$$
$$\left(x + \frac{b}{2a}\right)^2 = \frac{b^2 - 4ac}{4a^2}$$

Taking the square root of both sides, we have

$$x + \frac{b}{2a} = \pm\frac{\sqrt{b^2 - 4ac}}{2a}$$
$$x = \frac{-b}{2a} \pm \frac{\sqrt{b^2 - 4ac}}{2a} \qquad \text{Add } \frac{-b}{2a} \text{ to both sides}$$
$$x = \frac{-b \pm \sqrt{b^2 - 4ac}}{2a}$$

Our proof is now complete. What we have is this: If our equation is in the form $ax^2 + bx + c = 0$ (standard form), then the solutions can always be found by using the quadratic formula:

$$x = \frac{-b \pm \sqrt{b^2 - 4ac}}{2a}$$

Note This formula is called the quadratic formula. You will see it many times if you continue taking math classes. By the time you are finished with this section and the problems in the problem set, you should have it memorized.

▼ **Example 1** Solve $x^2 - 5x - 6 = 0$ by using the quadratic formula.

Solution To use the quadratic formula, we must make sure the equation is in standard form; identify a, b, and c; substitute them into the formula; and work out the arithmetic.

For the equation $x^2 - 5x - 6 = 0$, $a = 1$, $b = -5$, and $c = -6$:

$$x = \frac{-b \pm \sqrt{b^2 - 4ac}}{2a}$$

$$= \frac{-(-5) \pm \sqrt{(-5)^2 - 4(1)(-6)}}{2(1)}$$

$$= \frac{5 \pm \sqrt{49}}{2}$$

$$= \frac{5 \pm 7}{2}$$

$$x = \frac{5 + 7}{2} \quad \text{or} \quad x = \frac{5 - 7}{2}$$

$$x = \frac{12}{2} \qquad x = -\frac{2}{2}$$

$$x = 6 \qquad x = -1$$

The two solutions are 6 and -1. ▲

Note Whenever the solutions to our quadratic equations turn out to be rational numbers, as in Example 1, it means the original equation could have been solved by factoring. (We didn't solve the equation in Example 1 by factoring because we are trying to get some practice with the quadratic formula.)

▼ **Example 2** Solve for x: $2x^2 = -4x + 3$.

Solution Before we can identify a, b, and c, we must write the equation in standard form. To do so, we add $4x$ and -3 to each side of the equation:

$$2x^2 = -4x + 3$$

$$2x^2 + 4x - 3 = 0 \qquad \text{Add } 4x \text{ and } -3 \text{ to each side}$$

Now that the equation is in standard form, we see that $a = 2$, $b = 4$, and $c = -3$. Using the quadratic formula we have

$$x = \frac{-b \pm \sqrt{b^2 - 4ac}}{2a}$$

$$= \frac{-4 \pm \sqrt{4^2 - 4(2)(-3)}}{2(2)}$$

$$= \frac{-4 \pm \sqrt{40}}{4}$$

$$= \frac{-4 \pm 2\sqrt{10}}{4}$$

We can reduce this last expression to lowest terms by factoring 2 from the numerator and denominator and then dividing it out:

$$x = \frac{\cancel{2}(-2 \pm \sqrt{10})}{\cancel{2} \cdot 2}$$

$$= \frac{-2 \pm \sqrt{10}}{2}$$

Our two solutions are $\frac{-2 + \sqrt{10}}{2}$ and $\frac{-2 - \sqrt{10}}{2}$. ▲

▼ **Example 3** Solve for x: $(x - 2)(x + 3) = 5$.

Solution We must put the equation into standard form before we can use the formula:

$$(x - 2)(x + 3) = 5$$

$$x^2 + x - 6 = 5 \quad \text{Multiply out the left side}$$

$$x^2 + x - 11 = 0 \quad \text{Add } -5 \text{ to each side}$$

Now $a = 1$, $b = 1$, and $c = -11$; therefore

$$x = \frac{-1 \pm \sqrt{1^2 - 4(1)(-11)}}{2(1)}$$

$$= \frac{-1 \pm \sqrt{45}}{2}$$

$$= \frac{-1 \pm 3\sqrt{5}}{2}$$

The solution set is $\left\{\frac{-1 + 3\sqrt{5}}{2}, \frac{-1 - 3\sqrt{5}}{2}\right\}$. ▲

▼ **Example 4** Solve: $x^2 - 6x = -7$.

Solution We begin by writing the equation in standard form:

$$x^2 - 6x = -7$$

$$x^2 - 6x + 7 = 0 \quad \text{Add 7 to each side}$$

Using $a = 1$, $b = -6$, and $c = 7$ in the quadratic formula

$$x = \frac{-b \pm \sqrt{b^2 - 4ac}}{2a}$$

we have

$$x = \frac{-(-6) \pm \sqrt{(-6)^2 - 4(1)(7)}}{2(1)}$$
$$= \frac{6 \pm \sqrt{36 - 28}}{2}$$
$$= \frac{6 \pm \sqrt{8}}{2}$$
$$= \frac{6 \pm 2\sqrt{2}}{2}$$

The two terms in the numerator have a 2 in common. We reduce to lowest terms by factoring the 2 from the numerator and then dividing numerator and denominator by 2:

$$= \frac{\cancel{2}(3 \pm \sqrt{2})}{\cancel{2}}$$
$$= 3 \pm \sqrt{2}$$

The two solutions are $3 + \sqrt{2}$ and $3 - \sqrt{2}$. This time, let's check our solutions in the original equation $x^2 - 6x = -7$.

Checking $x = 3 + \sqrt{2}$, we have

$(3 + \sqrt{2})^2 - 6(3 + \sqrt{2}) \stackrel{?}{=} -7$	
$9 + 6\sqrt{2} + 2 - 18 - 6\sqrt{2} = -7$	Multiply
$11 - 18 + 6\sqrt{2} - 6\sqrt{2} = -7$	Add 9 and 2
$-7 + 0 = -7$	Subtraction
$-7 = -7$	A true statement

Checking $x = 3 - \sqrt{2}$, we have

$(3 - \sqrt{2})^2 - 6(3 - \sqrt{2}) \stackrel{?}{=} -7$	
$9 - 6\sqrt{2} + 2 - 18 + 6\sqrt{2} = -7$	Multiply
$11 - 18 - 6\sqrt{2} + 6\sqrt{2} = -7$	Add 9 and 2
$-7 + 0 = -7$	Subtraction
$-7 = -7$	A true statement

As you can see, both solutions yield true statements when used in place of the variable in the original equation. ▲

Problem Set 8.3

Solve the following equations by using the quadratic formula.

1. $x^2 + 3x + 2 = 0$
2. $x^2 - 5x + 6 = 0$
3. $x^2 + 5x + 6 = 0$
4. $x^2 - 7x - 8 = 0$
5. $x^2 + 6x + 9 = 0$
6. $x^2 - 10x + 25 = 0$

7. $x^2 + 6x + 7 = 0$
8. $x^2 - 4x - 1 = 0$
9. $2x^2 + 5x + 3 = 0$
10. $2x^2 + 3x - 20 = 0$
11. $4x^2 + 8x + 1 = 0$
12. $3x^2 + 6x + 2 = 0$
13. $x^2 - 2x + 1 = 0$
14. $x^2 + 2x - 3 = 0$
15. $x^2 - 5x = 7$
16. $2x^2 - 6x = 8$
17. $6x^2 - x - 2 = 0$
18. $6x^2 + 5x - 4 = 0$
19. $(x - 2)(x + 1) = 3$
20. $(x - 8)(x + 7) = 5$
21. $(2x - 3)(x + 2) = 1$
22. $(4x - 5)(x - 3) = 6$
23. $2x^2 - 3x = 5$
24. $3x^2 - 4x = 5$
25. $2x^2 = -6x + 7$
26. $5x^2 = -6x + 3$
27. $3x^2 = -4x + 2$
28. $3x^2 = 4x + 2$
29. $2x^2 - 5 = 2x$
30. $5x^2 + 1 = -10x$
31. Solve the equation $2x^3 + 3x^2 - 4x = 0$ by first factoring out the common factor, x, and then using the quadratic formula. There are three solutions to this equation.
32. Solve the equation $5y^3 - 10y^2 + 4y = 0$ by first factoring out the common factor, y, and then using the quadratic formula.
33. To apply the quadratic formula to the equation $3x^2 - 4x = 0$, you have to notice that $c = 0$. Solve the equation, using the quadratic formula.
34. Solve the equation $9x^2 - 16 = 0$ using the quadratic formula. (Notice $b = 0$.)
35. Solve the following equation by first multiplying both sides by the LCD and then applying the quadratic formula to the result.

$$\tfrac{1}{2}x^2 - \tfrac{1}{2}x - \tfrac{1}{6} = 0$$

36. Solve the following equation by first multiplying both sides by the LCD and then applying the quadratic formula to the result.

$$\tfrac{1}{2}y^2 - y - \tfrac{3}{2} = 0$$

Review Problems The following problems review the material we covered in Section 7.5. Reviewing these problems will help you with the material in the next section.

Multiply.

37. $(2\sqrt{3})(3\sqrt{5})$
38. $\sqrt{7}(\sqrt{7} + 4)$
39. $(\sqrt{6} + 2)(\sqrt{6} - 5)$
40. $(\sqrt{x} + 4)^2$
41. $(\sqrt{7} - \sqrt{2})(\sqrt{7} + \sqrt{2})$
42. $(\sqrt{a} + \sqrt{b})(\sqrt{a} - \sqrt{b})$

Rationalize the denominator.

43. $\dfrac{2}{3 + \sqrt{5}}$
44. $\dfrac{2 + \sqrt{3}}{2 - \sqrt{3}}$

Section 8.4 Complex Numbers

In order to solve quadratic equations such as $x^2 = -4$, we need to introduce a new set of numbers. If we try to solve $x^2 = -4$ using real numbers, we always get no solution. There is no real number whose square is -4.

The new set of numbers is called *complex numbers* and is based on the following definition.

DEFINITION The number i is a number such that $i = \sqrt{-1}$.

The first thing we notice about this definition is that i is not a real number. There are no real numbers that represent the square root of -1. The other observation we make about i is $i^2 = -1$. If $i = \sqrt{-1}$, then, squaring both sides, we must have $i^2 = -1$. The most common power of i is i^2. Whenever we see i^2, we can write it as -1. We are now ready for a definition of complex numbers.

DEFINITION A complex number is any number that can be put in the form $a + bi$, where a and b are real numbers and $i = \sqrt{-1}$.

Note The form $a + bi$ is called *standard form* for complex numbers. The definition indicates that if a number can be written in the form $a + bi$, then it is a complex number.

EXAMPLE The following are complex numbers:

$$3 + 4i \qquad \frac{1}{2} - 6i \qquad 8 + i\sqrt{2} \qquad \frac{3}{4} - 2i\sqrt{5}$$

EXAMPLE The number $4i$ is a complex number because $4i = 0 + 4i$.

EXAMPLE The number 8 is a complex number because $8 = 8 + 0i$.

From the last example we can see the real numbers are a subset of the complex numbers because any real number x can be written as $x + 0i$.

Addition and Subtraction of Complex Numbers

We add and subtract complex numbers according to the same procedure we used to add and subtract polynomials: we combine similar terms.

▼ **Example 1** Combine: $(3 + 4i) + (2 - 6i)$.

Solution

$$\begin{aligned}(3 + 4i) + (2 - 6i) &= (3 + 2) + (4i - 6i) && \text{Commutative and associative properties}\\ &= 5 + (-2i) && \text{Combine similar terms}\\ &= 5 - 2i\end{aligned}$$

▲

Example 2 Combine: $(2 - 5i) - (3 + 7i) + (2 - i)$.

Solution

$$\begin{aligned}(2 - 5i) - (3 + 7i) + (2 - i) &= 2 - 5i - 3 - 7i + 2 - i \\ &= (2 - 3 + 2) + (-5i - 7i - i) \\ &= 1 - 13i\end{aligned}$$

Multiplication of Complex Numbers

Multiplication of complex numbers is very similar to multiplication of polynomials. We can simplify many answers by using the fact that $i^2 = -1$.

Example 3 Multiply: $4i(3 + 5i)$.

Solution

$$\begin{aligned}4i(3 + 5i) &= 4i(3) + 4i(5i) && \text{Distributive property} \\ &= 12i + 20i^2 && \text{Multiplication} \\ &= 12i + 20(-1) && i^2 = -1 \\ &= -20 + 12i\end{aligned}$$

Example 4 Multiply: $(3 + 2i)(4 - 3i)$.

Solution

$$\begin{aligned}&(3 + 2i)(4 - 3i) \\ &= 3 \cdot 4 + 3(-3i) + 2i(4) + 2i(-3i) && \text{FOIL method} \\ &= 12 - 9i + 8i - 6i^2 \\ &= 12 - 9i + 8i - 6(-1) && i^2 = -1 \\ &= (12 + 6) + (-9i + 8i) \\ &= 18 - i\end{aligned}$$

Division of Complex Numbers

We divide complex numbers by applying the same process we used to rationalize denominators.

Example 5 Divide: $\dfrac{2i}{3 - 4i}$.

Solution We multiply numerator and denominator by the conjugate of the denominator, which is $3 + 4i$:

$$\begin{aligned}\left(\frac{2i}{3 - 4i}\right)\left(\frac{\mathbf{3 + 4i}}{\mathbf{3 + 4i}}\right) &= \frac{6i + 8i^2}{9 - 16i^2} \\ &= \frac{6i + 8(-1)}{9 - 16(-1)} \\ &= \frac{-8 + 6i}{25}\end{aligned}$$

Note The conjugate of $a + bi$ is $a - bi$. When we multiply complex conjugates the result is always a real number because

$$\begin{aligned}(a + bi)(a - bi) &= a^2 - (bi)^2 \\ &= a^2 - b^2i^2 \\ &= a^2 - b^2(-1) \\ &= a^2 + b^2\end{aligned}$$

which is a real number.

▼ **Example 6** Divide: $\dfrac{2 + i}{5 + 2i}$.

Solution The conjugate of the denominator is $5 - 2i$:

$$\begin{aligned}\left(\frac{2 + i}{5 + 2i}\right)\left(\frac{\mathbf{5 - 2i}}{\mathbf{5 - 2i}}\right) &= \frac{10 - 4i + 5i - 2i^2}{25 - 4i^2} \\ &= \frac{10 - 4i + 5i - 2(-1)}{25 - 4(-1)} \qquad i^2 = -1 \\ &= \frac{12 + i}{29}\end{aligned}$$

▲

Note If we were to write our answer to Example 6 in standard form for complex numbers, it would look like this:

$$\frac{12}{29} + \frac{1}{29}i$$

Problem Set 8.4

Combine the following complex numbers.

1. $(3 - 2i) + 3i$
2. $(5 - 4i) - 8i$
3. $(6 + 2i) - 10i$
4. $(8 - 10i) + 7i$
5. $(11 + 9i) - 9i$
6. $(12 + 2i) + 6i$
7. $(3 + 2i) + (6 - i)$
8. $(4 + 8i) - (7 + i)$
9. $(5 + 7i) - (6 + 8i)$
10. $(11 + 6i) - (3 + 6i)$
11. $(9 - i) + (2 - i)$
12. $(8 + 3i) - (8 - 3i)$
13. $(6 + i) - 4i - (2 - i)$
14. $(3 + 2i) - 5i - (5 + 4i)$
15. $(6 - 11i) + 3i + (2 + i)$
16. $(3 + 4i) - (5 + 7i) - (6 - i)$
17. $(2 + 3i) - (6 - 2i) + (3 - i)$
18. $(8 + 9i) + (5 - 6i) - (4 - 3i)$

Multiply the following complex numbers.

19. $3(2 - i)$
20. $4(5 + 3i)$
21. $2i(8 - 7i)$
22. $-3i(2 + 5i)$
23. $(2 + i)(4 - i)$
24. $(6 + 3i)(4 + 3i)$

25. $(2 + i)(3 - 5i)$
26. $(4 - i)(2 - i)$
27. $(3 + 5i)(3 - 5i)$
28. $(8 + 6i)(8 - 6i)$
29. $(2 + i)(2 - i)$
30. $(3 + i)(3 - i)$

Divide the following complex numbers.

31. $\dfrac{2}{3 - 2i}$
32. $\dfrac{3}{5 + 6i}$
33. $\dfrac{-3i}{2 + 3i}$
34. $\dfrac{4i}{3 + i}$
35. $\dfrac{6i}{3 - i}$
36. $\dfrac{-7i}{5 - 4i}$
37. $\dfrac{2 + i}{2 - i}$
38. $\dfrac{3 + 2i}{3 - 2i}$
39. $\dfrac{4 + 5i}{3 - 6i}$
40. $\dfrac{-2 + i}{5 + 6i}$
41. Use the FOIL method to multiply $(x + 3i)(x - 3i)$.
42. Use the FOIL method to multiply $(x + 5i)(x - 5i)$.
43. The opposite of i is $-i$. The reciprocal of i is $\dfrac{1}{i}$. Multiply the numerator and denominator of $\dfrac{1}{i}$ by i and simplify the result to see that the opposite of i and the reciprocal of i are the same number.
44. If $i^2 = -1$, what are i^3 and i^4? (*Hint:* $i^3 = i^2 \cdot i$.)

Review Problems The following problems review material we covered in Sections 8.1 and 7.3. Reviewing the problems from Section 8.1 will help you with the next section.

Solve each equation by taking the square root of each side. [8.1]

45. $(x - 3)^2 = 25$
46. $(x - 2)^2 = 9$
47. $(2x - 6)^2 = 16$
48. $(2x + 1)^2 = 49$
49. $(x + 3)^2 = 12$
50. $(x + 3)^2 = 8$

Put each expression into simplified form for radicals. [7.3]

51. $\sqrt{\dfrac{1}{2}}$
52. $\sqrt{\dfrac{5}{6}}$
53. $\sqrt{\dfrac{8x^2y^3}{3}}$
54. $\sqrt{\dfrac{45xy^4}{7}}$
55. $\sqrt[3]{\dfrac{1}{4}}$
56. $\sqrt[3]{\dfrac{2}{3}}$

Section 8.5 Complex Solutions To Quadratic Equations

The quadratic formula tells us solutions to equations of the form $ax^2 + bx + c = 0$ are always

$$x = \frac{-b \pm \sqrt{b^2 - 4ac}}{2a}$$

The part of the quadratic formula under the radical sign is called the discriminant:

$$\text{Discriminant} = b^2 - 4ac$$

When the discriminant is negative, we have to deal with the square root of a negative number. We handle square roots of negative numbers by using the definition $i = \sqrt{-1}$. To illustrate, suppose we want to simplify an expression that contains $\sqrt{-9}$, which is not a real number. We begin by writing $\sqrt{-9}$ as $\sqrt{9(-1)}$. Then, we write this expression as the product of two separate radicals; $\sqrt{9}\sqrt{-1}$. Applying the definition $i = \sqrt{-1}$ to this last expression, we have

$$\sqrt{9}\sqrt{-1} = 3i$$

As you may recall from the previous section, the number $3i$ is called a complex number. Here are some further examples.

▼ **Example 1** Write the following radicals as complex numbers.

(a) $\sqrt{-4} = \sqrt{4(-1)} = \sqrt{4}\sqrt{-1} = 2i$
(b) $\sqrt{-36} = \sqrt{36(-1)} = \sqrt{36}\sqrt{-1} = 6i$
(c) $\sqrt{-7} = \sqrt{7(-1)} = \sqrt{7}\sqrt{-1} = i\sqrt{7}$
(d) $\sqrt{-75} = \sqrt{75(-1)} = \sqrt{75}\sqrt{-1} = 5i\sqrt{3}$ ▲

In parts (c) and (d) of Example 1, we wrote i before the radical because it is less confusing that way. If we put i after the radical, it is sometimes mistaken for being under the radical.

Let's see how complex numbers relate to quadratic equations by looking at some examples of quadratic equations whose solutions are complex numbers.

▼ **Example 2** Solve for x: $(x + 2)^2 = -9$.

Solution We can solve this equation by expanding the left side, putting the results into standard form, and then applying the quadratic formula. It is faster, however, simply to take the square root of both sides:

$$(x + 2)^2 = -9$$
$$x + 2 = \pm\sqrt{-9} \qquad \text{Square root of both sides}$$
$$x + 2 = \pm 3i \qquad \sqrt{-9} = \sqrt{9}\sqrt{-1} = 3i$$
$$x = -2 \pm 3i \qquad \text{Add } -2 \text{ to both sides}$$

The solution set contains two complex solutions. Notice that the two solutions are conjugates.

The solution set is $\{-2 + 3i, -2 - 3i\}$. ▲

▼ **Example 3** Solve for x: $x^2 - 2x = -5$.

Solution The easiest way to solve this equation is by using the quadratic formula. We begin by adding 5 to both sides in order to put the equation into standard form:

$$\begin{aligned} x^2 - 2x &= -5 \\ x^2 - 2x + 5 &= 0 \qquad \text{Add 5 to both sides} \end{aligned}$$

Applying the quadratic formula with $a = 1$, $b = -2$, and $c = 5$, we have

$$\begin{aligned} x &= \frac{-(-2) \pm \sqrt{(-2)^2 - 4(1)(5)}}{2(1)} \\ &= \frac{2 \pm \sqrt{-16}}{2} \\ &= \frac{2 \pm 4i}{2} \end{aligned}$$

Dividing the numerator and denominator by 2, we have the two solutions:

$$x = 1 \pm 2i$$

The two solutions are $1 + 2i$ and $1 - 2i$. Let's check our first solution, $1 + 2i$, in the original equation:

$$\begin{aligned} (1 + 2i)^2 - 2(1 + 2i) &\stackrel{?}{=} -5 \\ 1 + 4i + 4i^2 - 2 - 4i &= -5 \\ 1 + 4i - 4 - 2 - 4i &= -5 \\ -5 &= -5 \qquad \text{A true statement} \end{aligned}$$

The second solution checks also. ▲

▼ **Example 4** Solve: $(2x - 3)(2x - 1) = -4$.

Solution We multiply the binomials on the left side and then add 4 to each side to write the equation in standard form. From there, we identify a, b, and c and apply the quadratic formula.

$$\begin{aligned} (2x - 3)(2x - 1) &= -4 \\ 4x^2 - 8x + 3 &= -4 \qquad \text{Multiply binomials on left side} \\ 4x^2 - 8x + 7 &= 0 \qquad \text{Add 4 to each side} \end{aligned}$$

Placing $a = 4$, $b = -8$, and $c = 7$ in the quadratic formula, we have:

$$x = \frac{-(-8) \pm \sqrt{(-8)^2 - 4(4)(7)}}{2(4)}$$

$$= \frac{8 \pm \sqrt{64 - 112}}{8}$$

$$= \frac{8 \pm \sqrt{-48}}{8}$$

$$= \frac{8 \pm 4i\sqrt{3}}{8} \qquad \sqrt{-48} = i\sqrt{48} = i\sqrt{16}\,\sqrt{3} = 4i\sqrt{3}$$

To reduce this last expression to lowest terms, we factor a 4 from the numerator and then divide numerator and denominator by 4:

$$= \frac{\cancel{4}(2 \pm i\sqrt{3})}{\cancel{4} \cdot 2}$$

$$= \frac{2 \pm i\sqrt{3}}{2}$$

▲

Note It would be a mistake to try to reduce this last expression further. Sometimes first year algebra students will try to divide the 2 in the denominator into the 2 in the numerator, which is a mistake. Remember, when we reduce to lowest terms we do so by dividing the numerator and denominator by any factors they have in common. In this case, 2 is not a factor of the numerator. This expression is in lowest terms.

This completes our work with solving quadratic equations. We can solve any quadratic equation we come across. Factoring is probably still the fastest method of solution, but again, factoring works only if the equation is factorable. Applying the quadratic formula always produces solutions, whether the equation is factorable or not.

Problem Set 8.5

Write the following radicals as complex numbers.

1. $\sqrt{-16}$ **2.** $\sqrt{-25}$ **3.** $\sqrt{-49}$ **4.** $\sqrt{-81}$

5. $\sqrt{-6}$ **6.** $\sqrt{-10}$ **7.** $\sqrt{-11}$ **8.** $\sqrt{-19}$

9. $\sqrt{-32}$ **10.** $\sqrt{-288}$ **11.** $\sqrt{-50}$ **12.** $\sqrt{-45}$

13. $\sqrt{-8}$ **14.** $\sqrt{-24}$ **15.** $\sqrt{-48}$ **16.** $\sqrt{-27}$

Solve the following quadratic equations. Use whatever method seems to fit the situation or is convenient for you.

17. $x^2 = 2x - 2$ **18.** $x^2 = 4x - 5$

19. $x^2 - 4x = -4$ **20.** $x^2 - 4x = 4$

21. $2x^2 + 5x = 12$
22. $2x^2 + 30 = 16x$
23. $(x - 2)^2 = -4$
24. $(x - 5)^2 = -25$
25. $(2x + 1)^2 = -9$
26. $(4x - 2)^2 = -8$
27. $(6x - 3)^2 = -27$
28. $(8x + 4)^2 = -32$
29. $x^2 + x + 1 = 0$
30. $x^2 - 3x + 4 = 0$
31. $x^2 - 5x + 6 = 0$
32. $x^2 + 2x + 2 = 0$
33. $3x^2 + 2x + 1 = 0$
34. $4x^2 + x + 5 = 0$
35. $2x^2 = -3x + 2$
36. $3x^2 = -2x + 1$
37. $(x + 2)(x - 3) = 5$
38. $(x - 1)(x + 1) = 6$
39. $(x - 5)(x - 3) = -10$
40. $(x - 2)(x - 4) = -5$
41. $(2x - 2)(x - 3) = 9$
42. $(x - 1)(2x + 6) = 9$
43. Is $x = 2 + 2i$ a solution to the equation $x^2 - 4x + 8 = 0$?
44. Is $x = 5 + 3i$ a solution to the equation $x^2 - 10x + 34 = 0$?
45. If one solution to a quadratic equation is $3 + 7i$, what do you think the other solution is?
46. If one solution to a quadratic equation is $4 - 2i$, what do you think the other solution is?

Review Problems The following problems review material we covered in Sections 3.2, 3.3, and 7.4. Reviewing the problems from Sections 3.2 and 3.3 will help you understand the next section.

Graph each line. [3.2, 3.3]

47. $y = x - 2$
48. $y = -x - 2$
49. $2x + 4y = 8$
50. $2x - 4y = 8$

Write each term in simplified form for radicals. Then combine similar terms. [7.4]

51. $3\sqrt{50} + 2\sqrt{32}$
52. $4\sqrt{18} + \sqrt{32} - \sqrt{2}$
53. $\sqrt{24} - \sqrt{54} - \sqrt{150}$
54. $\sqrt{72} - \sqrt{8} + \sqrt{50}$
55. $2\sqrt{27x^2} - x\sqrt{48}$
56. $5\sqrt{8x^3} + x\sqrt{50x}$

Section 8.6 Graphing Parabolas

In this section we will graph equations of the form $y = ax^2 + bx + c$ and equations that can be put into this form. The graphs of this type of equation all have similar shapes.

We will begin by graphing the simplest quadratic equation, $y = x^2$. To get the idea of the shape of this graph, we need to find some ordered pairs that are solutions. We can do this by setting up the following table:

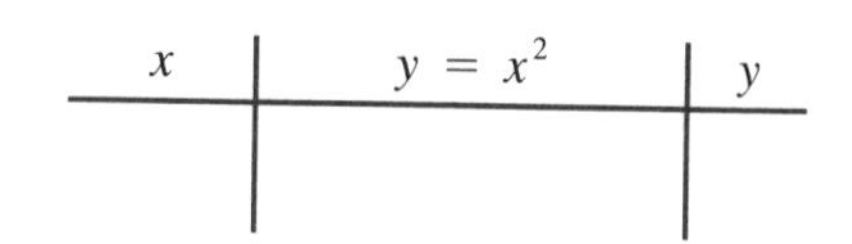

x	$y = x^2$	y

We can choose any convenient numbers for x and then use the equation $y = x^2$ to find the corresponding values for y. Let's use the values -3, -2, -1, 0, 1, 2, and 3 for x and find corresponding values for y. Here is how the table looks when we let x have these values:

x	$y = x^2$	y
-3	$y = (-3)^2 = 9$	9
-2	$y = (-2)^2 = 4$	4
-1	$y = (-1)^2 = 1$	1
0	$y = 0^2 = 0$	0
1	$y = 1^2 = 1$	1
2	$y = 2^2 = 4$	4
3	$y = 3^2 = 9$	9

The table gives us the solutions $(-3, 9)$, $(-2, 4)$, $(-1, 1)$, $(0, 0)$, $(1, 1)$, $(2, 4)$, and $(3, 9)$ for the equation $y = x^2$. We plot each of the points on a rectangular coordinate system and draw a smooth curve between them, as shown in Figure 1.

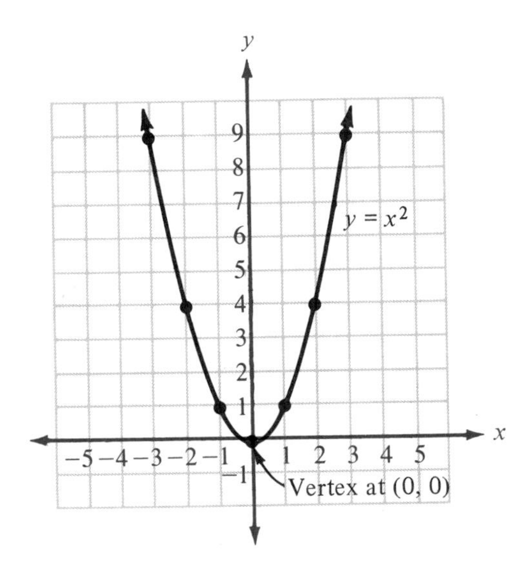

Figure 1

This graph is called a *parabola.* All equations of the form $y = ax^2 + bx + c\ (a \neq 0)$ produce parabolas when graphed.

Note that the point $(0, 0)$ is called the vertex of the parabola in Figure 1. It is the lowest point on the graph. Although all the parabolas in this section will open up, there are some parabolas that open downward. For those parabolas, the vertex is the highest point on the graph.

▼ **Example 1** Graph the equation: $y = x^2 - 3$.

Solution We begin by making a table using convenient values for x:

x	$y = x^2 - 3$	y
-2	$y = (-2)^2 - 3 = 4 - 3 = 1$	1
-1	$y = (-1)^2 - 3 = 1 - 3 = -2$	-2
0	$y = 0^2 - 3 = -3$	-3
1	$y = 1^2 - 3 = 1 - 3 = -2$	-2
2	$y = 2^2 - 3 = 4 - 3 = 1$	1

The table gives us the ordered pairs $(-2, 1)$, $(-1, -2)$, $(0, -3)$, $(1, -2)$, and $(2, 1)$ as solutions to $y = x^2 - 3$. The graph is shown in Figure 2.

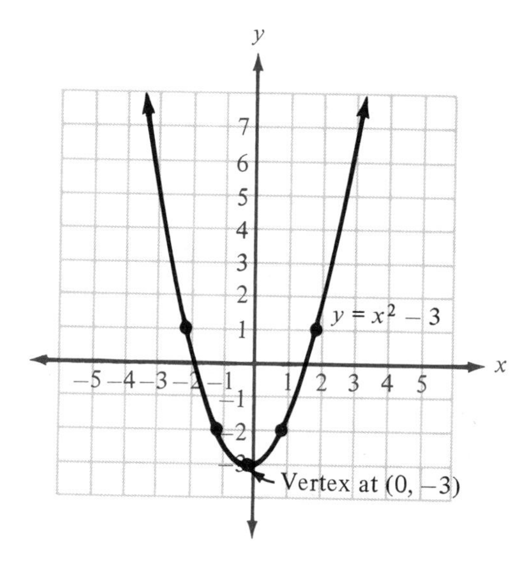

Figure 2 ▲

▼ **Example 2** Graph: $y = (x - 2)^2$.

Solution

x	$y = (x - 2)^2$	y
-1	$y = (-1 - 2)^2 = (-3)^2 = 9$	9
0	$y = (0 - 2)^2 = (-2)^2 = 4$	4
1	$y = (1 - 2)^2 = (-1)^2 = 1$	1
2	$y = (2 - 2)^2 = 0^2 = 0$	0

We can continue the table if we feel more solutions will make the graph clearer.

3	$y = (3 - 2)^2 = 1^2 = 1$	1
4	$y = (4 - 2)^2 = 2^2 = 4$	4
5	$y = (5 - 2)^2 = 3^2 = 9$	9

Putting the results of the table onto a coordinate system, we have the graph in Figure 3.

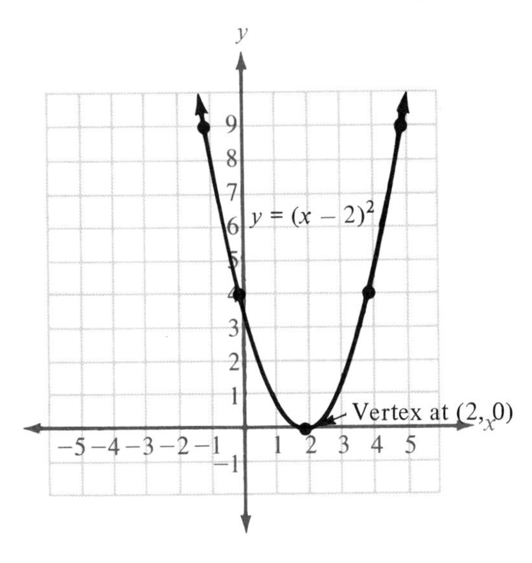

Figure 3 ▲

When graphing parabolas, it is sometimes easier to obtain the correct graph if you have some idea what the graph will look like before you start making your table.

If you look over the first two examples closely, you will see that the graphs in Figures 2 and 3 have the same shape as the graph of $y = x^2$. The difference is in the position of the vertex. For example, the graph of $y = x^2 - 3$, as shown in Figure 2, looks like the graph of $y = x^2$ with its vertex moved down three units vertically. Similarly, the graph of $y = (x - 2)^2$, shown in Figure 3, looks like the graph of $y = x^2$ with its vertex moved two units to the right.

Without showing the table necessary to graph them, here are four more parabolas and their corresponding equations. Can you see the correlation between the numbers in the equations and the position of graphs on the coordinate systems?

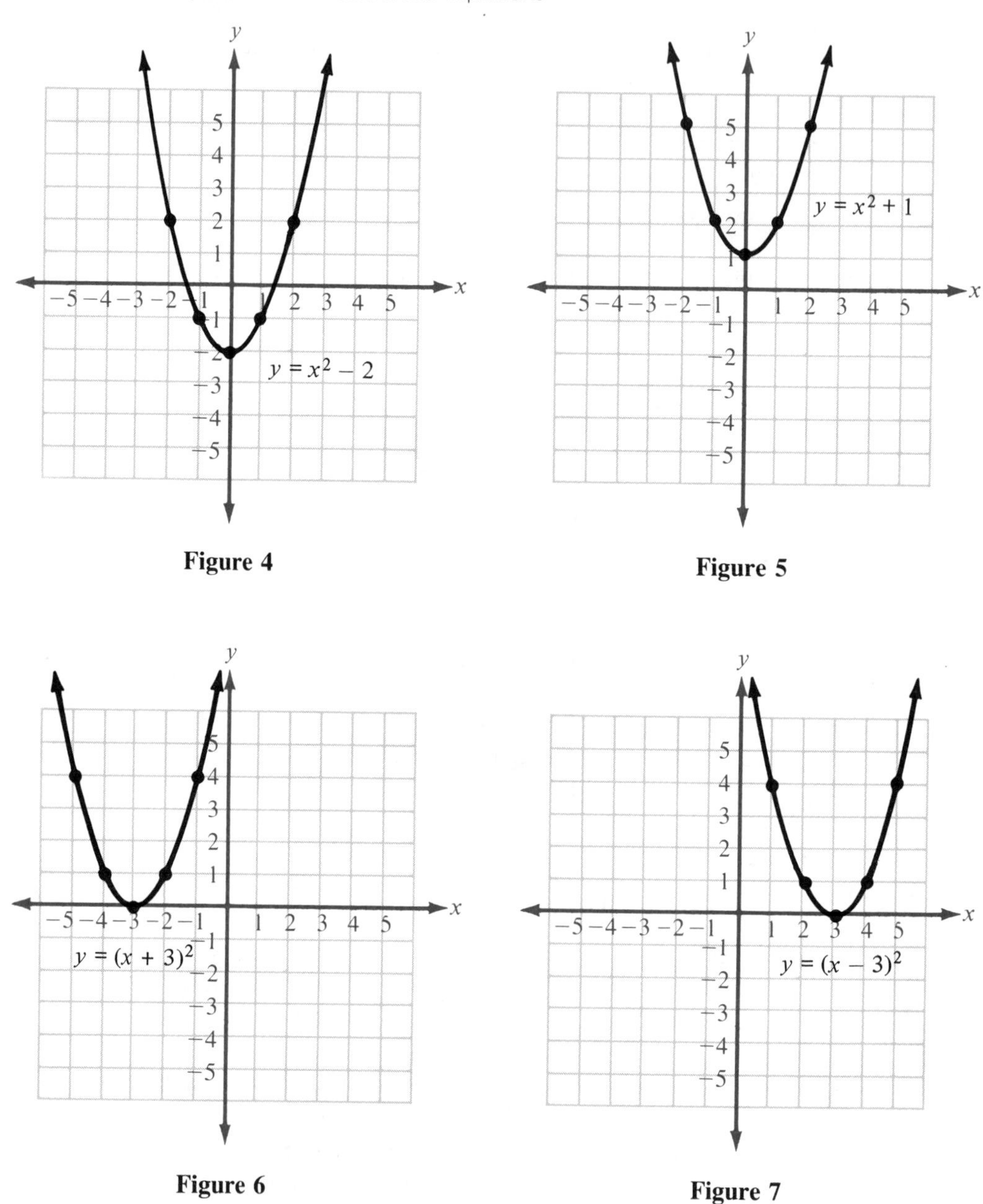

Figure 4

Figure 5

Figure 6

Figure 7

For our last example, we will graph a parabola in which the vertex has been moved both vertically and horizontally from the origin.

▼ **Example 3** Graph: $y = (x + 1)^2 - 3$.

Solution

x	$y = (x + 1)^2 - 3$	y
-4	$y = (-4 + 1)^2 - 3 = 9 - 3$	6
-3	$y = (-3 + 1)^2 - 3 = 4 - 3$	1
-1	$y = (-1 + 1)^2 - 3 = 0 - 3$	-3
1	$y = (1 + 1)^2 - 3 = 4 - 3$	1
2	$y = (2 + 1)^2 - 3 = 9 - 3$	6

Graphing the results of the table, we have Figure 8.

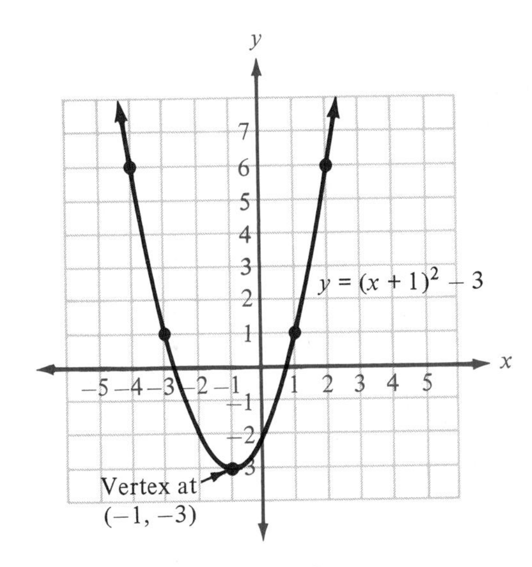

Figure 8 ▲

To summarize this section, we graph second-degree equations by finding as many points as necessary to ensure that the graph has the correct shape.

Problem Set 8.6

Graph each of the following equations by first making a table using the given values of x.

1. $y = x^2$ $\quad x = -4, -2, 0, 2, 4$

2. $y = -x^2$ $\quad x = -4, -2, 0, 2, 4$

3. $y = x^2 + 2$ $\quad x = -3, -2, -1, 0, 1, 2, 3$

4. $y = x^2 - 1$ $\quad x = -3, -1, 0, 1, 3$

5. $y = (x + 1)^2$ $\quad x = -3, -2, -1, 0, 1, 2$
6. $y = (x - 3)^2$ $\quad x = 0, 1, 2, 3, 4, 5, 6$

Graph each of the following equations by first making a table using convenient values for x.

7. $y = x^2 - 4$
8. $y = x^2 + 2$
9. $y = x^2 + 5$
10. $y = x^2 - 2$
11. $y = (x + 2)^2$
12. $y = (x + 5)^2$
13. $y = (x - 3)^2$
14. $y = (x - 2)^2$
15. $y = (x - 5)^2$
16. $y = (x + 3)^2$
17. $y = (x + 1)^2 - 2$
18. $y = (x - 1)^2 + 2$
19. $y = (x + 2)^2 - 3$
20. $y = (x - 2)^2 + 3$

The following equations have graphs that are also parabolas. However, the graphs of these equations will open downward; the vertex of each will be the highest point on the graph. Graph each equation by first making a table of ordered pairs using the given values of x.

21. $y = 4 - x^2$ $\quad x = -3, -2, -1, 0, 1, 2, 3$
22. $y = 3 - x^2$ $\quad x = -3, -2, -1, 0, 1, 2, 3$
23. $y = -1 - x^2$ $\quad x = -2, -1, 0, 1, 2$
24. $y = -2 - x^2$ $\quad x = -2, -1, 0, 1, 2$
25. Graph the line $y = x + 2$ and the parabola $y = x^2$ on the same coordinate system. Name the points where the two graphs intersect.
26. Graph the line $y = x$ and the parabola $y = x^2 - 2$ on the same coordinate system. Name the points where the two graphs intersect.
27. Graph the parabola $y = 2x^2$ and the parabola $y = \frac{1}{2}x^2$ on the same coordinate system. (Remember, when you substitute a value for x into $y = 2x^2$, you square it first and then multiply by 2. If $x = 3$, then $y = 2 \cdot 3^2 = 2 \cdot 9 = 18$.)
28. Graph the parabola $y = 3x^2$ and the parabola $y = \frac{1}{3}x^2$ on the same coordinate system.

Review Problems The following problems review material we covered in Section 7.6.

Solve each equation.

29. $\sqrt{x + 5} = 4$
30. $\sqrt{x - 3} = 2$
31. $\sqrt{2x + 3} = -3$
32. $\sqrt{3x - 3} = -5$
33. $\sqrt{3a + 2} - 3 = 5$
34. $\sqrt{2a - 4} + 3 = 7$
35. $\sqrt{x + 10} = x - 2$
36. $\sqrt{6x + 1} = x - 1$

Chapter 8 Summary and Review

Examples

SOLVING QUADRATIC EQUATIONS OF THE FORM $(ax + b)^2 = c$ [8.1]

We can solve equations of the form $(ax + b)^2 = c$ by taking the square root of both sides. We must remember to take both the positive and negative square roots of the right side when we do so.

1.
$$(x - 3)^2 = 25$$
$$x - 3 = \pm 5$$
$$x = 3 \pm 5$$
$$x = -2 \quad \text{or} \quad x = 8$$

COMPLETING THE SQUARE [8.2]

To complete the square on a quadratic equation as a method of solution we use the following steps:

Step 1: Move the constant term to one side and the variable terms to the other. Then, divide each side by the coefficient of x^2 if it is other than one.

Step 2: Take the square of half the coefficient of the linear term and add it to both sides of the equation.

Step 3: Write the left side as a binomial square and then take the square root of both sides.

Step 4: Solve the resulting equation.

2.
$$x^2 - 6x + 2 = 0$$
$$x^2 - 6x = -2$$
$$x^2 - 6x + \mathbf{9} = -2 + \mathbf{9}$$
$$(x - 3)^2 = 7$$
$$x - 3 = \pm\sqrt{7}$$
$$x = 3 \pm \sqrt{7}$$

THE QUADRATIC FORMULA [8.3]

Any equation that is in the form $ax^2 + bx + c = 0$, where $a \neq 0$, has as its solutions

$$x = \frac{-b \pm \sqrt{b^2 - 4ac}}{2a}$$

The expression under the square root sign, $b^2 - 4ac$, is known as the discriminant. When the discriminant is negative, the solutions are complex numbers.

3. If $2x^2 + 3x - 4 = 0$

then $x = \dfrac{-3 \pm \sqrt{9 - 4(2)(-4)}}{2(2)}$

$= \dfrac{-3 \pm \sqrt{41}}{4}$

COMPLEX NUMBERS [8.4]

Any number that can be put in the form $a + bi$, where $i = \sqrt{-1}$, is called a complex number.

4. The numbers, 5, $3i$, $2 + 4i$, and $7 - i$ are all complex numbers.

ADDITION AND SUBTRACTION OF COMPLEX NUMBERS [8.4]

We add (or subtract) complex numbers by using the same procedure we used to add (or subtract) polynomials: we combine similar terms.

5. $(3 + 4i) + (6 - 7i)$
$= (3 + 6) + (4i - 7i)$
$= 9 - 3i$

MULTIPLICATION OF COMPLEX NUMBERS [8.4]

We multiply complex numbers in the same way we multiply binomials. The result, however, can be simplified further by substituting -1 for i^2 whenever it appears.

6. $(2 + 3i)(3 - i)$
$= 6 - 2i + 9i - 3i^2$
$= 6 + 7i + 3$
$= 9 + 7i$

DIVISION OF COMPLEX NUMBERS [8.4]

Division with complex numbers is accomplished with the method for rationalizing the denominator we developed while working with radical expressions. If the denominator has the form $a + bi$, we multiply both the numerator and the denominator by its conjugate, $a - bi$.

7. $\dfrac{3}{2 + 5i}$

$= \dfrac{3}{2 + 5i} \cdot \dfrac{\mathbf{2 - 5i}}{\mathbf{2 - 5i}}$

$= \dfrac{6 - 15i}{29}$

GRAPHING PARABOLAS [8.6]

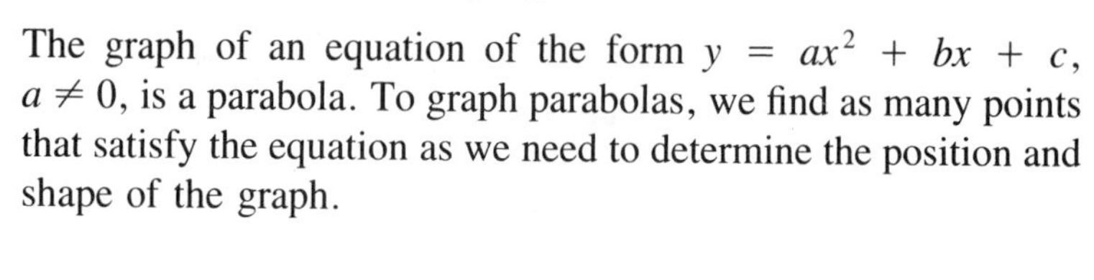
The graph of an equation of the form $y = ax^2 + bx + c$, $a \neq 0$, is a parabola. To graph parabolas, we find as many points that satisfy the equation as we need to determine the position and shape of the graph.

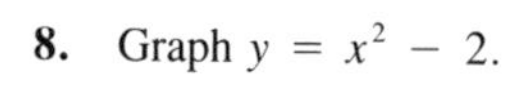
8. Graph $y = x^2 - 2$.

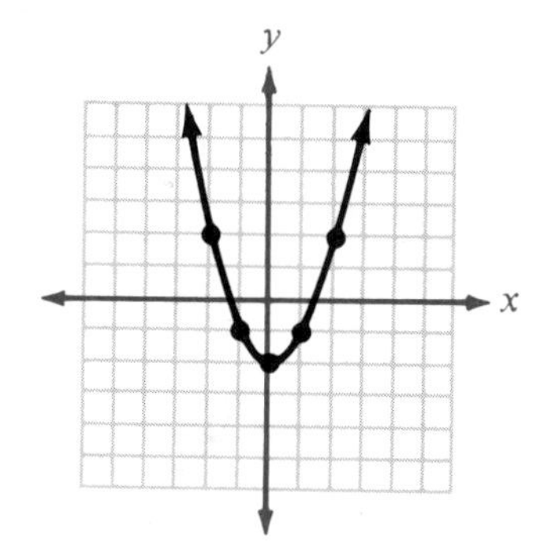

COMMON MISTAKES

1. The most common mistake when working with complex numbers is to say $i = -1$. It does not; i is the *square root* of -1, not -1 itself.

2. The most common mistake when working with the quadratic formula is to try to identify the constants a, b, and c before putting the equation into standard form.

Chapter 8 Test

Solve the following quadratic equations. [8.1, 8.3, 8.5]

1. $x^2 - 7x - 8 = 0$

2. $(x - 3)^2 = 12$

3. $(2x - 5)^2 = -75$

4. $2x^2 = 3x - 5$

5. $3x^2 = -2x + 1$

6. $(x + 2)(x - 1) = 6$

7. $9x^2 + 12x + 4 = 0$

8. Solve $x^2 - 6x - 6 = 0$ by completing the square. [8.2]

Write as complex numbers. [8.5]

9. $\sqrt{-9}$

10. $\sqrt{-121}$

11. $\sqrt{-72}$

12. $\sqrt{-18}$

Work the following problems involving complex numbers. [8.4]

13. $(3i + 1) + (2 + 5i)$

14. $(6 - 2i) - (7 - 4i)$

15. $(2 + i)(2 - i)$

16. $(3 + 2i)(1 + i)$

17. $\dfrac{i}{3 - i}$

18. $\dfrac{2 + i}{2 - i}$

Graph the following equations. [8.6]

19. $y = x^2 - 4$

20. $y = (x - 4)^2$

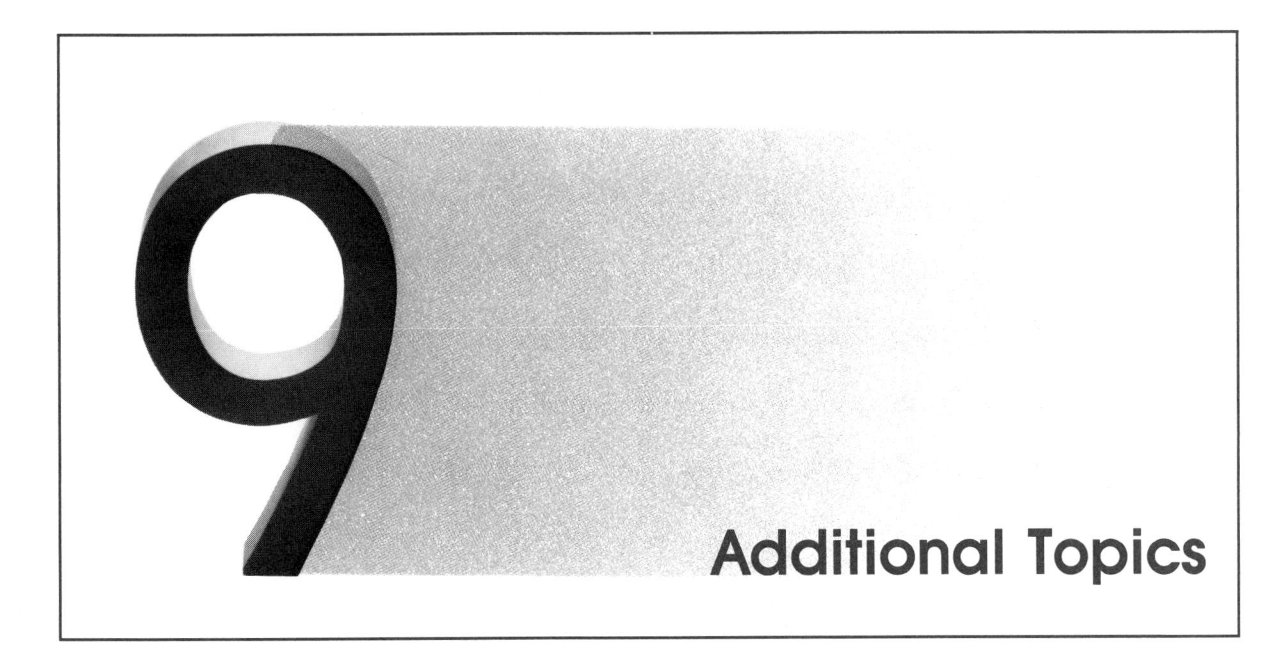

In this chapter we will extend some of the topics we have already covered and preview some other topics that are common to intermediate algebra. The sections in this chapter do not all fit together in a smooth progression. Many of them can be studied individually and taken up in any order.

The review problems in this chapter cover the first eight chapters of the book. If you are getting ready to take a final exam in this course, doing the review problems at the end of each section should give you a good idea which of the first eight chapters you need to work on most.

Section 9.1 Compound Inequalities

The *union* of two sets A and B is the set of all elements that are in A or in B. The word "or" is the key word in the definition. The intersection of two sets A and B is the set of elements contained in both A and B. The key word in this definition is the word "and." We can put the words "and" and "or" together with our methods of graphing inequalities to find the solution sets for compound inequalities.

DEFINITION A compound inequality is two or more inequalities connected by the words "and" or "or."

▼ **Example 1** Graph the solution set for the compound inequality

$$x < -1 \quad \text{or} \quad x \geq 3$$

Solution Graphing each inequality separately, we have:

$x < -1$

$x \geq 3$

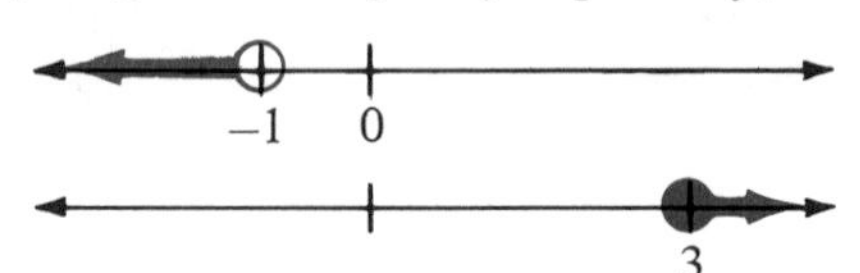

Since the two inequalities are connected by "or," we want to graph their union: that is, we graph all points that are on either the first graph or the second graph. Essentially, we put the two graphs together on the same number line:

$$x < -1 \quad \text{or} \quad x \geq 3$$

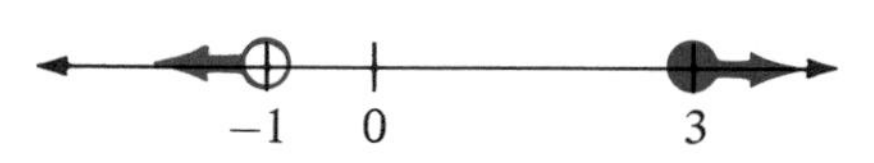

▲

▼ **Example 2** Graph the solution set for the compound inequality

$$x > -2 \quad \text{and} \quad x < 3$$

Solution Graphing each inequality separately, we have:

$x > -2$

$x < 3$

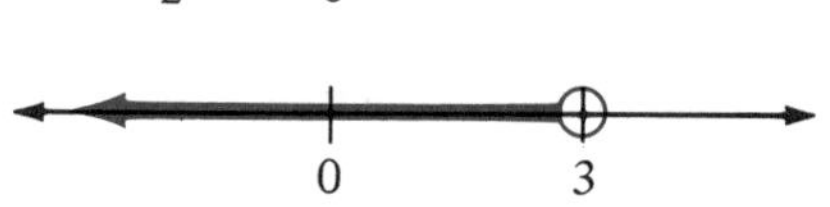

Since the two inequalities are connected by the word "and," we will graph their intersection, which consists of all points that are common to both graphs: that is, we graph the region where the two graphs overlap:

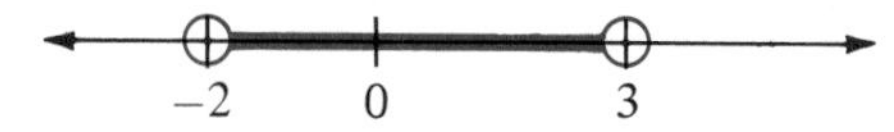

▲

▼ **Example 3** Solve and graph the solution set for

$$2x - 1 \geq 3 \quad \text{and} \quad -3x > -12$$

Solution Solving the two inequalities separately, we have:

$$\begin{aligned} 2x - 1 &\geq 3 & \text{and} \quad & -3x > -12 \\ 2x &\geq 4 & & \\ x &\geq 2 & \text{and} \quad & -\tfrac{1}{3}(-3x) < -\tfrac{1}{3}(-12) \\ & & & x < 4 \end{aligned}$$

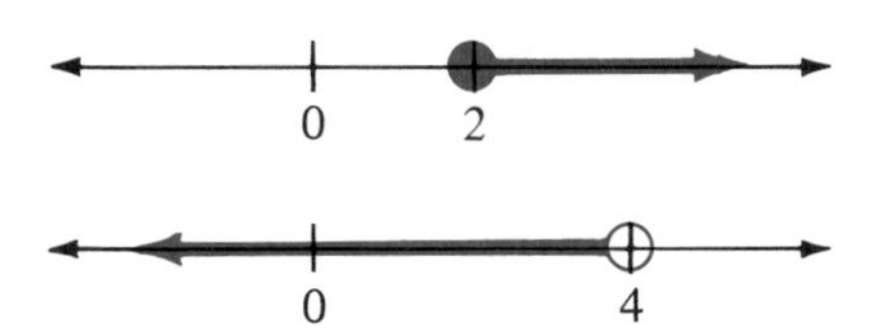

Since the word "and" connects the two graphs, we will graph their intersection—the points they have in common:

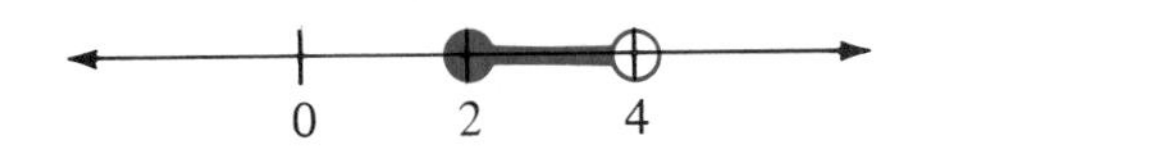

▲

Notation Sometimes compound inequalities that use the word "and" can be written in a shorter form. For example, the compound inequality $-2 < x$ and $x < 3$ can be written as $-2 < x < 3$. The word "and" does not appear when an inequality is written in this form. It is implied. The solution set for $-2 < x$ and $x < 3$ is

It is all the numbers between -2 and 3 on the number line. It seems reasonable, then, that this graph should be the graph of

$$-2 < x < 3$$

In both the graph and the inequality, x is said to be between -2 and 3.

▼ **Example 4** Solve and graph: $-3 \le 2x - 1 \le 9$.

Solution To solve for x we must add 1 to the center expression and then divide the result by 2. Whatever we do to the center expression, we must also do to the two expressions on the ends. In this way, we can be sure we are producing equivalent inequalities. The solution set will not be affected.

$-3 \le 2x - 1 \le 9$	
$-2 \le 2x \le 10$	Add 1 to each expression
$-1 \le x \le 5$	Multiply each expression by $\frac{1}{2}$

▲

▼ **Example 5** Solve and graph: $-5 < 3x + 1 < 10$.

Solution

$$-5 < 3x + 1 < 10$$

$$-6 < 3x < 9 \quad \text{Add } -1 \text{ to each member}$$

$$-2 < x < 3 \quad \text{Multiply through by } \tfrac{1}{3}$$

The solution set is all real numbers between -2 and 3. We must keep in mind that the notation $-2 < x < 3$ is equivalent to the expression $-2 < x$ and $x < 3$. This notation always implies the word "and"; it never implies the word "or." ▲

▼ **Example 6** The length of a rectangle is 3 feet longer than the width. If the perimeter must be between 14 feet and 30 feet, find all possible values for the width.

Solution If x = the width of the rectangle, then the length is $x + 3$. Since the perimeter is the sum of twice the width and twice the length, we can write it as $2x + 2(x + 3)$. This last expression must be between 14 feet and 30 feet. The inequality that describes the situation is

$$14 < 2x + 2(x + 3) < 30 \quad \text{The perimeter is between 14 and 30}$$

$$14 < 2x + 2x + 6 < 30 \quad \text{Multiply 2 and } (x + 3)$$

$$14 < 4x + 6 < 30 \quad \text{Add } 2x \text{ and } 2x$$

$$8 < 4x < 24 \quad \text{Add } -6 \text{ to each expression}$$

$$2 < x < 6 \quad \text{Multiply each expression by } \tfrac{1}{4}$$

The width is between 2 feet and 6 feet. ▲

Problem Set 9.1

Graph the following compound inequalities.

1. $x < -1$ or $x > 5$

2. $x \leq -2$ or $x \geq -1$

3. $x < -3$ or $x \geq 0$

4. $x < 5$ and $x > 1$

5. $x \leq 6$ and $x > -1$

6. $x \leq 7$ and $x > 0$

7. $x > 2$ and $x < 4$

8. $x < 2$ or $x > 4$

9. $x \geq -2$ and $x \leq 4$

10. $x \leq 2$ or $x \geq 4$

11. $x < 5$ and $x > -1$

12. $x > 5$ or $x < -1$

13. $-1 < x < 3$

14. $-1 \leq x \leq 3$

15. $-3 < x \leq -2$

16. $-5 \leq x \leq 0$

Solve the following. Graph the solution set in each case.

17. $3x - 1 < 5$ or $5x - 5 > 10$

18. $x + 1 < -3$ or $x - 2 > 6$

19. $x - 2 > -5$ and $x + 7 < 13$
20. $3x + 2 \le 11$ and $2x + 2 \ge 0$
21. $11x < 22$ or $12x > 36$
22. $-5x < 25$ and $-2x \ge -12$
23. $3x - 5 < 10$ and $2x + 1 > -5$
24. $5x + 8 < -7$ or $3x - 8 > 10$
25. $2x - 3 < 8$ and $3x + 1 > -10$
26. $11x - 8 > 3$ or $12x + 7 < -5$
27. $2x - 1 < 3$ and $3x - 2 > 1$
28. $3x + 9 < 7$ or $2x - 7 > 11$

Solve and graph each of the following.

29. $-1 \le x - 5 \le 2$
30. $0 \le x + 2 \le 3$
31. $-4 \le 2x \le 6$
32. $-5 < 5x < 10$
33. $-3 < 2x + 1 < 5$
34. $-7 \le 2x - 3 \le 7$
35. $0 \le 3x + 2 \le 7$
36. $2 \le 5x - 3 \le 12$
37. $-7 < 2x + 3 < 11$
38. $-5 < 6x - 2 < 8$
39. $-1 \le 4x + 5 \le 9$
40. $-8 \le 7x - 1 \le 13$

41. The sum of a number and 5 is between 10 and 20. Find the number.
42. The difference of a number and 2 is between 6 and 14. Find the number.
43. The difference of twice a number and 3 is between 5 and 7. Find the number.
44. The sum of twice a number and 5 is between 7 and 13. Find the number.
45. The length of a rectangle is 4 inches longer than the width. If the perimeter is between 20 inches and 30 inches, find all possible values for the width.
46. The length of a rectangle is 6 feet longer than the width. If the perimeter is between 24 feet and 36 feet, find all possible values for the width.

Review Problems The problems that follow review the material we covered in Chapter 1.

Simplify each expression.

47. $-|-5|$
48. $(-\frac{2}{3})^3$
49. $-3 - 4(-2)$
50. $2^4 + 3^3 \div 9 - 4^2$
51. $5|3 - 8| - 6|2 - 5|$
52. $7 - 3(2 - 6)$
53. $5 - 2[-3(5 - 7) - 8]$
54. $\dfrac{5 + 3(7 - 2)}{2(-3) - 4}$

55. Find the difference of -3 and -9.
56. If you add -4 to the product of -3 and 5, what number results?
57. Apply the distributive property to $\frac{1}{2}(4x - 6)$.
58. Use the associative property to simplify $-6(\frac{1}{3}x)$.

For the set $\{-3, -\frac{4}{5}, 0, \frac{5}{8}, 2, \sqrt{5}\}$, which numbers are

59. Integers
60. Rational numbers

Section 9.2 Linear Inequalities in Two Variables

A linear inequality in two variables is any expression that can be put in the form

$$ax + by < c$$

where a, b, and c are real numbers (a and b not 0). The inequality symbol can be any one of the following four: $<$, $\leq$, $>$, $\geq$.

Some examples of linear inequalities are

$$2x + 3y < 6 \qquad y \geq 2x + 1 \qquad x - y \leq 0$$

Although not all these inequalities have the form $ax + by < c$, each one can be put in that form.

The solution set for a linear inequality is a section of the coordinate plane. The boundary for the section is found by replacing the inequality symbol with an equal sign and graphing the resulting equation. The boundary is included in the solution set (and represented with a solid line) if the inequality symbol used originally is $\leq$ or $\geq$. The boundary is not included (and is represented with a dotted line) if the original symbol is $<$ or $>$.

Let's look at some examples.

▼ **Example 1** Graph the solution set for $x + y \leq 4$.

Solution The boundary for the graph is the graph of $x + y = 4$; we graph $x + y = 4$ by using the methods described in Chapter 3. The boundary is included in the solution set because the inequality symbol is $\leq$.

The graph of the boundary is shown in Figure 1.

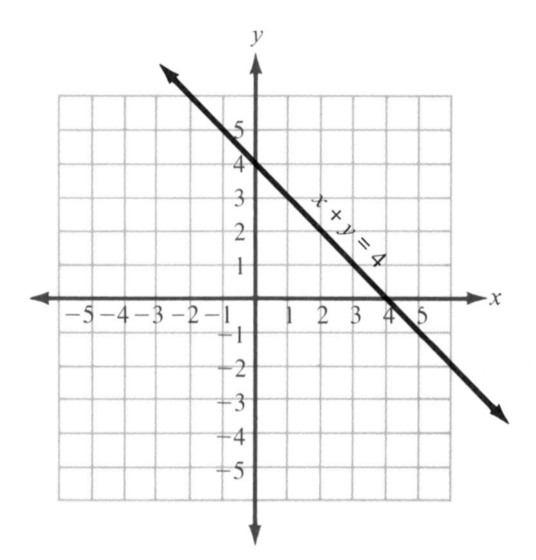

Figure 1

The boundary separates the coordinate plane into two sections, or regions—the region above the boundary and the region below the boundary. The solution set for $x + y \le 4$ is one of these two regions, along with the boundary. To find the correct region, we simply choose any convenient point that is *not* on the boundary. We then substitute the coordinates of the point into the original inequality $x + y \le 4$. If the point we choose satisfies the inequality, then it is a member of the solution set, and we can assume that all points on the same side of the boundary as the chosen point are also in the solution set. If the coordinates of our point do not satisfy the original inequality, then the solution set lies on the other side of the boundary.

In this example, a convenient point not on the boundary is the origin. Substituting (0, 0) into $x + y \le 4$ gives us

$$0 + 0 \overset{?}{\le} 4$$

$$0 \le 4 \qquad \text{A true statement}$$

Since the origin is a solution to the inequality $x + y \le 4$, and the origin is below the boundary, all other points below the boundary are also solutions.

The graph of $x + y \le 4$ is shown in Figure 2.

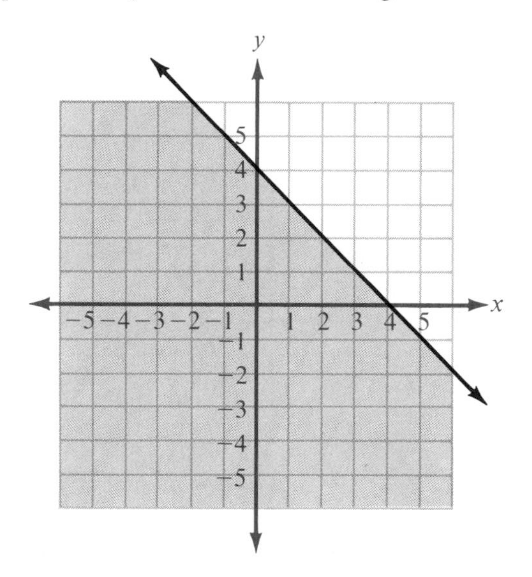

Figure 2

The region above the boundary is described by the inequality $x + y > 4$. ▲

Here is a list of steps to follow when graphing the solution set for linear inequalities in two variables.

To Graph the Solution Set for Linear Inequalities in Two Variables

Step 1: Replace the inequality symbol with an equal sign. The resulting equation represents the boundary for the solution set.

Step 2: Graph the boundary found in Step 1 using a *solid line* if the boundary is included in the solution set (that is, if the original inequality symbol was either $\leq$ or $\geq$). Use a *broken line* to graph the boundary if it is *not* included in the solution set. (It is not included if the original inequality was either $<$ or $>$.)

Step 3: Choose any convenient point not on the boundary and substitute the coordinates into the *original* inequality. If the resulting statement is *true*, the graph lies on the *same* side of the boundary as the chosen point. If the resulting statement is *false*, the solution set lies on the *opposite* side of the boundary.

▼ **Example 2** Graph the solution set for $y < 2x - 3$.

Solution The boundary is the graph of $y = 2x - 3$. The boundary is not included since the original inequality symbol is $<$. Therefore, we use a broken line to represent the boundary, as shown in Figure 3.

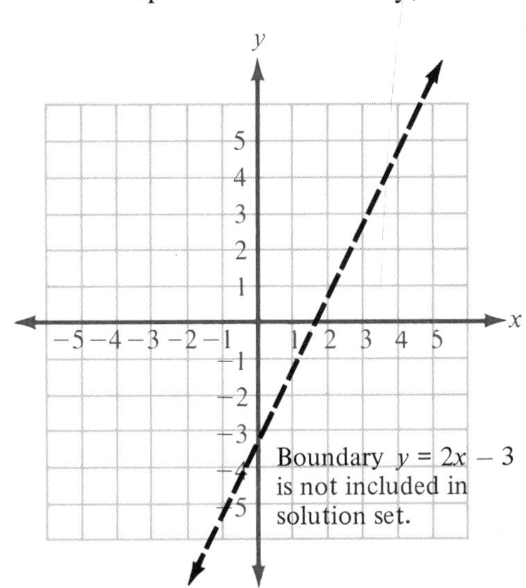

Figure 3

A convenient test point is again the origin. Using (0, 0) in $y < 2x - 3$, we have

$$0 \stackrel{?}{<} 2(0) - 3$$

$$0 < -3 \qquad \text{A false statement}$$

Since our test point gives us a false statement and it lies above the boundary, the solution set must lie on the other side of the boundary, as shown in Figure 4.

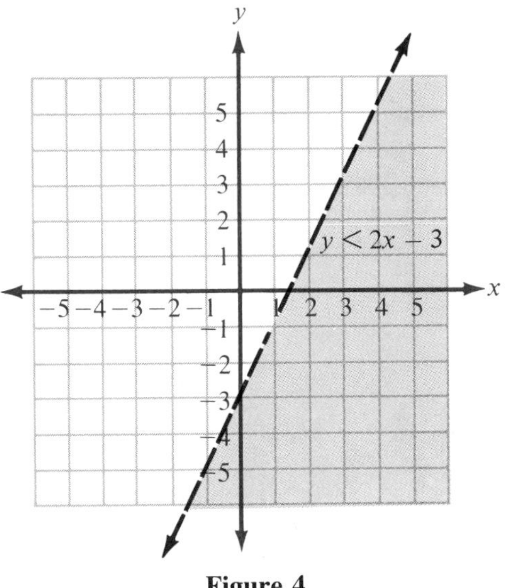

Figure 4

▲

▼ **Example 3** Graph the inequality: $2x + 3y \leq 6$.

Solution We begin by graphing the boundary $2x + 3y = 6$. The boundary is included in the solution because the inequality symbol is $\leq$.

If we use (0, 0) as our test point, we see that it yields a true statement when its coordinates are substituted into $2x + 3y \leq 6$. The graph, therefore, lies below the boundary.

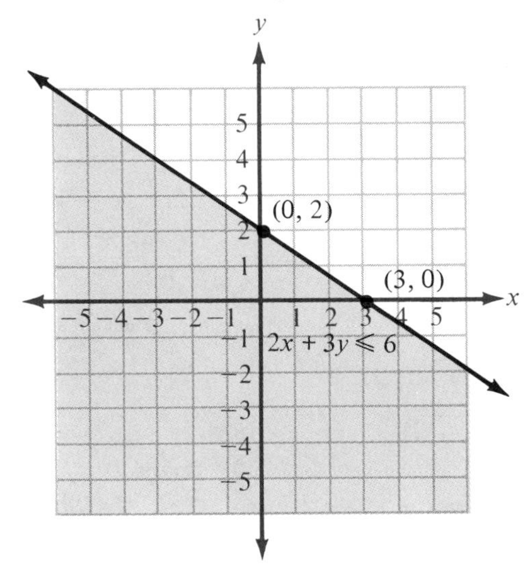

Figure 5

The ordered pair (0, 0) is a solution to $2x + 3y \leq 6$; all points on the same side of the boundary as (0, 0) must also be solutions to the inequality $2x + 3y \leq 6$. ▲

▼ **Example 4** Graph the solution set for $x \leq 5$.

Solution The boundary is $x = 5$, which is a vertical line. All points to the left have x-coordinates less than 5 and all points to the right have x-coordinates greater than 5, as shown in the following graph:

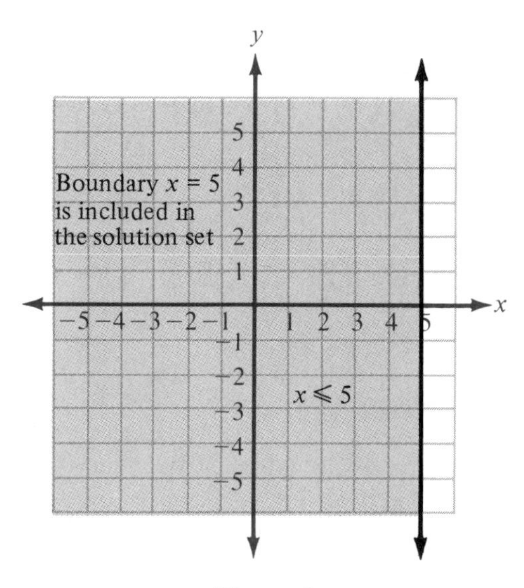

Figure 6 ▲

Problem Set 9.2

Graph the following linear inequalities.

1. $2x - 3y < 6$	**2.** $3x + 2y \geq 6$
3. $x - 2y \leq 4$	**4.** $2x + y > 4$
5. $x - y \leq 2$	**6.** $x - y \leq 1$
7. $3x - 4y \geq 12$	**8.** $4x + 3y < 12$
9. $5x - y \leq 5$	**10.** $4x + y > 4$
11. $2x + 6y \leq 12$	**12.** $x - 5y > 5$
13. $x \geq 1$	**14.** $x < 5$
15. $x \geq -3$	**16.** $y \leq -7$
17. $y < 2$	**18.** $3x - y > 1$
19. $2x + y > 3$	**20.** $5x + 2y < 2$
21. $y \leq 3x - 1$	**22.** $y \geq 3x + 2$
23. $y \geq x - 5$	**24.** $y > x + 3$
25. $y \leq -\frac{1}{2}x + 2$	**26.** $y < \frac{1}{3}x + 3$
27. $y < -x + 4$	**28.** $y \geq -x - 3$

Review Problems The following problems review material we covered in Chapter 2.

29. Simplify the expression $7 - 3(2x - 4) - 8$.
30. Find the value of $x^2 - 2xy + y^2$ when $x = 3$ and $y = -4$.

Solve each equation.

31. $-\frac{3}{2}x = 12$
32. $2x - 4 = 5x + 2$
33. $8 - 2(x + 7) = 2$
34. $3(2x - 5) - (2x - 4) = 6 - (4x + 5)$
35. Solve the formula $P = 2l + 2w$ for w.

Solve each inequality and graph the solution.

36. $-4x < 20$
37. $3 - 2x > 5$
38. $3 - 4(x - 2) \geq -5x + 6$
39. Solve the formula $3x - 2y \leq 12$ for y.
40. What number is 12% of 2,000?
41. The length of a rectangle is 5 inches more than three times the width. If the perimeter is 26 inches, find the length and width.
42. Patrick is eight years older than his cousin Stacey. In five years the sum of their ages will be 38. How old are they now?

Section 9.3 The Distance Formula and the Circle

Before we find the general equation of a circle, we must first derive what is known as the *distance formula*.

Suppose (x_1, y_1) and (x_2, y_2) are any two points in the first quadrant. (Actually, we could choose the two points to be anywhere on the coordinate plane. It is just more convenient to have them in the first quadrant.) We can name the points P_1 and P_2, respectively, and draw the diagram shown in Figure 1.

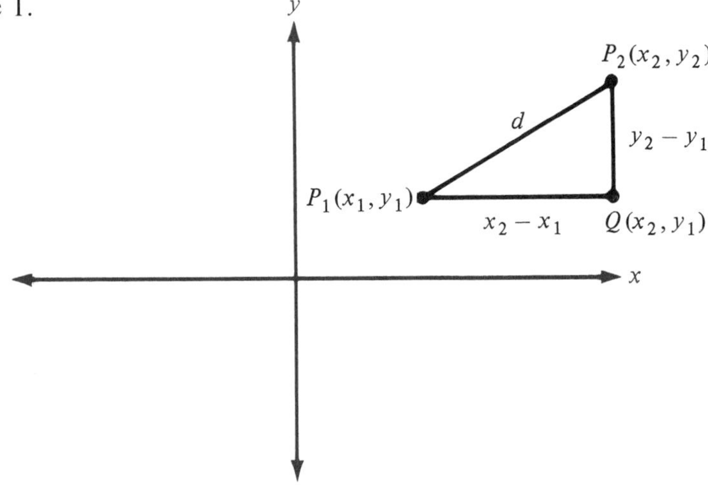

Figure 1

Notice the coordinates of point Q. The x-coordinate is x_2 since Q is directly below point P_2. The y-coordinate of Q is y_1 since Q is directly across from point P_1. It is evident from the diagram that the length of P_2Q is $y_2 - y_1$ and the length of P_1Q is $x_2 - x_1$. Using the Pythagorean theorem, we have

$$d^2 = (x_2 - x_1)^2 + (y_2 - y_1)^2$$

Taking the square root of both sides, we have

$$d = \sqrt{(x_2 - x_1)^2 + (y_2 - y_1)^2}$$

We know this is the positive square root, since d is the distance from P_1 to P_2 and must therefore be positive. This formula is called the *distance formula*.

▼ **Example 1** Find the distance between (3, 5) and (2, −1).

Solution If we let (3, 5) be (x_1, y_1) and (2, −1) be (x_2, y_2) and apply the distance formula, we have

$$\begin{aligned} d &= \sqrt{(2 - 3)^2 + (-1 - 5)^2} \\ &= \sqrt{(-1)^2 + (-6)^2} \\ &= \sqrt{1 + 36} \\ &= \sqrt{37} \end{aligned}$$

▲

▼ **Example 2** Find x if the distance from $(x, 5)$ to (3, 4) is $\sqrt{2}$.

Solution Using the distance formula, we have

$$\begin{aligned} \sqrt{2} &= \sqrt{(x - 3)^2 + (5 - 4)^2} \\ 2 &= (x - 3)^2 + 1^2 \\ 2 &= x^2 - 6x + 9 + 1 \\ 0 &= x^2 - 6x + 8 \\ 0 &= (x - 4)(x - 2) \\ x = 4 &\quad \text{or} \quad x = 2 \end{aligned}$$

The two solutions are 4 and 2, which indicates there are two points, (4, 5) and (2, 5), which are $\sqrt{2}$ units from (3, 4). ▲

We can use the distance formula to derive the equation of a circle.

Theorem 9.1 The equation of the circle with center at (a, b) and radius r is given by

$$(x - a)^2 + (y - b)^2 = r^2$$

PROOF By definition, all points on the circle are a distance r from the center (a, b). If we let (x, y) represent any point on the circle, then (x, y) is r units from (a, b). Applying the distance formula, we have

$$r = \sqrt{(x - a)^2 + (y - b)^2}$$

Squaring both sides of this equation gives the equation of the circle:

$$(x - a)^2 + (y - b)^2 = r^2$$

We can use Theorem 9.1 to find the equation of a circle, given its center and radius, or to find its center and radius, given the equation.

▼ **Example 3** Find the equation of the circle with center at $(-3, 2)$ having a radius of 5.

Solution We have $(a, b) = (-3, 2)$ and $r = 5$. Applying Theorem 9.1 yields

$$[x - (-3)]^2 + (y - 2)^2 = 5^2$$
$$(x + 3)^2 + (y - 2)^2 = 25$$

▲

▼ **Example 4** Give the equation of the circle with radius 3 whose center is at the origin.

Solution The coordinates of the center are $(0, 0)$ and the radius is 3. The equation must be

$$(x - 0)^2 + (y - 0)^2 = 3^2$$
$$x^2 + y^2 = 9$$

▲

We can see from Example 4 that the equation of any circle with its center at the origin and radius r will be

$$x^2 + y^2 = r^2$$

▼ **Example 5** Find the center and radius, and sketch the graph, of the circle whose equation is

$$(x - 1)^2 + (y + 3)^2 = 4$$

Solution Writing the equation in the form

$$(x - a)^2 + (y - b)^2 = r^2$$

we have

$$(x - 1)^2 + [y - (-3)]^2 = 2^2$$

The center is at $(1, -3)$ and the radius is 2.

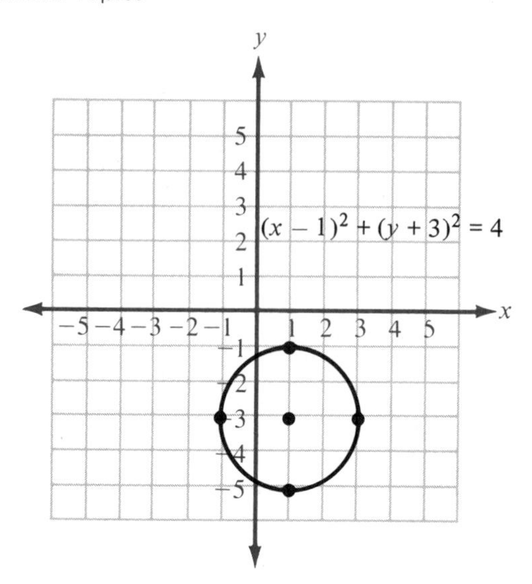

Figure 2

▲

▼ **Example 6** Sketch the graph of $x^2 + y^2 = 9$.

Solution Since the equation can be written in the form

$$(x - 0)^2 + (y - 0)^2 = 3^2$$

it must have its center at (0, 0) and a radius of 3.

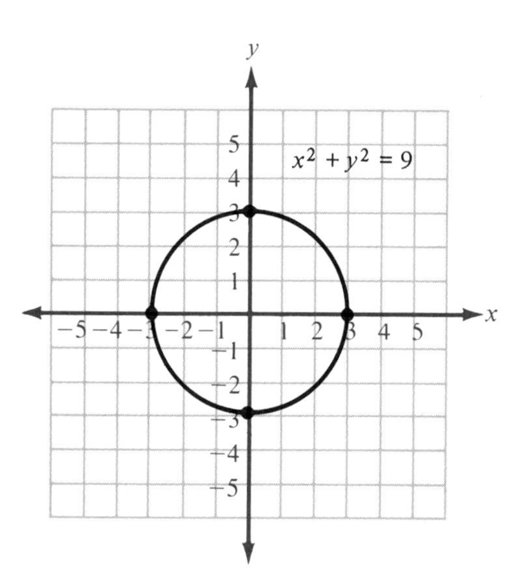

Figure 3

▲

Problem Set 9.3

Find the distance between the following points.

1. (3, 7) and (6, 3)
2. (4, 7) and (8, 1)
3. (0, 9) and (5, 0)
4. (−3, 0) and (0, 4)
5. (3, −5) and (−2, 1)
6. (−8, 9) and (−3, −2)
7. (−1, −2) and (−10, 5)
8. (−3, −8) and (−1, 6)
9. Find x so the distance between $(x, 2)$ and (1, 5) is $\sqrt{13}$.
10. Find x so the distance between (−2, 3) and $(x, 1)$ is 3.
11. Find y so the distance between $(7, y)$ and (8, 3) is 1.
12. Find y so the distance between (3, −5) and $(3, y)$ is 9.

Write the equation of the circle with the given center and radius.

13. Center (2, 3); $r = 4$
14. Center (3, −1); $r = 5$
15. Center (3, −2); $r = 3$
16. Center (−2, 4); $r = 1$
17. Center (−5, −1); $r = \sqrt{5}$
18. Center (−7, −6); $r = \sqrt{3}$
19. Center (0, −5); $r = 1$
20. Center (0, −1); $r = 7$
21. Center (0, 0); $r = 2$
22. Center (0, 0); $r = 5$

Give the center and radius, and sketch the graph, of each of the following circles.

23. $x^2 + y^2 = 4$
24. $x^2 + y^2 = 16$
25. $(x - 1)^2 + (y - 3)^2 = 25$
26. $(x - 4)^2 + (y - 1)^2 = 36$
27. $(x - 2)^2 + (y - 3)^2 = 9$
28. $(x - 3)^2 + (y + 1)^2 = 9$
29. $(x + 1)^2 + (y + 1)^2 = 1$
30. $(x + 3)^2 + (y + 2)^2 = 9$

Review Problems The following problems review material we covered in Chapter 3.

31. Fill in each ordered pair so that it is a solution to $y = \frac{1}{2}x + 3$.
    (−2,) (0,) (2,)
32. Graph the line $y = \frac{1}{2}x + 3$.
33. Graph the line $x = -2$.
34. Find the intercepts and slope for $2x - 3y = 6$. Then, graph the line.
35. Find the slope of the line through (−1, 2) and (5, −4).
36. Find x if the line through $(x, -2)$ and (3, 4) has a slope of 6.
37. Find the equation of the line with slope −3 that contains the point (−1, 4).
38. Solve this system by graphing: $\begin{aligned} x + y &= 4 \\ x - y &= 2 \end{aligned}$

Solve each system by the elimination method.

39. $\begin{aligned} 3x + 2y &= 1 \\ 2x + y &= 3 \end{aligned}$
40. $\begin{aligned} 2x + 3y &= -1 \\ 3x + 5y &= -2 \end{aligned}$

Solve each system by the substitution method.

41. $x + y = 20$
$y = 5x + 2$

42. $2x - 6y = 2$
$y = 3x + 1$

43. A total of \$1,200 is invested in two accounts. One of the accounts pays 8% interest annually and the other pays 10% interest annually. If the total amount of interest earned from both accounts for the year is \$104, how much is invested in each account?

44. Amy has \$1.85 in dimes and quarters. If she has a total of 11 coins, how many of each coin does she have?

Section 9.4 Fractional Exponents

Consider the expression $3^{1/2}$ and the expression $\sqrt{3}$. The second expression is the positive square root of 3. We have never encountered the first expression before. Assuming the properties of exponents apply to fractional exponents in the same way they apply to integer exponents, we can take a look at what happens when we square both of the preceding quantities:

$$\begin{array}{cc} (3^{1/2})^2 & (\sqrt{3})^2 \\ 3^{(1/2)2} & \sqrt{3}\cdot\sqrt{3} \\ 3^1 & \sqrt{9} \\ 3 & 3 \end{array}$$

The result is the same in both cases. The result allows us to define fractional exponents.

DEFINITION If x represents a nonnegative real number, then:

$$\sqrt{x} = x^{1/2}$$
$$\sqrt[3]{x} = x^{1/3}$$
$$\sqrt[4]{x} = x^{1/4}$$

and, in general,

$$\sqrt[n]{x} = x^{1/n}$$

We can use this definition in many ways. We can change any expression involving radicals to an expression that involves only exponents. We can avoid radical notation altogether if we choose.

▼ **Examples** Write each of the following as a radical and then simplify.

1. $25^{1/2}$

Solution The exponent $\frac{1}{2}$ indicates that we are to find the positive square root of 25:

$$25^{1/2} = \sqrt{25} = 5$$

2. $9^{1/2}$

Solution Again, an exponent of $\frac{1}{2}$ indicates that we are to find the positive square root of 9:

$$9^{1/2} = \sqrt{9} = 3$$

3. $8^{1/3}$

Solution The exponent is $\frac{1}{3}$, indicating that we are to find the cube root of 8:

$$8^{1/3} = \sqrt[3]{8} = 2$$

4. $81^{1/4}$

Solution An exponent of $\frac{1}{4}$ indicates that we are to find the fourth root:

$$81^{1/4} = \sqrt[4]{81} = 3$$

▲

Consider the expression $9^{3/2}$. If we assume that all our properties of exponents hold for fractional exponents, we can rewrite this expression as a power raised to another power:

$$9^{3/2} = (9^{1/2})^3$$

Since $9^{1/2}$ is equivalent to $\sqrt{9}$, we can simplify further:

$$(9^{1/2})^3 = (\sqrt{9})^3 = 3^3 = 27$$

We can summarize this discussion with the following definition.

DEFINITION If x represents a positive real number, and n and m are integers, then

$$x^{m/n} = \sqrt[n]{x^m} = (\sqrt[n]{x})^m$$

We can think of the exponent m/n as consisting of two parts: the numerator m is the power, and the denominator n is the root.

▼ **Example 5** Simplify: $8^{2/3}$.

Solution Using the preceding definition, we have

$8^{2/3} = (8^{1/3})^2$	Separate exponents
$= (\sqrt[3]{8})^2$	Write as cube root
$= 2^2$	$\sqrt[3]{8} = 2$
$= 4$	$2^2 = 4$

▲

▼ **Example 6** Simplify: $27^{4/3}$.

Solution

$$\begin{aligned} 27^{4/3} &= (27^{1/3})^4 && \text{Separate exponents} \\ &= (\sqrt[3]{27})^4 && \text{Write as cube root} \\ &= 3^4 && \sqrt[3]{27} = 3 \\ &= 81 && 3^4 = 81 \end{aligned}$$

▲

We can apply the properties of exponents to expressions that involve fractional exponents.

▼ **Examples** Use the properties of exponents to simplify each expression. Assume that all variables represent positive numbers.

7. $x^{1/3} \cdot x^{2/3}$

Solution To multiply with the same base, we add exponents. The property we use is $a^r a^s = a^{r+s}$:

$$x^{1/3} \cdot x^{2/3} = x^{1/3+2/3} = x^{3/3} = x$$

8. $(x^{1/4})^8$

Solution To raise a power to another power, we multiply exponents. The property we use is $(a^r)^s = a^{rs}$:

$$(x^{1/4})^8 = x^{(1/4)8} = x^2$$

9. $\dfrac{a^{5/6}}{a^{4/6}}$

Solution To divide with the same base, we subtract the exponent in the denominator from the exponent in the numerator:

$$\frac{a^{5/6}}{a^{4/6}} = a^{5/6-4/6} = a^{1/6}$$

10. $(8y^{12})^{1/3}$

Solution Distributing the exponent $\frac{1}{3}$ across the product, we have

$$(8y^{12})^{1/3} = 8^{1/3}(y^{12})^{1/3} = 2y^4$$

▲

Problem Set 9.4

Change each of the following to an expression involving roots and then simplify.

1. $4^{1/2}$ **2.** $9^{1/2}$ **3.** $16^{1/2}$ **4.** $25^{1/2}$

5. $27^{1/3}$ **6.** $8^{1/3}$ **7.** $125^{1/3}$ **8.** $16^{1/4}$

9. $81^{1/4}$ **10.** $36^{1/2}$ **11.** $81^{1/2}$ **12.** $144^{1/2}$

13. $8^{2/3}$ **14.** $25^{3/2}$ **15.** $125^{2/3}$ **16.** $36^{3/2}$

17. $16^{3/4}$
18. $9^{3/2}$
19. $16^{3/2}$
20. $8^{5/3}$
21. $4^{3/2}$
22. $4^{5/2}$
23. $(-8)^{2/3}$
24. $(-27)^{2/3}$
25. $(-32)^{1/5}$
26. $(-32)^{3/5}$
27. $4^{1/2} + 9^{1/2}$
28. $16^{1/2} + 25^{1/2}$
29. $16^{3/4} + 27^{2/3}$
30. $49^{1/2} + 64^{1/2}$
31. $4^{1/2} \cdot 27^{1/3}$
32. $8^{1/3} \cdot 25^{1/2}$

Use the properties of exponents to simplify each of the following expressions. Assume that all variables represent positive numbers.

33. $x^{1/4} \cdot x^{3/4}$
34. $x^{1/8} \cdot x^{3/8}$
35. $(x^{2/3})^3$
36. $(x^{1/4})^{12}$
37. $\frac{a^{3/5}}{a^{1/5}}$
38. $\frac{a^{5/7}}{a^{3/7}}$
39. $(27y^6)^{1/3}$
40. $(81y^8)^{1/4}$
41. $(9a^4b^2)^{1/2}$
42. $(25a^8b^4)^{1/2}$
43. $\frac{x^{3/5} \cdot x^{4/5}}{x^{2/5}}$
44. $\frac{x^{4/7} \cdot x^{6/7}}{x^{3/7}}$

Recall that exponents indicate reciprocals. That is, $a^{-r} = \frac{1}{a^r}$. Use this property of exponents, along with what you have learned in this section, to simplify the following expressions.

45. $25^{-1/2}$
46. $8^{-1/3}$
47. $8^{-2/3}$
48. $81^{-3/4}$
49. $27^{-2/3}$
50. $16^{-3/4}$

Review Problems The problems that follow review material we covered in Chapter 4.

Simplify each expression. (Write all answers with positive exponents only.)

51. $(5x^3)^2(2x^6)^3$
52. 2^{-3}
53. $\frac{x^4}{x^{-3}}$
54. $\frac{(20x^2y^3)(5x^4y)}{(2xy^5)10x^2y^3)}$
55. $(2 \times 10^{-4})(4 \times 10^5)$
56. $\frac{9 \times 10^{-3}}{3 \times 10^{-2}}$
57. $20ab^2 - 16ab^2 + 6ab^2$
58. Subtract $6x^2 - 5x - 7$ from $9x^2 + 3x - 2$

Multiply.

59. $2x^2(3x^2 + 3x - 1)$
60. $(2x + 3)(5x - 2)$
61. $(3y - 5)^2$
62. $(a - 4)(a^2 + 4a + 16)$
63. $(2a^2 + 7)(2a^2 - 7)$
64. Divide $15x^{10} - 10x^8 + 25x^6$ by $5x^6$
65. Divide, using long division. $\frac{x^2 - 2x + 6}{x - 4}$

Section 9.5 Factoring the Sum and Difference of Two Cubes

In Chapter 5 we factored a variety of polynomials. Among the polynomials we factored were polynomials that were the difference of two squares. The formula we used to factor the difference of two squares looks like this:

$$a^2 - b^2 = (a + b)(a - b)$$

If we ran across a binomial that had the form of the difference of two squares, we factored it by applying this formula. For example, to factor $x^2 - 25$, we simply notice that it can be written in the form $x^2 - 5^2$, which looks like the difference of two squares. According to the formula above, it factors into $(x + 5)(x - 5)$.

In this section we want to use two new formulas that will allow us to factor the sum and difference of two cubes. For example, we want to factor the binomial $x^3 - 8$, which is the difference of two cubes. (To see that it is the difference of two cubes, notice that it can be written as $x^3 - 2^3$.) We also want to factor $y^3 + 27$, which is the sum of two cubes. (To see this, notice that $y^3 + 27$ can be written as $y^3 + 3^3$.)

The formulas that allow us to factor the sum of two cubes and the difference of two cubes are not as simple as the formula for factoring the difference of two squares. Here is what they look like:

$$a^3 + b^3 = (a + b)(a^2 - ab + b^2)$$
$$a^3 - b^3 = (a - b)(a^2 + ab + b^2)$$

Let's begin our work with these two formulas by showing that they are true. To do so, we multiply out the right side of each formula.

▼ **Example 1** Verify the two formulas.

Solution We verify the formulas by multiplying the right sides and comparing the results with the left sides:

$$\begin{array}{rrrr} & a^2 & - ab & + b^2 \\ & & a & + b \\ \hline & a^2b & - ab^2 & + b^3 \\ a^3 & - a^2b & + ab^2 & \\ \hline a^3 & & & + b^3 \end{array}$$

The first formula is correct.

$$\begin{array}{rrrr} & a^2 & + ab & + b^2 \\ & & a & - b \\ \hline & - a^2b & - ab^2 & - b^3 \\ a^3 & + a^2b & + ab^2 & \\ \hline a^3 & & & - b^3 \end{array}$$

The second formula is correct. ▲

Here are some examples that use the formulas for factoring the sum and difference of two cubes.

▼ **Example 2** Factor: $x^3 - 8$.

Solution Since the two terms are perfect cubes, we write them as such and apply the formula:

$$\begin{aligned} x^3 - 8 &= x^3 - 2^3 \\ &= (x - 2)(x^2 + 2x + 2^2) \\ &= (x - 2)(x^2 + 2x + 4) \end{aligned}$$ ▲

▼ **Example 3** Factor: $y^3 + 27$.

Solution Proceding as we did in Example 2, we first write 27 as 3^3. Then, we apply the formula for factoring the sum of two cubes, which is $a^3 + b^3 = (a + b)(a^2 - ab + b^2)$:

$$\begin{aligned} y^3 + 27 &= y^3 + 3^3 \\ &= (y + 3)(y^2 - 3y + 3^2) \\ &= (y + 3)(y^2 - 3y + 9) \end{aligned}$$ ▲

▼ **Example 4** Factor: $8x^3 - y^3$.

Solution Since we can write $8x^3$ as $(2x)^3$, we have a binomial that is the difference of two cubes:

$$\begin{aligned} 8x^3 - y^3 &= (2x)^3 - y^3 \\ &= (2x - y)[(2x)^2 + (2x)y + y^2] \\ &= (2x - y)(4x^2 + 2xy + y^2) \end{aligned}$$ ▲

Note If you are wondering if the second factor $4x^2 + 2xy + y^2$ can be factored further, it cannot. If you think that it can be factored further, write down the factors you think it has and multiply them. Doing so will convince you that it does not factor.

▼ **Example 5** Factor: $27x^3 + 125y^3$.

Solution Since $27x^3$ can be written as $(3x)^3$ and $125y^3$ can be written as $(5y)^3$, we have a binomial that is the sum of two cubes:

$$\begin{aligned} 27x^3 + 125y^3 &= (3x)^3 + (5y)^3 \\ &= (3x + 5y)[(3x)^2 - (3x)(5y) + (5y)^2] \\ &= (3x + 5y)(9x^2 - 15xy + 25y^2) \end{aligned}$$

Again, the second factor, $9x^2 - 15xy + 25y^2$ cannot be factored further. ▲

Problem Set 9.5

Multiply.

1. $(x + 4)(x - 4)$
2. $(x + 7)(x - 7)$
3. $(2x + 9)(2x - 9)$
4. $(3x + 1)(3x - 1)$
5. $(x + 2)(x^2 - 2x + 4)$
6. $(x - 2)(x^2 + 2x + 4)$
7. $(x - 1)(x^2 + x + 1)$
8. $(x + 1)(x^2 - x + 1)$
9. $(x + 4)(x^2 - 4x + 16)$
10. $(x - 4)(x^2 + 4x + 16)$
11. $(2x - 3)(4x^2 + 6x + 9)$
12. $(2x + 3)(4x^2 - 6x + 9)$
13. $(x + y)(x^2 - xy + y^2)$
14. $(x - y)(x^2 + xy + y^2)$

Factor each of the following.

15. $x^3 - 2^3$
16. $x^3 + 2^3$
17. $x^3 + 7^3$
18. $x^3 - 7^3$
19. $(3x)^3 - (5y)^3$
20. $(5x)^3 - (3y)^3$
21. $x^3 + 8$
22. $x^3 - 27$
23. $x^3 - 125$
24. $x^3 - 64$
25. $x^3 + 1$
26. $x^3 - 1$
27. $8x^3 + y^3$
28. $x^3 + 8y^3$
29. $27x^3 - 8y^3$
30. $8x^3 - 27y^3$
31. $125x^3 + 8$
32. $8x^3 + 125$
33. $x^3 + 64y^3$
34. $x^3 - 64y^3$

Review Problems The problems below review material we covered in Chapter 5.

35. Factor 630 into the product of prime factors.
36. Factor by grouping: $3ax - 2a + 15x - 10$.

Factor completely.

37. $x^2 - 4x - 12$
38. $4x^2 - 20xy + 25y^2$
39. $x^4 - 16$
40. $2x^2 + xy - 21y^2$
41. $5x^3 - 25x^2 - 30x$

Solve each equation.

42. $x^2 - 9x + 18 = 0$
43. $x^2 - 6x = 0$
44. $8x^2 = -2x + 15$
45. $x(x + 2) = 80$
46. The product of two consecutive even integers is four more than twice their sum. Find the two integers.
47. The hypotenuse of a right triangle is 15 inches. One of the legs is 3 inches more than the other. Find the lengths of the two legs.
48. If the total cost, C, of manufacturing x hundred items is given by the equation $C = 600 + 1000x - 100x^2$, find x when the total cost is \$2200.

Section 9.6 Variation

Two variables are said to *vary directly* if one is a constant multiple of the other. For instance, y varies directly as x if $y = Kx$, where K is a constant. The constant K is called the *constant of variation*. The following table gives the relation between direct variation statements and their equivalent algebraic equations.

Statement	Equation (K = constant of variation)
y varies directly as x	$y = Kx$
y varies directly as the square of x	$y = Kx^2$
s varies directly as the square root of t	$s = K\sqrt{t}$
r varies directly as the cube of s	$r = Ks^3$

Any time we run across a statement similar to the ones in the table, we can immediately write an equivalent expression involving variables and a constant of variation K.

▼ **Example 1** Suppose y varies directly as x. When y is 15, x is 3. Find y when x is 4.

Solution From the first sentence we can write the relationship between x and y as

$$y = Kx$$

We now use the second sentence to find the value of K. Since y is 15 when x is 3, we have

$$15 = K(3) \qquad \text{or} \qquad K = 5$$

Now we can rewrite the relationship between x and y more specifically as

$$y = 5x$$

To find the value of y when x is 4, we simply substitute $x = 4$ into our last equation. Substituting

$$x = 4$$

into

$$y = 5x$$

we have

$$y = 5(4)$$
$$y = 20$$

▲

▼ **Example 2** Suppose y varies directly as the square of x. When x is 4, y is 32. Find x when y is 50.

Solution The first sentence gives us

$$y = Kx^2$$

Since y is 32 when x is 4, we have

$$\begin{aligned} 32 &= K(4)^2 \\ 32 &= 16K \\ K &= 2 \end{aligned}$$

The equation now becomes

$$y = 2x^2$$

When y is 50, we have

$$\begin{aligned} 50 &= 2x^2 \\ 25 &= x^2 \\ x &= \pm 5 \end{aligned}$$

There are two possible solutions, $x = 5$ or $x = -5$. ▲

▼ **Example 3** The cost of a certain kind of candy varies directly with the weight of the candy. If 12 ounces of the candy cost $1.68, how much will 16 ounces cost?

Solution Let $x =$ the number of ounces of candy and $y =$ the cost of the candy. Then $y = Kx$. Since y is 1.68 when x is 12, we have

$$\begin{aligned} 1.68 &= K \cdot 12 \\ K &= \frac{1.68}{12} \\ &= 0.14 \end{aligned}$$

The equation must be

$$y = 0.14x$$

When x is 16, we have

$$\begin{aligned} y &= 0.14(16) \\ &= 2.24 \end{aligned}$$

The cost of 16 ounces of candy is $2.24. ▲

Inverse Variation

Two variables are said to *vary inversely* if one is a constant multiple of the reciprocal of the other. For example, y varies inversely as x if $y = \frac{K}{x}$, where K is a real number constant. Again, K is called the constant of variation. The table that follows gives some examples of inverse variation statements and their associated algebraic equations.

Statement	Equation (K = constant of variation)
y varies inversely as x	$y = \frac{K}{x}$
y varies inversely as the square of x	$y = \frac{K}{x^2}$
F varies inversely as the square root of t	$F = \frac{K}{\sqrt{t}}$
r varies inversely as the cube of s	$r = \frac{K}{s^3}$

Every inverse variation statement has an associated inverse variation equation.

▼ **Example 4** Suppose y varies inversely as x. When y is 4, x is 5. Find y when x is 10.

Solution The first sentence gives us the relationship between x and y:

$$y = \frac{K}{x}$$

We use the second sentence to find the value of the constant K.

$$4 = \frac{K}{5}$$

or

$$K = 20$$

We can now write the relationship between x and y more specifically as

$$y = \frac{20}{x}$$

We use this equation to find the value of y when x is 10. Substituting

$$x = 10$$

into

$$y = \frac{20}{x}$$

we have

$$y = \frac{20}{10}$$

$$y = 2$$

▲

▼ **Example 5** The intensity (I) of light from a source varies inversely as the square of the distance (d) from the source. Ten feet away from the source the intensity is 200 candlepower. What is the intensity 5 feet from the source?

Solution

$$I = \frac{K}{d^2}$$

Since $I = 200$ when $d = 10$, we have

$$200 = \frac{K}{10^2}$$

$$200 = \frac{K}{100}$$

$$K = 20{,}000$$

The equation becomes

$$I = \frac{20{,}000}{d^2}$$

When $d = 5$, we have

$$I = \frac{20{,}000}{5^2}$$

$$= \frac{20{,}000}{25}$$

$$= 800 \text{ candlepower}$$

▲

Problem Set 9.6

For each of the following problems, y varies directly as x.

1. If $y = 10$ when $x = 5$, find y when x is 4.
2. If $y = 20$ when $x = 4$, find y when x is 11.
3. If $y = 39$ when $x = 3$, find y when x is 10.
4. If $y = -18$ when $x = 6$, find y when x is 3.
5. If $y = -24$ when $x = 4$, find x when y is -30.
6. If $y = 30$ when $x = -15$, find x when y is 8.
7. If $y = -7$ when $x = -1$, find x when y is -21.
8. If $y = 30$ when $x = 4$, find y when x is 7.

For each of the following, y varies directly as the square of x.

9. If $y = 75$ when $x = 5$, find y when x is 1.
10. If $y = -72$ when $x = 6$, find y when x is 3.
11. If $y = 48$ when $x = 4$, find y when x is 9.
12. If $y = 27$ when $x = 3$, find x when y is 75.

For each of the following problems, y varies inversely with x.

13. If $y = 5$ when $x = 2$, find y when x is 5.
14. If $y = 2$ when $x = 10$, find y when x is 4.
15. If $y = 2$ when $x = 1$, find y when x is 4.
16. If $y = 4$ when $x = 3$, find y when x is 6.
17. If $y = 5$ when $x = 3$, find x when y is 15.
18. If $y = 12$ when $x = 10$, find x when y is 60.
19. If $y = 10$ when $x = 10$, find x when y is 20.
20. If $y = 15$ when $x = 2$, find x when y is 6.

For each of the following, y varies inversely as the square of x.

21. If $y = 4$ when $x = 5$, find y when x is 2.
22. If $y = 5$ when $x = 2$, find y when x is 6.
23. If $y = 4$ when $x = 3$, find y when x is 2.
24. If $y = 9$ when $x = 4$, find y when x is 3.

Solve the following problems.

25. The tension t in a spring varies directly with the distance d the spring is stretched. If the tension is 42 pounds when the spring is stretched 2 inches, find the tension when the spring is stretched twice as far.

26. The time t it takes to fill a bucket varies directly with the volume g of the bucket. If it takes 1 minute to fill a 4-gallon bucket, how long will it take to fill a 6-gallon bucket?

27. The power P in an electric circuit varies directly with the square of the current I. If $P = 30$ when $I = 2$, find P when $I = 7$.

28. The resistance R in an electric circuit varies directly with the voltage V. If $R = 20$ when $V = 120$, find R when $V = 240$.

29. The amount of money M a person makes per week varies directly with the number of hours h he works per week. If he works 20 hours and earns $157, how much does he make if he works 30 hours?

30. The volume V of a gas varies directly as the temperature T. If $V = 3$ when $T = 150$, find V when T is 200.

31. The weight F of a body varies inversely with the square of the distance d between the body and the center of the earth. If a man weighs 150 pounds 4000 miles from the center of the earth, how much will he weigh at a distance of 5000 miles from the center of the earth?

32. The intensity I of a light source varies inversely with the square of the distance d from the source. Four feet from the source the intensity is 9 footcandles. What is the intensity 3 feet from the source?

33. The current I in an electric circuit varies inversely with the resistance R. If a current of 30 amps is produced by a resistance of 2 ohms, what current will be produced by a resistance of 5 ohms?

34. The pressure exerted by a gas on the container in which it is held varies inversely with the volume of the container. A pressure of 40 pounds per square inch is exerted on a container of volume 2 cubic feet. What is the pressure on a container whose volume is 8 cubic feet?

Review Problems The problems below review material we covered in Chapter 6.

35. Reduce to lowest terms: $\dfrac{x^2 - x - 6}{x^2 - 9}$.

36. Use factoring by grouping to reduce to lowest terms: $\dfrac{xy + 3y + 2x + 6}{xy + 3y + 5x + 15}$

Perform the indicated operations.

37. $\dfrac{x^2 - 25}{x + 4} \cdot \dfrac{2x + 8}{x^2 - 9x + 20}$

38. $\dfrac{3x + 6}{x^2 + 4x + 3} \div \dfrac{x^2 + x - 2}{x^2 + 2x - 3}$

39. $\dfrac{x}{x^2 - 16} + \dfrac{4}{x^2 - 16}$

40. $\dfrac{2}{x^2 - 1} - \dfrac{5}{x^2 + 3x - 4}$

41. $\dfrac{1 - \dfrac{25}{x^2}}{1 - \dfrac{8}{x} + \dfrac{15}{x^2}}$

Solve each equation.

42. $\dfrac{x}{2} - \dfrac{5}{x} = -\dfrac{3}{2}$

43. $\dfrac{x}{x^2 - 9} - \dfrac{3}{x - 3} = \dfrac{1}{x + 3}$

44. A boat travels 30 miles up a river in the same amount of time it takes to travel 50 miles down the same river. If the current is 5 mph, what is the speed of the boat in still water?

45. A pool can be filled by an inlet pipe in 8 hours. The drain will empty the pool in 12 hours. How long will it take to fill the pool if both the inlet pipe and the drain are open?

46. If 30 liters of a certain solution contains 2 liters of alcohol, how much alcohol is in 45 liters of the same solution?

Section 9.7 Functions and Function Notation

The idea of a function is very common in mathematics. Functions are studied in almost all branches of mathematics. They are the result of the process of working with quantities that vary with one another. That is, a change in one quantity brings about a change in another. The idea of a function is also a very simple concept mathematically.

DEFINITION A *function* is any set of ordered pairs in which no two different ordered pairs have the same first coordinate. The *domain* of a function is the set of all first coordinates. The *range* of a function is the set of all second coordinates.

EXAMPLE The set $\{(2, 3), (1, 5), (6, 3) (-3, 4)\}$ is a function because it is a set of ordered pairs and no two different ordered pairs have the same first coordinates. The domain of this function is the set $\{2, 1, 6, -3\}$. The range is the set $\{3, 5, 4\}$.

In this example we listed all the ordered pairs of the function. Usually the ordered pairs are given in the form of a rule instead of a list.

EXAMPLE The set $\{(x, y) | y = 2x + 1; x = 1, 2, 3, 4\}$ is an example of a function. The domain is the set $\{1, 2, 3, 4\}$. The range is the set $\{3, 5, 7, 9\}$, since these are the values of y that correspond to the x values 1, 2, 3, 4.

In this example we specified the function by giving the rule for obtaining ordered pairs, $y = 2x + 1$, and then listing the values that the variable x could assume, $x = 1, 2, 3, 4$. Most of the time, when a function is given by a rule, the domain is not specified. When this happens, we assume the domain includes all replacements for x except those that make the function undefined. For example, the domain of the function

$$y = \frac{3}{x}$$

is all real numbers except $x = 0$, because when x is 0 the expression $\frac{3}{0}$ is undefined. There is no value of y that corresponds to $x = 0$.

▼ **Example 1** Specify the domain for $y = \dfrac{3}{x - 2}$.

Solution We can replace x with any real number except 2, because $x = 2$ makes the denominator zero:

$$\text{Domain} = \{x | x \text{ is real}, x \neq 2\}$$ ▲

▼ **Example 2** Specify the domain for $y = \dfrac{3}{x^2 - x - 6}$.

Solution We are interested in eliminating any values of x that will make the denominator zero:

$$x^2 - x - 6 = 0$$
$$(x - 3)(x + 2) = 0$$
$$x = 3 \quad \text{or} \quad x = -2$$

Since $x = 3$ and $x = -2$ both make the denominator zero, we cannot use them as replacements for x:

$$\text{Domain} = \{x | x \text{ is real}, x \neq 3, x \neq -2\}$$ ▲

Function Notation

The notation $f(x)$ is read "f of x" and is defined to be the value of the function f at x. It is not to be read or interpreted as f times x. The equations $y = 2x + 3$ and $f(x) = 2x + 3$ are essentially the same; that is, $y = f(x)$.

Suppose we are given the equation $y = 2x + 3$ and we want to find the value of y when x is 4. We would have to say, "If $x = 4$, then $y = 2(4) + 3 = 11$." Using the new notation, $f(x) = 2x + 3$, we have $f(4) = 2(4) + 3 = 11$. The following table gives more examples of this concept.

Function Notation	y in Terms of x
$f(x) = 2x + 3$	$y = 2x + 3$
$f(1) = 2(1) + 3 = 5$	If $x = 1$, then $y = 2(1) + 3 = 5$
$f(0) = 2(0) + 3 = 3$	If $x = 0$, then $y = 2(0) + 3 = 3$
$f(-5) = 2(-5) + 3 = -7$	If $x = -5$, then $y = 2(-5) + 3 = -7$

Here are some additional examples of the use of function notation.

▼ **Example 3** If $f(x) = x^2 - 4$, find $f(0)$, $f(-1)$, $f(3)$, $f(5)$, $f(a)$.

Solution

$$\begin{aligned} f(x) &= x^2 - 4 \\ f(0) &= 0^2 - 4 = -4 \\ f(-1) &= (-1)^2 - 4 = -3 \\ f(3) &= 3^2 - 4 = 5 \\ f(5) &= 5^2 - 4 = 21 \\ f(a) &= a^2 - 4 \end{aligned}$$ ▲

▼ **Example 4** If $f(x) = 2x + 5$ and $g(x) = 2x^2 + 1$, find $f(2)$, $g(-3)$, $g(a)$, $f(z)$.

Solution

$$\begin{aligned} f(2) &= 2(2) + 5 = 4 + 5 = 9 \\ g(-3) &= 2(-3)^2 + 1 = 2(9) + 1 = 19 \\ g(a) &= 2a^2 + 1 \\ f(z) &= 2z + 5 \end{aligned}$$ ▲

Problem Set 9.7

Specify the domain and range of the following functions.

1. $\{(1, 2), (3, 4), (5, 6)\}$
2. $\{(-1, 1), (-2, 2), (-3, 3)\}$
3. $\{(1, a), (3, b), (4, c)\}$
4. $\{(a, -1), (b, c), (d, -2)\}$

5. $y = x, \quad x = 1, 2, 3, 4$

6. $y = 2x, \quad x = 2, 4, 6, 8$

7. $y = 3x - 1, \quad x = 1, 3, 5, 7, 9$

8. $y = 2x - 3, \quad x = 2, 4, 6, 8, 10$

9. $2x + 3y = 6, \quad x = 0, 2, 4$

10. $3x - 2y = 6, \quad y = 0, -2, -3$

11. $2x - y = 4, \quad y = 2, 4, 6, 8, 10$

12. $3x - 5y = 15, \quad x = 0, 5, 15$

Specify any restrictions on the domain of the following functions.

13. $y = \dfrac{1}{x - 2}$

14. $y = \dfrac{2}{x - 1}$

15. $y = \dfrac{3}{x + 3}$

16. $y = \dfrac{5}{x + 5}$

17. $y = \dfrac{4}{(x + 1)(x + 2)}$

18. $y = \dfrac{(x + 5)}{(x - 3)(x + 2)}$

19. $y = \dfrac{x}{(x + 5)(x - 1)}$

20. $y = \dfrac{2x}{(x - 7)(x + 6)}$

21. $y = \dfrac{3}{x^2 - 9}$

22. $y = \dfrac{2}{x^2 - 25}$

23. $y = \dfrac{3}{x^2 - 5x + 6}$

24. $y = \dfrac{2x}{x^2 + 4x + 4}$

25. $y = \dfrac{x}{2x^2 - 5x - 3}$

26. $y = \dfrac{3x}{3x^2 + x - 2}$

27. If $f(x) = 2x$, find $f(0), f(1), f(2)$.
28. If $f(x) = -3x$, find $f(-2), f(0), f(2)$.
29. If $f(x) = x - 5$, find $f(2), f(0), f(-1), f(a)$.
30. If $f(x) = x + 4$, find $f(-3), f(-2), f(0), f(b)$.
31. If $f(x) = 2x - 3$, find $f(1), f(2), f(\frac{1}{2}), f(z)$.
32. If $f(x) = 5x - 6$, find $f(-3), f(-5), f(\frac{2}{5}), f(c)$.
33. If $f(x) = x^2 + 1$, find $f(2), f(3), f(4), f(a)$.
34. If $f(x) = 2x^2 - 3$, find $f(0), f(-3), f(2), f(a + 1)$.
35. If $f(x) = 1/x$, find $f(-1), f(-2), f(3), f(\frac{1}{2}), f(\frac{4}{3})$.
36. If $f(x) = \sqrt{x}$, find $f(4), f(a), f(16), f(b)$.
37. If $f(x) = x^2 + 2x$, find $f(3), f(-2), f(a), f(a - 2)$.
38. If $f(x) = 2x^2 + 3x$, find $f(5), f(-3), f(b), f(a - b)$.
39. If $f(x) = x^3 - 1$, find $f(1), f(2), f(-3), f(a)$.
40. If $f(x) = x^3 + 4$, find $f(0), f(-3), f(-2), f(a + 1)$.

Review Problems The following problems review material we covered in Chapter 7 and Chapter 8.

Find each root.

41. $\sqrt{49}$

42. $\sqrt[3]{-8}$

Write in simplified form for radicals.

43. $\sqrt{50}$

44. $2\sqrt{18x^2y^3}$

45. $\sqrt{\frac{2}{5}}$

46. $\sqrt[3]{\frac{1}{2}}$

Perform the indicated operations.

47. $3\sqrt{12} + 5\sqrt{27}$

48. $\sqrt{3}(\sqrt{3} - 4)$

49. $(\sqrt{6} + 2)(\sqrt{6} - 5)$

50. $(\sqrt{x} + 3)^2$

Rationalize the denominator.

51. $\dfrac{8}{\sqrt{5} - \sqrt{3}}$

52. $\dfrac{\sqrt{5} - \sqrt{2}}{\sqrt{5} + \sqrt{2}}$

Solve for x.

53. $\sqrt{2x - 5} = 3$

54. $\sqrt{x + 15} = x + 3$

Solve each equation. Use any convenient method.

55. $(2x + 1)^2 = 25$

56. $x^2 - 2x - 8 = 0$

57. $3x^2 = 4x + 2$

58. $x^2 - 5x - 5 = 0$

59. $(2x - 3)(2x - 1) = -4$

60. Solve by completing the square: $x^2 + 4x - 3 = 0$

Perform the indicated operations.

61. $(4 + 3i) - (2 - 5i)$

62. $2i(1 - 3i)$

63. $(4 + 5i)(2 - 3i)$

64. $\dfrac{8}{2 - 3i}$

Chapter 9 Summary and Review

COMPOUND INEQUALITIES [9.1]

Two inequalities connecting by the word "and" or "or" form a compound inequality. If the connecting word is "or," we graph all points that are on either graph. If the connecting word is "and," we graph only those points that are common to both graphs. The inequality $-2 \leq x \leq 3$ is equivalent to the compound inequality $-2 \leq x$ and $x \leq 3$.

Examples

1. $x < -3$ or $x > 1$

-3 0 1

$-2 \leq x \leq 3$

-2 0 3

TO GRAPH A LINEAR INEQUALITY IN TWO VARIABLES [9.2]

Step 1: Replace the inequality symbol with an equal sign. The resulting equation represents the boundary for the solution set.

Step 2: Graph the boundary found in Step 1, using a *solid line* if the original inequality symbol was either $\leq$ or $\geq$. Use a *broken line* otherwise.

Step 3: Choose any convenient point not on the boundary and substitute the coordinates into the *original* inequality. If the resulting statement is *true,* the graph lies on the *same* side of the boundary as the chosen point. If the resulting statement is *false,* the solution set lies on the *opposite* side of the boundary.

2. Graph $x - y > 3$.

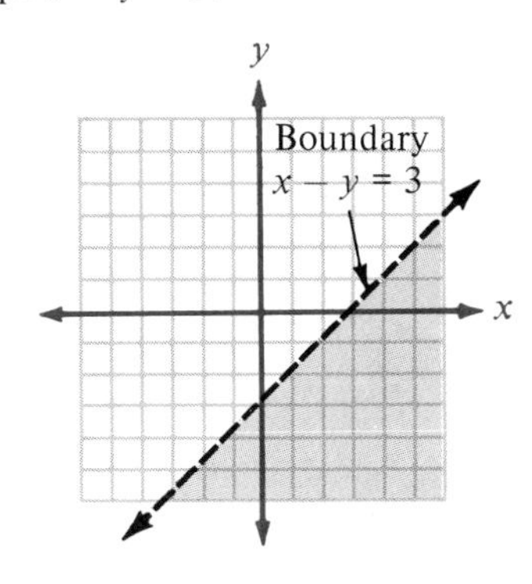

DISTANCE FORMULA [9.3]

The distance between the two points (x_1, y_1) and (x_2, y_2) is given by the formula

$$d = \sqrt{(x_2 - x_1)^2 + (y_2 - y_1)^2}$$

3. The distance between (5, 2) and (−1, 1) is

$$d = \sqrt{(5 + 1)^2 + (2 - 1)^2} = \sqrt{37}$$

THE CIRCLE [9.3]

The graph of any equation of the form

$$(x - a)^2 + (y - b)^2 = r^2$$

will be a circle having its center at (a, b) and a radius of r.

4. The graph of the circle $(x - 3)^2 + (y + 2)^2 = 25$ will have its center at $(3, -2)$ and the radius will be 5.

FRACTIONAL EXPONENTS [9.4]

Fractional exponents are used to specify roots and avoid radical notation. For any positive number a, we have

$$a^{m/n} = \sqrt[n]{a^m}$$

where m is the power and n represents the root.

5. $8^{2/3} = (\sqrt[3]{8})^2 = 2^2 = 4$
$9^{3/2} = (\sqrt{9})^3 = 3^3 = 27$

FACTORING THE SUM AND DIFFERENCE OF TWO CUBES [9.5]

We factor the sum of two cubes and the difference of two cubes according to the following two formulas:

$$a^3 + b^3 = (a + b)(a^2 - ab + b^2)$$
$$a^3 - b^3 = (a - b)(a^2 + ab + b^2)$$

6. $x^3 - 8 = x^3 - 2^3 = (x - 2)(x^2 + 2x + 4)$

$x^3 + 125 = x^3 + 5^3 = (x + 5)(x^2 - 5x + 25)$

DIRECT VARIATION [9.6]

The variable y is said to vary directly with the variable x if $y = Kx$, where K is a real number.

7. If y varies directly with the square of x, then

$$y = Kx^2$$

INVERSE VARIATION [9.6]

The variable y is said to vary inversely with the variable x if $y = K/x$, where K is a real number constant.

8. If y varies inversely with the cube of x, then

$$y = \frac{K}{x^3}$$

FUNCTIONS [9.7]

A *function* is any set of ordered pairs in which no two different ordered pairs have the same first coordinates. The *domain* of the function is the set of all first coordinates. The *range* is the set of all second coordinates.

The notation $f(x)$ is read "f of x" and is defined to be the value of the function f at x.

9. If $f(x) = 3x - 5$

then $f(0) = 3(0) - 5 = -5$

$f(2) = 3(2) - 5 = 1$

$f(3) = 3(3) - 5 = 4$

Chapter 9 Test

Graph each of the following inequalities. [9.1]

1. $x < 2$ or $x > 3$

2. $x \geq -2$ and $x \leq 3$

Solve each inequality and graph the solution. [9.1]

3. $3 - 4x \geq -5$ or $2x \geq 10$

4. $-7 < 2x - 1 < 9$

Graph each linear inequality in two variables. [9.2]

5. $y < -x + 4$

6. $3x - 4y \geq 12$

Find the distance between each pair of points. [9.3]

7. $(1, 2)$ $(4, 6)$

8. $(-4, 1)$ $(2, 5)$

9. Write the equation of the circle with center at $(2, -3)$ and radius 5. [9.3]

10. Graph the circle $(x - 2)^2 + (y - 3)^2 = 9$. [9.3]

Simplify each of the following expressions. [9.4]

11. $25^{1/2}$

12. $8^{2/3}$

13. $x^{3/4}x^{1/4}$

14. $(27x^6y^9)^{1/3}$

Factor each of the following. [9.5]

15. $x^3 - 7^3$

16. $x^3 + 27$

17. $(2x)^3 + (5y)^3$

18. $8x^3 - 27y^3$

19. Suppose y varies directly with the square of x. If x is 3 when y is 36, find y when x is 5. [9.6]

20. If y varies inversely with x, and x is 3 when y is 6, find y when x is 9. [9.6]

21. State the domain and range for the function $\{(1, 2), (3, 4), (5, 6)\}$. [9.7]

22. State the domain for the function $y = \frac{5}{x^2 - 25}$. [9.7]

If $f(x) = 3x^2 - 1$ and $g(x) = 5x + 2$, find the following: [9.7]

23. $f(3)$

24. $f(-2)$

25. $g(0)$

26. $g(-5)$

Answers to Odd-Numbered Exercises and Chapter Tests

CHAPTER 1

PROBLEM SET 1.1

1. The sum of 7 and 8 is 15 **3.** 7 is less than 10 **5.** The difference of 8 and 3 is not equal to 6.
7. The product of 2 and x is less than 20. **9.** The sum of x and 1 is equal to 5. **11.** $x + 5 = 14$ **13.** $5y < 30$
15. $x + 3 = 5 - y$ **17.** $3y \leq y + 6$ **19.** $\frac{x}{3} = x + 2$ **21.** T **23.** F **25.** F **27.** T **29.** 9
31. 49 **33.** 8 **35.** 64 **37.** 16 **39.** 100 **41.** 121 **43.** 11 **45.** 16 **47.** 17 **49.** 42
51. 30 **53.** 30 **55.** 24 **57.** 20 **59.** 27 **61.** 35 **63.** 13 **65.** 13 **67.** 16 **69.** 16
71. 81 **73.** 41 **75.** 345 **77.** 2345 **79.** 75 **81.** 2 **83.** 58 **85.** $2(10 + 4)$ **87.** $3(50) - 14$

PROBLEM SET 1.2

1.–9.

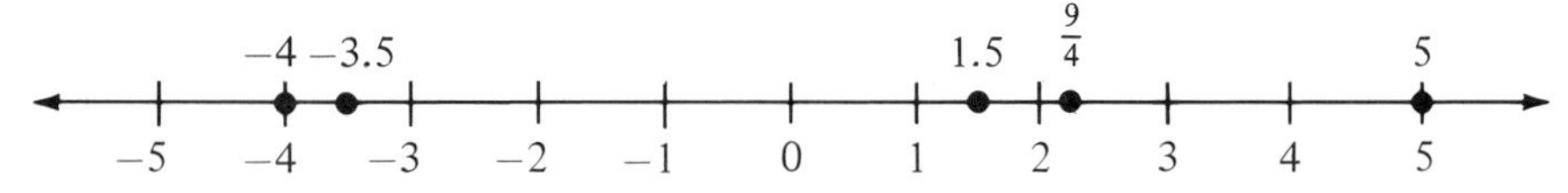

11. 0, 1 **13.** $-3, -2.5, 0, 1, \frac{3}{2}$ **15.** 9 **17.** $-10, -8, \frac{2}{3}, 4.5, 9$ **19.** $-10, \frac{1}{10}, 10$ **21.** $-\frac{3}{4}, \frac{4}{3}, \frac{3}{4}$
23. $-\frac{11}{2}, \frac{2}{11}, \frac{11}{2}$ **25.** $3, -\frac{1}{3}, 3$ **27.** $\frac{2}{5}, -\frac{5}{2}, \frac{2}{5}$ **29.** $-x, \frac{1}{x}$, the distance between x and 0 on the number line.
31. $<$ **33.** $>$ **35.** $>$ **37.** $>$ **39.** $<$ **41.** $<$ **43.** 6 **45.** 22 **47.** 3 **49.** 7 **51.** 3
53. T **55.** F **57.** F **59.** T **61.** F **63.** T **65.** F **67.** $\frac{8}{15}$ **69.** $\frac{3}{2}$ **71.** $\frac{5}{4}$ **73.** 1 **75.** 1
77. 1 **79.** $\frac{9}{16}$ **81.** $\frac{8}{27}$ **83.** $\frac{1}{10,000}$ **85.** $-8, -2$

PROBLEM SET 1.3

1. $3 + 5 = 8, 3 + (-5) = -2, -3 + 5 = 2, -3 + (-5) = -8$ **3.** $15 + 20 = 35, 15 + (-20) = -5, -15 + 20 = 5, -15 + (-20) = -35$ **5.** 3 **7.** -7 **9.** -14 **11.** -3 **13.** -25 **15.** -12 **17.** -19 **19.** -25 **21.** -8 **23.** -4 **25.** 6 **27.** 6 **29.** 8 **31.** -4 **33.** -14 **35.** -17 **37.** 4 **39.** 3 **41.** 15 **43.** -8 **45.** 12 **47.** 8 **49.** 4 **51.** 6 **53.** -7 **55.** -7 **57.** -10 **59.** -10 **61.** -25 **63.** -25 [A rule that summarizes the results of Problems 53–64 is: $-(a + b) = -a + (-b)$.] **65.** $5 + 9 = 14$ **67.** $[-7 + (-5)] + 4 = -8$ **69.** $[-2 + (-3)] + 10 = 5$ **71.** 3 **73.** -3 **75.** $-12 + 4$ **77.** $10 + (-6) + (-8) = -4$

PROBLEM SET 1.4

1. -3 **3.** -6 **5.** 0 **7.** -10 **9.** -16 **11.** -12 **13.** -7 **15.** 35 **17.** 0 **19.** -4 **21.** 4 **23.** -24 **25.** -28 **27.** 25 **29.** 4 **31.** 7 **33.** 17 **35.** 8 **37.** 4 **39.** 18 **41.** 10 **43.** 17 **45.** 1 **47.** 1 **49.** 27 **51.** -26 **53.** -2 **55.** 68 **57.** 3 **59.** -4 **61.** 0 **63.** $-7 - 4 = -11$ **65.** $12 - (-8) = 20$ **67.** $-5 - (-7) = 2$ **69.** $[4 + (-5)] - 17 = -18$ **71.** $8 - 5 = 3$ **73.** $-8 - 5 = -13$ **75.** $8 - (-5) = 13$ **77.** 10 **79.** -2 **81.** $1500 - 730$ **83.** $-35 + 15 - 20 = -40$

PROBLEM SET 1.5

1. Commutative **3.** Multiplicative inverse **5.** Commutative **7.** Distributive **9.** Commutative, associative **11.** Commutative, associative **13.** Commutative **15.** Commutative, associative **17.** Commutative **19.** Additive inverse **21.** $3x + 6$ **23.** $9a + 9b$ **25.** 0 **27.** 0 **29.** 10 **31.** $(4 + 2) + x = 6 + x$ **33.** $x + (2 + 7) = x + 9$ **35.** $(3 \cdot 5)x = 15x$ **37.** $(9 \cdot 6)y = 54y$ **39.** $(\frac{1}{2} \cdot 3)a = \frac{3}{2}a$ **41.** $(\frac{1}{3} \cdot 3)x = x$ **43.** $(\frac{1}{2} \cdot 2)y = y$ **45.** $(\frac{3}{4} \cdot \frac{4}{3})x = x$ **47.** $(\frac{6}{5} \cdot \frac{5}{6})a = a$ **49.** $8(x) + 8(2) = 8x + 16$ **51.** $8(x) - 8(2) = 8x - 16$ **53.** $4(y) + 4(1) = 4y + 4$ **55.** $3(6x) + 3(5) = 18x + 15$ **57.** $2(3a) + 2(7) = 6a + 14$ **59.** $9(6y) - 9(8) = 54y - 72$ **61.** $\frac{1}{2}(3x) - \frac{1}{2}(6) = \frac{3}{2}x - 3$ **63.** $\frac{1}{3}(3x) + \frac{1}{3}(6) = x + 2$ **65.** $3x + 3y$ **67.** $8a - 8b$ **69.** $12x + 18y$ **71.** $12a - 8b$ **73.** $3x + 2y$ **75.** $4a + 25$ **77.** $6x + 12$ **79.** $14x + 38$ **81.** No **83.** No, not commutative **85.** $8 \div 4 \neq 4 \div 8$

PROBLEM SET 1.6

1. -42 **3.** -21 **5.** -16 **7.** 3 **9.** 121 **11.** 6 **13.** -60 **15.** 24 **17.** 49 **19.** -27 **21.** 6 **23.** 10 **25.** 9 **27.** 45 **29.** 14 **31.** -2 **33.** 216 **35.** -2 **37.** -18 **39.** 29 **41.** -35 **43.** 9 **45.** 5 **47.** -38 **49.** -5 **51.** 37 **53.** 80 **55.** $-\frac{10}{21}$ **57.** -4 **59.** 1 **61.** 1 **63.** $\frac{9}{16}$ **65.** $-\frac{8}{27}$ **67.** $-8x$ **69.** $42x$ **71.** x **73.** x **75.** $-4a - 8$ **77.** $-\frac{3}{2}x + 3$ **79.** $-6x + 8$ **81.** $-15x - 30$ **83.** -25 **85.** $2(-4x) = -8x$ **87.** -26

PROBLEM SET 1.7

1. -2 **3.** -5 **5.** $-\frac{1}{3}$ **7.** -3 **9.** -9 **11.** $-\frac{1}{3}$ **13.** 3 **15.** $\frac{1}{7}$ **17.** $\frac{2}{5}$ **19.** 0 **21.** $\frac{16}{15}$ **23.** $\frac{4}{3}$ **25.** $-\frac{8}{13}$ **27.** -1 **29.** 1 **31.** $\frac{3}{5}$ **33.** $-\frac{5}{3}$ **35.** -2 **37.** -3 **39.** Undefined **41.** Undefined **43.** 5 **45.** $-\frac{7}{3}$ **47.** -1 **49.** -7 **51.** $\frac{15}{17}$ **53.** $-\frac{32}{17}$ **55.** $\frac{1}{3}$ **57.** 1 **59.** 1 **61.** -2 **63.** $\frac{9}{7}$ **65.** $\frac{16}{11}$ **67.** -1 **69.** -1 **71.** -7 **73.** $-\frac{4}{3}$ **75.** 3 **77.** -10 **79.** -3 **81.** -8

CHAPTER 1 TEST

1. The sum of 3 and 4 is 7. **2.** The difference of 8 and 2 is less than 10. **3.** 40 **4.** 16 **5.** Opposite 4, reciprocal $-\frac{1}{4}$, absolute value 4 **6.** Opposite $-\frac{3}{4}$, reciprocal $\frac{4}{3}$, absolute value $\frac{3}{4}$ **7.** -4 **8.** 17 **9.** -12 **10.** 0 **11.** c **12.** e **13.** d **14.** a **15.** -21 **16.** 64 **17.** -2 **18.** $-\frac{8}{27}$ **19.** 4 **20.** 204 **21.** 25 **22.** 52 **23.** 2 **24.** 2 **25.** $8 + 2x$ **26.** $-2 + x$ **27.** x **28.** $10x$ **29.** $6x + 10$ **30.** $-3x - 12$ **31.** $-10x + 5$ **32.** $-2x + 1$ **33.** $1, -8$ **34.** $1, 1.5, \frac{3}{4}, -8$ **35.** $\sqrt{2}$ **36.** all of them **37.** $8 + (-3) = 5$ **38.** $-24 - 2 = -26$ **39.** $-5(-4) = 20$

40. $\dfrac{-24}{-2} = 12$

CHAPTER 2

PROBLEM SET 2.1

1. $-3x$ **3.** $-a$ **5.** $12x$ **7.** $6a$ **9.** $6x - 3$ **11.** $7a + 5$ **13.** $5x - 5$ **15.** $4a + 2$ **17.** $-9x - 2$ **19.** $12a + 3$ **21.** $10x - 1$ **23.** $21y + 6$ **25.** $-6x + 8$ **27.** $-2a + 3$ **29.** $-4x + 26$ **31.** $4y - 16$ **33.** $-6x - 1$ **35.** $5x - 12$ **37.** $7y - 39$ **39.** $-21x - 24$ **41.** 5 **43.** 6 **45.** -9 **47.** 4 **49.** 4 **51.** -5 **53.** -5 **55.** 16 **57.** -28 **59.** -37 **61.** -41 **63.** -19 **65.** -8 **67.** 64 **69.** 64 **71.** 144 **73.** 144 **75.** Three times the quantity $5x$ plus 1, minus 6 **77.** Four times the quantity $2x$ plus $3y$, minus 1 **79.** Four times the quantity $2x$ plus $3y$ minus 1 **81.** 10 **83.** 17 **85.** -17 **87.** 2

PROBLEM SET 2.2

1. 11 **3.** 4 **5.** -9 **7.** -8 **9.** -17 **11.** 3 **13.** -4 **15.** -1 **17.** 1 **19.** 5 **21.** 3 **23.** 13 **25.** 21 **27.** 7 **29.** 6 **31.** 22 **33.** -6 **35.** 0 **37.** -2 **39.** -16 **41.** -3 **43.** 10 **45.** -12 **47.** -1 **49.** 4 **51.** 2 **53.** -5 **55.** -1 **57.** -3 **59.** 8 **61.** -8 **63.** 5 **65.** -11 **67.** $18x$ **69.** x **71.** y **73.** x **75.** a

PROBLEM SET 2.3

1. 2 **3.** 4 **5.** $-\frac{1}{2}$ **7.** -2 **9.** 3 **11.** 4 **13.** 0 **15.** 0 **17.** 6 **19.** -50 **21.** 6 **23.** 12 **25.** -20 **27.** 32 **29.** -8 **31.** $\frac{1}{2}$ **33.** 4 **35.** 3 **37.** -4 **39.** 4 **41.** -15 **43.** $-\frac{1}{2}$ **45.** 3 **47.** 1 **49.** $\frac{1}{4}$ **51.** -3 **53.** 3 **55.** 2 **57.** $-\frac{3}{2}$ **59.** 1 **61.** -2 **63.** Any method of solution results in a false statement. **65.** All methods of solution result in true statements. **67.** 10 **69.** 240 **71.** 25% **73.** 35% **75.** 64 **77.** 2000 **79.** $10x - 43$ **81.** $-3x - 13$ **83.** $-6y + 4$ **85.** $-5x + 7$

PROBLEM SET 2.4

1. 3 **3.** -2 **5.** -1 **7.** 2 **9.** -4 **11.** -2 **13.** 0 **15.** 1 **17.** $\frac{1}{2}$ **19.** 7 **21.** 8 **23.** -3 **25.** $-\frac{1}{3}$ **27.** 0 **29.** 2 **31.** 6 **33.** 8 **35.** 0 **37.** $\frac{3}{7}$ **39.** 1 **41.** All real numbers **43.** No solution **45.** All real numbers **47.** No solution **49.** $\frac{3}{2}$ **51.** 4 **53.** 1 **55.** $6x - 10$ **57.** $\frac{3}{2}x + 3$ **59.** $-x + 2$

PROBLEM SET 2.5

1. 100 feet **3.** 0 **5.** 2 **7.** 7 meters **9.** $\frac{3}{2}$ or 1.5 inches **11.** 15 **13.** 10 **15.** 132 feet **17.** $\frac{2}{9}$ centimeters **19.** 4 **21.** 2 **23.** $l = \dfrac{A}{w}$ **25.** $r = \dfrac{d}{t}$ **27.** $h = \dfrac{V}{lw}$ **29.** $R = \dfrac{V}{I}$ **31.** $P = \dfrac{nRT}{V}$ **33.** $h = \dfrac{V}{\pi r^2}$ **35.** $a^2 = c^2 - b^2$ **37.** $a = P - b - c$ **39.** $x = 3y - 1$ **41.** $y = 3x + 6$ **43.** $y = -\frac{2}{3}x + 2$ **45.** $y = -2x + 4$ **47.** $y = \frac{5}{2}x - \frac{3}{2}$ **49.** $w = \dfrac{P - 2l}{2}$ **51.** $C = \frac{5}{9}(F - 32)$ **53.** $v = \dfrac{h - 16t^2}{t}$ **55.** $h = \dfrac{A - \pi r^2}{2\pi r}$ **57.** $b = \dfrac{2A}{h} - B$ **59.** $y = \frac{3}{4}x - 3$ **61.** $x = -2y + 7$ **63.** 25 **65.** 376.8 square centimeters **67.** 552.64 cubic centimeters **69.** $2(6 + 4) = 20$ **71.** $2 \cdot 6 + 4 = 16$ **73.** $x = .12(500) = 60$ **75.** $x = .10(2000) = 200$ **77.** $160 = .08x$; $x = 2000$ **79.** $220 = .11x$; $x = 2000$

PROBLEM SET 2.6

Along with the answers to the odd-numbered problems in this problem set we are including the equations used to solve each problem. Be sure that you try the problems on your own before looking here to see what the correct equations are.

1. $x + 5 = 13$; 8 **3.** $2x + 4 = 14$; 5 **5.** $5(x + 7) = 30$; -1
7. The two numbers are x and $x + 2$; $x + (x + 2) = 8$; 3 and 5

9. The two numbers are x and $3x - 4$; $x + (3x - 4) + 5 = 25$; 6 and 14
11. Barney's age is x, Fred's age is $x + 4$; $(x - 5) + (x - 1) = 48$; Barney is 27, Fred 31.
13. Lacy's age is x, Jack's age is $2x$; $(x + 3) + (2x + 3) = 54$; Lacy is 16, Jack is 32.
15. Let x = Patrick's age now; then $x + 20$ = Pat's age now; $x + 22 = 2(x + 2)$; Patrick is 18, Pat is 38.
17. The width is x, the length is $x + 5$; $2x + 2(x + 5) = 34$; 6 inches and 11 inches
19. The length of a side is x; $4x = 48$; 12 meters
21. The width is x, the length is $2x - 3$; $2x + 2(2x - 3) = 54$; 10 inches and 17 inches
23. If the number of nickels is x, then the number of dimes is $x + 9$; $5x + 10(x + 9) = 210$; 8 nickels, 17 dimes.
25. If you have x dimes and $2x$ quarters, then $10x + 25(2x) = 900$; 15 dimes, 30 quarters.
27. If she has x nickels, then she has $x + 3$ dimes and $x + 5$ quarters; $5x + 10(x + 3) + 25(x + 5) = 435$; 7 nickels, 10 dimes, 12 quarters.
29. If x = the amount of money invested at 8%, then $x + 2000$ is the amount invested at 9%; $.08x + .09(x + 2000) = 860$; \$4,000 invested at 8%, \$6,000 invested at 9%.
31. If x = the amount invested at 10%, then $x + 500$ is the amount invested at 12%; $.10x + .12(x + 500) = 214$; \$700 invested at 10%, \$1200 invested at 12%.
33. Let x = the amount invested at 8%, $2x$ = the amount invested at 9%, and $3x$ = the amount invested at 10%; $.08x + .09(2x) + .10(3x) = 280$; \$500 at 8%, \$1,000 at 9%, \$1500 at 10%.
35. 4 is less than 10 **37.** 9 is greater than or equal to -5 **39.** $<$ **41.** $<$ **43.** 2 **45.** 12

PROBLEM SET 2.7

1. $x < 12$

3. $a \leq 12$

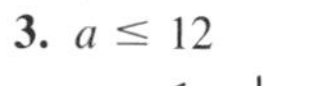

5. $x > 13$

7. $y \geq 4$

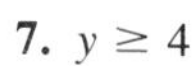

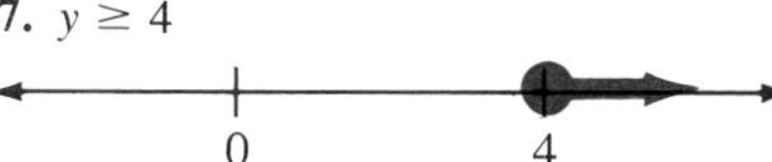

9. $x > 9$

11. $x < 2$

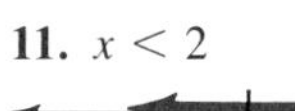

13. $a \leq 5$

15. $x > 15$

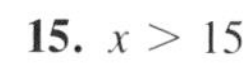

17. $x < -3$

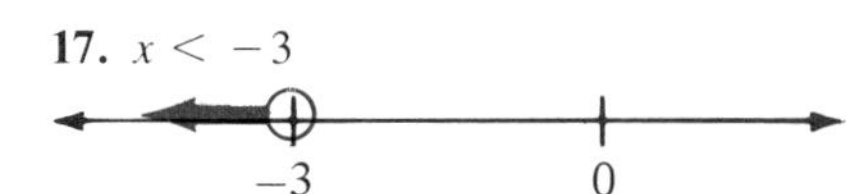

19. $x \leq 6$

21. $x \geq -50$

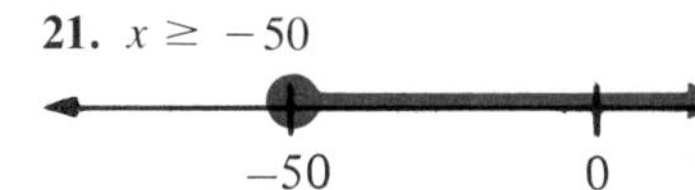

23. $y < -6$

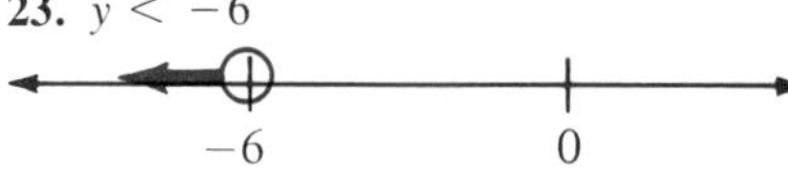

25. $x < 6$

27. $y \geq -5$

29. $x < 3$

31. $x \le 18$

33. $a < -20$

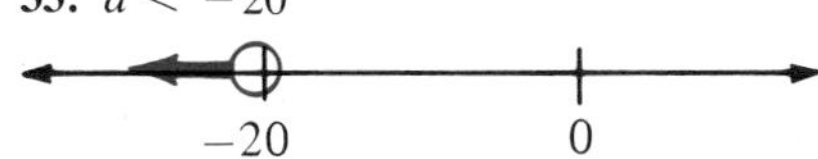

35. $y < 25$

37. $a \le 3$

39. $x > \frac{15}{2}$

41. $x < -1$

43. $y \ge -2$

45. $x < -1$

47. $m \le -6$

49. $x \le -5$

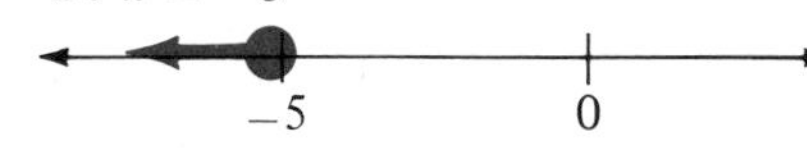

51. $y < -\frac{3}{2}x + 3$ **53.** $y < \frac{2}{5}x - 2$ **55.** $y \le \frac{3}{7}x + 3$ **57.** $y \le \frac{1}{2}x + 1$ **59.** $2x + 6 < 10; x < 2$
61. $4x > x - 8; x > -\frac{8}{3}$ **63.** $2(x + 5) \le 12; x \le 1$ **65.** $3x - 5 < x + 7; x < 6$
67. $2x + 2(3x) \ge 48; x \ge 6$; the width is at least 6 meters
69. $x + (x + 2) + (x + 4) > 24; x > 6$; the shortest side is even and greater than 6 inches
71. $x \ge .12(3000); x \ge \$360$ **73.** $x \ge .08(2000) + .10(3000); x \ge \$460$ **75.** 0, 2
77. $-3, -\frac{5}{3}, 0, 2, \frac{15}{2}$ **79.** *B* **81.** *A* **83.** *B* and *C*

CHAPTER 2 TEST

1. $-4x + 5$ **2.** $3a - 4$ **3.** $-3y - 12$ **4.** $11x + 28$ **5.** 22 **6.** 25 **7.** 6 **8.** $\frac{4}{3}$ **9.** 2
10. $-\frac{9}{2}$ **11.** -3 **12.** 2 **13.** -3 **14.** -1 **15.** 5.7 **16.** 2000 **17.** 3 **18.** $\frac{28}{3}$
19. $y = -\frac{2}{5}x + 4$ **20.** $v = \dfrac{h - x - 16t^2}{t}$ **21.** Rick is 20; Dave is 40. **22.** Width is 10 inches, length is 20 inches **23.** 8 quarters, 15 dimes **24.** \$800 at 7%, \$1400 at 9%

25. $x < 1$

26. $a < -4$

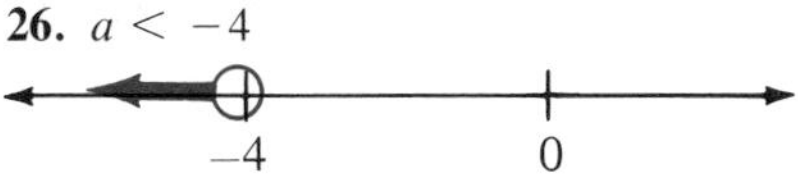

27. $x \le -3$

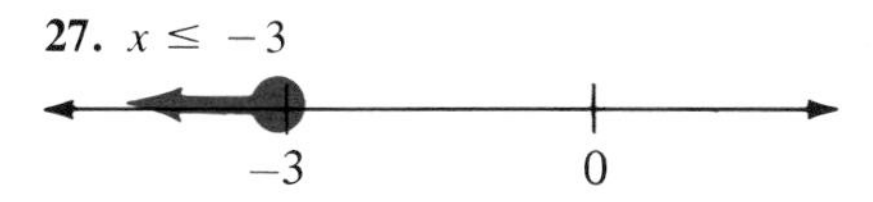

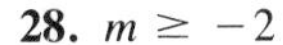

28. $m \ge -2$

CHAPTER 3

PROBLEM SET 3.1

1. (0, 6), (3, 0), (6, −6) **3.** (0, 3), (4, 0), (−4, 6) **5.** (1, 1), ($\frac{3}{4}$, 0), (5, 17) **7.** (2, 13), (1, 6), (0, −1)
9. (−5, 4)(−5, −3)(−5, 0)

11.

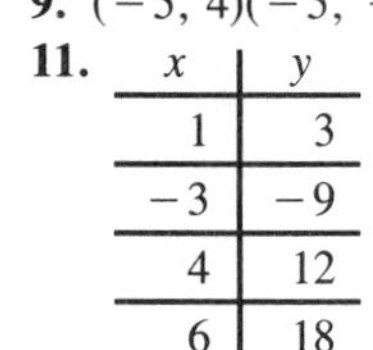

x	y
1	3
−3	−9
4	12
6	18

13.

x	y
0	0
$-\frac{1}{2}$	−2
−3	−12
3	12

15.

x	y
2	3
3	2
5	0
9	−4

17.

x	y
2	0
3	2
1	−2
−3	−10

19.

x	y
0	−1
−1	−7
−3	−19
$\frac{3}{2}$	8

21. (0, −2) **23.** (1, 5), (0, −2), (−2, −16) **25.** (1, 6), (−2, −12), (0, 0) **27.** (2, −2)
29. (3, 0), (3, −3) **31.** 12 inches **33.** (5, 10) **35.** (3, 3), (−3, 3), (5, 5), (−5, 5)
37. −3 **39.** 2 **41.** 0 **43.** $y = -5x + 4$ **45.** $y = \frac{3}{2}x - 3$

PROBLEM SET 3.2

1–17.

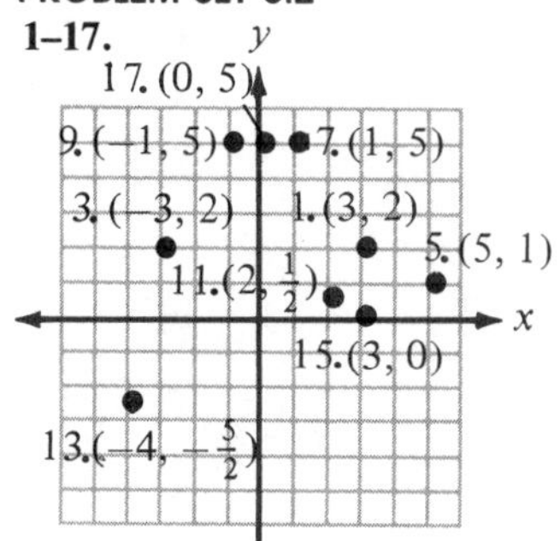

19.

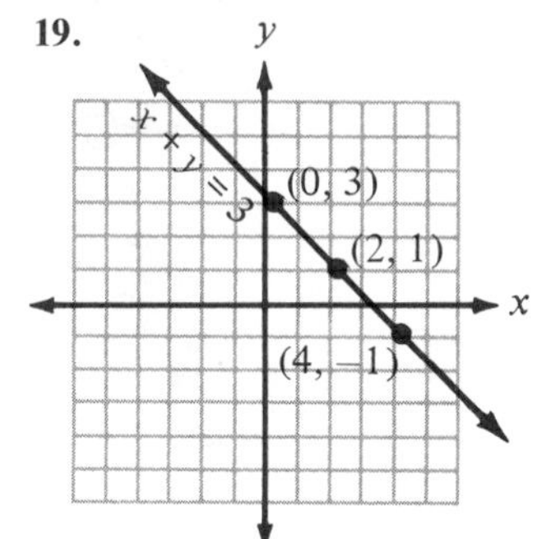

21.

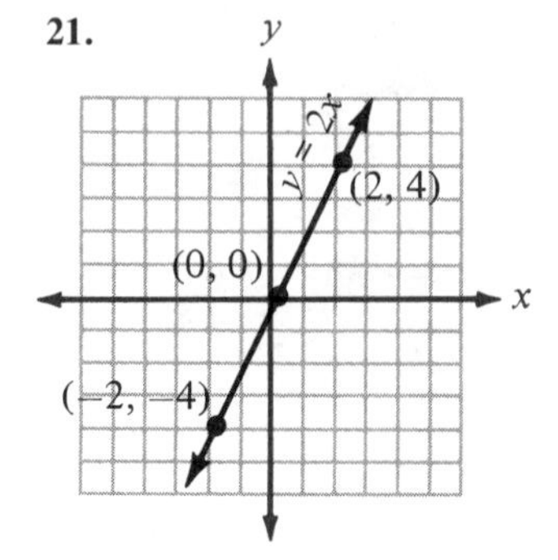

23.

(3, 1)
(−3, −1)
(0, 0)
y = 1/3 x

25.

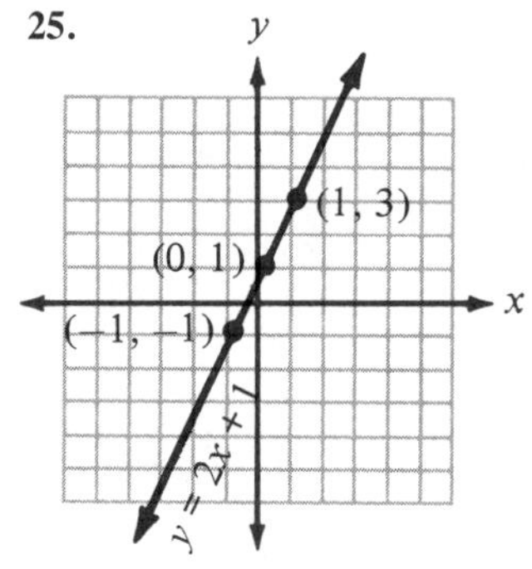

27.

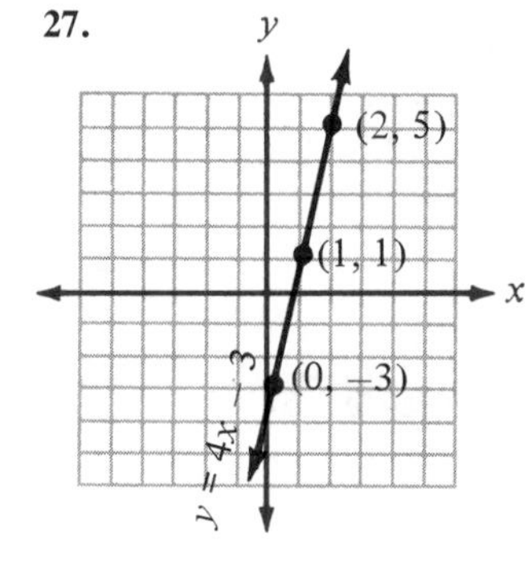

29.

y = 1/2 x + 3
(2, 4)
(0, 3)
(−2, 2)

31.

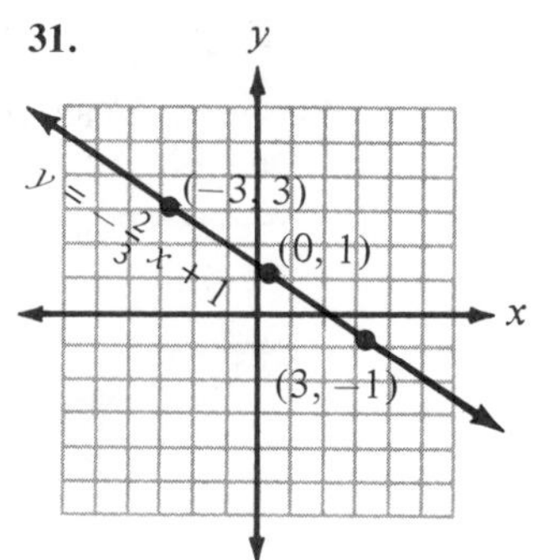

33.

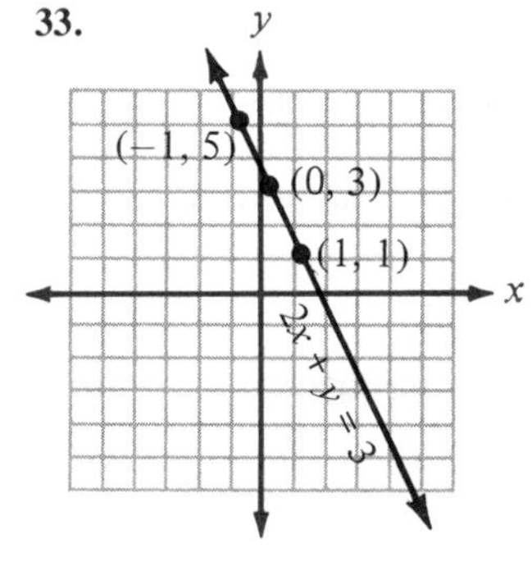

35.

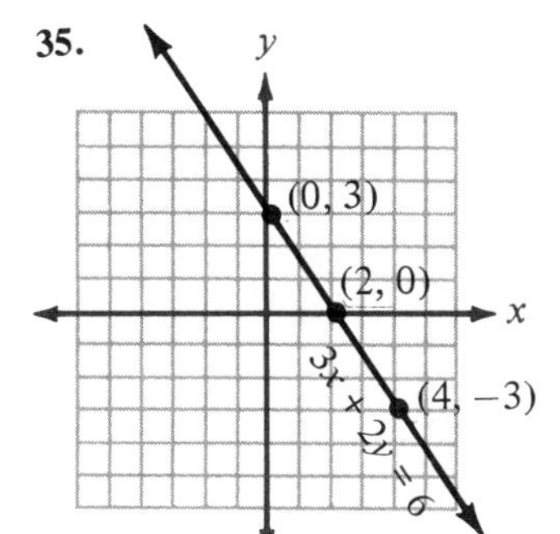

37.

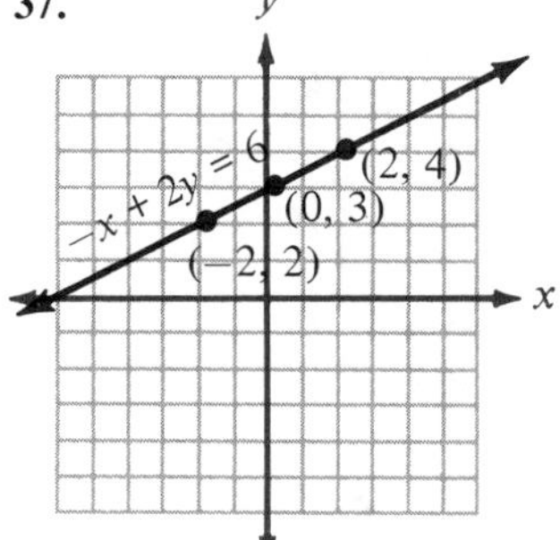

39.

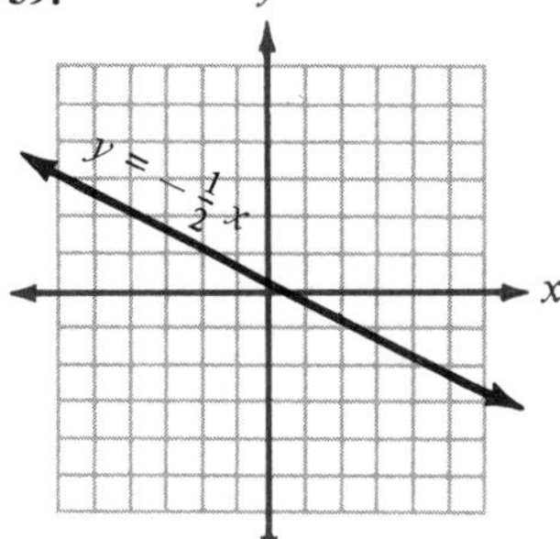

41.

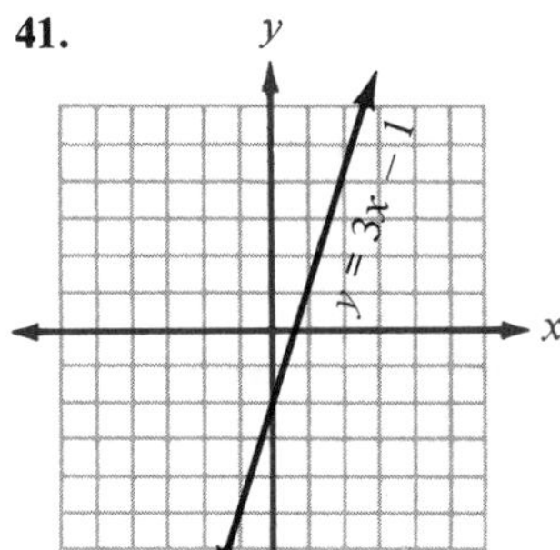

43.

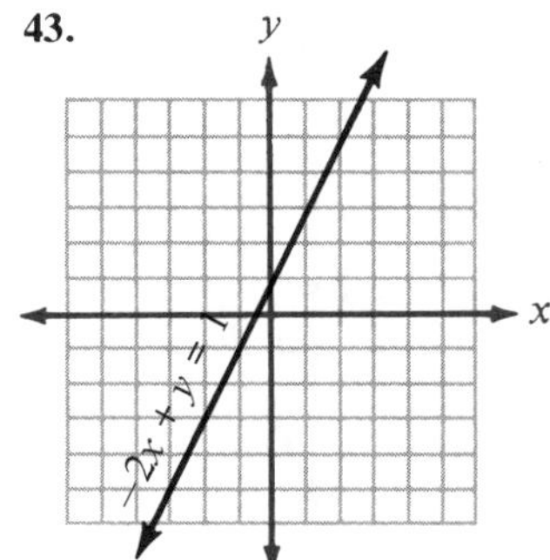

45.

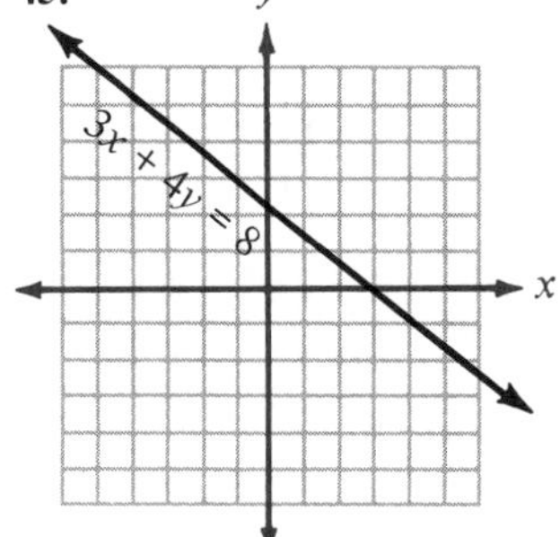

47.

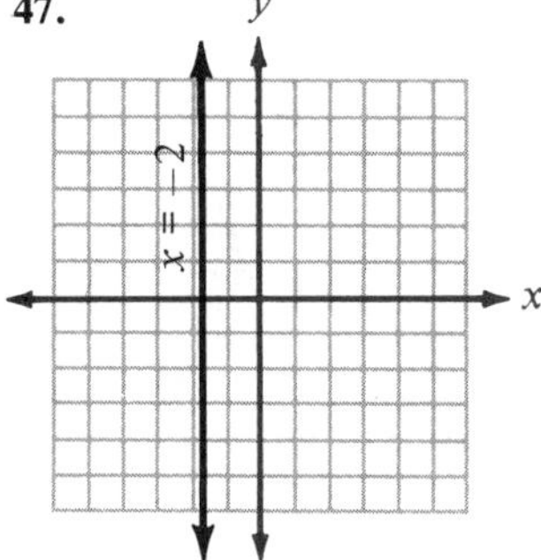

49.

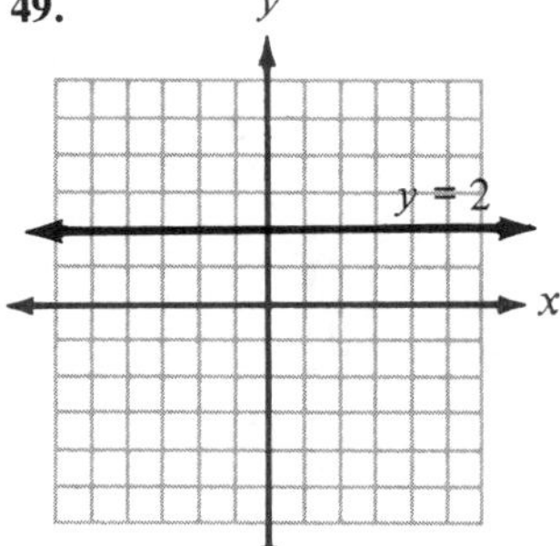

51.

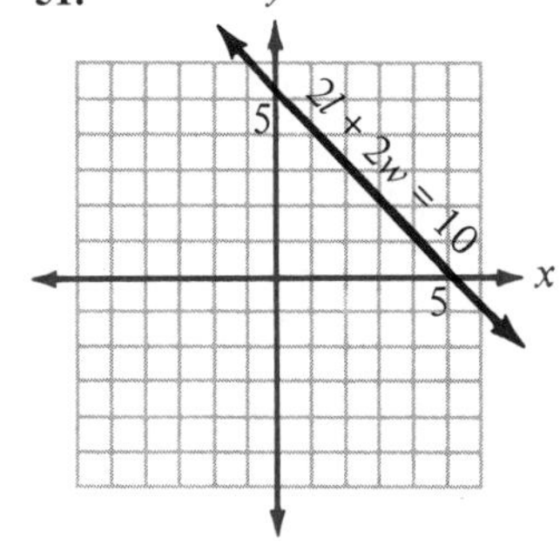

53.

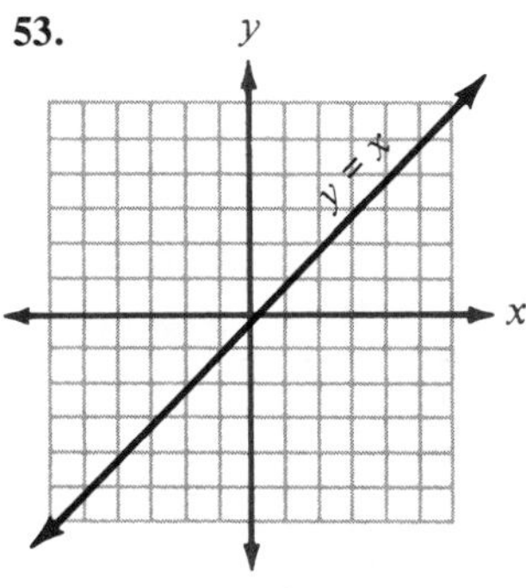

55.

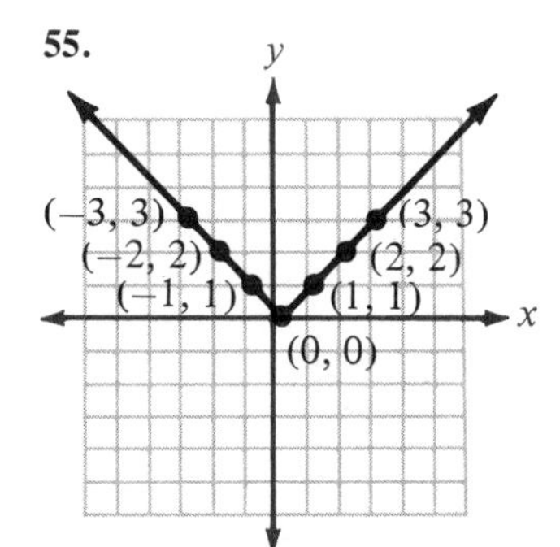

57.

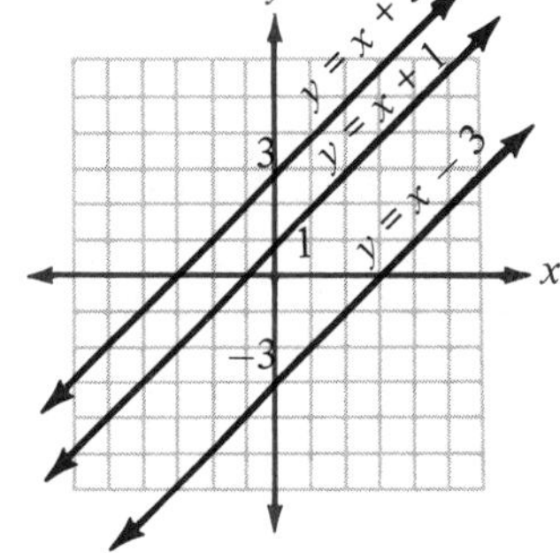

59. 5 **61.** -2 **63.** 3

PROBLEM SET 3.3

1.

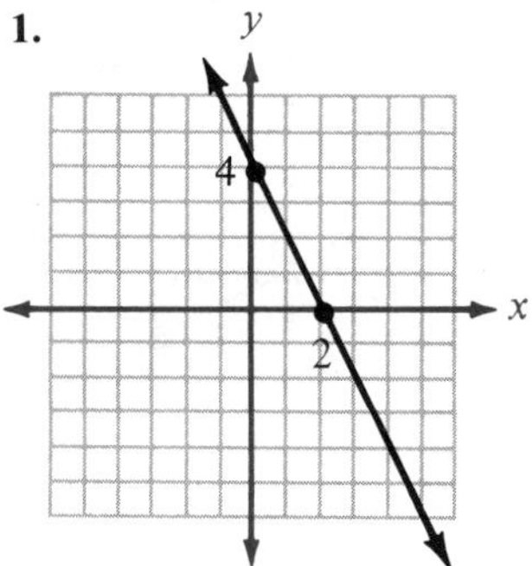

3.

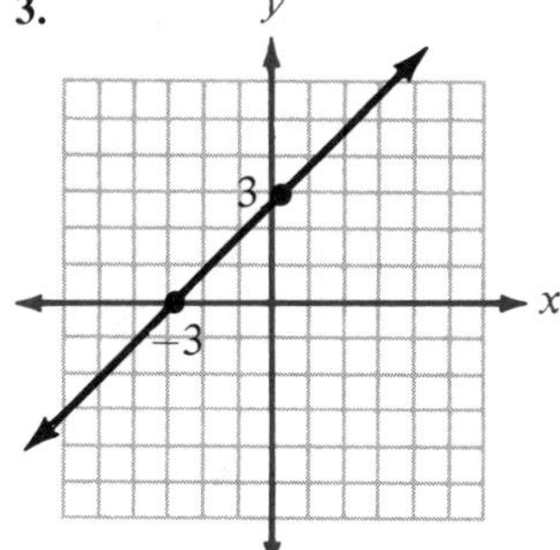

5.

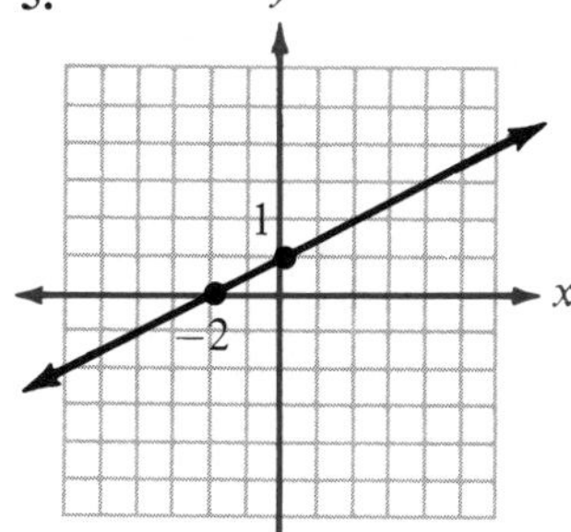

7.

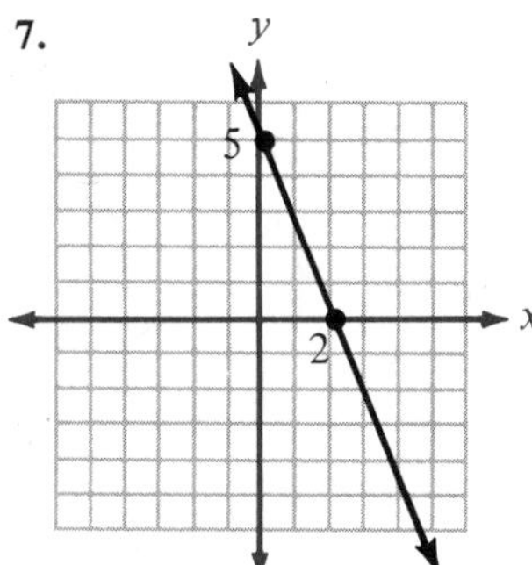

9.

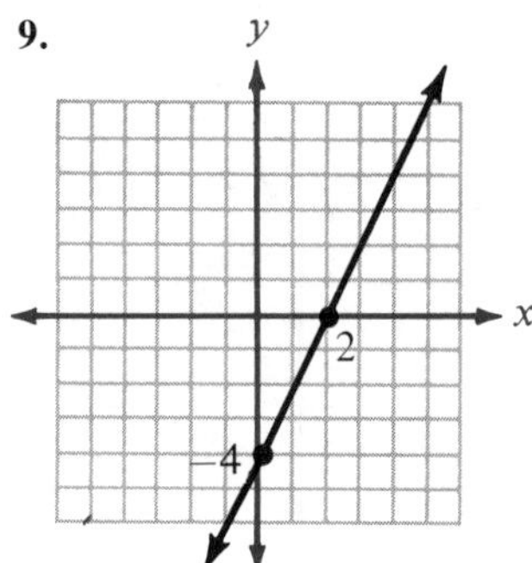

11.

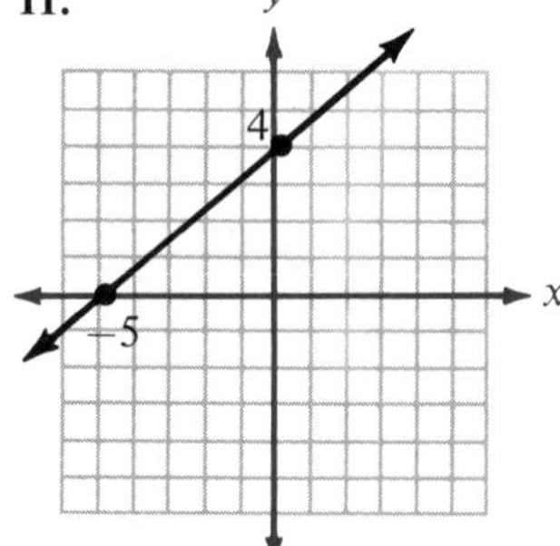

13.

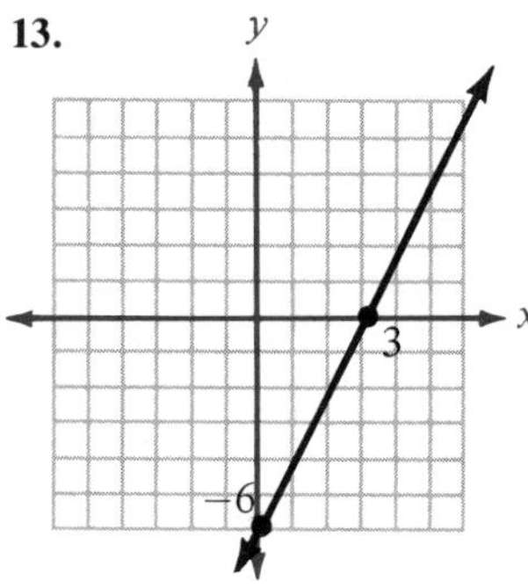

15.

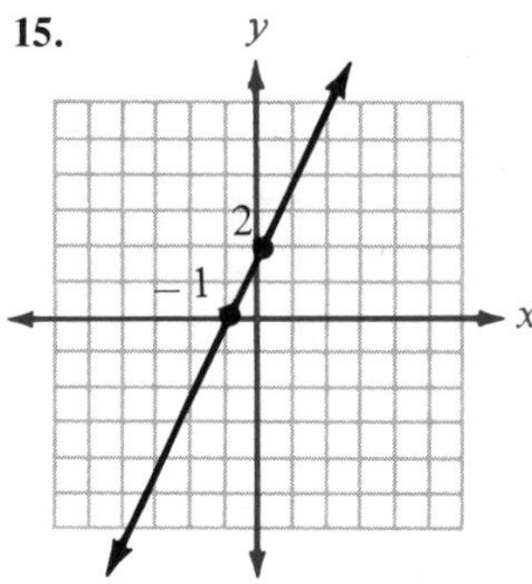

17.

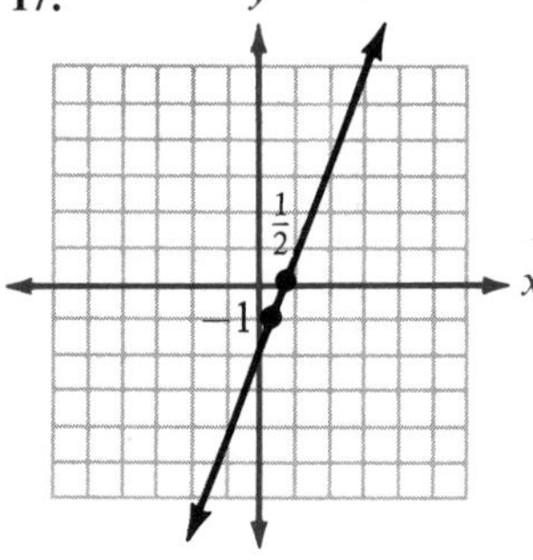

19.

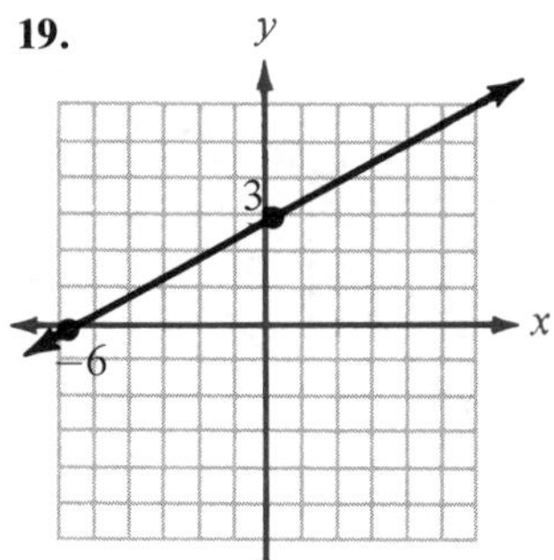

21.

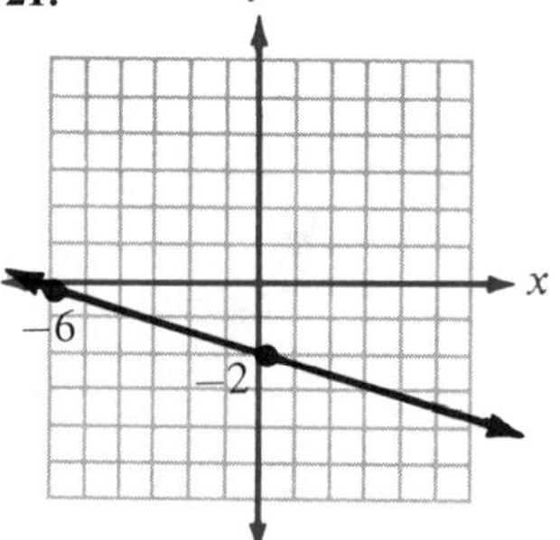

23.

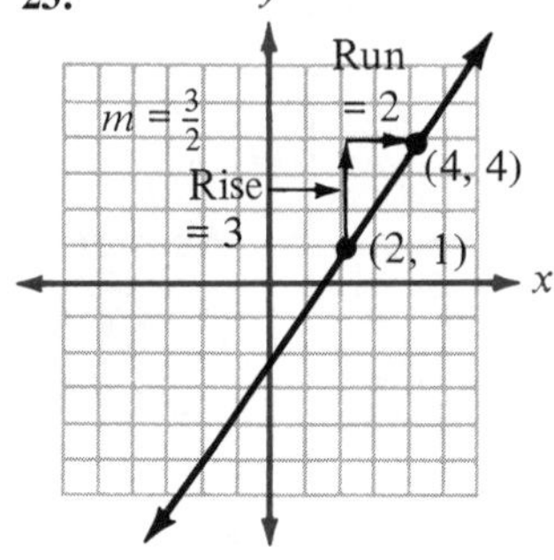

25.
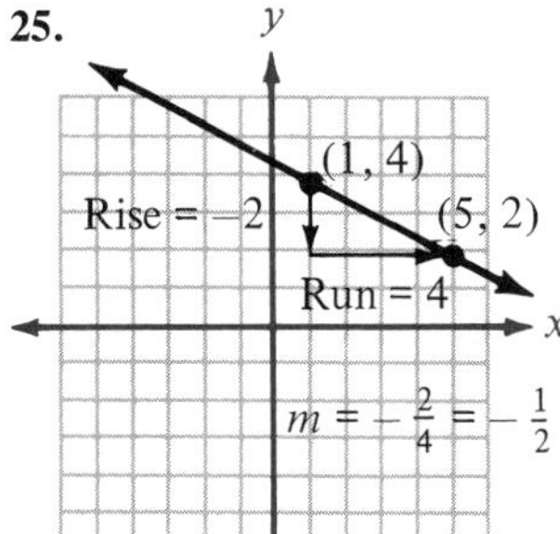

27.
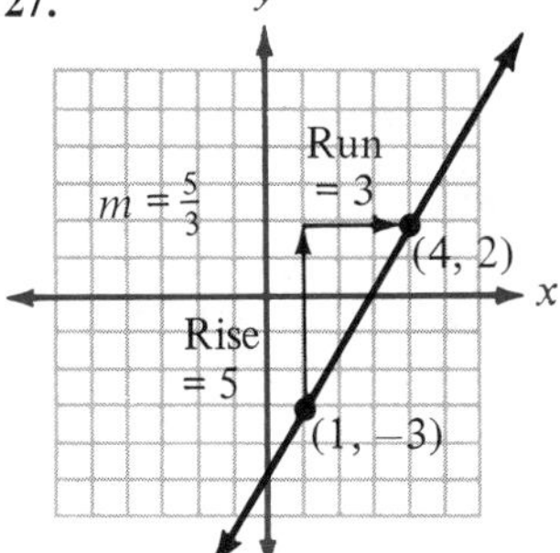

29.
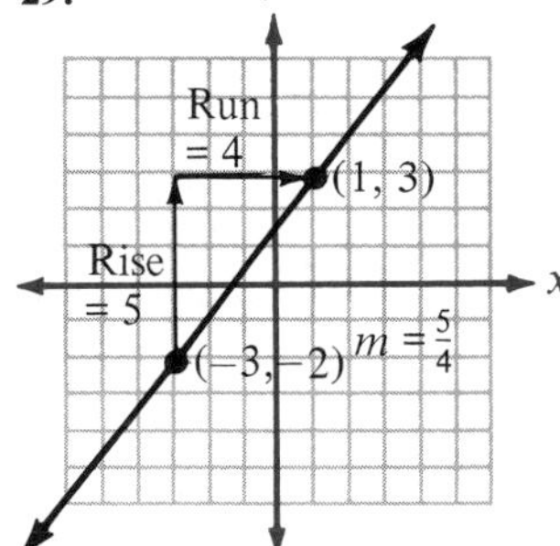

31.
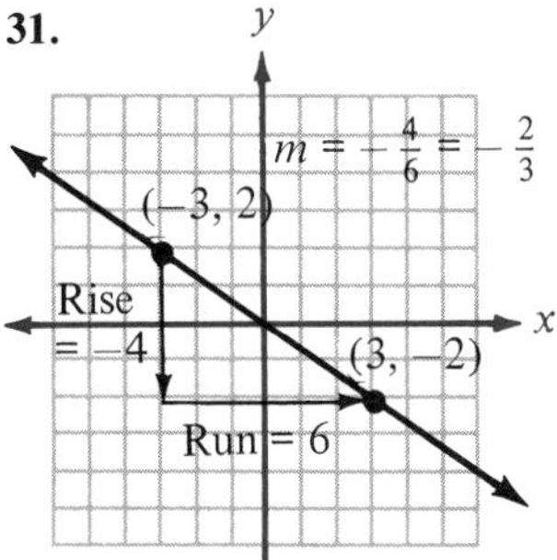

33.
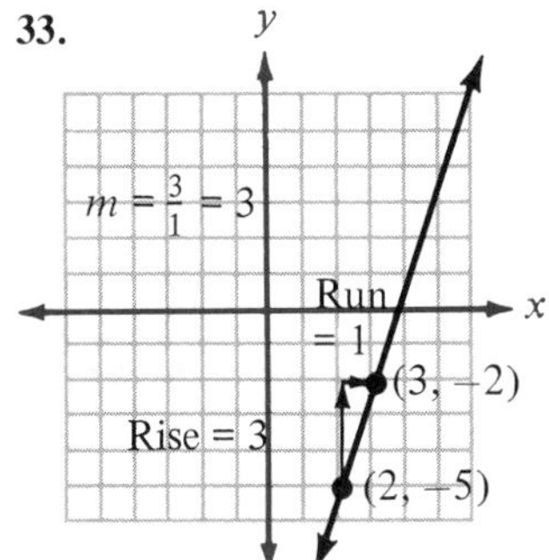

35.
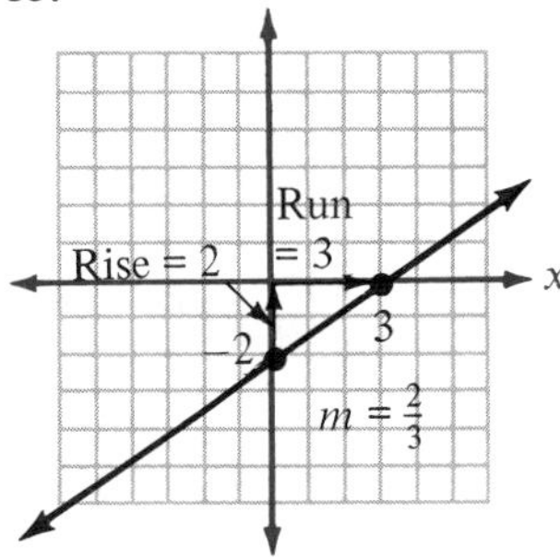

37.
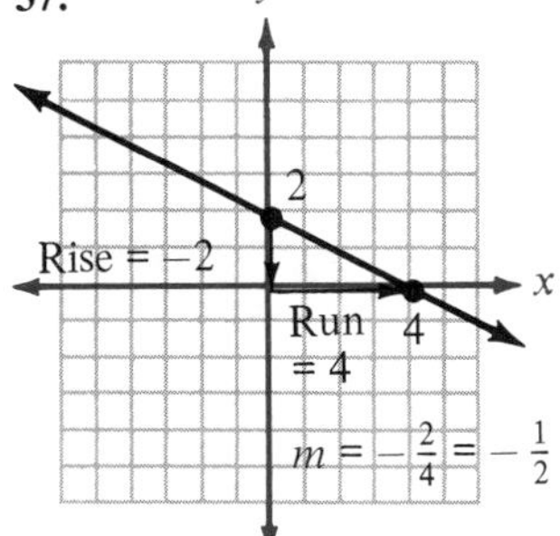

39.
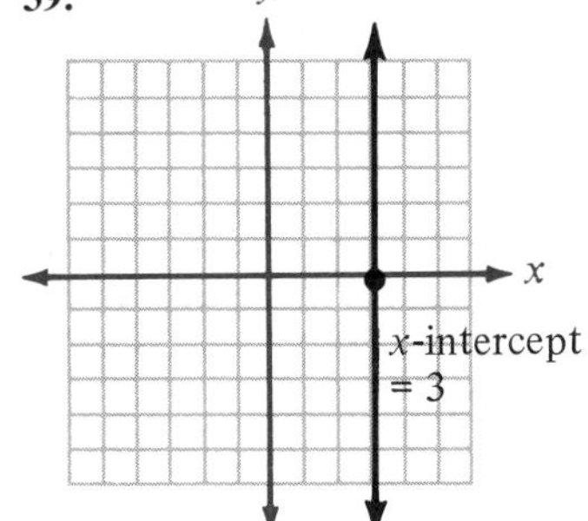

41.
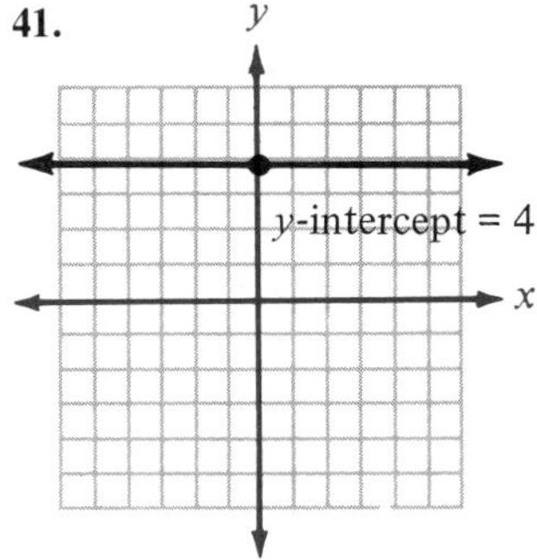

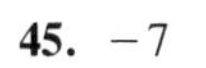

43. 49 **45.** −7 **47.** −12 **49.** −18 **51.** −17

PROBLEM SET 3.4

1.
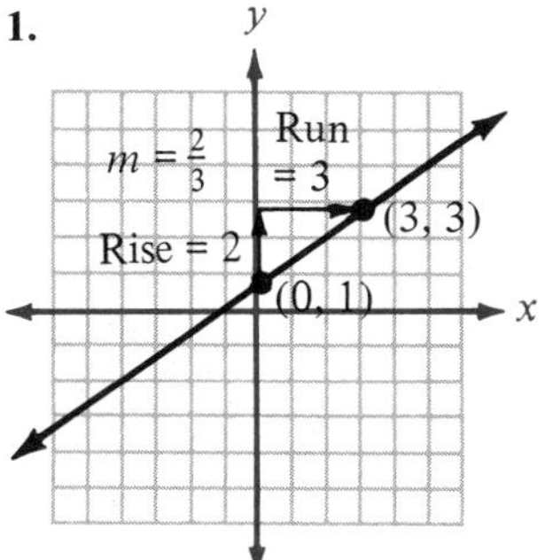

3.
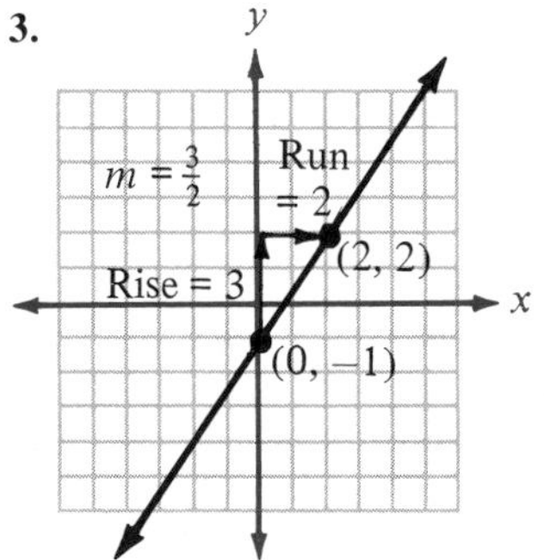

5.
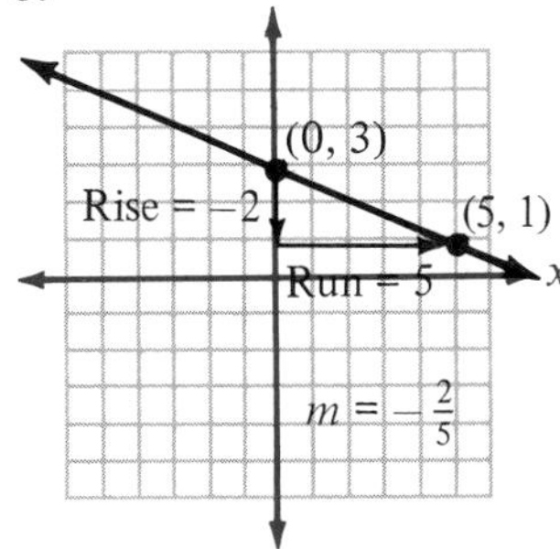

7.

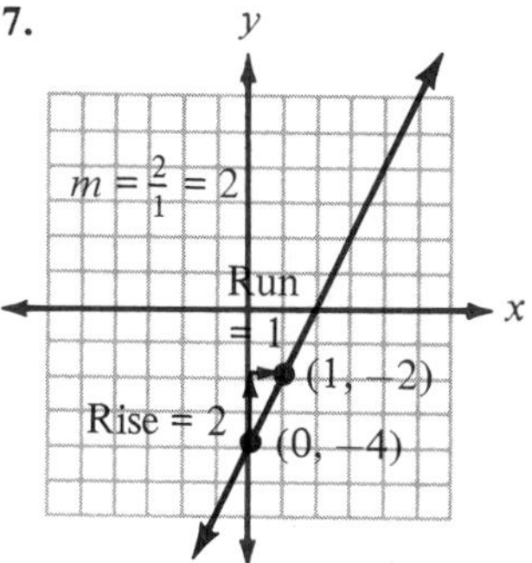

9.

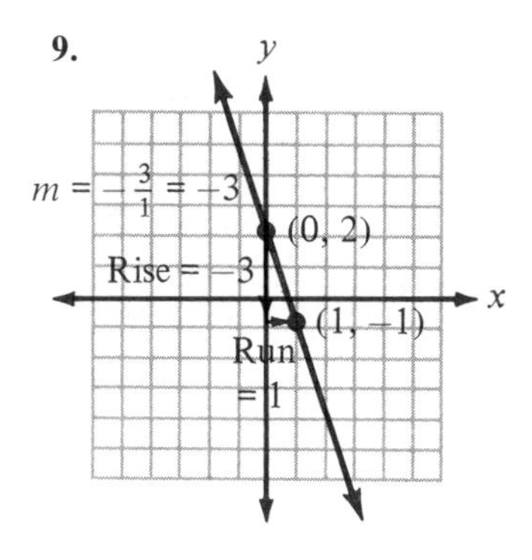

11.

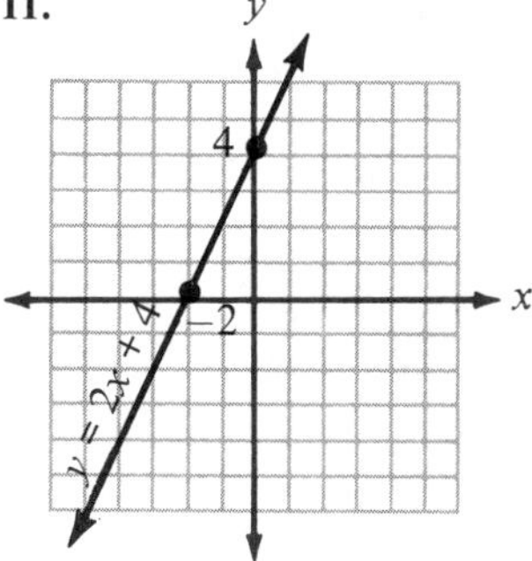

13.

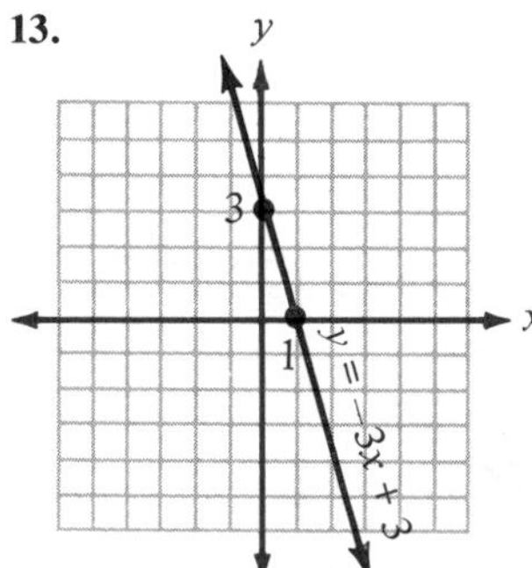

15.

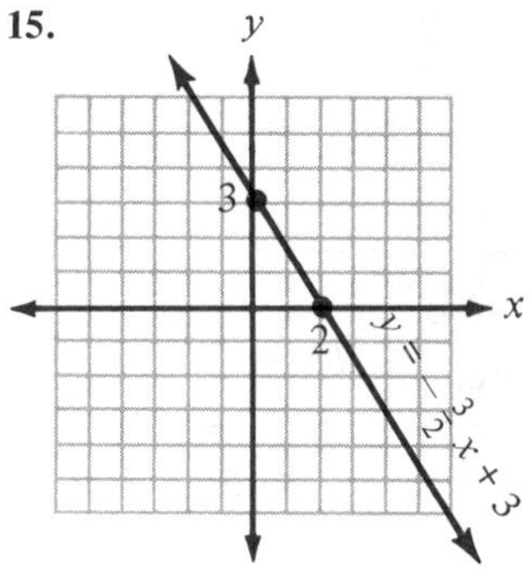

17.

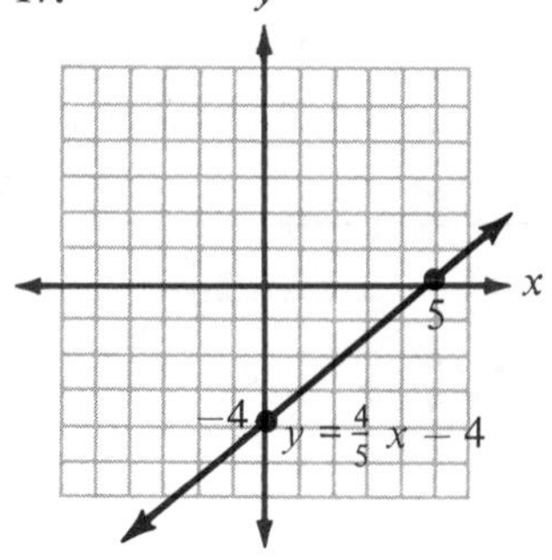

19.

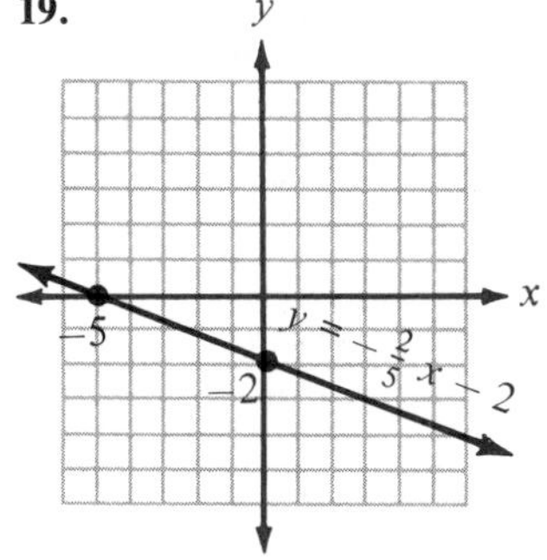

21. $y = 2x - 1$ **23.** $y = -\frac{1}{2}x - 1$ **25.** $y = \frac{3}{2}x - 6$ **27.** $y = -3x + 1$ **29.** $y = x + 2$
31. $y = x - 2$ **33.** $y = 2x - 3$ **35.** $y = \frac{4}{3}x + 2$ **37.** $y = -\frac{2}{3}x - 3$ **39.** $y = -x - 1$
41. $y = -\frac{2}{3}x + 2$ **43.** $y = -\frac{5}{2}x - 5$ **45.** $x = 3$ **47.** $y = 3$ **49.** 75 **51.** 20%
53. 400 **55.** $2240

PROBLEM SET 3.5

1. (2, 1) **3.** (−1, 2) **5.** (3, 5) **7.** (4, 3) **9.** (0, −6) **11.** (1, 0) **13.** (0, 0) **15.** (−5, −6)
17. (−1, −1) **19.** (−3, 2) **21.** (−3, 5) **23.** (−4, 6) **25.** Lines are parallel; there is no solution to the system. **27.** Lines coincide; any solution to one equation is a solution to the other. **29.** −15 **31.** 27
33. −27 **35.** 13

PROBLEM SET 3.6

1. (2, 1) **3.** (3, 7) **5.** (2, −5) **7.** (−1, 0) **9.** Lines coincide. **11.** (4, 8) **13.** (2, 3) **15.** (1, 0) **17.** (−1, −2) **19.** (5, −1) **21.** (−4, 5) **23.** (−3, −10) **25.** (3, 2) **27.** (6, 5) **29.** (4, −3) **31.** (2, 2) **33.** Lines are parallel. **35.** (1, 1) **37.** Lines are parallel. **39.** (7, 5) **41.** (10, 12) **43.** $x < 5$ **45.** $x \leq -4$ **47.** $x \geq 3$ **49.** $x > -4$

PROBLEM SET 3.7

1. (4, 7) **3.** (3, 17) **5.** (−1, −2) **7.** (2, 4) **9.** (0, 4) **11.** (−1, 3) **13.** (1, 1) **15.** (2, −3) **17.** (8, −2) **19.** (−3, 5) **21.** Lines are parallel. **23.** (3, 1) **25.** (3, −3) **27.** (2, 6) **29.** (4, 4) **31.** (5, −2) **33.** (18, 10) **35.** Lines coincide. **37.** 4 **39.** 3 meters; 9 meters **41.** 12 nickels, 15 dimes **43.** 480 **45.** \$800 at 8%, \$1600 at 10%

PROBLEM SET 3.8

As you can see, in addition to the answers to the problems we have sometimes included the system of equations used to solve the problems. Remember, you should attempt the problem on your own before looking here to check your answers or equations.

1. $x + y = 25$, $y = x + 5$ The two numbers are 10 and 15. **3.** 3, 12

5. $x - y = 5$, $x = 2y + 1$ The two numbers are 9 and 4. **7.** 6, 29

9. Let x = the amount invested at 6% and y = the amount invested at 8%.

$x + y = 20{,}000$
$0.06x + 0.08y = 1380$
He has \$9,000 at 8% and \$11,000 at 6%.

11. \$8000 at 5%, \$2000 at 6% **13.** 8 quarters, 6 nickels

15. Let x = the number of dimes and y = the number of quarters.

$x + y = 21$
$0.10x + 0.25y = 3.45$
He has 12 dimes and 9 quarters.

17. Let x = the number of liters of 50% solution and y = the number of liters of 20% solution.

$x + y = 18$
$0.05x + 0.20y = 0.30(18)$
6 liters of 50% solution
12 liters of 20% solution

19. 10 gallons of 10% solution, 20 gallons of 7% solution **21.** $x < -4$ **23.** $x \leq -3$ **25.** $x \leq -1$ **27.** $x > 1$

CHAPTER 3 TEST

1. (0, −2), (5, 0), (10, 2), $(-\frac{5}{2}, -3)$ **2.** (2, 5), (0, −3)

3.

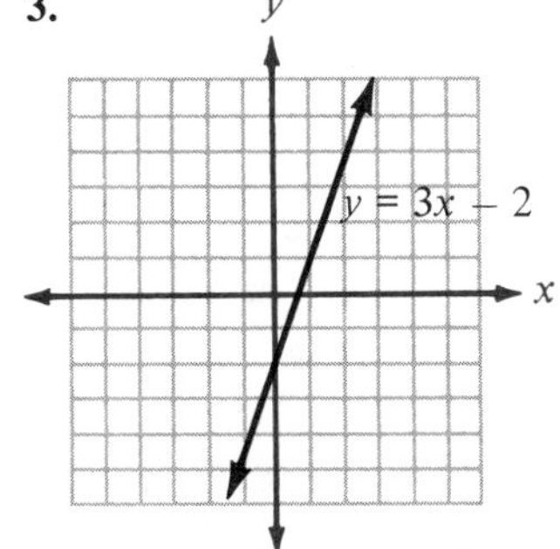

4.

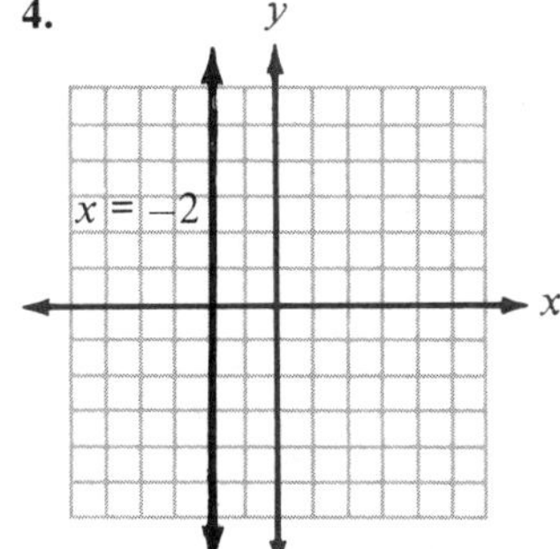

5.

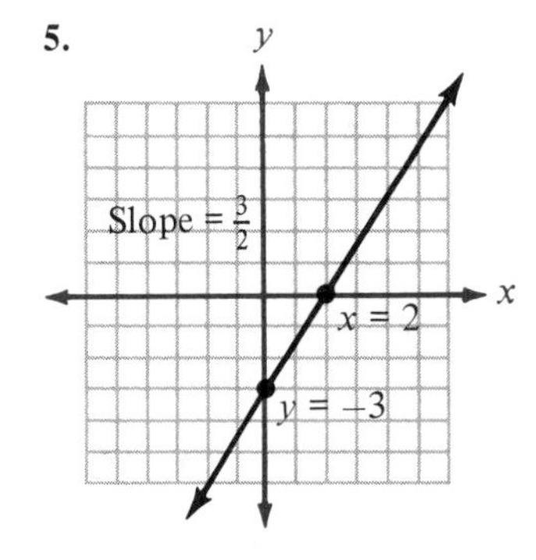

6.

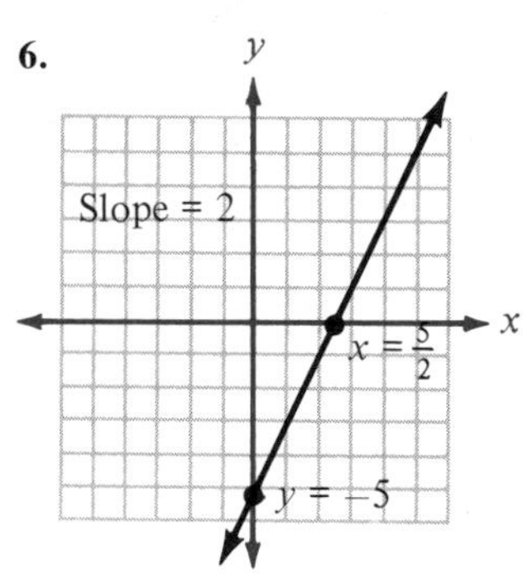

7. $y = 4x + 8$ **8.** $y = 3x + 10$

9.

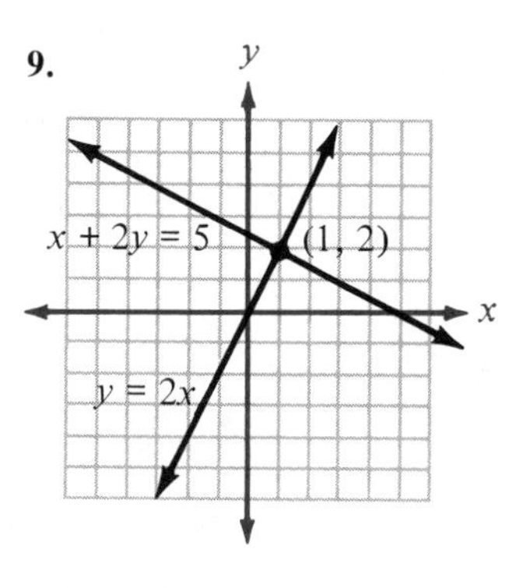

10. $(-3, -4)$ **11.** $(5, -3)$ **12.** $(2, -2)$ **13.** Lines coincide. **14.** $(2, 7)$ **15.** $(4, 3)$ **16.** $(2, 9)$
17. 5, 7 **18.** 3, 12 **19.** \$6,000 **20.** 7 nickels, 5 quarters

CHAPTER 4

PROBLEM SET 4.1

1. Base 4; exponent 2; 16 **3.** Base 7; exponent 2; 49 **5.** Base 4; exponent 3; 64 **7.** Base -5; exponent 2; 25 **9.** Base 2; exponent 3; -8 **11.** Base 3; exponent 4; 81 **13.** Base $\frac{2}{3}$; exponent 2; $\frac{4}{9}$ **15.** Base $\frac{1}{2}$; exponent 4; $\frac{1}{16}$ **17.** x^9 **19.** 7^7 **21.** y^{30} **23.** 2^{12} **25.** x^{28} **27.** x^{10} **29.** 5^{12} **31.** y^9 **33.** 2^{50} **35.** a^{3x} **37.** b^{xy} **39.** $16x^2$ **41.** $32y^5$ **43.** $81x^4$ **45.** $49a^2b^2$ **47.** $64x^3y^3z^3$ **49.** $8x^{12}$ **51.** $16a^6$ **53.** x^{14} **55.** a^{11} **57.** $128x^7$ **59.** $432x^{10}$ **61.** $16x^4y^6$ **63.** $8a^{12}b^3$ **65.** $30x^9$ **67.** $256x^8y^5$ **69.** 4.32×10^4 **71.** 5.7×10^2 **73.** 2.38×10^5 **75.** 2,490 **77.** 352 **79.** 28,000 **81.** 6.5×10^8 **83.** \$740,000 **85.** 100,000; the exponent and the number of zeros are the same. **87.** $(2^3)^2 = 2^6 = 64$; $2^{3^2} = 2^9 = 512$ **89.** b **91.** -3 **93.** 11 **95.** -5 **97.** 5

PROBLEM SET 4.2

1. $\frac{1}{9}$ **3.** $\frac{1}{36}$ **5.** $\frac{1}{64}$ **7.** $\frac{1}{125}$ **9.** $\frac{2}{x^3}$ **11.** $\frac{1}{8x^3}$ **13.** $\frac{1}{25y^2}$ **15.** $\frac{1}{100}$ **17.** 25 **19.** $\frac{1}{25}$ **21.** x^6 **23.** 64 **25.** $8x^3$ **27.** 6^{10} **29.** $\frac{1}{6^{10}}$ **31.** $\frac{1}{2^8}$ **33.** 2^8 **35.** $27x^3$ **37.** $\frac{16}{x^{12}}$ **39.** 1 **41.** $2a^2b$ **43.** $\frac{1}{49y^6}$ **45.** $\frac{1}{x^8}$ **47.** $\frac{1}{y^3}$ **49.** x^2 **51.** a^6 **53.** $\frac{1}{y^9}$ **55.** $\frac{1}{x^3}$ **57.** $\frac{1}{x}$ **59.** x^{18} **61.** a^{16} **63.** $\frac{1}{a^4}$ **65.** x **67.** 3.57×10^{-4} **69.** 3.57×10^4 **71.** 4.8×10^{-3} **73.** 2.5×10^1 **75.** 9×10^{-6} **77.** 0.00423 **79.** 56,000 **81.** 0.00008 **83.** 78.9 **85.** 4.2 **87.** 0.002 **89.** 6×10^{-3} **91.** 2.5×10^4 **93.** $7x$ **95.** $2a$ **97.** $10y$

PROBLEM SET 4.3

1. $12x^7$ **3.** $-16y^{11}$ **5.** $32x^2$ **7.** $200a^6$ **9.** $-24a^3b^3$ **11.** $24x^6y^8$ **13.** $3x$ **15.** $\frac{6}{y^3}$ **17.** $\frac{1}{2a}$ **19.** $-\frac{3a}{b^2}$ **21.** $\frac{x^2}{9z^2}$ **23.** $-\frac{12}{xy^6}$ **25.** 6×10^8 **27.** 1.75×10^{-1} **29.** 1.21×10^{-6} **31.** 4.2×10^3 **33.** 3×10^{10} **35.** 5×10^{-3} **37.** $8x^2$ **39.** $-11x^5$ **41.** 0 **43.** $4x^3$ **45.** $31ab^2$ **47.** $-14abc$ **49.** $\frac{7}{5}$ **51.** $\frac{14}{25}$ **53.** -1 **55.** $\frac{7}{12}$ **57.** $4x^3$ **59.** $\frac{1}{b^2}$ **61.** $\frac{6y^{10}}{x^4}$ **63.** 2×10^6 **65.** 1×10^1 **67.** 4.2×10^{-6} **69.** $9x^3$ **71.** $-20a^2$ **73.** $6x^5y^2$ **75.** 2 **77.** 4 **79.** $(4 + 5)^2 = 9^2 = 81$; $4^2 + 5^2 = 16 + 25 = 41$ **83.** -8 **85.** 9 **87.** 0

89.

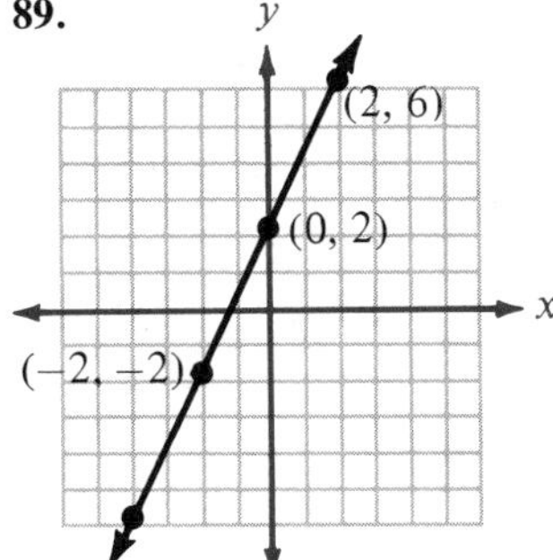

91.

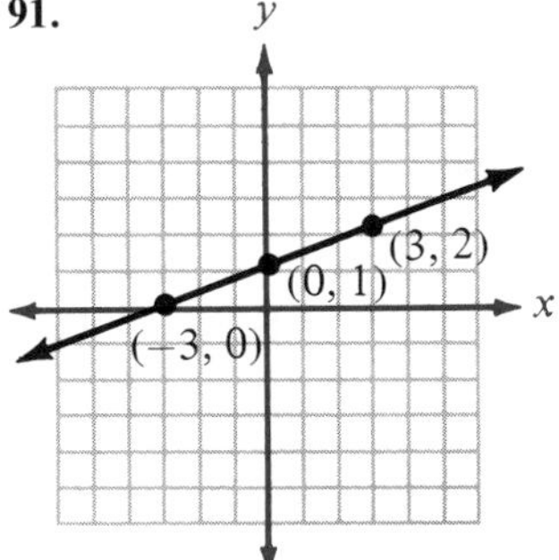

PROBLEM SET 4.4

1. Trinomial, 3 **3.** Trinomial, 3 **5.** Binomial, 1 **7.** Binomial, 2 **9.** Monomial, 2 **11.** Monomial, 0 **13.** $5x^2 + 5x + 9$ **15.** $5a^2 - 9a + 7$ **17.** $x^2 + 6x + 8$ **19.** $6x^2 - 13x + 5$ **21.** $x^2 - 9$ **23.** $3y^2 - 11y + 10$ **25.** $6x^3 + 5x^2 - 4x + 3$ **27.** $8x^2 - 2$ **29.** $2a^2 - 2a - 2$ **31.** $-x^2 - 26x + 4$ **33.** $-4y^2 + 15y - 22$ **35.** $x^2 - 33x + 63$ **37.** $8y^2 + 4y + 26$ **39.** $75x^2 - 150x - 75$ **41.** $12x + 2$ **43.** 4 **45.** 25 **47.** 16 **49.** $10x^2$ **51.** $-15x^2$ **53.** $6x^3$ **55.** $6x^4$

57.

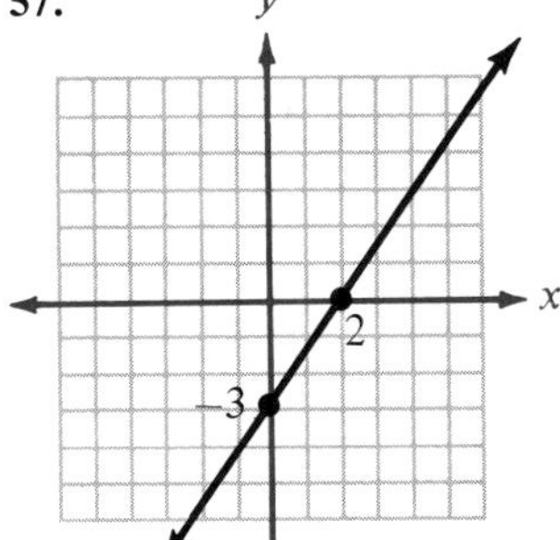

59.

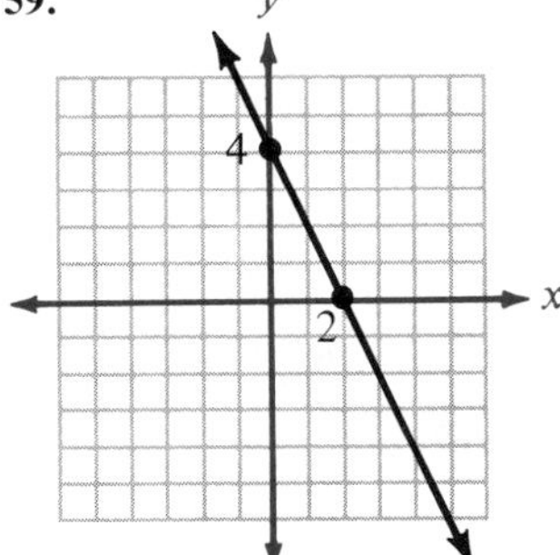

61. $\frac{4}{3}$

PROBLEM SET 4.5

1. $6x^2 + 2x$ **3.** $6x^4 - 4x^3 + 2x^2$ **5.** $2a^3b - 2a^2b^2 + 2ab$ **7.** $3y^4 + 9y^3 + 12y^2$ **9.** $8x^5y^2 + 12x^4y^3 + 32x^2y^4$ **11.** $x^2 + 7x + 12$ **13.** $x^2 + 7x + 6$ **15.** $xy + 4x + 2y + 8$ **17.** $a^2 + 2a - 15$ **19.** $xy + bx - ay - ab$ **21.** $x^2 - 36$ **23.** $y^2 - 4$ **25.** $2x^2 - 11x + 12$ **27.** $2a^2 + 3a - 2$ **29.** $6x^2 - 19x + 10$ **31.** $2ax + 8x + 3a + 12$ **33.** $25x^2 - 16$ **35.** $8x^2 - 13x - 6$ **37.** $3 - 10a + 8a^2$ **39.** $56 - 83x + 30x^2$ **41.** $x^3 + 4x^2 - x - 4$ **43.** $a^3 - 6a^2 + 11a - 6$ **45.** $x^3 + 8$ **47.** $2x^3 + 17x^2 + 26x + 9$ **49.** $5x^4 - 13x^3 + 20x^2 + 7x + 5$ **51.** $A = x(2x + 5) = 2x^2 + 5x$ **53.** $A = x(x + 1) = x^2 + x$ **55.** $R = xp = (1200 - 100p)p = 1200p - 100p^2$ **57.** $R = xp = (1700 - 100p)p = 1700p - 100p^2$ **59.**

y
x
Run = 3
Rise = 2
y-intercept = −1

61. $y = \frac{1}{2}x + 2$; slope $= \frac{1}{2}$, y-intercept $= 2$ **63.** $y = 3x$ **65.** $y = x - 2$

PROBLEM SET 4.6

1. $x^2 - 4x + 4$ **3.** $a^2 + 6a + 9$ **5.** $x^2 - 10x + 25$ **7.** $a^2 - 14a + 49$ **9.** $x^2 + 20x + 100$ **11.** $a^2 + 2ab + b^2$ **13.** $4x^2 - 4x + 1$ **15.** $16a^2 + 40a + 25$ **17.** $9x^2 - 12x + 4$ **19.** $9a^2 + 30ab + 25b^2$ **21.** $16x^2 - 40xy + 25y^2$ **23.** $49m^2 + 28mn + 4n^2$ **25.** $36x^2 - 120xy + 100y^2$ **27.** $x^4 + 10x^2 + 25$ **29.** $a^4 + 2a^2 + 1$ **31.** $x^6 - 14x^3 + 49$ **33.** $x^2 - 9$ **35.** $a^2 - 25$ **37.** $y^2 - 1$ **39.** $81 - x^2$ **41.** $4x^2 - 25$ **43.** $16x^2 - 1$ **45.** $4a^2 - 49$ **47.** $36 - 49x^2$ **49.** $x^4 - 9$ **51.** $a^4 - 16$ **53.** $(50 - 1)(50 + 1) = 2500 - 1 = 2499$ **55.** Both equal 25. **57.** $100 + 200r + 100r^2$ **59.** $x^2 + (x + 1)^2 = 2x^2 + 2x + 1$ **61.** $x^2 + (x + 1)^2 + (x + 2)^2 = 3x^2 + 6x + 5$ **63.** $2x^2$ **65.** $5x$

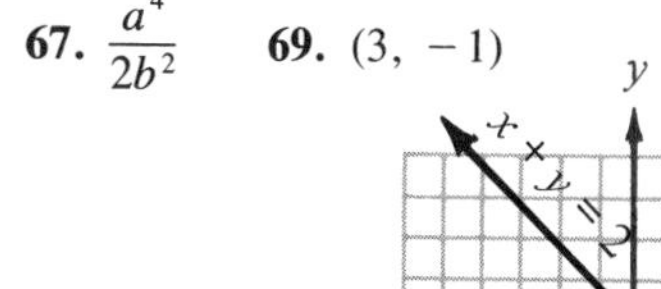

67. $\dfrac{a^4}{2b^2}$ **69.** $(3, -1)$

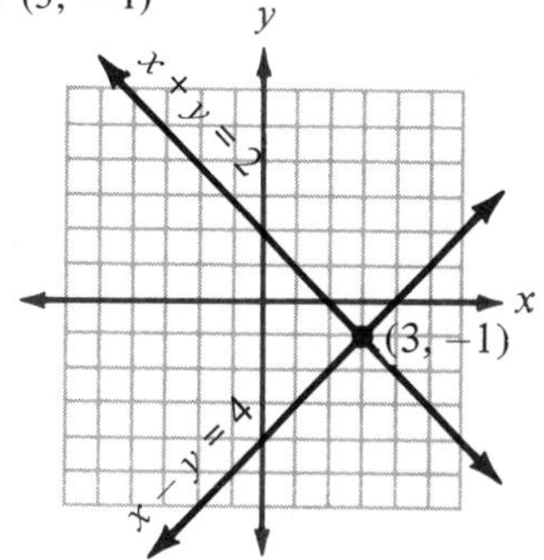

71. $(-1, 1)$

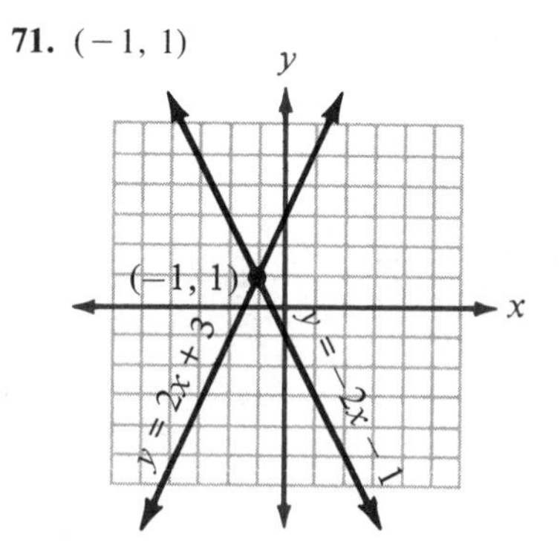

PROBLEM SET 4.7

1. $x - 2$ **3.** $3 - 2x^2$ **5.** $5xy - 2y$ **7.** $7x^4 - 6x^3 + 5x^2$ **9.** $10x^4 - 5x^2 + 1$ **11.** $-4a + 2$ **13.** $-8a^4 - 12a^3$ **15.** $-4b - 5a$ **17.** $-6a^2b + 3ab^2 - 7b^3$ **19.** $-\dfrac{a}{2} - b - \dfrac{b^2}{2a}$ **21.** $3x + 4y$ **23.** $-y + 3$ **25.** $5y - 4$ **27.** $xy - x^2y^2$ **29.** $-1 + xy$ **31.** $-a + 1$ **33.** $x^2 - 3xy + y^2$ **35.** $2 - 3b + 5b^2$ **37.** $-2xy + 1$ **39.** $xy - \frac{1}{2}$ **41.** $\dfrac{1}{4x} - \dfrac{1}{2a} + \dfrac{3}{4}$ **43.** $\dfrac{4x^2}{3} + \dfrac{2}{3x} + \dfrac{1}{x^2}$ **45.** Both equal 7. **47.** $\dfrac{3(10) + 8}{2} = \dfrac{38}{2} = 19$; $3(10) + 4 = 34$ **49.** $(7, -1)$ **51.** $(2, 3)$ **53.** $(1, 1)$ **55.** Lines coincide

PROBLEM SET 4.8

1. $x - 2$ **3.** $a + 4$ **5.** $x - 3$ **7.** $x + 3$ **9.** $a - 5$ **11.** $x + 2 + \dfrac{2}{x + 3}$ **13.** $a - 2 + \dfrac{12}{a + 5}$ **15.** $x + 4 + \dfrac{9}{x - 2}$ **17.** $x + 4 + \dfrac{-10}{x + 1}$ **19.** $a + 1 + \dfrac{-1}{a + 2}$ **21.** $x - 3 + \dfrac{17}{2x + 4}$ **23.** $3a - 2 + \dfrac{7}{2a + 3}$ **25.** $2a^2 - a - 3$ **27.** $x^2 - x + 5$ **29.** $x^2 + x + 1$ **31.** $x^2 + 2x + 4$ **33.** 5, 20 **35.** \$800 at 8%, \$400 at 9% **37.** 8 five-dollar bills, 12 ten-dollar bills. **39.** 10 gallons of 20%, 6 gallons of 60%.

CHAPTER 4 TEST

1. 81 **2.** $\frac{9}{16}$ **3.** $72x^{18}$ **4.** $\frac{1}{9}$ **5.** 1 **6.** a^2 **7.** x **8.** 2.78×10^{-2} **9.** 243,000 **10.** $\dfrac{y^2}{2x^4}$ **11.** $3a$ **12.** $10x^5$ **13.** 9×10^9 **14.** 2 **15.** $7x^5$ **16.** $8x^2 + 2x + 2$ **17.** $3x^2 + 4x + 6$

18. $3x - 4$ **19.** 10 **20.** $6a^4 - 10a^3 + 8a^2$ **21.** $x^2 + 5x + 6$ **22.** $8x^2 + 2x - 15$ **23.** $x^3 - 27$ **24.** $x^2 + 10x + 25$ **25.** $9a^2 - 12ab + 4b^2$ **26.** $9x^2 - 16y^2$ **27.** $a^4 - 9$ **28.** $2x^2 + 3x - 1$ **29.** $4x + 3 + \frac{4}{2x - 3}$ **30.** $3x^2 + 9x + 25 + \frac{76}{x - 3}$

CHAPTER 5

PROBLEM SET 5.1

1. Composite **3.** Prime **5.** Composite **7.** Composite **9.** Prime **11.** Composite **13.** Prime **15.** Composite **17.** Composite **19.** Composite **21.** $5 \cdot 7$ **23.** 2^7 **25.** $2^4 \cdot 3^2$ **27.** $2 \cdot 19$ **29.** $3 \cdot 5 \cdot 7$ **31.** $2^2 \cdot 3^2 \cdot 5$ **33.** $5 \cdot 7 \cdot 11$ **35.** 11^2 **37.** $2^2 \cdot 3 \cdot 5 \cdot 7$ **39.** $2^2 \cdot 5 \cdot 31$ **41.** $\frac{7}{11}$ **43.** $\frac{5}{7}$ **45.** $\frac{11}{13}$ **47.** $\frac{14}{15}$ **49.** $\frac{5}{9}$ **51.** $\frac{5}{8}$ **53.** $6^3 = (2 \cdot 3)^3 = 2^3 \cdot 3^3$ **55.** $9^4 \cdot 16^2 = (3^2)^4(2^4)^2 = 3^8 \cdot 2^8$ **57.** $3 \cdot 8 + 3 \cdot 7 + 3 \cdot 5 = 24 + 21 + 15 = 60 = 2^2 \cdot 3 \cdot 5$ **59.** $3x - 15$ **61.** $5x^3 - 15x^2$ **63.** $6x^3y - 18x^2y^2 + 12xy^3$ **65.** $xy + 3x + 2y + 6$ **67.** $xa + xb + ya + yb$ **69.** $2ax + 8x + 3a + 12$

PROBLEM SET 5.2

1. $5(3x + 5)$ **3.** $3(2a + 3)$ **5.** $4(x - 2y)$ **7.** $3(x^2 - 2x - 3)$ **9.** $3(a^2 - a - 20)$ **11.** $4(6y^2 - 13y + 6)$ **13.** $x^2(9 - 8x)$ **15.** $13a^2(1 - 2a)$ **17.** $7xy(3x - 4y)$ **19.** $11ab^2(2a - 1)$ **21.** $7x(x^2 + 3x - 4)$ **23.** $11(11y^4 - x^4)$ **25.** $25x^2(4x^2 - 2x + 1)$ **27.** $8(a^2 + 2b^2 + 4c^2)$ **29.** $4ab(a - 4b + 8ab)$ **31.** $11a^2b^2(11a - 2b + 3ab)$ **33.** $12x^2y^3(1 - 6x^3 - 3x^2y)$ **35.** $(y + 5)(x + 3)$ **37.** $(y + 6)(x + 2)$ **39.** $(b + 7)(a + 3)$ **41.** $(a - b)(x + y)$ **43.** $(a + 3)(2x + 5)$ **45.** $(3x - 4)(b + 2)$ **47.** $(x + a)(x + 2)$ **49.** $(x + a)(x + b)$ **51.** 6 **53.** $3(4x^2 + 2x + 1)$ **55.** $h = 16t(3 - t)$ when $t = 3$, $h = 0$ **57.** $A = 1{,}000(1 + r)$ when $r = .12$, $A = 1000(1.12) = \$1120$ **59.** $x^2 + 7x + 12$ **61.** $x^2 + 5x - 14$ **63.** $x^2 - 5x - 14$ **65.** $x^2 - x - 6$ **67.** $x^3 + 27$ **69.** $2x^3 + 9x^2 - 2x - 3$

PROBLEM SET 5.3

1. $(x + 3)(x + 4)$ **3.** $(x + 1)(x + 2)$ **5.** $(a + 3)(a + 7)$ **7.** $(x - 2)(x - 5)$ **9.** $(y - 3)(y - 7)$ **11.** $(x - 4)(x + 3)$ **13.** $(y + 4)(y - 3)$ **15.** $(x + 7)(x - 2)$ **17.** $(r - 9)(r + 1)$ **19.** $(x - 6)(x + 5)$ **21.** $(a + 7)(a + 8)$ **23.** $(y - 7)(y + 6)$ **25.** $(x + 7)(x + 6)$ **27.** $2(x + 1)(x + 2)$ **29.** $3(a - 5)(a + 4)$ **31.** $100(x - 2)(x - 3)$ **33.** $100(p - 8)(p - 5)$ **35.** $x^2(x - 4)(x + 3)$ **37.** $2r(r - 3)(r + 5)$ **39.** $2y^2(y - 4)(y + 1)$ **41.** $x^3(x + 2)(x + 2)$ **43.** $3y^2(y - 5)(y + 1)$ **45.** $4x^2(x - 4)(x - 9)$ **47.** $(x + 2y)(x + 3y)$ **49.** $(x - 4y)(x - 5y)$ **51.** $(a + 4b)(a - 2b)$ **53.** $(a - 5b)(a - 5b)$ **55.** $(a + 5b)(a + 5b)$ **57.** $(x - 6a)(x + 8a)$ **59.** $(x - 9b)(x + 4b)$ **61.** $(x + 16)$ **63.** $4x^2 - x - 3$ **65.** $h = 16(4 - t)(1 + t)$, when $t = 4$, $h = 16(4 - 4)(1 + 4) = 0$ **67.** $C = 100(8 - x)(1 + x)$, when $x = 8$, $C = 100(8 - 8)(1 + 8) = 0$ when $x = 3$, $C = 2000$ **69.** $6a^2 + 13a + 2$ **71.** $6a^2 + 7a + 2$ **73.** $6a^2 + 8a + 2$ **75.** $2x^2 + 7x - 11$ **77.** $(7x + 3) - (4x - 5) = 3x + 8$ **79.** $(5x^2 - 5) - (2x^2 - 4x) = 3x^2 + 4x - 5$

PROBLEM SET 5.4

1. $(2x + 1)(x + 3)$ **3.** $(2a - 3)(a + 1)$ **5.** $(3x + 5)(x - 1)$ **7.** $(3y + 1)(y - 5)$ **9.** $(2x + 3)(3x + 2)$ **11.** $(2x - 3y)(2x - 3y)$ **13.** $(4y + 1)(y - 3)$ **15.** $(4x - 5)(5x - 4)$ **17.** $(10a - b)(2a + 5b)$ **19.** $(4x - 5)(5x + 1)$ **21.** $(2m + 3)(6m - 1)$ **23.** $(4x + 5)(5x + 3)$ **25.** $(3a - 4b)(4a - 3b)$ **27.** $(3x - 7y)(x + 2y)$ **29.** $(7x - 3)(2x + 5)$ **31.** $(3x - 5)(2x - 11)$ **33.** $(5t - 19)(3t - 2)$ **35.** $2(2x + 3)(x - 1)$ **37.** $2(4a - 3)(3a - 4)$ **39.** $x(5x - 4)(2x - 3)$ **41.** $x^2(2x - 5)(3x + 2)$ **43.** $2a(5a + 2)(a - 1)$ **45.** $3x(5x + 1)(x - 7)$ **47.** $5y(7y + 2)(y - 2)$ **49.** $a^2(3a - 1)(5a + 1)$ **51.** $3y(4x + 5)(2x - 3)$ **53.** $2y(3x - 7y)(2x - y)$ **55.** Both equal 25. **57.** $4x^2 - 9$ **59.** $x^4 - 81$ **61.** $x^2 - 9$ **63.** $36a^2 - 1$ **65.** $x^2 + 8x + 16$ **67.** $4x^2 + 12x + 9$ **69.** $2x^4 - 3x^2 + 4x$ **71.** $a - a^2b^3$

PROBLEM SET 5.5

1. $(x - 3)(x + 3)$ **3.** $(a - 6)(a + 6)$ **5.** $(x - 7)(x + 7)$ **7.** $4(a - 2)(a + 2)$ **9.** No factors **11.** $(5x - 13)(5x + 13)$ **13.** $(3a - 4b)(3a + 4b)$ **15.** $(3 - m)(3 + m)$ **17.** $(5 - 2x)(5 + 2x)$ **19.** $2(x - 3)(x + 3)$ **21.** $32(a - 2)(a + 2)$ **23.** $2y(2x - 3)(2x + 3)$ **25.** $(a^2 + b^2)(a - b)(a + b)$ **27.** $(4m^2 + 9)(2m - 3)(2m + 3)$ **29.** $3xy(x - 5y)(x + 5y)$ **31.** $(x - 1)^2$ **33.** $(x + 1)^2$ **35.** $(a - 5)^2$ **37.** $(y + 2)^2$ **39.** $(x - 2)^2$ **41.** $(m - 6)^2$ **43.** $(2a + 3)^2$ **45.** $(7x - 1)^2$ **47.** $(3y - 5)^2$ **49.** $(x + 5y)^2$ **51.** $(3a + b)^2$ **53.** $3(a + 3)^2$ **55.** $2(x + 5y)^2$ **57.** $5x(x + 3y)^2$ **59.** 14 **61.** 25 **63.** $A = 1000(1 + r)^2$ when $r = .12$, $A = 1000(1.12)^2 = 1000(1.2544) = \$1254.40$ **65.** $x - 2 + \dfrac{2}{x - 3}$ **67.** $x + 3$ **69.** $3x - 2 + \dfrac{9}{2x + 3}$ **71.** $x^2 + 2x + 4$

PROBLEM SET 5.6

1. $(x + 9)(x - 9)$ **3.** $(x + 5)(x - 3)$ **5.** $(x + 3)^2$ **7.** $(y - 5)^2$ **9.** $2ab(a^2 + 3a + 1)$ **11.** Cannot be factored further **13.** $3(2a + 5)(2a - 5)$ **15.** $(3x - 2y)^2$ **17.** $4x(x^2 + 4y^2)$ **19.** $2y(y + 5)^2$ **21.** $a^4(a^2 + 4b^2)$ **23.** $(y + 3)(x + 4)$ **25.** $(x + 2)(x - 2)(x^2 + 4)$ **27.** $(y - 5)(x + 2)$ **29.** $5(a + b)^2$ **31.** Cannot be factored further. **33.** $3(x + 2y)(x + 3y)$ **35.** $(2x + 19)(x - 2)$ **37.** $100(x - 2)(x - 1)$ **39.** $(x + 8)(x - 8)$ **41.** $(x + 3)(x + a)$ **43.** $a^5(7a + 3)(7a - 3)$ **45.** Cannot be factored further. **47.** $a(5a + 1)(5a + 3)$ **49.** $(a - b)(x + y)$ **51.** $3a^2b(4a + 1)(4a - 1)$ **53.** $5x^2(2x + 3)(2x - 3)$ **55.** $(3x + 41y)(x - 2y)$ **57.** $2x^3(4x - 5)(2x - 3)$ **59.** $(x + a)(2x + 3)$ **61.** $(y + 1)(y - 1)(y^2 + 1)$ **63.** $3x^2y^2(2x + 3y)^2$ **65.** 5 **67.** $-\frac{3}{2}$ **69.** $-\frac{3}{4}$ **71.** x^{15} **73.** $72x^{18}$ **75.** 5.76×10^4

PROBLEM SET 5.7

1. $-2, 1$ **3.** 4, 5 **5.** $0, -1, 3$ **7.** $-\frac{2}{3}, -\frac{3}{2}$ **9.** $0, -\frac{4}{3}, \frac{4}{3}$ **11.** $0, -\frac{1}{3}, -\frac{3}{5}$ **13.** $-1, -2$ **15.** 4, 5 **17.** $6, -4$ **19.** 2, 3 **21.** -3 **23.** $4, -4$ **25.** $-4, \frac{3}{2}$ **27.** $-\frac{2}{3}$ **29.** 5 **31.** $-\frac{5}{2}, 4$ **33.** $\frac{5}{3}, -4$ **35.** $\frac{7}{2}, -\frac{7}{2}$ **37.** $0, -6$ **39.** 0, 3 **41.** 0, 4 **43.** 0, 5 **45.** 2, 5 **47.** $\frac{1}{2}, -\frac{4}{3}$ **49.** $-\frac{5}{2}, 4$ **51.** $-10, 8$ **53.** 5, 8 **55.** 6, 8 **57.** -4 **59.** 5, 8 **61.** $6, -8$ **63.** $0, -\frac{3}{2}, -4$ **65.** $0, 3, -\frac{5}{2}$ **67.** $0, \frac{1}{2}, -\frac{5}{2}$ **69.** $0, \frac{3}{5}, -\frac{3}{2}$ **71.** \$15 for the suit, \$75 for the bicycle **73.** \$600 for the lot, \$2400 for the house **75.** $\frac{1}{8}$ **77.** x^8 **79.** x^{18} **81.** 5.6×10^{-3}

PROBLEM SET 5.8

1. Two consecutive even integers are x and $x + 2$; $x(x + 2) = 80$; 8, 10 and $-10, -8$ **3.** 9, 11 and $-11, -9$ **5.** $x(x + 2) = 5(x + x + 2) - 10$; 8, 10 and 0, 2 **7.** 8, 6 **9.** The numbers are x and $5x + 2$; $x(5x + 2) = 24$; 2, 12 and $-\frac{12}{5}, -10$ **11.** 20, 5 and 0, 0 **13.** Let $x =$ the width; $x(x + 1) = 12$; width 3 inches, length 4 inches **15.** Let $x =$ the base; $\frac{1}{2}(x)(2x) = 9$; base 3 inches, height 6 inches **17.** $x^2 + (x + 2)^2 = 10^2$; 6 inches and 8 inches **19.** 12 meters **21.** $1400 = 400 + 700x - 100x^2$; 2 hundred items and 5 hundred items **23.** 2 hundred items and 8 hundred items **25.** $R = xp = (1200 - 100p)p = 3200$; \$4 or \$8 **27.** \$7 and \$10 **29.** $42a^5b^3$ **31.** $2x^2y^2$ **33.** $25x^5$ **35.** 6×10^{-3} **37.** $2a^3$

CHAPTER 5 TEST

1. $5 \cdot 7$ **2.** $7 \cdot 11$ **3.** 2^7 **4.** $2^2 \cdot 3^4$ **5.** $(x - 2)(x - 3)$ **6.** $(x - 3)(x + 2)$ **7.** $(a - 4)(a + 4)$ **8.** $x^2 + 25$ **9.** $(x^2 + 9)(x - 3)(x + 3)$ **10.** $3(3x - 5y)(3x + 5y)$ **11.** $(y + 4)(x + 7)$ **12.** $(x - b)(x + 5)$ **13.** $2(2a + 1)(a + 5)$ **14.** $3(m - 3)(m + 2)$ **15.** $(2y - 1)(3y + 5)$ **16.** $2x(2x + 1)(3x - 5)$ **17.** $-3, -4$ **18.** 2 **19.** $-6, 6$ **20.** $5, -4$ **21.** 5, 6 **22.** $0, 4, -4$ **23.** $3, -\frac{5}{2}$ **24.** $0, 1, -\frac{1}{3}$ **25.** 4, 16 **26.** 3, 5 and $-3, -1$ **27.** 3, 14 feet **28.** 5 meters and 12 meters **29.** 2 hundred or 3 hundred items **30.** \$3 or \$6

CHAPTER 6

PROBLEM SET 6.1

1. $\frac{1}{x-2}, x \neq 2$ **3.** $\frac{-1}{x+2}, x \neq -2$ **5.** $\frac{1}{a+3}, a \neq -3, a \neq 3$ **7.** $\frac{1}{x-5}, x \neq 5, x \neq -5$ **9.** $\frac{1}{5a}, a \neq 0$ **11.** $\frac{x^2-4}{2}$ **13.** $\frac{5}{m^2}$ **15.** $\frac{2x-10}{3x-6}$ **17.** 2 **19.** $\frac{5(x-1)}{4}$ **21.** $\frac{1}{x-3}$ **23.** $\frac{1}{y-3}$ **25.** $\frac{1}{x+3}$ **27.** $\frac{1}{a-5}$ **29.** $\frac{1}{3x+2}$ **31.** $\frac{x+5}{x+2}$ **33.** $\frac{2m(m+2)}{m-2}$ **35.** $\frac{3x}{2}$ **37.** $\frac{x-1}{x-4}$ **39.** $\frac{a-2}{3a+2}$ **41.** $\frac{2x-3}{2x+3}$ **43.** $\frac{1}{(x-3)(x^2+9)}$ **45.** $\frac{x+2}{x+5}$ **47.** $\frac{x+a}{x+b}$ **49.** $\frac{x+a}{x+3}$ **51.** $5 + 4 = 9$ **53.** $\frac{2205}{704} \approx 3.1$ lumens/square foot **55.** $P = \frac{1254.40}{(1+r)^2}$, when $r = .12$, $P = \frac{1254.40}{(1.12)^2} = \$1{,}000$ **57.** $\frac{15}{16}$ **59.** $\frac{1}{6}$ **61.** 2 **63.** $\frac{8}{9}$ **65.** $3 \cdot 5^2$ **67.** $3^2 \cdot 5^2$

PROBLEM SET 6.2

1. 2 **3.** $\frac{x}{2}$ **5.** $\frac{3}{2}$ **7.** $\frac{1}{2(x-3)}$ **9.** $\frac{4a(a+5)}{7(a+4)}$ **11.** $\frac{y-2}{4}$ **13.** $\frac{2(x+4)}{x-2}$ **15.** $\frac{x+3}{(x-3)(x+1)}$ **17.** 1 **19.** $\frac{y-5}{(y+2)(y-2)}$ **21.** $\frac{x+5}{x-5}$ **23.** 1 **25.** $\frac{a+3}{a+4}$ **27.** $\frac{2y-3}{y-6}$ **29.** 1 **31.** 15 **33.** $2(x-3)$ **35.** $2a$ **37.** $(x+2)(x+1)$ **39.** $-2x(x-5)$ **41.** $\frac{2(x+5)}{x(y+5)}$ **43.** $\frac{2x}{x-y}$ **45.** $\frac{9}{2}$ **47.** 1 **49.** $\frac{2}{3}$ **51.** $9x^2$ **53.** $6ab^2$

PROBLEM SET 6.3

1. $\frac{7}{x}$ **3.** $\frac{4}{a}$ **5.** 1 **7.** $y+1$ **9.** $x+2$ **11.** $x-2$ **13.** $\frac{6}{x+6}$ **15.** $\frac{(y+2)(y-2)}{2y}$ **17.** $\frac{3+2a}{6}$ **19.** $\frac{7x+3}{4(x+1)}$ **21.** $\frac{17}{5(x-2)}$ **23.** $\frac{4-3x}{3(x+4)}$ **25.** $\frac{3}{x-2}$ **27.** $\frac{4}{a+3}$ **29.** $\frac{2x-9}{(x+5)(x-5)}$ **31.** $\frac{1}{x+3}$ **33.** $\frac{3a-4}{(a-3)(a+2)}$ **35.** $\frac{x-2}{(x-3)(x+3)}$ **37.** $\frac{7a+5}{a(a+1)}$ **39.** $\frac{1}{(x+4)(x+3)}$ **41.** $\frac{y}{(y+4)(y+5)}$ **43.** $x + \frac{2}{x} = \frac{x^2+2}{x}$ **45.** $\frac{1}{x} + \frac{1}{2x} = \frac{3}{2x}$ **47.** 3 **49.** -2 **51.** $\frac{1}{5}$ **53.** $-2, -3$ **55.** $-2, 3$ **57.** 0, 5

PROBLEM SET 6.4

1. -3 **3.** 20 **5.** -1 **7.** 5 **9.** -2 **11.** 4 **13.** 3, 5 **15.** -8, 1 **17.** 2 **19.** 1 **21.** 8 **23.** 5 **25.** Possible solution 2, which does not check; $\emptyset$ **27.** 3 **29.** Possible solution -3, which does not check; $\emptyset$ **31.** 0 **33.** Possible solutions 2 and 3, but only 2 checks; 2 **35.** -4 **37.** -1 **39.** $-6, -7$ **41.** Possible solutions -3 and -1, but only -1 checks; -1 **43.** 7 **45.** Width is 4 inches, length is 13 inches **47.** $-8, -6$ or 6, 8 **49.** 6 inches and 8 inches

PROBLEM SET 6.5

1. $\frac{1}{x} + \frac{1}{3x} = \frac{16}{3}$; $\frac{1}{4}$ and $\frac{3}{4}$ **3.** $x + \frac{1}{x} = \frac{13}{6}$; $\frac{2}{3}$ and $\frac{3}{2}$ **5.** $\frac{7+x}{9+x} = \frac{5}{7}$; -2 **7.** $\frac{1}{x} + \frac{1}{x+2} = \frac{5}{12}$; 4 and 6

9. Let x = the speed of the boat in still water.

	d	r	t
upstream	26	$x - 3$	$\frac{26}{x-3}$
downstream	38	$x + 3$	$\frac{38}{x+3}$

The equation is $\frac{26}{x-3} = \frac{38}{x+3}$; $x = 16$ mph.

11. 300 mph **13.** 170 mph and 190 mph **15.** $\frac{1}{12} - \frac{1}{15} = \frac{1}{x}$; 60 hours **17.** $\frac{60}{11}$ minutes **19.** $(2, -1)$ **21.** $(4, 3)$ **23.** $(1, 1)$ **25.** $(5, 2)$

PROBLEM SET 6.6

1. 6 **3.** $\frac{1}{6}$ **5.** xy^2 **7.** $\frac{xy}{2}$ **9.** $\frac{y}{x}$ **11.** $\frac{a+1}{a-1}$ **13.** $\frac{x+1}{2(x-3)}$ **15.** $a - 3$ **17.** $\frac{y+3}{y+2}$ **19.** $x + y$ **21.** $\frac{a+1}{a}$ **23.** $\frac{1}{x}$ **25.** $\frac{2a+3}{3a+4}$ **27.** $x < 1$ **29.** $x \geq -7$ **31.** $x < 6$ **33.** $x \leq 2$

PROBLEM SET 6.7

1. $\frac{4}{3}$ **3.** $\frac{4}{5}$ **5.** $\frac{8}{1}$ **7.** $\frac{3}{5}$ **9.** $\frac{8}{9}$ **11.** $\frac{5}{x}$ **13.** $\frac{12}{11}$ **15.** $\frac{22}{125}$ **17.** $\frac{1}{6}$ **19.** 1 **21.** 10 **23.** 40 **25.** $\frac{5}{4}$ **27.** $\frac{7}{3}$ **29.** $-4, 3$ **31.** $-4, 4$ **33.** $-4, 2$ **35.** $-1, 6$ **37.** 15 hits **39.** 21 milliliters alcohol **41.** 45.5 grams of fat **43.** 14.7 inches **45.** 343 miles **47.** $\frac{x+2}{x+3}$ **49.** $\frac{2(x+5)}{x-4}$ **51.** $\frac{1}{x-4}$

CHAPTER 6 TEST

1. $\frac{x+4}{x-4}$ **2.** $\frac{2}{a+2}$ **3.** $\frac{y+7}{x+a}$ **4.** 3 **5.** $(x-7)(x-1)$ **6.** $\frac{x+4}{3x-4}$ **7.** $(x-3)(x+2)$ **8.** $\frac{-3}{x-2}$ **9.** $\frac{2x+3}{(x+3)(x-3)}$ **10.** $\frac{3x}{(x+1)(x-2)}$ **11.** $\frac{11}{5}$ **12.** 1 **13.** 6 **14.** $\frac{4}{3}$ or $\frac{3}{4}$ **15.** 15 mph **16.** 60 hours **17.** $\frac{1}{2}$ and $\frac{1}{3}$ **18.** 132 **19.** $\frac{x+1}{x-1}$ **20.** $\frac{x+4}{x+2}$

CHAPTER 7

PROBLEM SET 7.1

1. 3 **3.** -3 **5.** Not a real number **7.** -12 **9.** 25 **11.** Not a real number **13.** -8 **15.** -10 **17.** 35 **19.** 1 **21.** -2 **23.** -5 **25.** -1 **27.** -3 **29.** -2 **31.** x **33.** $3x$ **35.** xy **37.** $a + b$ **39.** $7xy$ **41.** x **43.** $2x$ **45.** x^2 **47.** $6a^3$ **49.** $5a^4b^2$ **51.** x^2 **53.** $3a^4$ **55.** x^2 **57.** $3 + 4 = 7$ **59.** $\sqrt{25} = 5$ **61.** $12 + 5 = 17$ **63.** $\sqrt{169} = 13$ **65.** $x + 3$ **67.** 5 and 7 **69.** $r = \frac{\sqrt{65} - \sqrt{50}}{\sqrt{50}} \approx .14 = 14\%$ **71.** approximately .095 or 9.5% **73.** $x - 4$ **75.** $\frac{2}{a-2}$ **77.** $\frac{2x+1}{x}$ **79.** $\frac{x+2}{x+a}$

PROBLEM SET 7.2

1. $2\sqrt{2}$ **3.** $2\sqrt{3}$ **5.** $2\sqrt[3]{3}$ **7.** $5x\sqrt{2}$ **9.** $3ab\sqrt{5}$ **11.** $3x\sqrt[3]{2}$ **13.** $4x^2\sqrt{2}$ **15.** $4\sqrt{5}$ **17.** $2x\sqrt{7x}$ **19.** $2x\sqrt[3]{x}$ **21.** $3a\sqrt[3]{a^2}$ **23.** $15a\sqrt{5a}$ **25.** $15y\sqrt{2x}$ **27.** $14x\sqrt{3y}$ **29.** $\frac{4}{5}$ **31.** $\frac{2}{3}$ **33.** $\frac{2}{3}$ **35.** $\frac{2}{3}$ **37.** $2x$ **39.** $3ab$ **41.** $\dfrac{3x}{2y}$ **43.** $\dfrac{5\sqrt{2}}{3}$ **45.** $\sqrt{3}$ **47.** $\dfrac{8\sqrt{2}}{7}$ **49.** $\dfrac{12\sqrt{2x}}{5}$ **51.** $\dfrac{3a\sqrt{6}}{5}$ **53.** $\dfrac{15\sqrt{2}}{2}$ **55.** $\dfrac{14y\sqrt{7}}{3}$ **57.** $5ab\sqrt{2}$ **59.** $6x\sqrt{2y}$ **61.** $\dfrac{8xy\sqrt{3y}}{5}$ **63.** $\sqrt{12} \approx 3.464$; $2\sqrt{3} \approx 2(1.732) = 3.464$ **65.** $\frac{5}{4}$ seconds **67.** $\dfrac{2(x+5)}{x(x+1)}$ **69.** $\dfrac{x+1}{x-1}$ **71.** $(x+6)(x+3)$

PROBLEM SET 7.3

1. $\dfrac{\sqrt{2}}{2}$ **3.** $\dfrac{\sqrt{3}}{3}$ **5.** $\dfrac{\sqrt{10}}{5}$ **7.** $\dfrac{\sqrt{6}}{2}$ **9.** $\dfrac{2\sqrt{15}}{3}$ **11.** $\dfrac{\sqrt{30}}{2}$ **13.** 2 **15.** $\sqrt{7}$ **17.** $\sqrt{5}$ **19.** $2\sqrt{5}$ **21.** $2\sqrt{3}$ **23.** $\dfrac{\sqrt{7}}{2}$ **25.** $xy\sqrt{2}$ **27.** $\dfrac{x\sqrt{15y}}{3}$ **29.** $\dfrac{4a^2\sqrt{5}}{5}$ **31.** $\dfrac{6a^2\sqrt{10a}}{5}$ **33.** $\dfrac{2xy\sqrt{15y}}{3}$ **35.** $\dfrac{4xy\sqrt{5y}}{3}$ **37.** $\dfrac{18ab\sqrt{6b}}{5}$ **39.** $18x\sqrt{x}$ **41.** $\dfrac{\sqrt[3]{4}}{2}$ **43.** $\dfrac{\sqrt[3]{3}}{3}$ **45.** $\dfrac{\sqrt[3]{12}}{2}$ **47.** $\dfrac{\sqrt[3]{6}}{2}$ **49.** Both are 0.707 **51.** 6 miles **53.** $10x$ **55.** $23x$ **57.** $10a$ **59.** $x+5$ **61.** $\dfrac{5a+6}{15}$ **63.** $\dfrac{1}{(a+3)(a+2)}$

PROBLEM SET 7.4

1. $7\sqrt{2}$ **3.** $2\sqrt{5}$ **5.** $7\sqrt{3}$ **7.** $5\sqrt{5}$ **9.** $13\sqrt{13}$ **11.** $6\sqrt{10}$ **13.** $6\sqrt{5}$ **15.** $4\sqrt{2}$ **17.** 0 **19.** $-30\sqrt{3}$ **21.** $-6\sqrt{3}$ **23.** $4\sqrt{3}$ **25.** $16\sqrt{2} + 25\sqrt{3}$ **27.** 0 **29.** $9\sqrt{2}$ **31.** $-3\sqrt{7}$ **33.** $35\sqrt{3}$ **35.** $-6\sqrt{2} - 6\sqrt{3}$ **37.** $2x\sqrt{x}$ **39.** $4a\sqrt{3}$ **41.** $15x\sqrt{2x}$ **43.** $13x\sqrt{3xy}$ **45.** $-b\sqrt{5a}$ **47.** $27x\sqrt{2x} - 8x\sqrt{3x}$ **49.** $23x\sqrt{2y}$ **51.** 3.968 is not equal to the decimal approximation of $\sqrt{8}$, which is 2.828. **53.** $9\sqrt{3}$ **55.** $9x^2 + 6xy + y^2$ **57.** $9x^2 - 16y^2$ **59.** 9 **61.** 2, 3 **63.** Possible solutions 2 and 4; only 2 checks.

PROBLEM SET 7.5

1. $\sqrt{6}$ **3.** $2\sqrt{3}$ **5.** $10\sqrt{21}$ **7.** $24\sqrt{2}$ **9.** $\sqrt{6} - \sqrt{2}$ **11.** $\sqrt{6} + 2$ **13.** $2\sqrt{6} + 3$ **15.** $6 - \sqrt{15}$ **17.** $2\sqrt{6} + 2\sqrt{15}$ **19.** $3 + 2\sqrt{2}$ **21.** $x + 6\sqrt{x} + 9$ **23.** $27 - 10\sqrt{2}$ **25.** $a - 10\sqrt{a} + 25$ **27.** $16 + 6\sqrt{7}$ **29.** $11 + 5\sqrt{5}$ **31.** $-28 + \sqrt{2}$ **33.** $\sqrt{6} + 3\sqrt{3} + \sqrt{2} + 3$ **35.** $x - 36$ **37.** $a - 2\sqrt{a} - 8$ **39.** 1 **41.** $30 + \sqrt{7}$ **43.** $6x + 16\sqrt{x} + 8$ **45.** $49a - 4b$ **47.** $\dfrac{\sqrt{15} + \sqrt{6}}{3}$ **49.** $\dfrac{5 - \sqrt{10}}{3}$ **51.** $6 + 2\sqrt{5}$ **53.** $5 + 2\sqrt{6}$ **55.** $\dfrac{5 - \sqrt{21}}{2}$ **57.** $\dfrac{x + 4\sqrt{x} + 4}{x - 4}$ **59.** $\dfrac{5 - \sqrt{15} - \sqrt{10} + \sqrt{6}}{2}$ **61.** $6\sqrt{5}$ **63.** $52 + 14\sqrt{3}$ **65.** $-6, 1$ **67.** 0, 3 **69.** $-9, 9$ **71.** $-5, 3$ **73.** 500 miles

PROBLEM SET 7.6

1. 3 **3.** 44 **5.** $\emptyset$ **7.** $\emptyset$ **9.** 8 **11.** 4 **13.** $\emptyset$ **15.** $-\frac{2}{3}$ **17.** 25 **19.** 4 **21.** 7 **23.** 8 **25.** $\emptyset$ **27.** Possible solutions 1 and 6; only 6 checks **29.** $-2, -1$ **31.** -4 **33.** Possible solutions 5 and 8; only 8 checks **35.** 0, 3 **37.** $x + 2 = \sqrt{8x}$; $x = 2$ **39.** $x - 3 = 2\sqrt{x}$; possible solutions 1 and 9; only 9 checks. **41.** $\dfrac{392}{121} \approx 3.2$ feet **43.** $\frac{3}{2}$ **45.** $\dfrac{a-1}{a+1}$ **47.** $\frac{1}{3}$ and 3 **49.** Walk 4 mph and jog 8 mph **51.** $\frac{12}{7}$ minutes

CHAPTER 7 TEST

1. 4 **2.** -6 **3.** 7, -7 **4.** 3 **5.** -2 **6.** -3 **7.** $5\sqrt{3}$ **8.** $4\sqrt{2}$ **9.** $\frac{\sqrt{6}}{3}$ **10.** $\frac{\sqrt[3]{2}}{2}$
11. $15x\sqrt{2}$ **12.** $\frac{2xy\sqrt{15y}}{5}$ **13.** $4\sqrt{3}$ **14.** $11x\sqrt{2}$ **15.** $\sqrt{15} - 2\sqrt{3}$ **16.** $-51 - \sqrt{5}$ **17.** $x - 36$
18. $8 - 2\sqrt{15}$ **19.** $\frac{5 - \sqrt{21}}{2}$ **20.** $\frac{x - 5\sqrt{x}}{x - 25}$ **21.** -1 **22.** 12 **23.** $\emptyset$ **24.** Possible solutions 2 and 6; only 6 checks. **25.** Possible solutions 1 and 16; only 16 checks.

CHAPTER 8

PROBLEM SET 8.1
1. ± 3 **3.** ± 5 **5.** $\pm 2\sqrt{2}$ **7.** $\pm 5\sqrt{2}$ **9.** $\pm 3\sqrt{2}$ **11.** 0, -4 **13.** 4, -6 **15.** $5 \pm 5\sqrt{3}$
17. $-1 \pm 5\sqrt{2}$ **19.** 2, -3 **21.** $\frac{11}{4}$, $-\frac{1}{4}$ **23.** $\frac{1 \pm 2\sqrt{3}}{3}$ **25.** $\frac{-2 \pm 3\sqrt{3}}{6}$ **27.** $3 \pm \sqrt{3}$
29. $-2 \pm \sqrt{5}$ **31.** $2 \pm \sqrt{2}$ **35.** $(x + 3)^2 = 16$; -7, 1 **37.** $r = \frac{\sqrt{A}}{10} - 1$ **39.** $x^2 + 6x + 9$
41. $x^2 - 10x + 25$ **43.** $(x - 6)^2$ **45.** $(x + 2)^2$ **47.** 2 **49.** 2

PROBLEM SET 8.2
1. 9 **3.** 1 **5.** 16 **7.** 1 **9.** 64 **11.** $\frac{9}{4}$ **13.** $\frac{49}{4}$ **15.** $\frac{1}{4}$ **17.** $\frac{9}{16}$ **19.** -6, 2 **21.** 8, -2
23. -3, 1 **25.** 0, 10 **27.** 3, -5 **29.** $-2 \pm \sqrt{7}$ **31.** $2 \pm 2\sqrt{2}$ **33.** 8, -1 **35.** $-1 \pm \sqrt{2}$
37. -2, 1 **39.** $\frac{-2 \pm \sqrt{3}}{2}$ **41.** $\frac{1 \pm \sqrt{3}}{2}$ **43.** $\frac{1}{2}$ **45.** $\frac{9 \pm \sqrt{105}}{6}$ **47.** $-2 + 2.646 = 0.646$; $-2 - 2.646 = -4.646$ **51.** $(-2 + \sqrt{7}) + (-2 - \sqrt{7}) = -4$ **53.** 4 **55.** -24 **57.** $2\sqrt{10}$
59. $2\sqrt{3}$ **61.** $2xy\sqrt{5y}$ **63.** $\frac{9}{5}$

PROBLEM SET 8.3

1. -1, -2 **3.** -2, -3 **5.** -3 **7.** $-3 \pm \sqrt{2}$ **9.** $-\frac{3}{2}$, -1 **11.** $\frac{-2 \pm \sqrt{3}}{2}$ **13.** 1
15. $\frac{5 \pm \sqrt{53}}{2}$ **17.** $\frac{2}{3}$, $-\frac{1}{2}$ **19.** $\frac{1 \pm \sqrt{21}}{2}$ **21.** $\frac{-1 \pm \sqrt{57}}{4}$ **23.** $\frac{5}{2}$, -1 **25.** $\frac{-3 \pm \sqrt{23}}{2}$
27. $\frac{-2 \pm \sqrt{10}}{3}$ **29.** $\frac{1 \pm \sqrt{11}}{2}$ **31.** 0, $\frac{-3 \pm \sqrt{41}}{4}$ **33.** 0, $\frac{4}{3}$ **35.** $\frac{3 \pm \sqrt{21}}{6}$ **37.** $6\sqrt{15}$
39. $-4 - 3\sqrt{6}$ **41.** 5 **43.** $\frac{3 - \sqrt{5}}{2}$

PROBLEM SET 8.4
1. $3 + i$ **3.** $6 - 8i$ **5.** 11 **7.** $9 + i$ **9.** $-1 - i$ **11.** $11 - 2i$ **13.** $4 - 2i$ **15.** $8 - 7i$
17. $-1 + 4i$ **19.** $6 - 3i$ **21.** $14 + 16i$ **23.** $9 + 2i$ **25.** $11 - 7i$ **27.** 34 **29.** 5 **31.** $\frac{6 + 4i}{13}$
33. $\frac{-9 - 6i}{13}$ **35.** $\frac{-3 + 9i}{5}$ **37.** $\frac{3 + 4i}{5}$ **39.** $\frac{-6 + 13i}{15}$ **41.** $x^2 + 9$ **45.** -2, 8 **47.** 1, 5
49. $-3 \pm 2\sqrt{3}$ **51.** $\frac{\sqrt{2}}{2}$ **53.** $\frac{2xy\sqrt{6y}}{3}$ **55.** $\frac{\sqrt[3]{2}}{2}$

PROBLEM SET 8.5

1. $4i$ **3.** $7i$ **5.** $i\sqrt{6}$ **7.** $i\sqrt{11}$ **9.** $4i\sqrt{2}$ **11.** $5i\sqrt{2}$ **13.** $2i\sqrt{2}$ **15.** $4i\sqrt{3}$ **17.** $1 \pm i$ **19.** 2 **21.** $-4, \frac{3}{2}$ **23.** $2 \pm 2i$ **25.** $\frac{-1 \pm 3i}{2}$ **27.** $\frac{1 \pm i\sqrt{3}}{2}$ **29.** $\frac{-1 \pm i\sqrt{3}}{2}$ **31.** 2, 3 **33.** $\frac{-1 \pm i\sqrt{2}}{3}$ **35.** $\frac{1}{2}, -2$ **37.** $\frac{1 \pm 3\sqrt{5}}{2}$ **39.** $4 \pm 3i$ **41.** $\frac{4 \pm \sqrt{22}}{2}$ **43.** Yes **45.** $3 - 7i$

47.

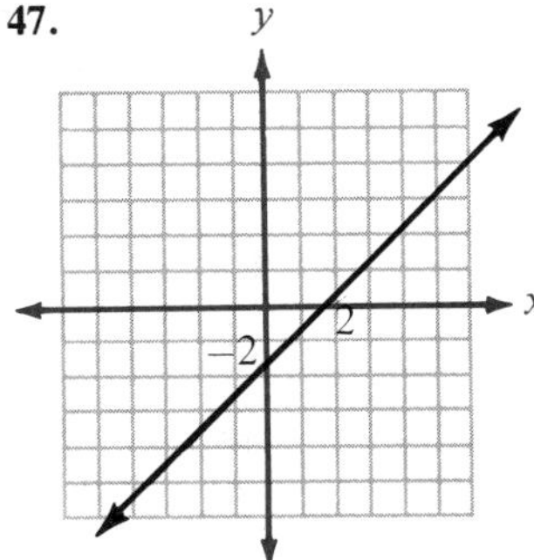

49.

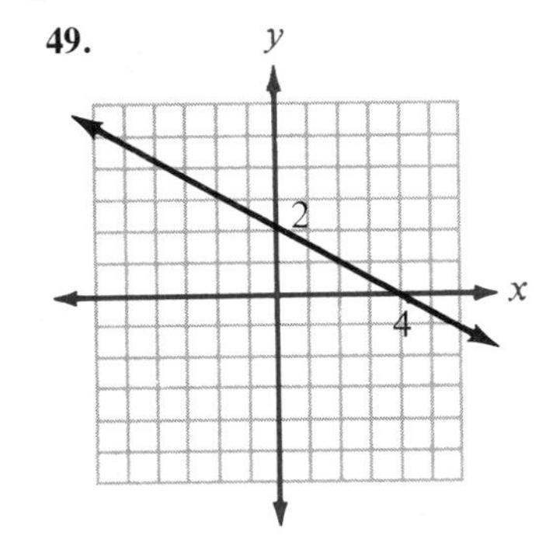

51. $23\sqrt{2}$ **53.** $-6\sqrt{6}$ **55.** $2x\sqrt{3}$

PROBLEM SET 8.6

1.

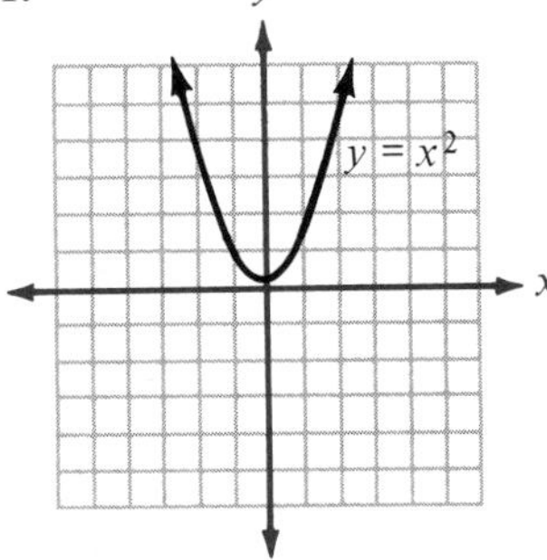

3.

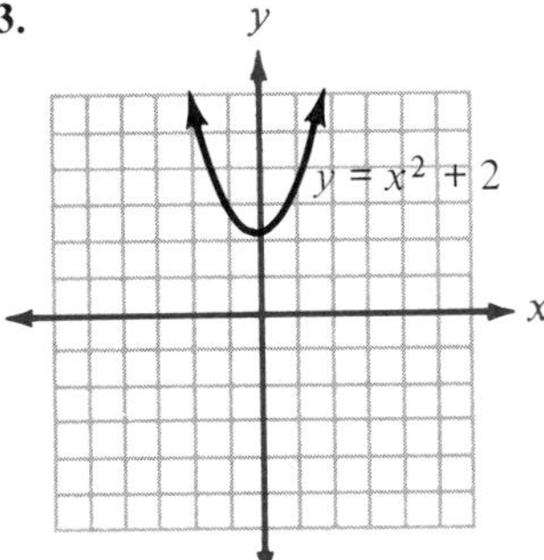

5.

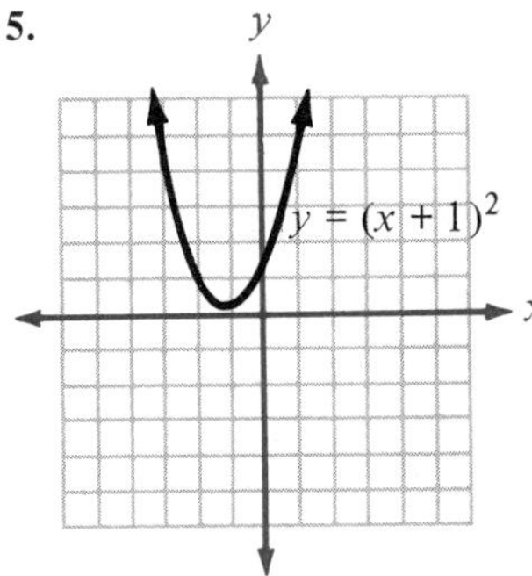

7.

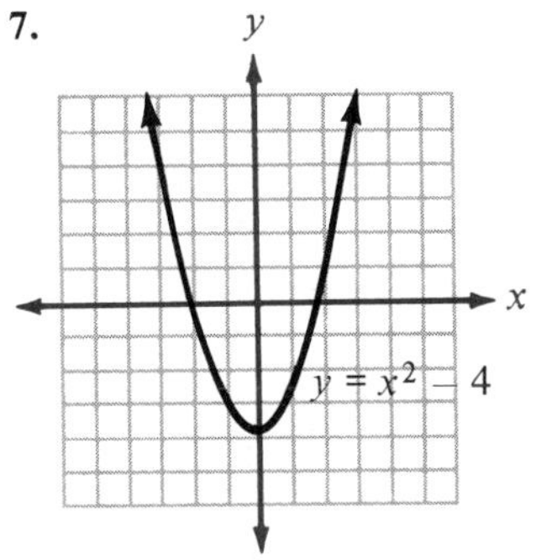

9.

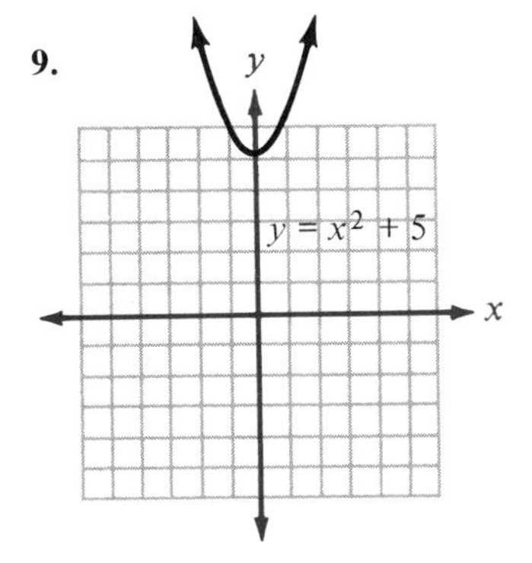

11.

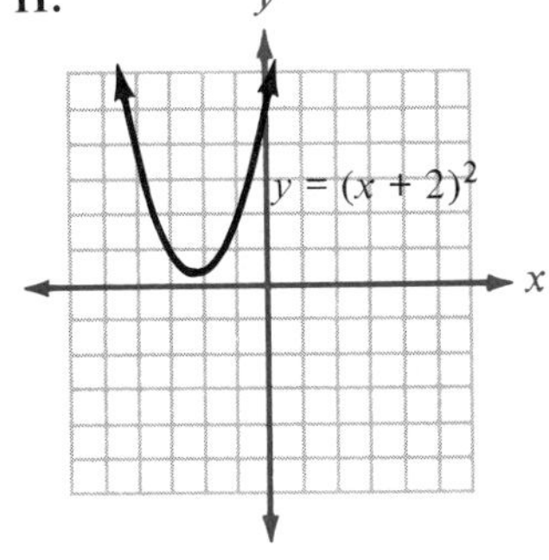

13.
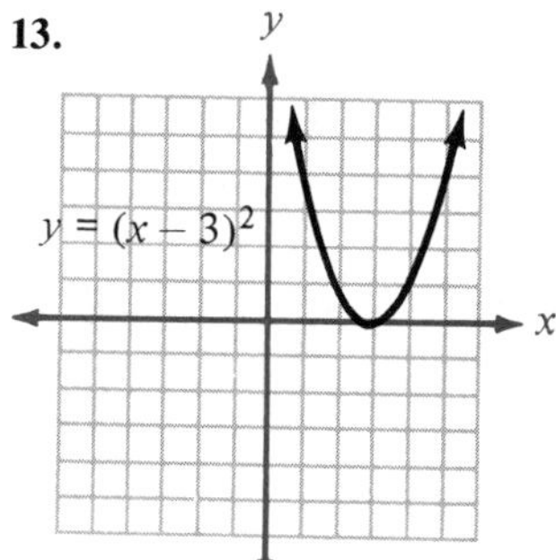

15.
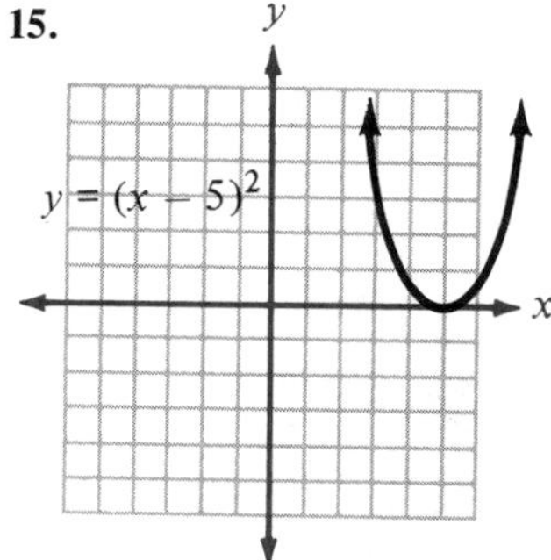

17.
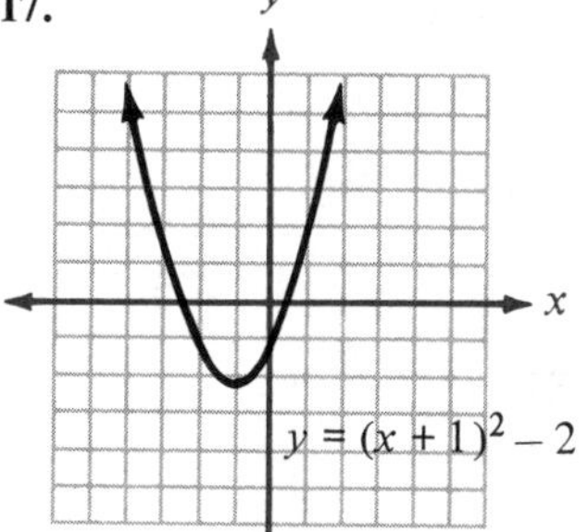

19.
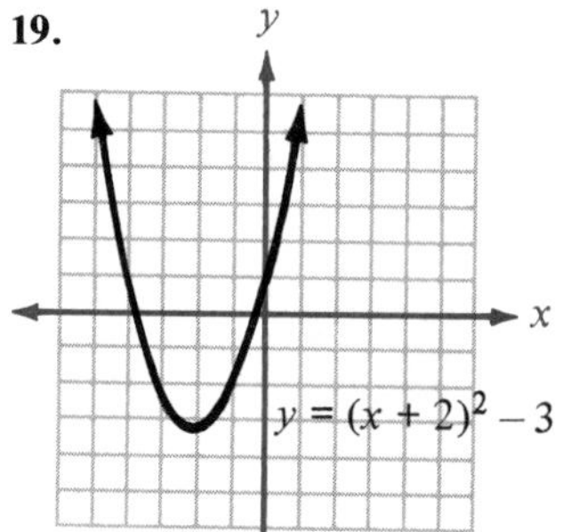

21.
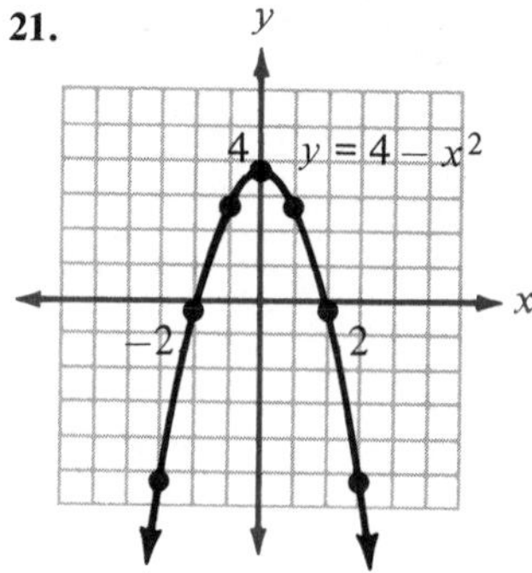

23.
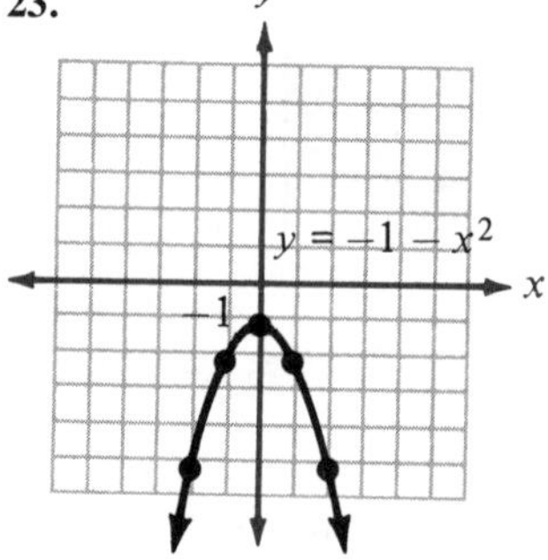

25.
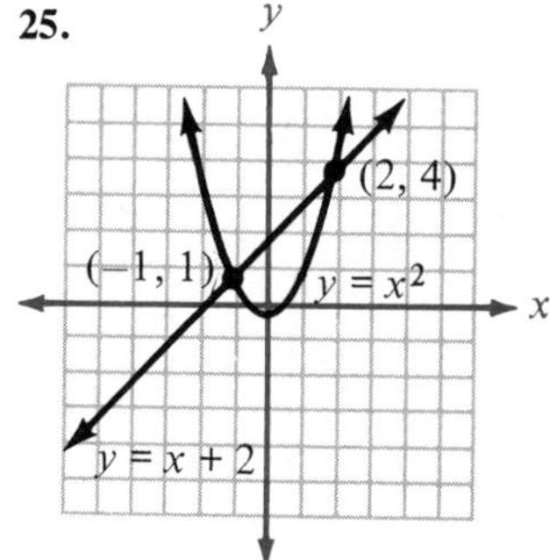

27.
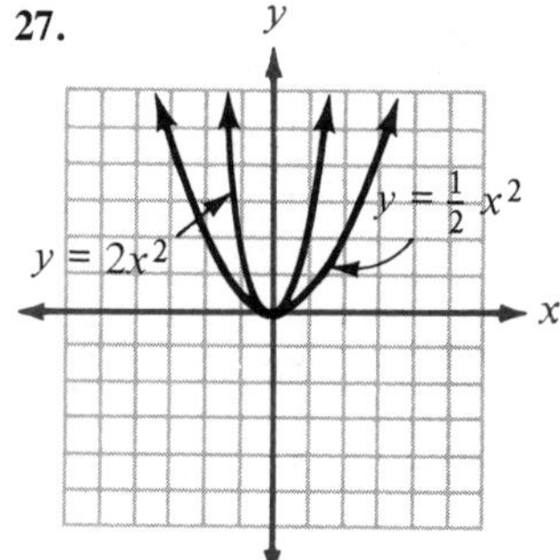

29. 11 **31.** $\emptyset$ **33.** $\frac{62}{3}$ **35.** Possible solutions -1 and 6; only 6 checks.

CHAPTER 8 TEST

1. $-1, 8$ **2.** $3 \pm 2\sqrt{3}$ **3.** $\dfrac{5 \pm 5i\sqrt{3}}{2}$ **4.** $\dfrac{3 \pm i\sqrt{31}}{4}$ **5.** $\frac{1}{3}, -1$ **6.** $\dfrac{-1 \pm \sqrt{33}}{2}$ **7.** $-\frac{2}{3}$
8. $3 \pm \sqrt{15}$ **9.** $3i$ **10.** $11i$ **11.** $6i\sqrt{2}$ **12.** $3i\sqrt{2}$ **13.** $3 + 8i$ **14.** $-1 + 2i$ **15.** 5
16. $1 + 5i$ **17.** $\dfrac{-1 + 3i}{10}$ **18.** $\dfrac{3 + 4i}{5}$

19.

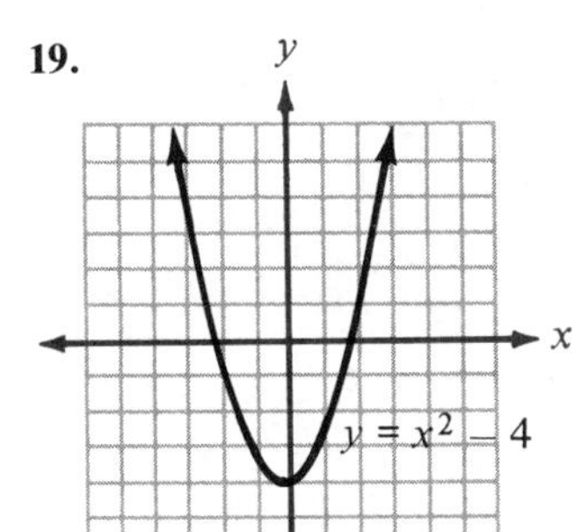

20.

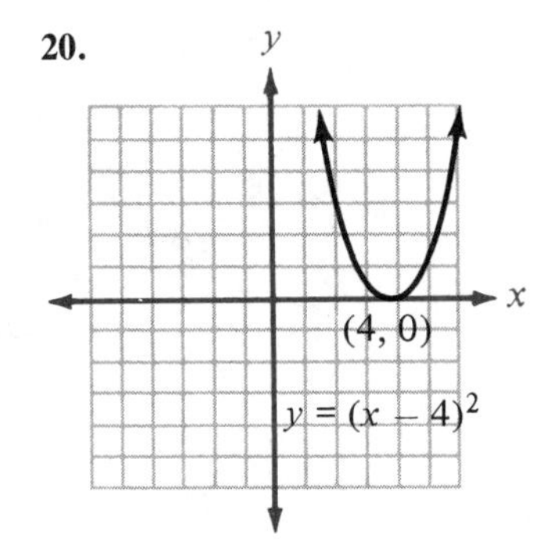

CHAPTER 9

PROBLEM SET 9.1

1.

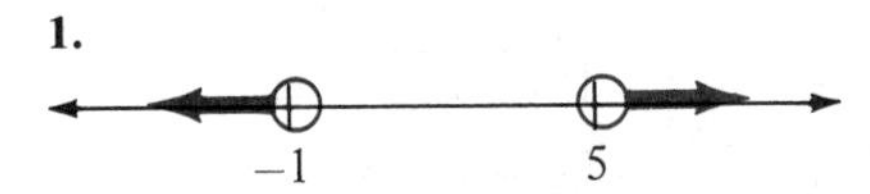

3.

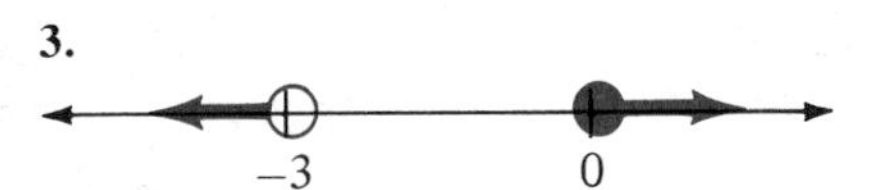

5. -1 $\quad$ 6

7. 2 $\quad$ 4

9. -2 $\quad$ 4

11. -1 $\quad$ 5

13. -1 $\quad$ 3

15. -3 $\quad$ -2

17. 2 $\quad$ 3

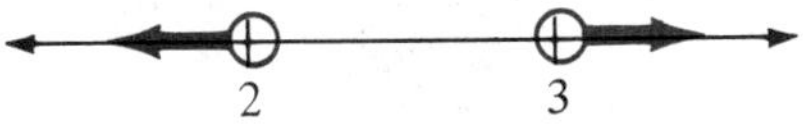

19. -3 $\quad$ 6

21. 2 $\quad$ 3

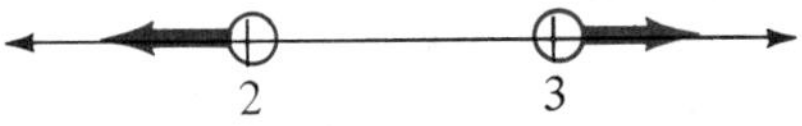

23. -3 $\quad$ 5

25. $-\frac{11}{3}$ $\quad$ $\frac{11}{2}$

27. 1 $\quad$ 2

29. 4 $\quad$ 7

31. -2 $\quad$ 3

33.

35.

37.

−5 4

39.

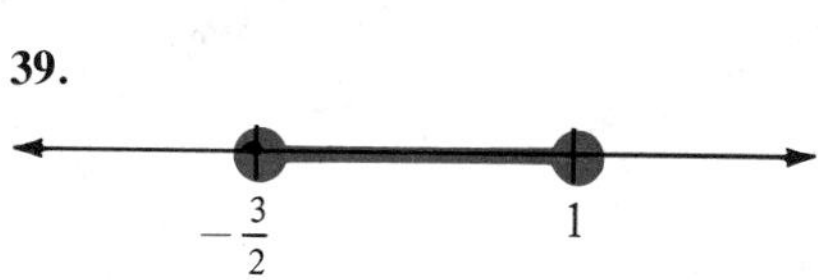

41. $5 < x < 15$ **43.** $4 < x < 5$ **45.** The width is between 3 inches and $\frac{11}{2}$ inches. **47.** -5 **49.** 5
51. 7 **53.** 9 **55.** 6 **57.** $2x - 3$ **59.** $-3, 0, 2$

PROBLEM SET 9.2

1.

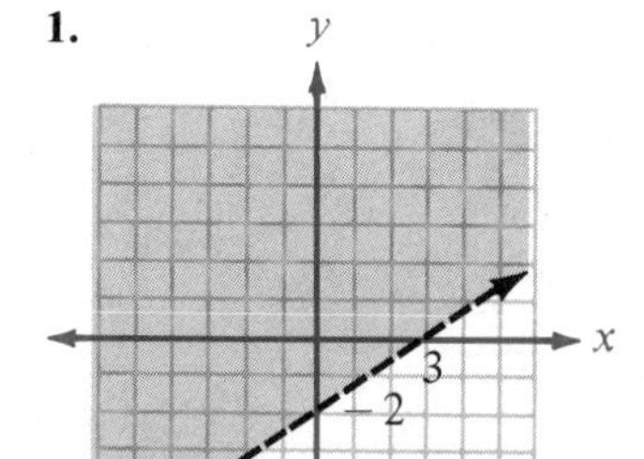

3.

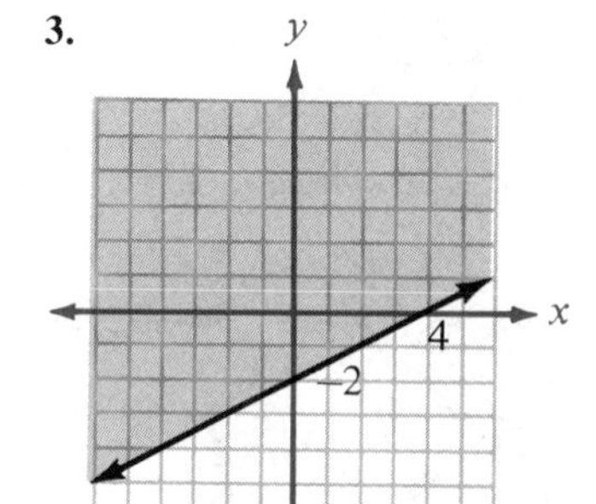

5.

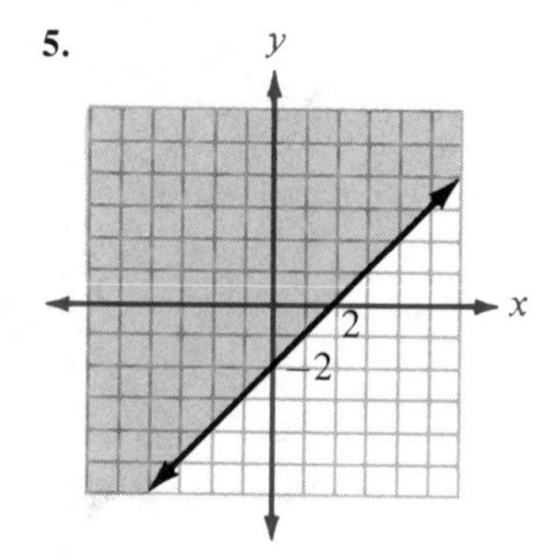

7.

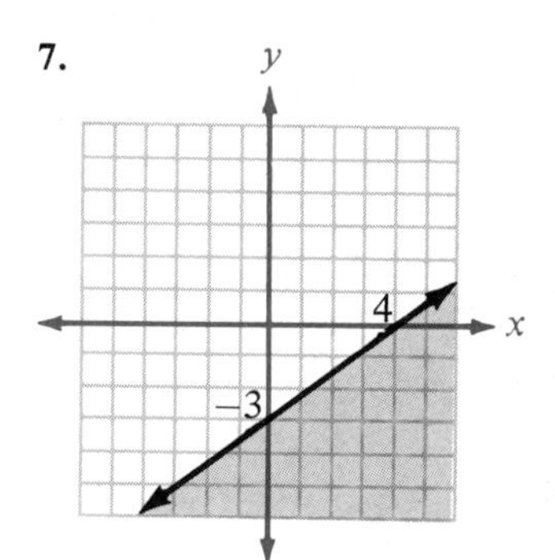

9.

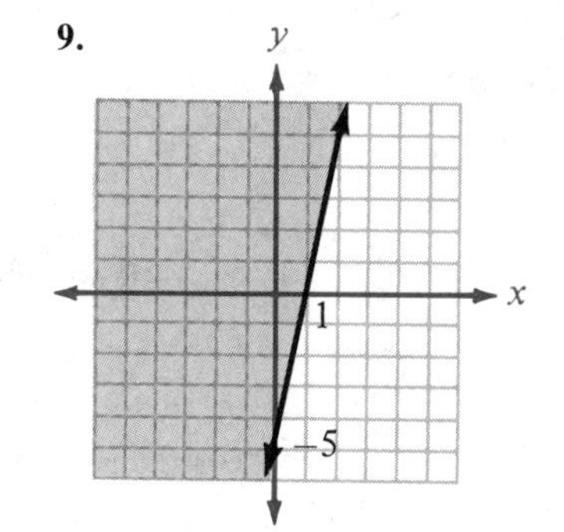

11.

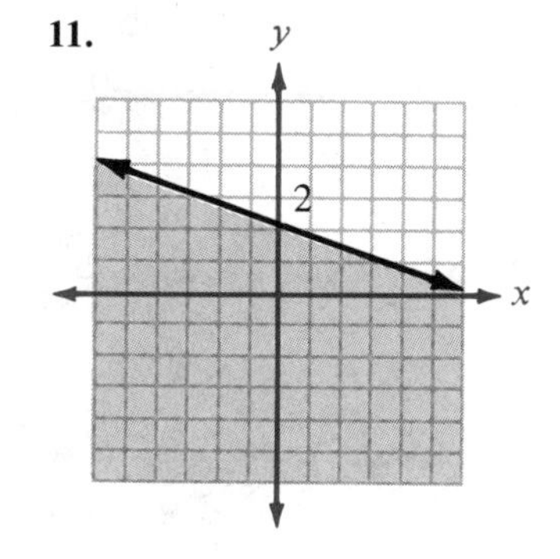

13.

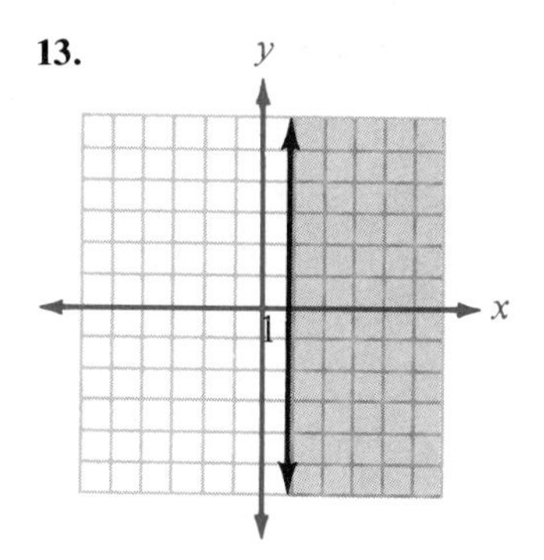

15.

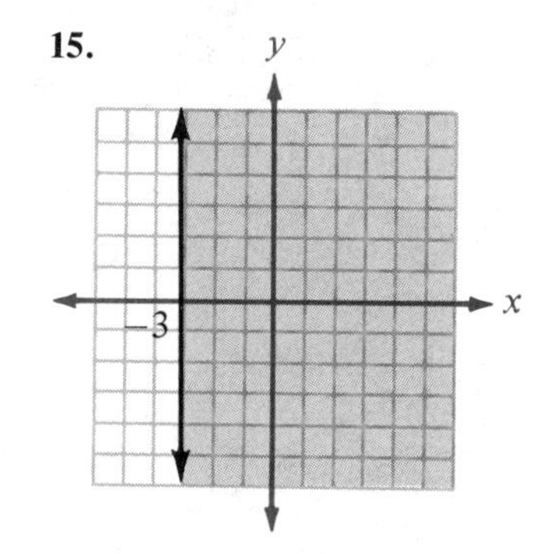

17.

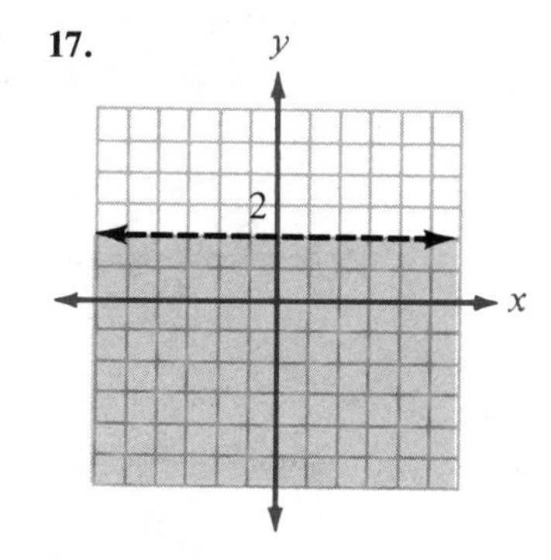

19.

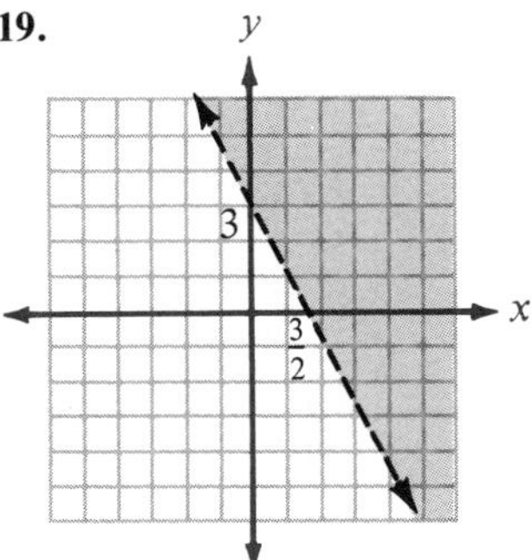

21.

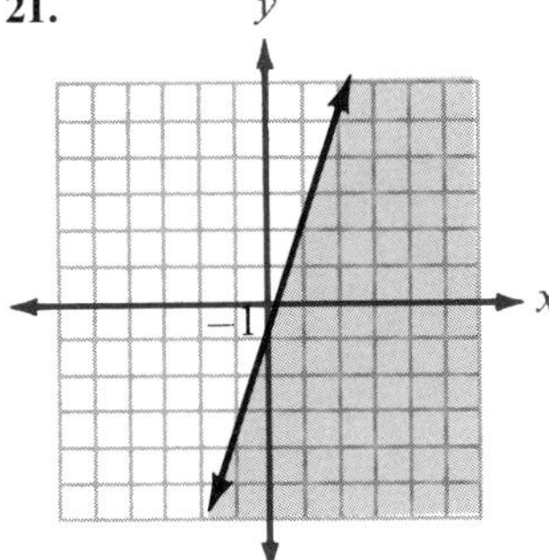

23.

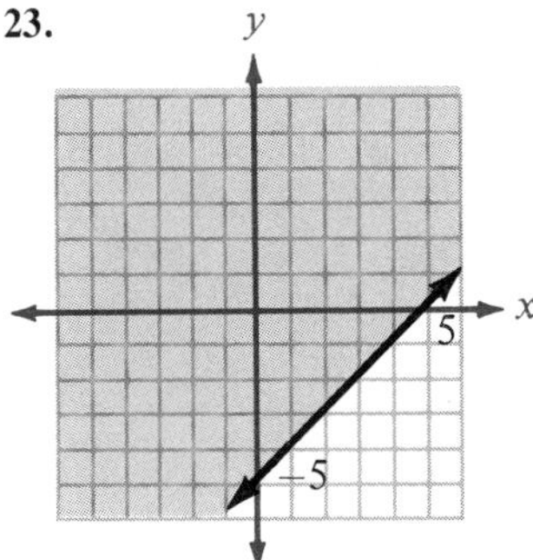

25.

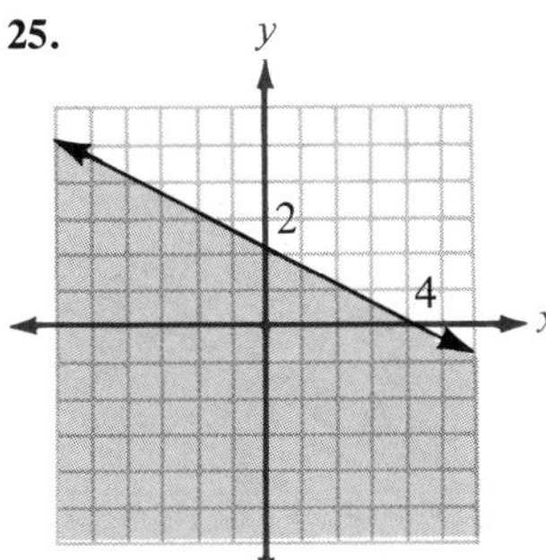

27.

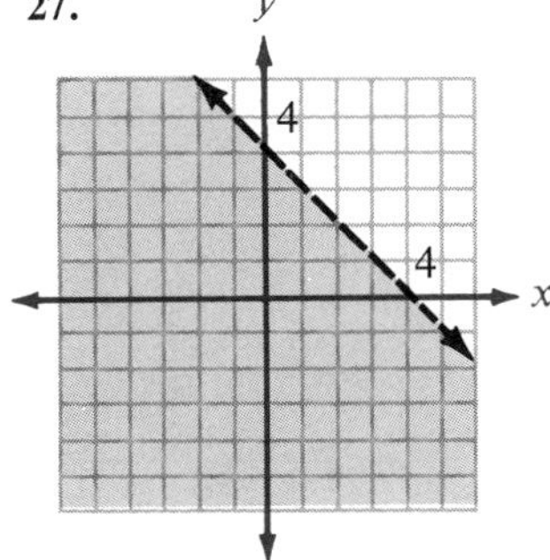

29. $-6x + 11$ **31.** -8 **33.** -4 **35.** $w = \dfrac{P - 2l}{2}$ **37.** $x < -1$

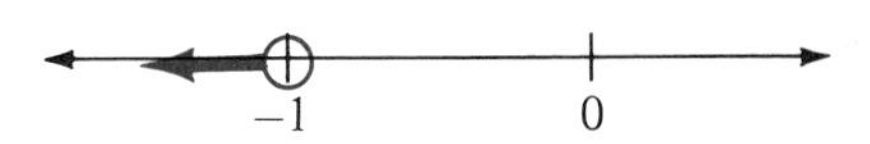

39. $y \geq \frac{3}{2}x - 6$ **41.** Width is 2 inches, length is 11 inches.

PROBLEM SET 9.3

1. 5 **3.** $\sqrt{106}$ **5.** $\sqrt{61}$ **7.** $\sqrt{130}$ **9.** 3 or -1 **11.** 3 **13.** $(x - 2)^2 + (y - 3)^2 = 16$
15. $(x - 3)^2 + (y + 2)^2 = 9$ **17.** $(x + 5)^2 + (y + 1)^2 = 5$ **19.** $x^2 + (y + 5)^2 = 1$ **21.** $x^2 + y^2 = 4$

23. center = (0, 0), radius = 2

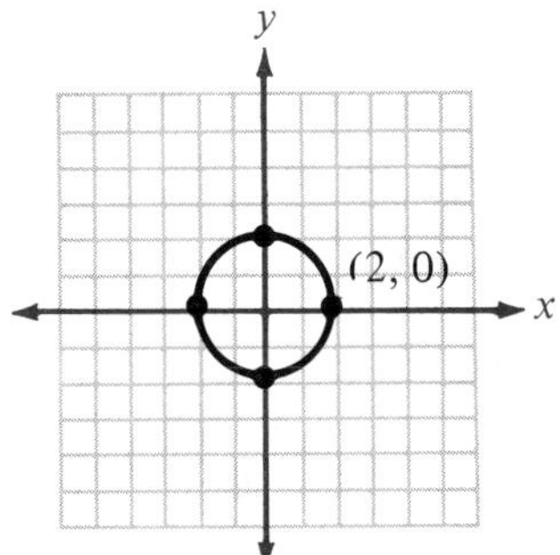
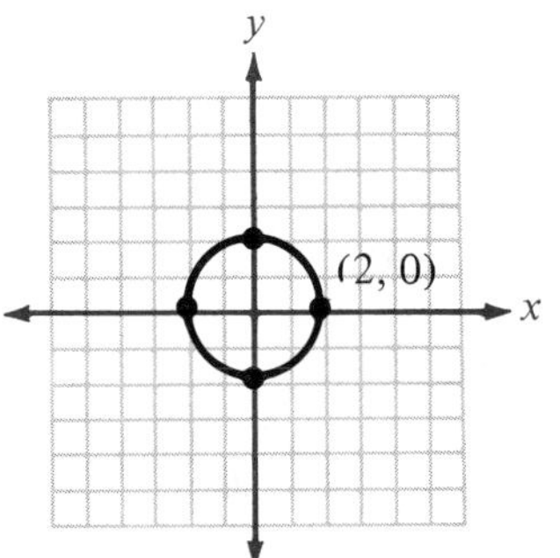

25. center = (1, 3), radius = 5

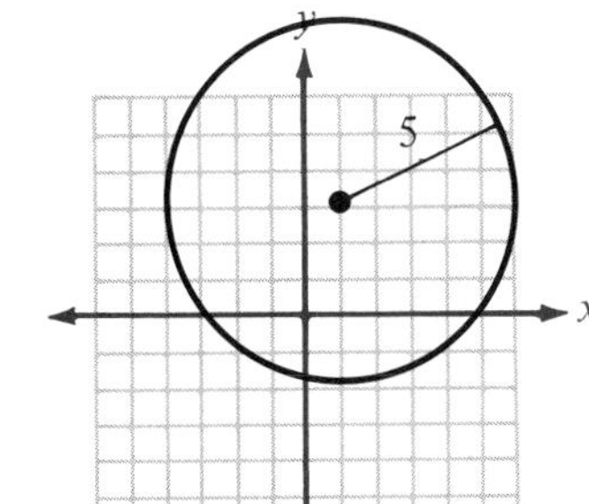

27. center = (2, 3),
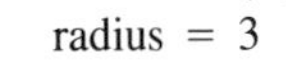
radius = 3

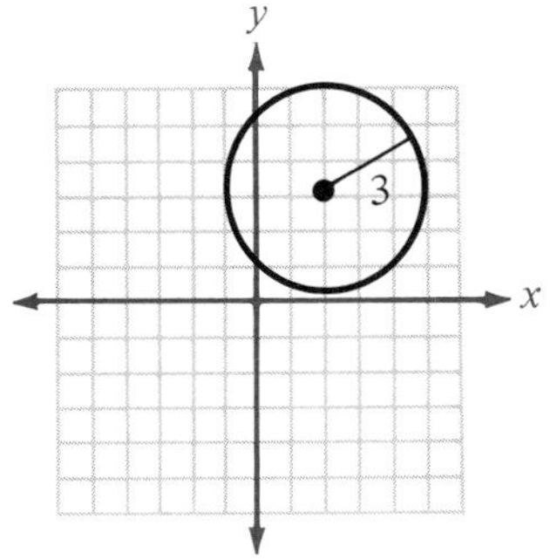

29. center $= (-1, -1)$, radius $= 1$

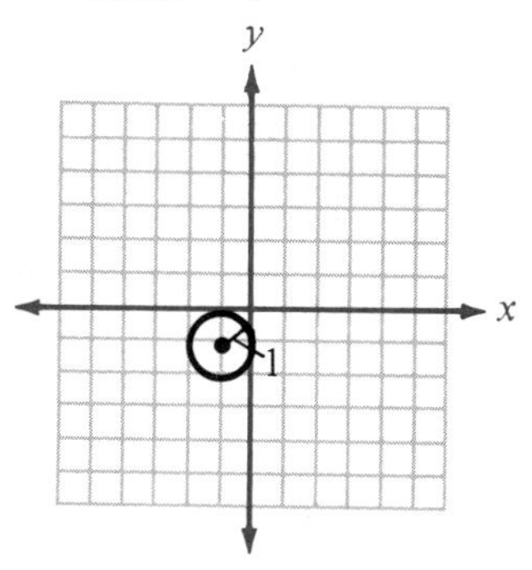

31. $(-2, 2)$ $(0, 3)$ $(2, 4)$

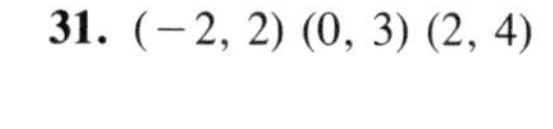

33.

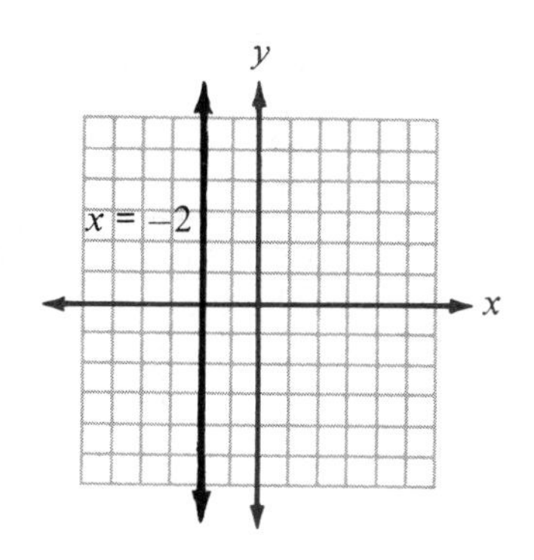

35. -1 **37.** $y = -3x + 1$ **39.** $(5, -7)$ **41.** $(3, 17)$ **43.** \$800 at 8%, \$400 at 10%.

PROBLEM SET 9.4

1. 2 **3.** 4 **5.** 3 **7.** 5 **9.** 3 **11.** 9 **13.** 4 **15.** 25 **17.** 8 **19.** 64 **21.** 8 **23.** 4 **25.** -2 **27.** 5 **29.** 17 **31.** 6 **33.** x **35.** x^2 **37.** $a^{2/5}$ **39.** $3y^2$ **41.** $3a^2b$ **43.** x **45.** $\frac{1}{5}$ **47.** $\frac{1}{4}$ **49.** $\frac{1}{9}$ **51.** $200x^{24}$ **53.** x^7 **55.** 8×10^1 **57.** $10ab^2$ **59.** $6x^4 + 6x^3 - 2x^2$ **61.** $9y^2 - 30y + 25$ **63.** $4a^4 - 49$ **65.** $x + 2 + \dfrac{14}{x - 4}$

PROBLEM SET 9.5

1. $x^2 - 16$ **3.** $4x^2 - 81$ **5.** $x^3 + 8$ **7.** $x^3 - 1$ **9.** $x^3 + 64$ **11.** $8x^3 - 27$ **13.** $x^3 + y^3$ **15.** $(x - 2)(x^2 + 2x + 4)$ **17.** $(x + 7)(x^2 - 7x + 49)$ **19.** $(3x - 5y)(9x^2 + 15xy + 25y^2)$ **21.** $(x + 2)(x^2 - 2x + 4)$ **23.** $(x - 5)(x^2 + 5x + 25)$ **25.** $(x + 1)(x^2 - x + 1)$ **27.** $(2x + y)(4x^2 - 2xy + y^2)$ **29.** $(3x - 2y)(9x^2 + 6xy + 4y^2)$ **31.** $(5x + 2)(25x^2 - 10x + 4)$ **33.** $(x + 4y)(x^2 - 4xy + 16y^2)$ **35.** $2 \cdot 3^2 \cdot 5 \cdot 7$ **37.** $(x - 6)(x + 2)$ **39.** $(x + 2)(x - 2)(x^2 + 4)$ **41.** $5x(x - 6)(x + 1)$ **43.** 0, 6 **45.** -10, 8 **47.** 9 inches and 12 inches

PROBLEM SET 9.6

1. 8 **3.** 130 **5.** 5 **7.** -3 **9.** 3 **11.** 243 **13.** 2 **15.** $\frac{1}{2}$ **17.** 1 **19.** 5 **21.** 25 **23.** 9 **25.** 84 pounds **27.** 367.5 **29.** \$235.50 **31.** 96 pounds **33.** 12 amps **35.** $\dfrac{x + 2}{x + 3}$ **37.** $\dfrac{2(x + 5)}{x - 4}$ **39.** $\dfrac{1}{x - 4}$ **41.** $\dfrac{x + 5}{x - 3}$ **43.** -2 **45.** 24 hours

PROBLEM SET 9.7

1. $D = \{1, 3, 5\}$, $R = \{2, 4, 6\}$ **3.** $D = \{1, 3, 4\}$, $R = \{a, b, c\}$ **5.** $D = \{1, 2, 3, 4\}$, $R = \{1, 2, 3, 4\}$ **7.** $D = \{1, 3, 5, 7, 9\}$, $R = \{2, 8, 14, 20, 26\}$ **9.** $D = \{0, 2, 4\}$, $R = \{2, \frac{2}{3}, -\frac{2}{3}\}$ **11.** $D = \{3, 4, 5, 6, 7\}$, $R = \{2, 4, 6, 8, 10\}$ **13.** 2 **15.** -3 **17.** $-1, -2$ **19.** $-5, 1$ **21.** $3, -3$ **23.** 2, 3 **25.** $-\frac{1}{2}$, 3 **27.** 0, 2, 4 **29.** $-3, -5, -6, (a - 5)$ **31.** $-1, 1, -2, (2z - 3)$ **33.** $5, 10, 17, (a^2 + 1)$ **35.** $-1, -\frac{1}{2}, \frac{1}{3}, 2, \frac{3}{4}$ **37.** $15, 0, (a^2 + 2a), (a^2 - 2a)$ **39.** $0, 7, -28, (a^3 - 1)$ **41.** 7 **43.** $5\sqrt{2}$ **45.** $\dfrac{\sqrt{10}}{5}$ **47.** $21\sqrt{3}$ **49.** $-4 - 3\sqrt{6}$ **51.** $4\sqrt{5} + 4\sqrt{3}$ **53.** 7 **55.** $-3, 2$ **57.** $\dfrac{2 \pm \sqrt{10}}{3}$ **59.** $\dfrac{2 \pm i\sqrt{3}}{2}$ **61.** $2 + 8i$ **63.** $23 - 2i$

CHAPTER 9 TEST

1.

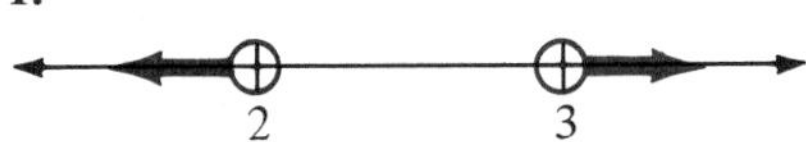

2.

3.

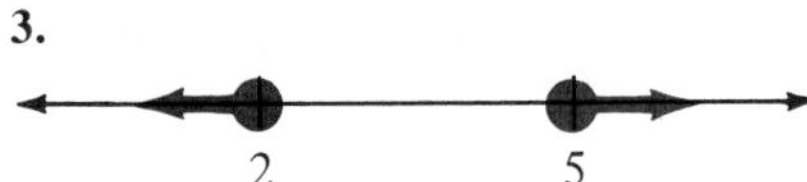

4.

5.

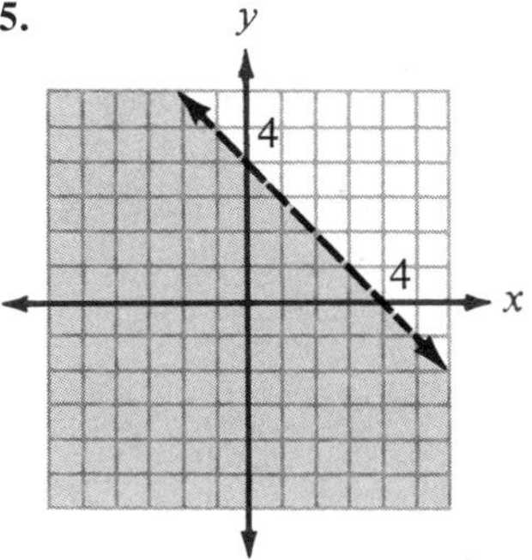

6.

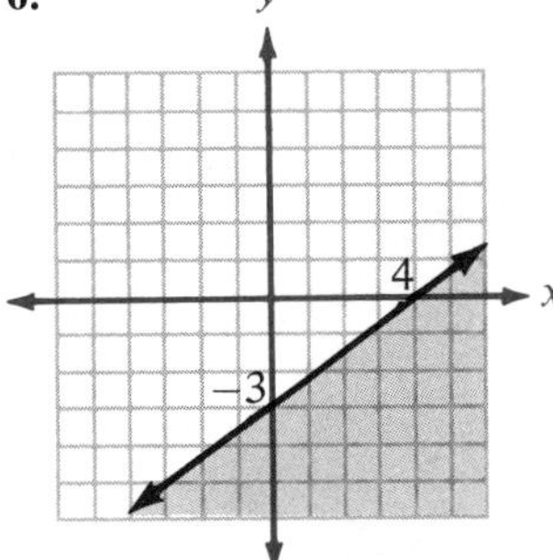

7. 5 **8.** $2\sqrt{13}$ **9.** $(x - 2)^2 + (y + 3)^2 = 25$

10.

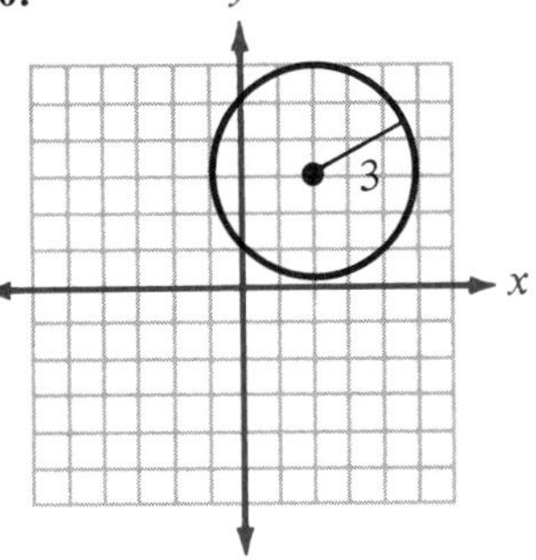

11. 5 **12.** 4 **13.** x **14.** $3x^2y^3$ **15.** $(x - 7)(x^2 + 7x + 49)$ **16.** $(x + 3)(x^2 - 3x + 9)$
17. $(2x + 5y)(4x^2 - 10xy + 25y^2)$ **18.** $(2x - 3y)(4x^2 + 6xy + 9y^2)$ **19.** 100 **20.** 2
21. $D = \{1, 3, 5\}$, $R = \{2, 4, 6\}$ **22.** $x \neq 5$ and $x \neq -5$ **23.** 26 **24.** 11 **25.** 2 **26.** -23

Index

8
9
E 0

FORMULAS FROM GEOMETRY

RECTANGLE

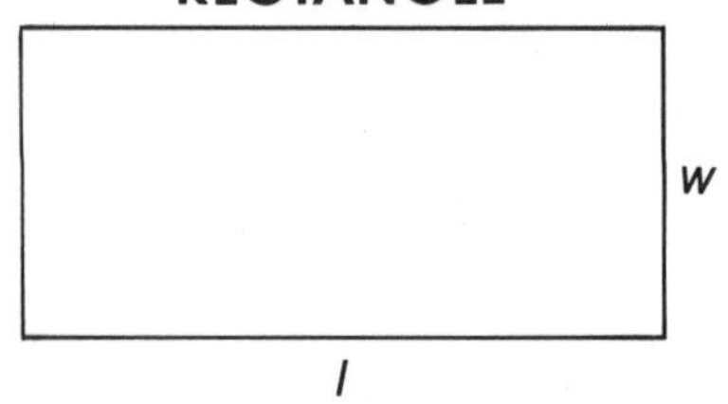

Area $= A = lw$
Perimeter $= P = 2l + 2w$

PARALLELOGRAM

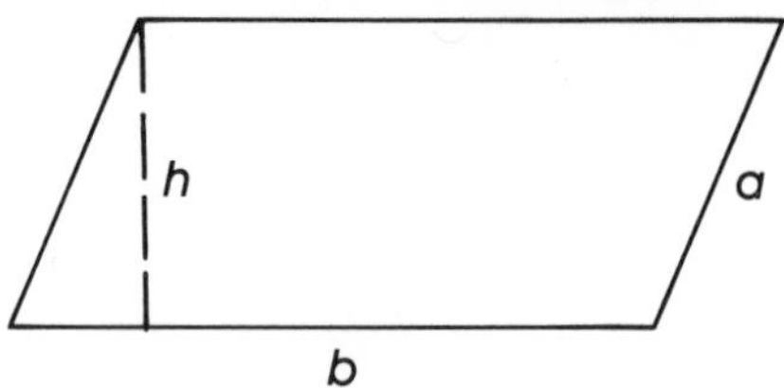

Area $= A = bh$
Perimeter $= P = 2a + 2b$

TRAPEZOID

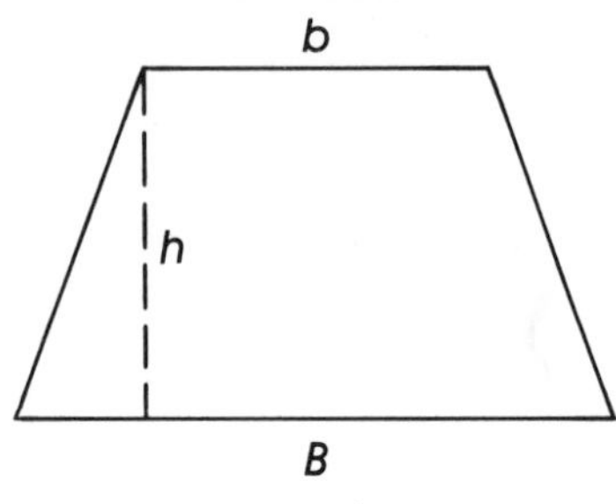

Area $= A = \frac{1}{2}(b + B)h$

CIRCLE

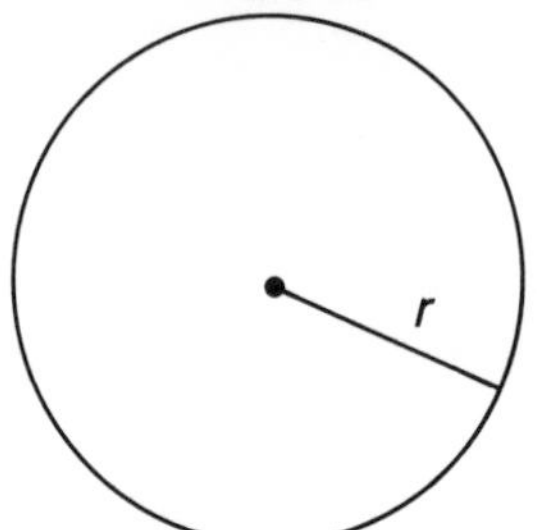

Area $= A = \pi r^2$
Circumference $= C = 2\pi r$

TRIANGLE

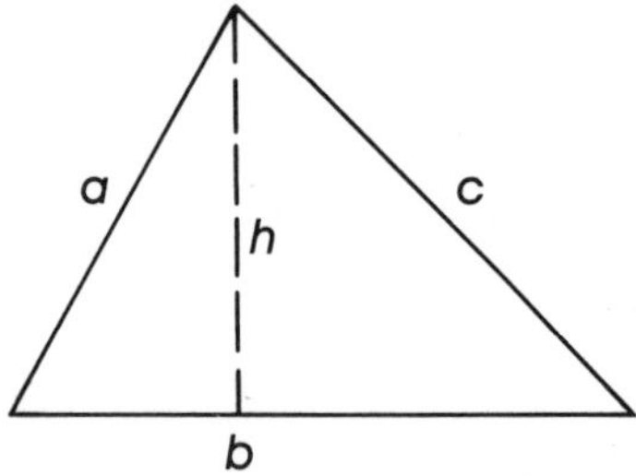

Area $= A = \frac{1}{2}bh$
Perimeter $= P = a + b + c$

RIGHT TRIANGLE

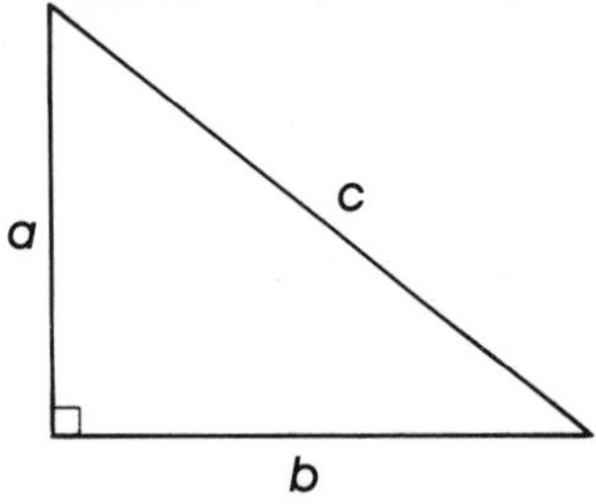

Pythagorean Theorem
$c^2 = a^2 + b^2$